Analysis of Drugs and Metabolites by Gas Chromatography– Mass Spectrometry

ANALYSIS OF DRUGS AND METABOLITIES
BY GAS CHROMATOGRAPHY-MASS SPECTROMETRY

VOLUME 1: Respiratory Gases, Volatile Anesthetics, Ethyl Alcohol, and Related Toxicological Materials

VOLUME 2: Hypnotics, Anticonvulsants, and Sedatives

VOLUME 3: Antipsychotic, Antiemetic, and Antidepressant Drugs

VOLUME 4: Central Nervous System Stimulants

VOLUME 5: Analgesics, Local Anesthetics, and Antibiotics

VOLUME 6: Cardiovascular, Antihypertensive, Hypoglycemic, and Thyroid-Related Agents

VOLUME 7: Natural, Pyrolytic, and Metabolic Products of Tobacco and Marijuana

OTHER VOLUMES IN PREPARATION

ANALYSIS OF DRUGS AND METABOLITES BY GAS CHROMATOGRAPHY-MASS SPECTROMETRY

VOLUME 7
Natural, Pyrolytic, and Metabolic Products of Tobacco and Marijuana

Benjamin J. Gudzinowicz
Department of Pathology
Rhode Island Hospital
Providence, Rhode Island

Michael J. Gudzinowicz
Center in Toxicology
Department of Biochemistry
School of Medicine
Vanderbilt University
Nashville, Tennessee

With the Assistance of

Horace F. Martin
Department of Pathology, Rhode Island Hospital
Providence, Rhode Island
and
Division of Biological and Medical Sciences
Brown University, Providence, Rhode Island

and

James L. Driscoll
Joanne Hologgitas
Department of Pathology, Rhode Island Hospital
Providence, Rhode Island

MARCEL DEKKER, INC.,   New York and Basel

Library of Congress Cataloging in Publication Data

Gudzinowicz, Benjamin J
  Analysis of drugs and metabolites by gas chromatography--mass spectrometry.

  Includes bibliographical references and indexes.
  CONTENTS: v.1. Respiratory gases, volatile anesthetics, ethyl alcohol, and related toxicological materials.--v.2. Hypnotics, anti-convulsants, and sedatives.--[etc.]--v.7. Natural, pyrolytic, and metabolic products of tobacco and marijuana.
  1. Drugs--Analysis. 2. Gas chromatography. 3. Mass spectrometry. 4. Drug metabolism. I. Gudzinowicz, Michael J., joint author. II. Title. [DNLM: 1. Drugs--Analysis. 2. Chromatography, Gas. 3. Spectrum analysis, Mass. QV25 G923a]
RS189.G83   615'.1901   76-56481
ISBN 0-8247-6576-1 (v. 1)

COPYRIGHT © 1980 by MARCEL DEKKER, INC. ALL RIGHTS RESERVED.

Neither this book nor any part may be reproduced or transmitted in any form or by any means, electronic or mechanical, including photocopying, microfilming, and recording, or by any information storage and retrieval system, without permission in writing from the publisher.

MARCEL DEKKER, INC.
270 Madison Avenue, New York, New York 10016

Current printing (last digit):
10 9 8 7 6 5 4 3 2 1

PRINTED IN THE UNITED STATES OF AMERICA

Dedicated to

HELEN L. GUDZINOWICZ
a devoted and understanding wife and mother

PREFACE

In the past two decades, remarkable progress has been made in the analysis of drugs, pharmaceuticals, and related toxicological materials. In great measure, these notable advances can be attributed to technological advancements in two specific types or areas of analytical instrumentation; namely, gas chromatography and integrated gas chromatography-mass spectrometry.

Since James and Martin revealed to the scientific community their gas chromatographic technique which permitted the separation of fatty acid mixtures into their individual components, the rapid growth of gas chromatography has been very evident. This remarkable progress can be directly correlated with the improvements that we have witnessed over the years in gas chromatographic stationary phase, carrier gas, column, and temperature- and pressure-controlling technology. Furthermore, it has assumed a position of even greater analytical significance since the advent of highly specific, rapid, sensitive detection systems.

On the other hand, the integrated GC-MS analytical system is rather unique and exceptional in that it combines the mass spectrometer's unexcelled identification potential with the gas chromatograph's separation capabilities. Although the integration of GC and MS was first reported in 1957 by Holmes and Morrell, it nevertheless remained a dormant, costly, and seemingly unappreciated technique until 1970. Since then, with improved instrumentation at a more reasonable price and newly developed operating techniques, numerous publications have appeared in the literature showing its applicability to a wide variety of difficult analytical problems, thus opening up new horizons for analytical research in toxicology, biochemistry, pharmacology, forensics, medicine, etc. To be able to monitor a drug, its persistence and metabolic fate in biological fluids of man via mass fragmentography at picogram concentration levels provides the researcher with a tool of immeasurable significance.

Because much has been written over the years about the analysis of drugs and their metabolites by either or both techniques, the objectives of these volumes are several-fold: (1) to compile from existing literature in a chronological manner the various GC and/or GC-MS procedures available for the analysis of specific drugs and their metabolites, (2) to describe with as much detail as possible all procedures (qualitative and quantitative) in order that they might be reproduced faithfully in one's laboratory, and (3) to indicate, wherever possible, not only the results, precision, accuracy, and limits of detection achieved by a given procedure, but also its applicability to pharmacokinetic studies. For this reason, in addition to the text, which is well referenced in each section, many illustrations of actual applications and tables of data for each instrumental technique are included as aids to the analyst for his greater appreciation and understanding of the limitations as well as potentials ascribed to each method. As stated in the past, from an analytical chemist's point of view, it is hoped that this deliberately combined visual and factual approach will find acceptance by the reader who would otherwise rely only on his interpretation of the written word relative to some published procedure.

Without wishing to be repetitious, in retrospect it must be again stated that this volume really represents the end result of many tedious and arduous investigations by numerous eminent scientists whose research efforts have appeared in the literature throughout the world. We are indeed humbly indebted to them, and to those journals, publishers, and organizations that granted special copyright permission to the authors.

Benjamin J. Gudzinowicz
Michael J. Gudzinowicz

CONTENTS

Preface	v
Contents of Other Volumes	ix

Chapter 1 Natural, Pyrolytic, and Carcinogenic Products
of Tobacco 1

I.	Chemical Composition of Tobacco Leaf	2
	A. Nicotine and Other Tobacco Alkaloids	2
	B. N'-Nitrosonornicotine	41
	C. Hydrocarbons	50
	D. Alcohols of Tobacco	59
	E. Sterols	68
	F. Fatty Acids and Esters	73
	G. Ammonia and Hydrazine	87
II.	Chemical Composition of Tobacco Smoke	88
	A. Tobacco Alkaloids in Smoke and Biological Fluids	103
	B. Low-Boiling Volatiles (General)	136
	C. Hydrocarbons	146
	D. Ketones	159
	E. Phenolic Constituents	159
	F. Acidic Components	171
	G. Polynuclear Aromatic Hydrocarbons and Related Compounds	180
	H. Sulfur-Containing Compounds	232
	I. Analysis of Specific Miscellaneous Compounds	247
	References	262

Chapter 2 Natural, Pyrolytic, and Metabolic Products
             of Marijuana   271

        I. Cannabinoid Patterns and Their Use in Determining
           Chemical Race and Origin   273
       II. Chemical Composition of the Marijuana Plant   325
      III. Some Chemical Constituents of Marijuana Smoke   407
       IV. Cannabinoids and Metabolic Products in
           Biological Media   432
       References   479

Author Index   491

Subject Index   509

CONTENTS OF OTHER VOLUMES

Volume 1 RESPIRATORY GASES, VOLATILE ANESTHETICS, ETHYL ALCOHOL, AND RELATED TOXICOLOGICAL MATERIALS

  Chapter 1. Respiratory Gases, Volatile Anesthetics, and Related Toxicological Materials

  Chapter 2. Ethyl Alcohol and Volatile Trace Components in Breath, Body Fluids, and Body Tissues

Volume 2 HYPNOTICS, ANTICONVULSANTS, AND SEDATIVES

  Chapter 1. Hypnotics, Anticonvulsants, and Sedatives: Barbiturate Compounds

  Chapter 2. Hypnotics, Anticonvulsants, and Sedatives: Nonbarbiturate Compounds

  Chapter 3. Hypnotics, Anticonvulsants, and Sedatives: Nonbarbiturate Compounds (Continued)

Volume 3 ANTIPSYCHOTIC, ANTIEMETIC, AND ANTIDEPRESSANT DRUGS

  Chapter 1. Antipsychotic and Antiemetic Drugs: Phenothiazine, Butyrophenone, and Thioxanthene Derivatives

  Chapter 2. Antidepressant Drugs: Monoamine Oxidase Inhibitors, Tricyclic Antidepressants, and Several Related Compounds

Volume 4 CENTRAL NERVOUS SYSTEM STIMULANTS

    Chapter 1. Amphetamines, Xanthines, and Related Compounds

    Chapter 2. Phenylethylamine-, Tryptamine-, and Propranolol-Related Compounds

Volume 5 ANALGESICS, LOCAL ANESTHETICS, AND ANTIBIOTICS

    Chapter 1. Narcotics, Narcotic Antagonists, and Synthetic Opiate-like Drugs

    Chapter 2. Antipyretic, Antiinflammatory, and Antihyperuricemic Agents; Local Anesthetics; and Antibiotics

Volume 6 CARDIOVASCULAR, ANTIHYPERTENSIVE, HYPOGLYCEMIC, AND THYROID-RELATED AGENTS

    Chapter 1. Cardiovascular Drugs

    Chapter 2. Antihypertensive, Hypoglycemic, and Thyroid-Related Drugs

Analysis of Drugs
and Metabolites
by Gas Chromatography-
Mass Spectrometry

Chapter 1

NATURAL, PYROLYTIC, AND CARCINOGENIC
PRODUCTS OF TOBACCO

The natural, pyrolytic, and carcinogenic products of tobacco will be discussed in Chapter 1 of this volume, and the natural constituents of marijuana and products isolated and identified as smoke by-products in Chapter 2.

With regard to the so-called active ingredients of tobacco and marijuana, Jaffe [1], in his discussion of drug addiction and drug abuse, noted that:

> Next to caffeine, nicotine (tobacco) is the substance most widely used for its effects on mood. That some persons develop a considerable psychological dependence on the smoking of tobacco is a matter of common experience. Some investigators feel that in chronic smokers distinct withdrawal phenomena follow the abrupt cessation of smoking. It is now generally accepted that the smoking of tobacco is linked to cardiovascular, pulmonary, and neoplastic disease. The inability of many persons to give up smoking in spite of these serious consequences makes it logical to include tobacco-using behavior as another form of compulsive drug use. There is already a vast literature on the pharmacological effects of chronic cigarette use, and technics for modifying the patterns of compulsive tobacco use are under study.

In addition to the above, the carcinogenic nature of tobacco smoke components is extremely important to personal health. Consequently, the GC and/or GC-MS conditions required to separate and identify such carcinogenics will be discussed in this chapter. Furthermore, GC methods for the analysis of polynuclear aromatics from sources other than tobacco smoke having potential carcinogenic activity using both packed and capillary columns, will be included.

In Chapter 2, the pharmacology, absorption, metabolism, and excretion of delta-9-tetrahydrocannabinol (Δ^9-THC), the major psychoactive ingredient of marijuana, will be explored, as well as the GC/GC-MS analysis of its natural and pyrolytic products. With regard to marijuana use, Jaffe notes that "the remarkable similarity between the descriptions of the subjective effects of Δ^9-THC and those of LSD is the basis for the classification of the former as a psychedelic agent. Yet cannabis has sedative effects not seen with LSD, does not produce the sympathomimetic effects characteristic of other psychedelics, induces a low order of tolerance compared to the high degree seen with LSD, and exhibits no cross-tolerance with LSD. These features indicate a distinctly different mechanism of action."

I. CHEMICAL COMPOSITION OF TOBACCO LEAF

In 1968, Stedman [2] prepared a comprehensive review of the constituents of tobacco and tobacco smoke, surveying only those compounds for which claims of identity appear to be reasonably justified. The various constituents found in tobacco leaf in this 1968 survey are listed in Table 1.1. The list of known components in tobacco and smoke has risen since 1959 from about 400 to more than 1200, not including the individual components in complex substances such as the brown pigments and resins, which, at that time, had not been resolved. In the following text, the GC and/or GC-MS analysis of some of these organics that have been reported in the literature will be discussed, with special emphasis directed to the determination of nicotine and related alkaloids in tobacco leaf.

A. Nicotine and Other Tobacco Alkaloids

In Figure 1.1 are shown the structures of nicotine and related tobacco alkaloids. The metabolites of nicotine formed by oxidation, demethylation, and pyridine-N-methylation are illustrated in Figure 1.2.

As early as 1958, Quin [3] reported preliminary results clearly indicating the usefulness of gas chromatography in the study of tobacco alkaloids. Using a Perkin-Elmer model 154-B gas chromatograph equipped with a thermal conductivity detector and 1-m by 6-mm-o.d. U-shaped glass columns packed with polyglycol liquid stationary phases (20% by weight) coated on alkali-washed Firebrick, the retention times of the various compounds studied are listed in Table 1.2. As noted by Quin, the polyglycol columns exhibited good selectivity for the alkaloids, making possible the

TABLE 1.1
Chemical Composition of Tobacco Leaf[a]

1. Alkanes
    Normal C_8–C_{35}
    Iso $\quad C_{27}$–C_{34}
    Anteiso C_{28}, C_{30}, C_{32}, C_{34}

2. Isoprenoid hydrocarbons
    α-Carotene
    β-Carotene
    Neo-β-carotene
    Neophytadiene
    Phytoene
    Phytofluene

3. Sterols
    Campesterol
    Cholesterol
    Ergosterol
    β-Sitosterol
    Stigmasterol

4. Monoterpenes
    Borneol
    1-Linalool

5. Diterpenes
    3,8,13-Duvatriene-1,5-diol (α-, β-)
    4,8,13-Duvatriene-1,3-diol (α-, β-)
    12α-Hydroxy-13-epimanoyl oxide
    α_2-Levantanolide
    Levantenolide (α-, β-)
    α-5,8-Oxido-3,9,13-duvatrien-1-ol
    α-5,8-Oxido-3,9(17),13-duvatrien-1-ol
    β-5,8-Oxido-3,9(17),13-duvatrien-1-ol

6. Triterpenes
    β-Amyrin

7. Tetraterpenes
    Cryptoxanthin
    Flavoxanthin
    Lutein
    Neoxanthin
    Violaxanthin
    Xeaxanthin

(continued)

TABLE 1.1 (continued)

8. Trisesquiterpene
   Solanesol
9. Related isoprenoids
   6,8-Dihydroxy-11-isopropyl-4,8-dimethyl-14-oxo-4,9-pentadecadienoic acid
   Hexahydrofarnesylacetone
   Solanochromene
   Solanone
   Tocopherols
   Vitamin K_1 (2-methyl-3-phytyl-1,4-naphthoquinone)
10. Alcohols
    a. Aliphatic
       1-Docosanol
       1-Eicosanol
       Ethyl alcohol
       1-Heneicosanol
       1-Heptadecanol
       Methanol
       3-Methyl-1-pentanol
       1-Nonadecanol
       1-Octadecanol
       1-Tricosanol
    b. Aromatic
       Benzyl alcohol
       β-Phenethyl alcohol
    c. Polyols
       Diethylene glycol
       Glycerol
       Propylene glycol
       Triethylene glycol
    d. Cyclic
       Furfuryl alcohol
       Inositol
       Menthol
11. Esters
    β-Amyrenyl esters
    Benzyl acetate
    Dibutyl phthalate
    Di(2-ethylhexyl) phthalate
    Dipropyl phthalate
    Ethyl acetate
    Ethyl butyrate

TABLE 1.1 (continued)

    Ethyl caproate
    Ethyl isovalerate
    Ethyl β-methylvalerate
    Ethyl propionate
    Ethyl valerate
    Glycerides
    Methyl and ethyl esters of higher fatty acids
    Methyl salicylate
    β-Phenethyl acetate
    Solanesyl esters
    Steryl esters
    Undecyl acetate

12. Aldehydes
    Acetaldehyde
    Acrolein
    p-Anisaldehyde
    Benzaldehyde
    Butyraldehyde
    Crotonaldehyde
    Formaldehyde
    Furfural
    Glycolaldehyde
    Glyoxal
    5-Hydroxymethylfurfural
    Isobutyraldehyde
    Isovaleraldehyde
    Mesoxaldialdehyde
    5-Methylfurfural
    Methylglyoxal
    Methylreductone
    Propionaldehyde
    Reductone
    m-Tolualdehyde
    Valeraldehyde

13. Ketones
    Acetone
    2-Butanone
    4-Methyl-2-pentanone
    Methyl α-pyrryl ketone
    2-Pentanone

14. Quinones
    9,10-Anthraquinone

(continued)

TABLE 1.1 (continued)

15. Acids
    Acetic
    Adipic
    Arachidic
    Arachidonic
    Auxin and indoleacetic acid
    Azelaic
    Benzoic
    Butyric
    C_{10}-C_{23} (saturated)
    C_{10}-C_{34} (normal)
    C_{15}-C_{26} (iso, anteiso)
    C_{16}, C_{18} (hydroxy)
    C_{22}-C_{25} (cyclohexyl)
    Caproic
    Citric
    Crotonic
    Decanoic
    Formic
    Fumaric
    Furoic
    D-Glyceric
    Glycolic
    Glyoxylic
    Heptanoic
    α-Hydroxyisocaproic
    β-Hydroxyisocaproic
    α-Hydroxy-β-methylvaleric
    β-Hydroxy-β-methylvaleric
    Hydroxypyruvic
    α-Hydroxyvaleric
    Isobutyric
    Isocaproic
    2-Isopropylmalic
    Isovaleric
    α-Ketoglutaric
    Lactic
    Lauric
    Linoleic
    Linolenic
    Maleic
    Malic
    Malonic

TABLE 1.1 (continued)

α-Methylbutyric
β-Methylvaleric
Myristic
Nonanoic
Octanoic
Oleic
Oxalacetic
Oxalic
Palmitic
Phenylacetic
α-Phenyllactic
Phenylpyruvic
Propionic
Pyruvic
Stearic
Succinic
Terephthalic
Valeric

16. Phenols and related compounds
    4-Allylcatechol
    p-Anisaldehyde
    Caffeic acid
    1-O-Caffeoylglucose
    4-Caffeoylquinic acid
    Catechol
    Chlorogenic acid
    p-Coumaric acid
    p-Coumarylquinic acid
    m-Cresol
    Esculetin
    Esculetin 7-glucoside
    Eugenol
    Ferulic acid
    1-O-Feruloylglucose
    3-Feruloylquinic acid
    Guaiacol
    Hydrocaffeic acid
    Hydroxyacetophenone (o-, m-, p-)
    Hydroxybenzaldehyde (o-, m-, p-)
    3-Hydroxybenzoic acid
    4-Hydroxybenzoic acid
    2-Hydroxyphenylacetic acid
    3-Hydroxyphenylacetic acid

(continued)

TABLE 1.1 (continued)

4-Hydroxyphenylacetic acid
3-Hydroxyphenylpropionic acid
4-Hydroxyphenylpropionic acid
Isoeugenol
Isoquercetrin
Kaempferol 3-rhamnoglucoside
Melilotic acid
Methyl salicylate
Naringin
Naringenin
Neochlorogenic acid
Phenol
Protocatechuic acid
Protocatechuic aldehyde
Quercetin methyl ethers
Quercimeritrin
Quinic acid
Rutin
Salicylaldehyde
Scopoletin
Scopoletin 7-glucoside
Scopoletin rhamnoglucoside
Shikimic acid
Syringaldehyde
Syringic acid
1,2,3-Trimethoxybenzene
Vanillic acid
Vanillin

17. Alkaloids and other bases
    Adenine
    Ammonia
    Anabasine
    Anatabine
    Cotinine
    Ethylamine
    Guanine
    Harmane
    Isoamylamine
    Isobutylamine
    Isonicotein (2,3'-bipyridyl)
    Methylamine
    N-Methylanabasine
    N-Methylanatabine

TABLE 1.1 (continued)

N-Methylnicotinamide
N-Methyl-2-phenylethylamine
2-Methylpyrrolidine
N-Methylpyrrolidine
N-Methyl-3-pyrroline
Myosmine
Nicotelline
Nicotinamide
Nicotine
Nicotine-N-oxide
Nicotinic acid
Nicotyrine
Norharmane
Nornicotine
2-Phenylethylamine
Piperidine
Pyridine
3-Pyridyl ethyl ketone
3-Pyridyl propyl ketone
Pyrrolidine
1,2,5,6-Tetrahydropyridine
Thymine
Trimethylamine

18. Volatile bases and alkaloids in brown leaf pigments
    Cotinine
    2-Methylpyridine
    3-Methylpyridine
    N-Methylpyrrolidine
    Nicotine
    Pyridine
    Pyrrole
    3-Vinylpyridine

19. Carbohydrates
    Arabinose
    Arabogalactan
    Cellulose
    1-Deoxy-1-L-alanino-D-fructose
    1-Deoxy-1-(N-γ-aminobutyric acid)-D-fructose
    1-Deoxy-1-L-proline-D-fructose
    Deoxyribose
    Erythrose
    Fructose

(continued)

TABLE 1.1 (continued)

Galactan
Galactosamine
Galactose
Galacturonic acid
Glucosamine
Glucose
Salt of gumlike polysaccharide
Lignin
Maltose
Mannose
Pectins
Pentoses
Planteose
Raffinose
Rhamnose
Ribose
Rutinose
Sorbitol
Stachyose
Starch
Sucrose
Xylan
Xylose

20. Amino acids and related compounds
    α-Alanine
    β-Alanine
    α-Aminoadipic acid
    α-Aminobutyric acid
    γ-Aminobutyric acid
    Arginine
    Asparagine
    Aspartic acid
    Betaine
    Choline
    Citrulline
    Cysteic acid
    Cysteine
    Cystine
    Glutamic acid
    Glutamine
    Glutathione
    Glycine
    Histidine

TABLE 1.1 (continued)

Homocystine
Homoserine
Hydroxyproline
Isoleucine
Leucine
Lysine
Methionine
Methionine sulfone
1-Methylhistidine
Norleucine
Ornithine
Phenylalanine
Pipecolic acid
Proline
Pyrrolidine-2-acetic acid
Serine
Taurine
Threonine
Tryptophan
Tyramine
Tyrosine
Valine

21. Miscellaneous components
    $C_{10}H_{14}O$
    Chlorophyll
    Nucleic acids
    Phosphatides
    Resins
    Saponins
    Silicones

[a] Adapted from Stedman [2].

separation of most of the members of complex mixtures. As indicated in Table 1.2, the elution sequence is identical on the polypropylene and polybutylene columns, although some relative differences in retention times are observable. A slightly different sequence is observed with the polyethylene glycol column. These effects are considered valuable in that a separation of alkaloids that is difficult on one column may be feasible on another.

Shortly thereafter, in 1961, Kobashi [4] investigated the separation and identification of the bases of nicotine and pyridine by gas chromatography

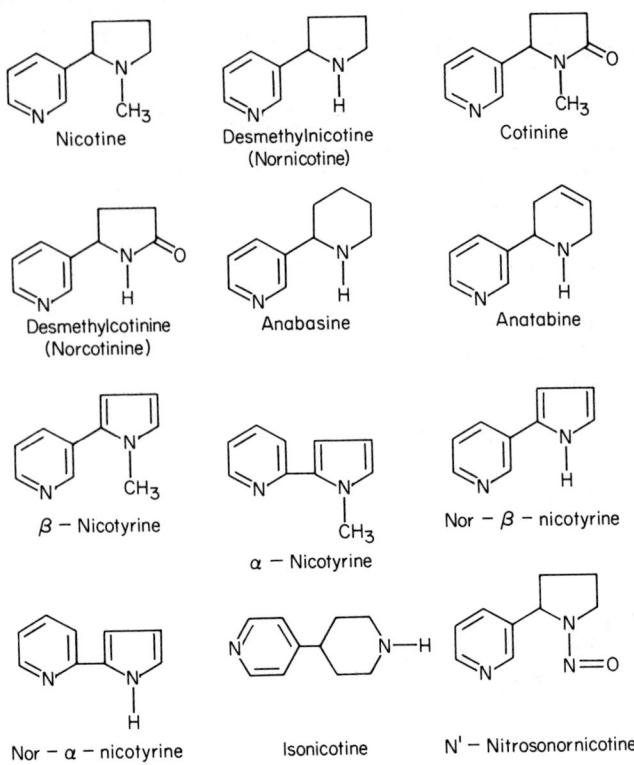

Figure 1.1. Nicotine and related tobacco alkaloid structures.

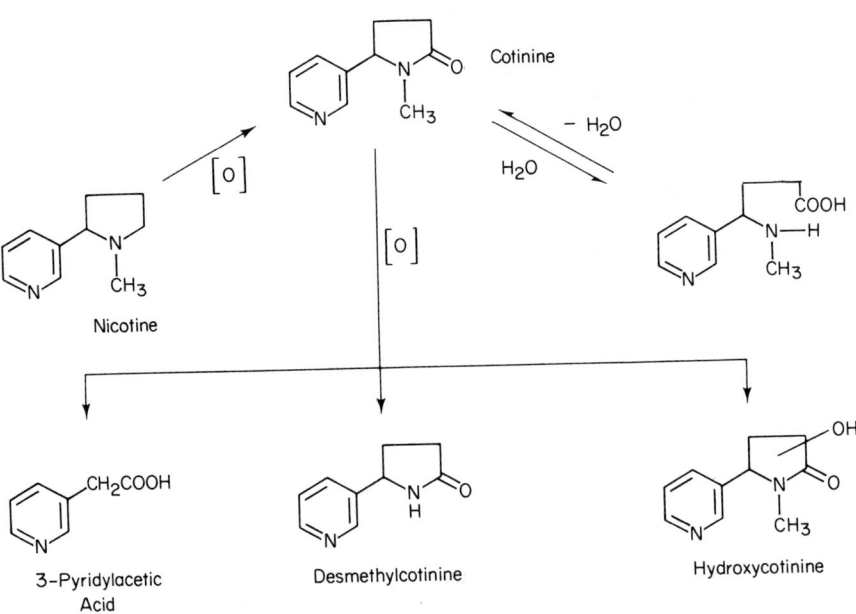

Figure 1.2. Metabolites of nicotine formed by oxidation, demethylation, and pyridine-N-methylation.

TABLE 1.2
Retention Times of Some Tobacco Alkaloids[a,b]

	Retention time (min)		
Compound	Column A	Column B	Column C
3-Pyridyl methyl ketone	4.3	4.3	3.1
3-Pyridyl ethyl ketone	5.3	6.1	5.0
3-Pyridyl n-propyl ketone	6.6	8.1	7.0
Nicotine	5.2	8.6	8.2
Nornicotine	12.3	16.1	14.3
Myosmine	13.4	16.4	14.7
Anabasine	13.8	19.4	18.1
Metanicotine	16.5	23.5	20.9
Nicotyrine	19.4	21.0	18.3
Cotinine	85	79	63
Helium flow rate (ml/min)	48	45	50
Column temp. (°C)	190	190	180

[a] Adapted from Quin [3].
[b] Liquid stationary phases: column A, polyethylene glycol (MW = 20,000); column B, polypropylene glycol (MW = 1,025); column C, polybutylene glycol (MW = 1,500).

for the purpose of studying the thermal decomposition of tobacco alkaloids. Pyridine bases (pyridine, α-picoline, β-picoline, γ-picoline, 2-ethyl pyridine, 3-ethyl pyridine, 4-ethyl pyridine, 2-vinyl pyridine, 3-vinyl pyridine, 4-vinyl pyridine, 2-methyl-5-ethyl pyridine, 2-methyl-5-vinyl pyridine, 2,4-lutidine, 2,5-lutidine, pyrrole, pyrrolidine, and piperidine) were eluted in the order of their boiling points using a dioctyl phthalate column, and a linear relationship was obtained between their relative retention times (RRT) and boiling points. Among the bases, in which the same positions were substituted by various groups, a similar linear relationship was found to exist between relative retention times and boiling points by use of a polyethylene glycol 1500 column treated in advance with sodium hydroxide to prevent tailing. These linear relationships for the phthalate and glycol columns are shown in Figures 1.3 and 1.4, respectively, and unknown bases could be tentatively identified accurately from a two-dimensional diagram of the

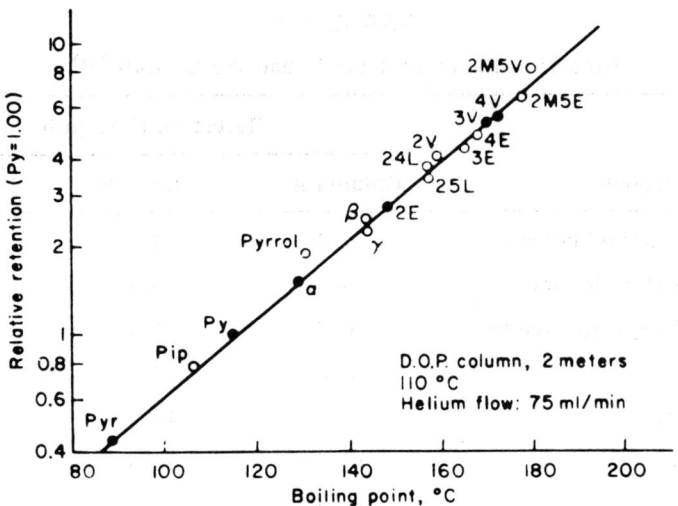

Figure 1.3. Relationship between RRT and boiling point (dioctyl phthalate column). Adapted from Kobashi [4].

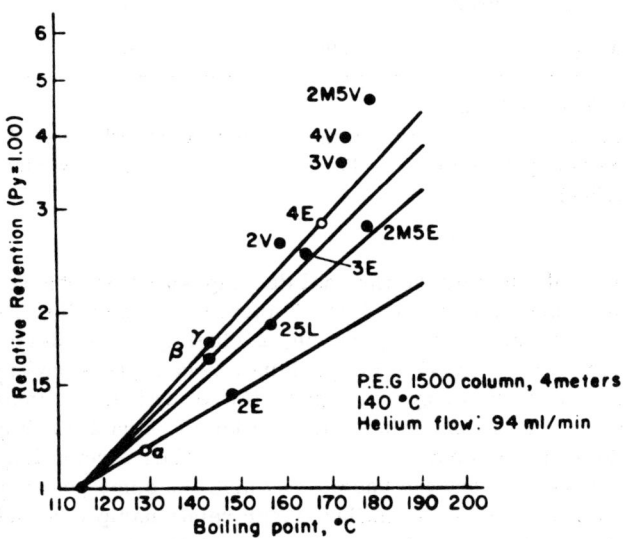

Figure 1.4. Relationship between RRT and boiling point (PEG 1500 column). Adapted from Kobashi [4].

relative retention times obtained both from dioctyl phthalate and from polyethylene glycol columns.

Nicotine bases and related high-boiling pyridine bases (nicotine, nornicotine, anabasine, anatabine, myosmine, nicotyrine, cotinine, dihydrometanicotine, N-methyl anabasine, metanicotine, α,β-dipyridyl, β-pyridyl methyl ketone, β-pyridyl n-propyl ketone, α-aminopyridine, β-aminopyridine, N-methyl nicotinamide, and α,α-dipyridyl) were separated and identified by the use of silicone grease and polyethylene glycol 6000 columns treated with sodium hydroxide. The retention times of these compounds relative to nicotine with the PEG 6000 and silicone columns are listed in Table 1.3, which also includes the boiling point of each component. However, a distinct linear relationship between the relative retention times and boiling points was not observed by Kobashi.

On the other hand, Kobashi and Watanabe [5] determined by means of gas chromatography the homologs of pyridine and nicotine that were expected to be produced by the pyrolysis of tobacco alkaloids. Mixtures of pyridine bases (pyridine, β-picoline, and 3-ethyl pyridine) were separated completely without tailing with a 4-m polyethylene glycol 1500 column which had been treated in advance with a small amount of sodium hydroxide at 140°C with a helium carrier-gas flow rate of 94 ml/min. For the determination of the bases, calibration curves showing the relation between a ratio of peak areas and a ratio of weights of bases were prepared within 1% error; 2,4-lutidine was used as an internal standard.

Nicotine, nornicotinine, and anabasine were separated by the use of a 2-m polyethylene glycol 6000 column pretreated with sodium hydroxide at 210°C under a helium flow of 40 ml/min. The determination of these alkaloids was carried out within 1.5% error by the use of calibration curves prepared in a manner similar to that described above, except that quinoline was used as the internal standard.

In 1962, Quin and Pappas [6] developed a GC method for the determination of nicotine and nornicotine in tobacco that is reportedly useful when the amount of alkaloid exceeds about 0.2%. The analytical extraction of the tobacco alkaloids consisted of the following:

Two grams of cured tobacco passing a 40-mesh sieve was accurately weighed and then wetted with 10 ml of a 5% acetic acid solution; exactly 20 ml of benzene-chloroform solution, 1 to 1 (v/v), saturated with a 36% sodium hydroxide solution was added, followed by 10 ml of 36% NaOH. The mixture was shaken for 20 min and then one spoonful of Celite was added. About 2 ml of the organic solution was collected by filtration and used directly, in 50-μl aliquots, for gas chromatography. If the tobacco contains less than about 0.5% of nicotine, a concentration step is necessary.

TABLE 1.3

Relative Retention Times of Nicotine Bases and High-Boiling Pyridine Bases[a]

No.	Compound	BP (°C)	Relative retention times	
			PEG 6000	Silicone grease
1	α-Aminopyridine	211	0.89	0.27
2	β-Pyridyl methyl ketone	218	0.88	0.39
3	β-Pyridyl n-propyl ketone	245	1.24	0.77
4	Nicotine	247	1.00 (10.6 min)	1.00 (11.3 min)
5	N-methyl anabasine	121/7 mm Hg	1.55	1.44
6	β-Aminopyridine	250	2.00	0.39
7	Dihydrometanicotine	141/15 mm Hg	2.00	1.46
8	Nornicotine	267	2.30	1.24
9	Myosmine	118/3.2 mm Hg	2.47	1.26
10	Anabasine	146/15 mm Hg	2.59	1.72
11	Anatabine	145/10 mm Hg	4.60	1.60
12	α,α'-Dipyridyl	272	2.78	1.37
13	Metanicotine	110/1.5 mm Hg	2.87	1.64
14	Nicotyrine	280	3.47	1.45
15	α,β'-Dipyridyl	137/4 mm Hg	5.50	1.83
16	N-methyl nicotinamide		10.30	1.73
17	Cotinine	210/6 mm Hg	14.20	3.10

[a] From Kobashi [4], courtesy of Nippon Kagaku Zasshi.

Using a Perkin-Elmer model 154-B gas chromatograph equipped with a thermal conductivity detector, two different column systems and GC conditions were employed for the analysis of nicotine and nornicotine.

For the determination of nicotine, the various parameters used were as follows:

Column: 1-m by 6-mm-o.d. glass U-shaped packed with 20% polypropylene glycol 1025 coated on 30-60 mesh, KOH-washed Firebrick

Column temp.: 180 or 190°C

Helium CG flow rate (ml/min): 75 (at 190°C) and 50 (at 180°C)

With the above conditions, nicotine was eluted in approximately 5 and 10 min, respectively, at column temperatures of 190 and 180°C. By adding 40 mg of nicotine (in benzene solution) to two 2.0-g samples of burley tobacco, the recovery was 101.2 ± 0.5%.

On the other hand, the GC method for nornicotine determination was as follows:

Column: 1-m by 6-mm-o.d. glass U-shaped packed with 20% polybutylene glycol 1500 coated on 30-60 mesh, KOH-washed Firebrick

Column temp.: 180°C

Helium CG flow rate (ml/min): 50 ml/min

With these GC operating parameters, nornicotine had a retention time of about 10 min. As noted, this column was satisfactory for resolving nornicotine from nicotine and anabasine; these latter two components eluted in 6.2 and 12.5 min, respectively. However, with the polypropylene column, nornicotine could be separated from nicotine, but not from anabasine.

Using calibration curves prepared by plotting peak height versus nicotine or nornicotine concentration, results for several replicate samples of several tobaccos are given in Table 1.4 which are compared with values for total tertiary alkaloids obtained with a revised Cundiff-Markunas procedure [7].

Since it is well known that only a part of the nicotine present in tobacco is transported into tobacco smoke and that more than 30 bases structurally related to nicotine have been identified in smoke, Jarboe and Rosene [8] used packed reactors and helium as an inert atmosphere in order to study the pyrolysis of nicotine. At 600 to 900°C the pyrolysis produced a variety of heterocyclic nitrogen and aromatic hydrocarbons; the volatiles were separated and characterized as shown in Table 1.5.

TABLE 1.4

Alkaloid Content of Several Tobaccos[a]

Sample	Nicotine (wt. %)							
	Burley		Bright		Commercial cigarette		Cigar filler	
	GC[b]	C-M[c]	GC[b]	C-M[c]	GC[b]	C-M[c]	GC[b]	C-M[c]
1	2.04	2.07	1.68	1.70	1.28	1.42	0.59	0.59
2	2.06	2.07	1.72	1.62	1.34	1.42	0.62	0.59
3	2.06	2.12	1.68	1.65	1.24	1.42	0.56	0.59
4	1.92		1.67		1.25		0.59	
5	1.93		1.67		1.31		0.62	
Av.	2.00	2.09	1.68	1.66	1.29	1.42	0.60	0.59
Std. dev.	0.07	0.03	0.02	0.04	0.04	0.00	0.03	0.00

	Nornicotine (wt. %)			
	N. sylvestris		Cherry red N. tabacum	
	GC[b]	C-M[d]	GC[b]	C-M[d]
1	1.53	1.62	1.41	1.31
2	1.56		1.41	1.34
3			1.41	
Av.	1.55	1.62	1.41	1.33

[a] From Quin and Pappas [6], courtesy of Journal of Agricultural and Food Chemistry.

[b] Gas chromatographic method, conditions as described in experimental section.

[c] Cundiff-Markunas method for total tertiary alkaloids, expressed as nicotine [7].

[d] Cundiff-Markunas method for total secondary alkaloids, expressed as nornicotine.

TABLE 1.5

Pyrolysis Products of Nicotine[a,b]

Compound	Retention time[c]		Temp. (°C)			
	Ucon	Carbowax	600	700	800	900
Pyridine	1.00	1.00	0.50	3.40	6.00	4.70
3-Picoline	1.64	1.46	1.40	4.00	2.60	0.05
3-Ethyl pyridine	2.15	1.92	1.10	0.46	0.03	0.03
3-Vinyl pyridine	2.82	2.42	10.00	8.00	2.60	0.03
Metanicotine		2.46	0.01	0.01	0.01	
Benzonitrile	4.10	3.61			0.01	0.44
3-Cyanopyridine	5.50	5.43	0.64	1.80	1.10	0.05
Naphthalene	6.85	5.46			0.01	0.45
Compound A		4.24	0.11	0.10	0.13	
3-(Buta-1,3-dienyl)pyridine		5.96	0.17	0.01		
2-Cyanopyridine	9.40	8.31			0.01	0.01
Compound B	10.60	8.65		0.41		
Quinoline	12.10	9.82	0.01	1.60	1.50	0.82
Nicotine	11.20	9.28	34.00			
Isoquinoline	13.90	11.30	0.01	0.20	0.40	0.20
Nornicotyrine		12.00	0.01	0.01	0.01	
Myosmine	22.20	23.00	11.00			
1,7-Diazaindene	76.30	61.80			0.01	0.01

[a] From Jarboe and Rosene [8], courtesy of <u>Journal of the Chemical Society</u>.
[b] Yields (wt. %) based on initial nicotine.
[c] Gas-liquid chromatography, with pyridine as internal standard.

In their experiments, no evidence was found for the production of nornicotine, nicotyrine, N-methylmyosmine, nicotinamide, or nicotinic acid, all of which had been found when air was the carrier gas. Based on the compounds isolated, activation energies, bond strengths, and analogy with similar systems, Jarboe and Rosene postulated a mechanism involving formation of the radical (1) form of nicotine which subsequently may react by elimination to form myosmine or rearrange to radical (2), which is free to form metanicotine by elimination.

The formation of 3-vinyl pyridine in relatively large quantities at lower temperatures is the most attractive source of 3-ethyl pyridine and 3-picoline. These compounds could most logically arise by reduction, formation of radical (3) by disproportionation, and hydrogen abstraction to form

3-picoline. The investigators also noted that the isolation of 1,7-diazaindene was of particular interest, as it follows the possibility of thermally initiated cyclization involving a three-atom rather than a four-atom side chain as in the formation of quinoline, isoquinoline, and naphthalene.

In their study of the mechanism of the demethylation of nicotine, Craig, Mary, Goldman, and Wolf [9], by the combined use of thin-layer and gas chromatography, showed that the metal complex-catalyzed rearrangement of nicotine-N-oxide yielded formaldehyde, nicotine, N-methylmyosmine, nornicotine, myosmine, nicotyrine, and cotinine. Although nicotine, metabolized in animals and plants to a variety of products, yields only nornicotine by a simple demethylation, the other metabolic products include nicotine-1'-oxide or oxynicotine; 3-methylaminopropyl-3'-pyridyl ketone; 3-nicotinoylpropionic acid, N-methylmyosmine; nicotyrine; γ-3-pyridyl-γ-hydroxybutyric acid; desmethylcotinine; γ-3-pyridyl-γ-methylaminobutyric acid; myosmine, a naturally occurring nicotine alkaloid which can be formed by autoxidation of nicotine; and γ-3-pyridyl-γ-oxo-N-methylbutyramide.

Whereas the rearrangement of N-benzyldimethylamine oxides was shown by Craig, Mary, and Wolf [10] to proceed by two pathways, both of which yield formaldehyde and benzaldehyde in a ratio determined by their relative acidities and the available α-protons adjacent to nitrogen, nicotine-1'-oxide may rearrange in three possible ways: (1) loss of proton from the N-methyl group leading to methylolamine, which in turn gives formaldehyde and nornicotine; (2) loss of the proton from the α-position adjacent to the pyridine ring giving the tertiary hydroxynicotine, the cyclic ketal of 3-methylaminopropyl-3'-pyridyl ketone, whose expected ease of dehydration

Chemical Composition of Tobacco Leaf

Pathway 1

nicotine-1'-oxide ⟶ R—[pyrrolidine with N-CH$_2$OH] ⟶ nornicotine + CH$_2$O

Pathway 2

nicotine-1'-oxide ⟶ R—[pyrrolidine with OH and N-Me] ⟶ R—[dihydropyrrole with N-Me]

Pathway 3

nicotine-1'-oxide ⟶ R—[pyrrolidine with H, OH and N] ⟶ R—[open chain with NH-Me and COOH] ⟶ R—[pyrrolidinone with N-Me, =O]

Where R = (3-pyridyl).

would give N-methylmyosmine; and (3) proton removal from the other α-position of the pyrrolidine ring to give the secondary hydroxynicotine which, by dehydrogenation and ring closure, can lead to γ-3-pyridyl-γ-methyl-aminobutyric acid and continine, respectively.

Using the metal complex-catalyzed rearrangement of nicotine-N-oxide with the iron(III) tartaric acid system as described [10] for the model system N-p-nitrobenzyldimethylamine oxide and for trimethylamine oxide, the postulated simple unified mechanism, depending only on an attack on the α-proton in the rearrangement of the N-oxide, was tested with gas chromatography playing a major role in the identification and determination of the alkaloidal products of rearrangement.

The retention times of the alkaloids are listed in Table 1.6, and the separation shown in Figure 1.5 was achieved with a U-shaped glass column, 6 ft by 5 mm, packed with 5.6% polyethylene glycol 20M on C-22 Firebrick, an argon flow of 43 ml/min, and a column temperature of 200°C. A 5-μl sample of the benzene extract of alkaloidal products from the reaction mixture was injected into the chromatograph, and the chromatogram obtained showed the presence of 10 peaks as indicated in Figure 1.6.

The individual and total amount of the alkaloids in the reaction mixture are shown in Table 1.7. Peaks W, X, Y, and Z remained unidentified, with retention times of 18.40, 22.88, 53.28, and 73.76, respectively. The preponderance of simple N-demethylation was expected in view of the enhanced acidity of the α-proton adjacent to the pyridine ring but, as seen in Table 1.7, 77 wt.% of the reaction product was found to be nornicotine.

TABLE 1.6

Retention Time of Individual Alkaloids[a]

Alkaloid	Time (min)	Ratio	Quin [11] ratio	Kobashi and Watanabe ratio
3-Pyridyl ethyl ketone	17.28	1.00	1.01	
3-Pyridyl n-propyl ketone	21.92	1.26	1.27	1.24
Nicotine	17.28	1.00	1.00	1.00
N-Methylmyosmine	25.76	1.49		
Nornicotine	39.68	2.30	2.37	2.30
Myosmine	42.88	2.49	2.57	2.47
Anabasine	43.52	2.52	2.65	2.59
Metanicotine	45.92	2.66	3.17	2.87
Nicotyrine	58.88	3.42	3.73	3.47
N-Methylnicotinamide	179.36	10.47	12.30	10.30
Cotinine	239.20	13.84	16.40	14.20

[a] From Craig et al. [9], courtesy of the Journal of the American Chemical Society.

However, the investigators postulated that the explanation may be found in the stereochemistry of the N-oxide, which can exist in two possible conformations.

In addition to the above studies, the GC analysis and separation of nicotine and other tobacco alkaloids have been performed by many investigators [12-34] over the past 20 years. For example, Kanazawa and Sato [12] described in 1964 a method for the determination of nicotine in nicotine sulfate. In their procedure, a Shimadzu GC-2A gas chromatograph was used and equipped with a 2-m by 4-mm-o.d. column packed with 10% PEG 6000 coated on 32-48 mesh Celite 545. For the elution of nicotine, the column was held isothermally at 170°C and the helium carrier-gas flow rate was maintained at 50 ml/min.

In 1964, Alworth et al. [13] studied the biosynthesis of nicotine in Nicotiana glutinosa from carbon-14 dioxide. As noted by these investigators, plants of Nicotiana glutinosa were grown in an atmosphere containing $^{14}CO_2$ for periods varying from 2 hr, the shortest time at which incorporation of

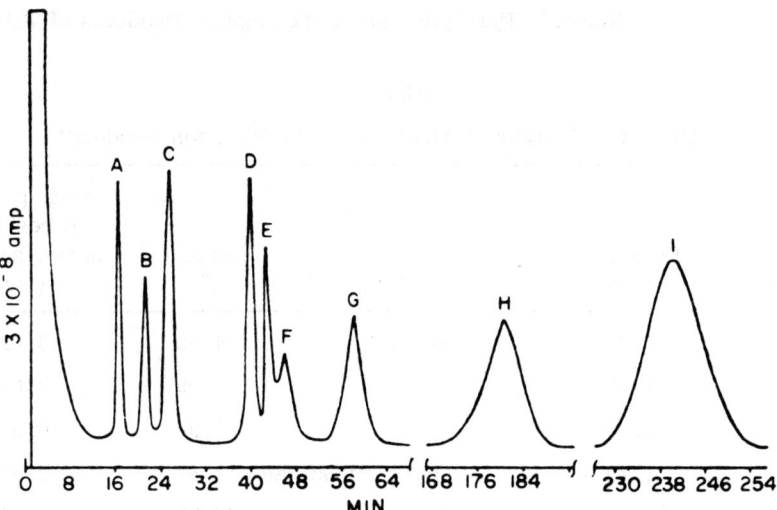

Figure 1.5. Gas chromatographic behavior of nicotine alkaloids. Peak A, nicotine (or 3-pyridyl ethyl ketone); B, 3-pyridyl n-propyl ketone; C, N-methylmyosmine; D, nornicotine; E, myosmine (or anabasine); F, meta-nicotine; G, nicotyrine; H, N-methylnicotinamide; I, cotinine. From Craig et al. [9], courtesy of the Journal of the American Chemical Society.

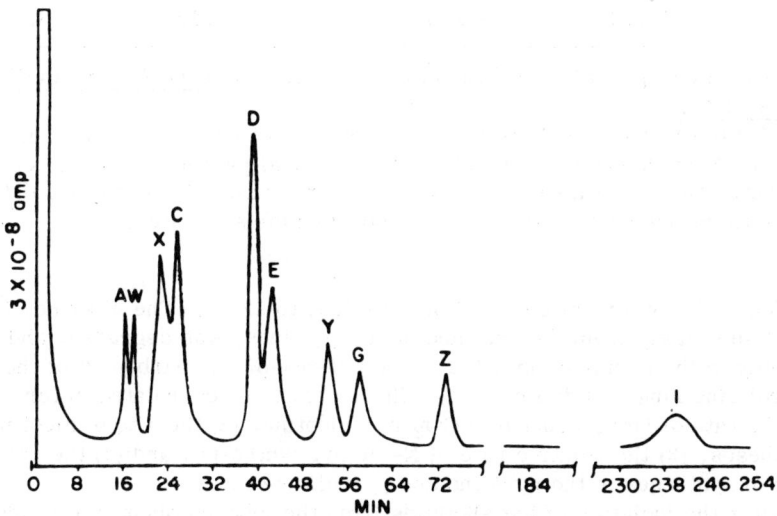

Figure 1.6. Gas chromatogram of benzene extract of nicotine N-oxide reaction mixture. Peak A, nicotine, sensitivity 1; W, unidentified, sensitivity 1; X, unidentified, sensitivity 30; C, N-methylmyosmine, sensitivity 30; D, nornicotine, sensitivity 3; E, myosmine, sensitivity 30; Y, unidentified, sensitivity 100; G, nicotyrine, sensitivity 100; Z, unidentified, sensitivity 100; I, cotinine, sensitivity 100. From Craig et al. [9], courtesy of the Journal of the American Chemical Society.

TABLE 1.7

Amount of Individual Alkaloids in the Reaction Product[a]

Peak	Retention time ratio	Alkaloid	Weight (%)	Amt. per ml of reaction prod. soln.[b] (mg)
A	1.00	Nicotine	4.09	3.62
W	1.07	?	4.09	3.62
X	1.35	?	0.96	0.85
C	1.49	N-Methylmyosmine	2.87	2.54
D	2.30	Nornicotine	77.15	68.24
E	2.49	Myosmine	9.50	8.40
Y	3.09	?	0.17	0.15
G	3.42	Nicotyrine	0.58	0.52
Z	4.28	?	0.16	0.14
I	13.84	Cotinine	0.40	0.36

[a] From Craig et al. [9], courtesy of the Journal of the American Chemical Society.
[b] Total amount = 88.44 per ml of reaction product solution; 180 mg of nicotine N-oxide gave 130 mg of crude alkaloidal product from the benzene extract. This was dissolved to make 1.30 ml of reaction product solution, which therefore contained 115 mg of total alkaloids isolated.

radioactivity into nicotine could be detected, to 12 hr. The nicotine, isolated separately from the root and aerial portions, was degraded, and the activity in the pyridine ring, the N-methyl group, and carbon-2' of the pyrrolidine ring was determined. These data were correlated in terms of (1) the rate of incorporation of CO_2 into nicotine, (2) the site of nicotine syntheses, (3) the relative rate of N-methyl syntheses, and (4) the relative rate of syntheses of the pyridine and pyrrolidine rings.

After the isolation of the alkaloids from the tobacco plant, the crude alkaloid fraction was chromatographed with a GC instrument equipped with a 5-ft by 0.5-in. stainless steel column packed with 10% by weight polybutylene glycol coated on KOH-treated, 60-80 mesh Firebrick. With this column maintained isothermally at 169°C and the helium flow rate held at 300 ml/min, the retention times for nicotine, nornicotine, anabasine, and anatabine were 6.25, 10.25, 13.00, and 16.00 min, respectively. Using a calibration curve prepared by plotting peak area versus nicotine concentration, the amount of nicotine in the alkaloid fraction could be determined. For further GC studies, the alkaloids corresponding to each of the four compounds cited above were collected by cooling the effluent in liquid nitrogen.

The radioactivity of the nicotine fraction was also determined with an apparatus consisting of an Aerograph gas chromatograph, a model 182 Nuclear-Chicago scaler, and a modified flow-through proportional counter. For this investigation, a 5-ft by 0.25-in. stainless steel column was used, packed as above, and held at approximately 198° C with a helium flow rate of 75 ml/min. The retention time of nicotine using these operating conditions was 6.20 min. The proportional counter tube was heated to 175 to 185° C by means of heating tape and, in order to provide a gas mixture that was suitable for ^{14}C counting, it was necessary to add a methane flow of about 35 ml/min to the carrier gas just before it entered the counter tube. For comparative purposes, values obtained by GC and scintillation counting are listed in Table 1.8, where it is noted that the specific activity of nicotine can be determined by the GC-proportional counter method to an accuracy of better than 15%.

In a subsequent study, the GC-proportional counting method was also employed for the determination of the specific activity of the N-methyl group of nicotine. As noted by the investigators:

> Samples which had been counted as described were collected on cotton as they were eluted from the counter tube. The nicotine was then removed and subjected to a Herzig-Meyer determination, collecting the methyl iodide generated in 5 ml of toluene cooled in a Dry Ice-acetone bath. The solution of methyl iodide in toluene could now be injected into the GC-proportional counter apparatus and the mass and activity determined by the procedure described for nicotine.

TABLE 1.8[a]

Specific Activity of Nicotine

Sample	Activity (μCi/mmole)	
	GC	Scintillation
A	40.70	47.40
B	5.85	6.23
C	0.86	0.95

[a] Adapted from Alworth et al. [13].

TABLE 1.9[a]

Specific Activity of N-Methyl Group of Nicotine

Sample	Activity (μCi/mmole)	
	GC	Scintillation
I	1.5	2.0
II	0.60	0.95
III	0.210	0.385
IV	0.062	0.130

[a] Adapted from Alworth et al. [13].

For CH_3I determination, Alworth et al. used a 10-ft by 0.25-in. stainless steel column packed with a mixed liquid stationary phase (7.5% Apiezon L: 7.5% polyamine 6) coated on 60-80 mesh Firebrick. Using a column temperature of 55°C and a methane carrier-gas flow rate of 40 ml/min, the elution time for CH_3I was approximately 4 min. As shown in Table 1.9, one notes that the GC-proportional counting method is less precise when applied to $^{14}CH_3I$ than when applied to nicotine.

With regard to the biosynthesis of nicotine alkaloids in Nicotiana glutinosa and using the instrumentation and methodology described above, Alworth and Rapoport [15] reported in 1965 that:

Biosyntheses from $^{14}CO_2$ carried out with both intact Nicotiana glutinosa plants and seedlings have demonstrated that nornicotine is not the precursor of nicotine in either the root or the aerial portions of the plant. This conclusion contrasts with that reached on the basis of ^3H incorporations into the alkaloids of Nicotiana rustica. The most probable relationship, based on $^{14}CO_2$ biosyntheses and supporting some previous conclusions, is that nornicotine arises by demethylation of nicotine in both the root and aerial portions. The relative specific activities of anabasine and anatabine after the $^{14}CO_2$ biosyntheses indicated that anabasine is not the precursor of anatabine. In the case of nicotine, nornicotine, anatabine, and anabasine, higher specific activities were noted in the samples isolated from the root than in those isolated from the aerial portions. This shows that these four alkaloids all can be formed in the Nicotiana root, and furthermore, since the ^{14}C in these experiments first enters into the metabolism of the aerial portion of the plants, these results indicate that the N. glutinosa root is the most active site of formation of each of these alkaloids. Finally, a comparison of the specific activities of anabasine, anatabine, and nicotine suggests that if a precursor-product relationship exists among these alkaloids it must be nicotine, anatabine or anabasine, or both.

In contrast to the use of packed columns for gas-liquid chromatography of alkaloids, Massingill and Hodgkins [16] evaluated three capillary for 25 alkaloids as well as four new packed columns as shown in Table 1.10, which also includes operating conditions used with a Barber-Colman model 5000 gas chromatograph equipped with a linear temperature programmer, a stream splitter, and a high-temperature flame ionization detector. Included among the alkaloids evaluated with these packed/capillary columns were anabasine and nicotine, whose retention times in minutes were determined on six of the seven columns listed in Table 1.10; these values were as follows:

Packed columns			Capillary columns			
1% SE-52	1% JXR	1% XE-60	SE-30	Apiezon L	QF-1	Compound
1.67	2.42	1.92	4.17	3.33	2.00	Nicotine
2.83	3.58	3.42		10.08	4.08	Anabasine

In 1968, Harke and Drews [20] also reported the separation of nicotine alkaloids by capillary gas chromatography. In their method, a Carlo Erba Fractovap D gas chromatograph equipped with a flame ionization detector and a 50-m by 0.5-mm glass capillary coated with polypropylene glycol (UCON LB 550 X) and KOH was used. Using a helium carrier-gas column inlet pressure of 0.5 atm, a column temperature of 195°C, and a FID temperature of 240°C, the retention times for 3-pyridyl-n-propylketone, nicotine, nornicotine, myosmine, anabasine, and nicotyrine were approximately 10.00, 10.65, 17.10, 17.10, 20.20, and 21.15 min, respectively. As noted, with this column nornicotine and myosmine could not be resolved.

Jacin et al. [18] described a GC method for the determination of nicotine in tobacco which was extracted using a modification of the Cundiff and Markunas [35] procedure. The extraction from tobacco was as follows:

A tobacco sample (0.3-0.5 g) was blended with 0.5 g of reagent-grade barium hydroxide octahydrate in a glass-stoppered 125-ml Erlenmeyer flask. Saturated barium hydroxide solution (10 ml) was then added to the blend and the contents were mixed manually. A benzene-chloroform (9:1) mixture (50 ml) was quantitatively added from a burette to the Erlenmeyer. The contents were shaken for 20

TABLE 1.10

Operating Conditions[a]

	Packed columns				Capillary columns		
	1% JXR	1% SE-52	1% XE-60	0.5% Epon 1001	QF-1	SE-30	Apiezon L
Percent liquid phase							
Column dimensions	6 ft × 1/8-in. o.d.	6 ft × 1/8-in. o.d.	6 ft × 1/8-in. o.d.	2 ft × 1/8-in. o.d.	100 ft × 0.01-in. i.d.	200 ft × 0.01-in. i.d.	100 ft × 0.01-in. i.d.
Column temperature (°C)[b]	100–300°C 12°C/min	100–300°C 12°C/min	100–250°C 12°C/min	100–250°C 12°C/min	100–200°C 12°C/min	100–250°C 12°C/min	100–250°C 12°C/min
Flow rate	58 ml/min	70 ml/min	70 ml/min	91 ml/min	15 ml/min	8.5 ml/min	15 ml/min
Inlet pressure (psi)[c]	40	40	40	40	80	80	80
Sample size[d]	0.1–1 μl	0.1–1 μl	0.1–1 μl	0.1–1 μl	0.5 μl	1 μl	0.5 μl
Split ratio					1/10	1/10	1/10

[a] From Massingill and Hodgkins [16], courtesy of Analytical Chemistry.
[b] Flash heater temperature, 290°C; detector temperature, 310°C.
[c] Outlet pressure, atmospheric.
[d] Sample concentration, 1 to 5 μg alkaloid/μl solvent (usually acetone).

min on a wrist-action shaker. The flask was then removed from the shaker and aliquots of the benzene-chloroform layer were injected into the gas chromatograph for nicotine determination.

For their investigation, Jacin et al. used a Perkin-Elmer model 800 gas chromatograph equipped with a flame ionization detector and a 6-ft by 1/8-in. column packed with 10% NGA (neopentyl glycol adipate) coated on 80-100 mesh Gas Chrom Q. With the specified chromatographic conditions (injection temperature, 235°C; column temperature, 168°C; nitrogen carrier-gas flow rate, 33 ml/min), nicotine, nornicotine, and anabasine were separated and eluted in 5.5, 14.0, and 15.5 min, respectively.

Using a calibration curve of peak height versus nicotine concentration over the range of 0.1 to 0.6 mg of nicotine per microliter, the recovery of nicotine added to Maryland stem, bright stem, bright lamella, and reconstituted tobacco sheet was 101.75 ± 2.75, 104.0 ± 1.0, 98.33 ± 1.44, and $102.67 \pm 0.44\%$, respectively. When compared with a spectrophotometric method used for determining total alkaloids in tobacco as nicotine, the GC values for 11 different tobacco specimens reported as a percentage of the spectral data were $94.41 \pm 3.67\%$. This was anticipated since the GC procedure determines nicotine only, whereas the spectrophotometric procedure includes other alkaloids.

Whereas Anastasov et al. [19] separated nicotine, nornicotine-myosmine, and anabasine into three respective peaks with 20% Apiezon L on Chromosorb W with retention times of 30 to 70 min, Kobashi and Watanabe [14] used 2-m column packed with 24% polyethylene glycol, a carrier-gas flow rate of 80 ml/min, and a column temperature of 180°C to partially resolve myosmine from nornicotine, and Yasumatsu [17] obtained relative retention times of 1.00 (15.50 min), 1.57, 1.97, and 2.36 for nicotine, nornicotine, anabasine, and anatabine, respectively, using a 2.25-m column of 25% DC 550, a column temperature of 200°C, and a carrier-gas flow rate of 60 ml/min, Weeks et al. [21] compared the analyses of nicotine and four of the minor tobacco alkaloids on three GC column substrates for retention times, effective plate number, and resolution.

These separations were performed with a Barber-Colman model 5000 gas chromatograph equipped with a dual flame ionization detector and 2.44-m columns packed either with 10% DC 550, 5% SE-30, or 10% Veramid 900 coated on 60-80 mesh, acid-washed, DMCS-treated Chromosorb W. Using injector and detector temperatures of 225 and 370°C, respectively, and helium as carrier gas, the retention times and resolution of the tobacco alkaloids studied are listed in Table 1.11 in addition to the carrier-gas flow rates and column temperatures employed. Under the operating parameters specified, the DC 550 and Versamid 900 columns were superior to the SE-30 column. With both these columns, nicotine, nornicotine, anabasine, and anatabine were 99% resolved from each other.

TABLE 1.11

Retention Times (r, min) and Resolution (R) of Tobacco Alkaloids[a]

Alkaloid	DC 550		SE-30		Versamid 900	
	r	R	r	R	r	R
Nicotine	9.07		2.05		6.48	
		5.12		1.57		
Nornicotine	14.05		2.48		13.92	
		1.24				0.52
Myosmine	16.95			1.20	12.77	
		0.64				
Anabasine	18.40		2.89		16.85	
		2.18				3.28
Anatabine	22.60				21.20	
CG flow rate (ml/min)	40		30		40	
Column temp. (°C)	170		190		170	

[a] Adapted from Weeks et al. [21].

Jenden et al. [24] described in 1972 a new liquid phase which permits the efficient separation and analysis of amines and amino esters by gas chromatography. As pointed out by Jenden et al., the phase is a polyamide containing tertiary amine groups. It may be coated neat or with a silicone oil on a deactivated support such as Gas Chrom Q and may be used as a modifier for porous polystyrene solid phases such as Porapak P. Although it was developed specifically for the analysis of N-demethylated choline esters, the new phase has proved very satisfactory for a variety of basic compounds, giving high efficiencies (400 to 600 plates/foot), symmetrical peaks, and bleed sufficiently low for satisfactory performance in an integrated GC-MS system. The brown viscous liquid stationary phase is believed to be a linear polymer of 4-dodecyldiethylenetriamine succinimide (DDTS), terminated by succinimide groups.

In their investigation, the retention times (measured after injection of 5 nmoles in methylene chloride) were determined with a F&M model 5705A gas chromatograph equipped with flame ionization detectors and 10-ft by 1/8-in. columns packed with either (1) 5% DDTS plus 5% OV-101, (2) 0.3%

DDTS plus 1% OV-101, or (3) 1% DDTS. Using a nitrogen carrier-gas flow rate of 45 ml/min, the 0.3% DDTS plus 1% OV-101 column held isothermally at 150°C, and injector and detector temperatures maintained about 50°C above the indicated column temperature, nicotine was eluted in nearly 4.5 min. The retention times of other amines and alkaloids of neurochemical interest are given in Table 2.5 of Volume 3.

Bush [25] developed a quantitative procedure for the four most important tobacco alkaloids (nicotine, nornicotine, anabasine, and anatabine) which was based primarily on the separation of the tobacco alkaloids as previously reported by Weeks et al. [21].

Using a modified version of the procedure of Keller et al. [36] for the extraction and isolation of alkaloids from tobacco samples with quinoline added as internal standard, their separation and quantitation were performed with a Varian model 1740 gas chromatograph equipped with flame ionization detectors and 3.05-m by 2-mm-i.d. glass coiled columns packed with 10% DC 550 coated on acid-washed, DMCS-treated, 60-80 mesh Chromosorb W. With quinoline added as internal standard, the injector, column, and detector temperatures maintained at 250, 185, and 300°C, respectively, and the nitrogen carrier-gas flow rate adjusted to 30 ml/min, the weight responses (RWR) relative to quinoline (RWR = 1.00) with the flame detector for nicotine, nornicotine, anabasine, and anatabine were 0.81, 0.31, 0.55, and 0.58, respectively. Tested on several tobacco samples spiked with two levels of alkaloids, the GC summations were 3.0 to 6.6% greater than total alkaloids by steam distillation. The largest standard deviation for the individual alkaloids was observed for nornicotine (3.6% average) and the smallest for anatabine (1.8%).

For a series of central nervous stimulant drugs, Stead et al. [26] compared GC retention indices on support-coated open tubular (SCOT) columns and on packed columns. The packed columns were (1) 2-m, 2% Apiezon L/5% KOH; (2) 1-m, 1% Carbowax 20M/5% KOH; both coated on acid-washed, DMCS-treated 80-100 mesh Chromosorb G. With nitrogen as the carrier gas, either a Pye 104 or Perkin-Elmer F11 gas chromatograph was used; the column operated isothermally in the range 110 to 160°C.

On the other hand, the capillary columns used were those developed and described by Caddy et al. [29].

The retention indices for nicotine on SCOT and packed columns using two stationary phases obtained by Stead et al. were as follows:

Apiezon L/KOH		Carbowax 20M/KOH	
SCOT	Packed	SCOT	Packed
1373	1382	1901	1658

In 1973, Caddy et al. [27-29] discussed the comparison of open tubular columns for the analysis of central nervous system stimulant drugs [27], and described a system based on Kovats' retention indices for the identification of CNS stimulants recovered from body fluids, making use of the difference in retention times found on polar and nonpolar liquid-phase packed GC columns (ΔI values) and the effect of operating temperature [29], and the limitations imposed by the use of a flame ionization detector in the quantitative analysis of CNS drugs on support-coated open tubular columns as well as the use of an alkali flame ionization detector which was advocated as a means for improving sensitivity [28].

In their study of open tubular columns, they proposed the use of SCOT columns for the routine analysis of amphetamines; the details of the two most favorable SCOT columns prepared are given in Volume 3, Table 2.6. Using the SCOT columns A and B (Table 2.6 in Volume 3), retention time data at 150 and 190°C at prescribed helium flow-rate settings for some 33 CNS stimulant drugs and metabolites (nicotine included) are listed in Table 1.36 of Volume 4. Using their proposed system, the CNS drugs examined could be determined at the nanogram-per-milliliter level. For nicotine, the retention times on two SCOT columns at 150 and 190°C were as follows:

Retention Time (min)

Column A		Column B	
150°C	190°C	150°C	190°C
5.33	1.80	3.76	0.77

As for the use of Kovats' retention indices for the identification of compounds, Caddy et al. [29] adopted a two-dimensional plot using information from two column types (see Table 2.6 in Volume 3) and expressed index limits as "boxes" to assist in peak identification. The mean retention indices of the CNS stimulants studied on polar (column A) and nonpolar (column B) columns, each operated at 150 and 190°C, are listed in Table 1.37 of Volume 4; these values were then used to construct identification charts by plotting a retention index found for each drug on the nonpolar column against a corresponding index value found on the polar column. The retention indices of nicotine on two SCOT columns at 150 and 190°C are noted below:

Retention Indices (I Values)

Column A		Column B	
150°C	190°C	150°C	190°C
1876	1926	1353	1392

Although the quantitative analysis of CNS stimulant drugs with SCOT columns using a flame ionization detector is possible, Caddy et al. [28] obtained a tenfold improvement in detection limits (down to and possibly lower than 200 ng/ml in some instances) when employing an alkali flame ionization detector.

Donike and Stratmann [30] also showed that, in drug control as well as in pharmacokinetic studies, it was possible to obtain reproducible analyses on nitrogen-containing drugs in the concentration range of 1 to 200 µg/ml with the aid of a nitrogen-specific alkali flame detector and a steep temperature gradient. Using a Hewlett-Packard model 7600 gas chromatograph equipped with an automatic sampler, a nitrogen-specific alkali flame ionization detector, and a 1.06-m by 2.5-mm-i.d. glass column packed with 2% Igepal CO-880 and 12.5% Apiezon L on 60-80 mesh, acid-base washed Chromosorb W, nitrogen-containing volatile drugs were chromatographed employing the following GC conditions: column temperature, programmed from 130 to 270°C at 20°C/min, then held isothermally at 270°C for 2 min; detector temperature, 380 to 400°C; injector temperature, 250°C; helium carrier-gas flow rate, 60/ml. With diphenylamine as an internal standard, a ten-component drug mixture (nicotine included) was separated; the absolute and relative retention times of the individual compounds are shown in Table 1.38 of Volume 4. With the automatic sampler, the standard deviation attained was 0.1% for absolute and relative retention times, but varied for quantitative determinations with the polarity and the concentration of the active agents from 0.5 to 10%.

For the determination of nicotine in subpicomole quantities, Neelakantan and Kostenbauder [31] reported a unique method for the preparation of a nicotine derivative for electron-capture studies which would also be amenable to mass spectrometric analysis. The procedure was based on the catalytic hydrogenation of nicotine (I) to yield N-methyl-4-(3'-piperidyl)-n-butylamine (octahydronicotine) (II), the secondary amino function of which could then be treated with a perfluoroacid anhydride to provide the electron-capturing derivative. Using the reaction sequence, the dipentafluoropropionyl derivative (III) of octahydronicotine could be easily quantitated in amounts corresponding to 0.03 pmole of nicotine.

Following a micro catalytic hydrogenation (10% Pd/C catalyst), the derivatization of nicotine was carried out in the following manner:

A 2-mg sample of nicotine dihydrochloride was hydrogenated, the octahydronicotine dihydrochloride solution was collected in a 15-ml centrifuge tube and evaporated to dryness at 55°C under reduced pressure, and the residue dried at 70°C under vacuum for 1 hr. To this tube was then added 0.4 ml of perfluoropropionic anhydride, the tube was then sealed with a Teflon-lined screw cap and placed in an oven at 70°C for 2 hr. The excess anhydride was then removed by directing a stream of nitrogen into the tube. The residue was dissolved in 2 ml of heptane, which was then washed first with 2 ml of water (the acidic aqueous wash was saved) and then with 2 ml of 5% sodium bicarbonate solution. The heptane layer was then transferred to another 15-ml centrifuge tube and the solvent was evaporated with a stream of nitrogen. The residue was further dried at 40°C under high vacuum (0.1 mm Hg) and redissolved in 1 ml of n-heptane. One microliter of this solution was withdrawn for GC analysis.

For the GC determination of nicotine, Neelakantan and Kostenbauder used a Varian model 2700 gas chromatograph equipped with both a flame ionization and electron-capture detector and 2-m by 2-mm-i.d. glass columns packed with 1.25% OV-17 on 100-120 mesh Chromosorb G. For FID analysis, the following operating conditions were employed, injector temperature, 245°C; column temperature, 136°C; detector temperature, 200°C; nitrogen carrier-gas flow rate, 30 ml/min. With electron-capture detection, the nitrogen flow rate was also 30 ml/min with the injector, column, and detector temperatures maintained at 245, 185, and 215°C, respectively.

With the FID instrument and operating conditions cited above, underivatized nicotine and octahydronicotine had retention times of 3.00 and 4.75 min, respectively. On the other hand, the pentafluoropropionyl derivative of octahydronicotine was eluted in 9.50 min using the ECD instrument. For quantitative studies, diphenylamine was used as internal standard, its PFP derivative having a retention time of approximately 3.65 min as shown in Figure 1.7.

Standard curves obtained using the PFP derivative of diphenylamine as internal standard indicate that a plot of peak height ratio versus weight ratio is linear over a range of at least 10 pg to 1 ng of OHNPFPD (corresponding to 0.03 to 3 pmoles of nicotine), as shown in Table 1.12.

Foltz et al. [32] developed a rapid method for the detection of drugs in body fluids from suspected overdose cases. Similar to that of Finkle, Foltz, and Taylor [37], the method (described in detail in Volume 2, Chapter 1) utilizes a computerized GC-MS system with initial identification based on the chemical ionization mass spectrum of the drug or the drug's metabolite. Using methane as reactant gas, CI mass spectra were obtained

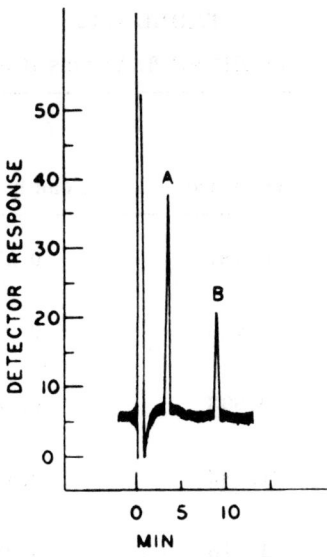

Figure 1.7. Gas Chromatogram showing EC response: A, 30.30 pg of pentafluoropropionyl diphenylamine as internal standard; B, 41.98 of OHNPFPD. From Neelakantan and Kostenbauder [31], courtesy of Analytical Chemistry.

for 375 drugs, drug metabolites, and other compounds encountered in body fluid extracts.

For tobacco- and marijuana-related compounds studied by Foltz et al. by chemical ionization, the following CI information was provided:

Methane CI-MS Spectral Data

Compound	MW	MH^+ (RI)	Prominent fragment ions > m/e 50
Nicotine	162	163 (100)	161 (27)
Cotinine	176	177 (100	
Tetrahydrocannabinol	314	315 (100)	313 (55), 299 (15), 81 (11), 135 (7)
11-Hydroxy-Δ^9-tetra-hydrocannabinol	330	331 (48	313 (100), 330 (47), 329 (25), 190 (9)

TABLE 1.12

Standard Curve for OHNPFPD Versus $(C_6H_5)_2NCOC_2F_5$[a]

Amount injected in pg, $\dfrac{\text{OHNPFPD}}{\text{marker}}$	Wt. ratio	Peak ht. ratio	$\dfrac{\text{Wt. ratio}}{\text{Peak ht. ratio}}$
$\dfrac{13.12}{16.68}$	0.787	0.271	2.90
$\dfrac{10.49}{13.35}$	0.786	0.273	2.88
$\dfrac{20.99}{30.30}$	0.693	0.235	2.95
$\dfrac{41.98}{30.30}$	1.385	0.464	2.98
$\dfrac{107.1}{38.00}$	2.818	0.986	2.86
$\dfrac{141.9}{38.00}$	3.734	1.317	2.83
$\dfrac{209.9}{151.5}$	1.385	0.503	2.75
$\dfrac{428.1}{303.0}$	1.413	0.486	2.91
$\dfrac{571.0}{303.0}$	1.884	0.625	3.01
$\dfrac{1135}{303.0}$	3.746	1.246	3.01

[a] From Neelakantan and Kostenbauder [31], courtesy of <u>Analytical Chemistry</u>.

Using SE-30 as a stationary phase for the gas-liquid chromatography of drugs, Moffat [38] suggested its use for the identification of drugs based on retention indices. In compiled retention index data for 480 drugs, the following tobacco-related compounds were included whose retention indices were reported as follows: cotinine, 1670; nicotine, 1340.

In 1977, Baker [34] described the use of a gas chromatographic system equipped with dual flame ionization and nitrogen-selective rubidium bead detectors in the identification of drugs. With the method, drugs were chromatographed along with a caffeine standard with a Perkin-Elmer model 900 gas chromatograph equipped with a flame ionization detector and a rubidium-bead nitrogen-phosphorus selective detector and a 183-cm by 2-mm glass column packed with 3% OV-17 on 110-120 mesh Anakrom ABS support. Following sample injection, the column was temperature programmed from 100 to 250°C at 4°C/min, then held at 250°C for 15 min, whereas the helium carrier gas was maintained at 27 ml/min and split equally between both detection systems. In addition to characterizing the drug in terms of absolute elution times and retention times relative to caffeine, the drug can also be characterized by the ratio of the nitrogen/FID response of the drug to the nitrogen/FID response of the caffeine standard. The response index and absolute and relative retention time of 71 nitrogen-containing drugs (nicotine included) commonly encountered in forensic and toxicology applications are listed in Table 1.13.

As noted by Baker, it was also found that the response index generally increased if the number of nitrogens in the compounds increased (Fig. 1.8).

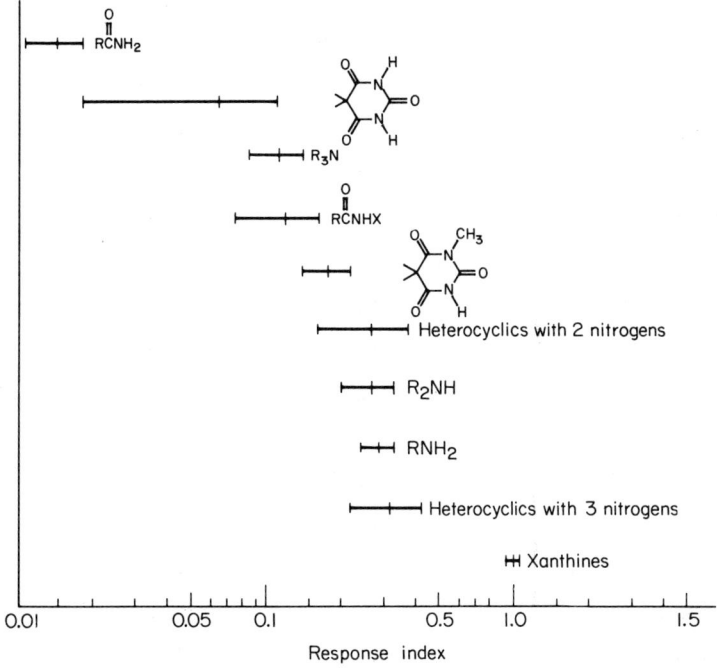

Figure 1.8. Effect of chemical structure on the response index. The vertical bars represent the mean and the standard deviation. From Baker [34], courtesy of Analytical Chemistry.

TABLE 1.13

Retention Times (Absolute and Relative) and Response Index of Selected Drugs[a]

Drug	RT (min)	RRT	Response index
Cyclopentamine	2.20	0.10	0.28
Propylhexidrine	3.08	0.14	0.27
Amphetamine	3.30	0.15	0.33
Methamphetamine	3.96	0.18	0.42
Mephenteramine	5.07	0.23	0.23
Carbromal	5.28	0.24	0.10
Acetylcarbromal	5.28	0.24	0.09
Nicotine	7.70	0.35	0.36
Ephedrine	8.15	0.37	0.34
Phenylpropanolamine	8.35	0.38	0.31
Ethinamate	9.90	0.45	0.02
Phendimetrazine	10.32	0.47	0.20
Phenmetrazine	10.57	0.48	0.24
Metharbital	10.78	0.49	0.24
Barbital	12.75	0.58	0.11
Salicylamide	13.40	0.61	0.22
Nikethamide	13.85	0.63	0.42
Methyprylon	14.08	0.64	0.12
Allobarbital	15.18	0.69	0.04
Aprobarbital	15.40	0.70	0.04
Butalbital	16.30	0.74	0.04
Butabarbital	16.30	0.74	0.04
Amobarbital	16.93	0.77	0.04
Methylphenidate	17.60	0.80	0.16
Pentobarbital	17.80	0.81	0.04
Phenacetin	18.25	0.83	0.10

TABLE 1.13 (continued)

Drug	RT (min)	RRT	Response index
Methohexital	18.48	0.84	0.16
Mescaline	18.70	0.85	0.23
Secobarbital	18.90	0.86	0.03
Meperidine	18.90	0.86	0.15
N,N-Dimethyltryptamine	19.10	0.87	0.30
Phencyclidine	19.33	0.88	0.12
Thiopental	20.00	0.91	0.21
Dimenhydrinate	20.20	0.92	0.14
Meprobamate	20.90	0.95	0.01
Thioamyal	21.10	0.96	0.18
Hexobarbital	21.10	0.96	0.18
Glutethimide	21.10	0.96	0.02
Carbisoprodol	21.55	0.98	0.20
Doxylamine	21.80	0.99	0.28
Mephobarbital	21.80	0.99	0.15
Caffeine	22.00	1.00	1.00
Methapyriline	23.55	1.07	0.47
Cyclobarbital	24.00	1.09	0.07
Phenobarbital	24.70	1.10	0.07
Tybamate	25.10	1.14	0.15
Procaine	25.10	1.14	0.33
Methadone	25.50	1.16	0.10
Dextromethorphan	25.50	1.16	0.11
Theophylline	25.95	1.18	0.95
Proxyphene	26.20	1.19	0.10
Methaqualone	27.95	1.27	0.24
Disulfiram	28.20	1.28	0.48

(continued)

TABLE 1.13 (continued)

Drug	RT (min)	RRT	Response index
Cocaine	28.20	1.28	0.13
Pentazocine	28.40	1.29	0.09
Promethazine	28.60	1.30	0.24
Levallorphan	30.15	1.37	0.09
Oxazepam	31.00	1.41	0.24
Chlordiazepoxide	31.00	1.41	0.23
Oxymetazoline	31.00	1.41	0.26
Dihydrocodeine	31.70	1.44	0.13
Codeine	31.70	1.44	0.13
Diphenylhydantoin	32.60	1.48	0.08
Morphine	33.00	1.50	0.12
Diazepam	33.20	1.51	0.27
Oxycodone	35.80	1.63	0.11
Oxymorphone	36.50	1.66	0.12
Fentanyl	36.50	1.66	0.19
Phenazocine	38.70	1.76	0.10
Heroin	39.40	1.79	0.11
Flurazepam	42.00	1.91	0.29

[a] Adapted from Baker [34].

However, this was found to be only a general trend, and many examples can be found in Table 1.13, where a compound containing two nitrogens had a lower response index than a compound with only one nitrogen. It was also observed that tertiary amines had a much lower response index than secondary and primary amines. A possible explanation proposed by Baker for the observed differences may be that primary and secondary amines readily lost hydrogen under the pyrolytic conditions, then were rapidly converted to a cyan radical, while with tertiary amines a carbon-nitrogen bond would have to be broken first.

Chemical Composition of Tobacco Leaf

In 1977, Burns and Collin [33] developed a rapid GC method for the determination of certain alkaloids, other than nicotine, in tobacco. Following a Soxhlet extraction of tobacco with methanol, after addition of quinaldine (2-methylquinoline) as internal standard, the alkaloid extract is analyzed using a Perkin-Elmer model F11 gas chromatograph equipped with a flame ionization detector and a 4-m by 3-mm-i.d. glass column packed with 6% Carbowax 20M:1% KOH coated on 80-100 mesh Diatomite M. With the injector and column temperatures held at 275 and 180°C, respectively, and nitrogen used as the carrier gas (flow rate not specified), the retention data for various alkaloids present in a typical tobacco extract are given in Table 1.14.

Using this GC procedure, the recovery of nornicotine and anabasine added to tobacco was 96.1 ± 4.4% and 97.2 ± 3.9%, respectively.

B. N'-Nitrosonornicotine

N'-Nitrosonornicotine, a known carcinogen, has been found to be present in rather high concentrations in tobacco [39-41]. As reported by Hoffmann et al. [39], various types of tobacco products contain N'-nitrosonornicotine (NNN) at 2 to 90 μg/g (dry weight of the tobacco) (2 to 90 ppm). The manner in which tobacco specimens were prepared for NNN analysis, following the outline given below, consisted of the following steps:

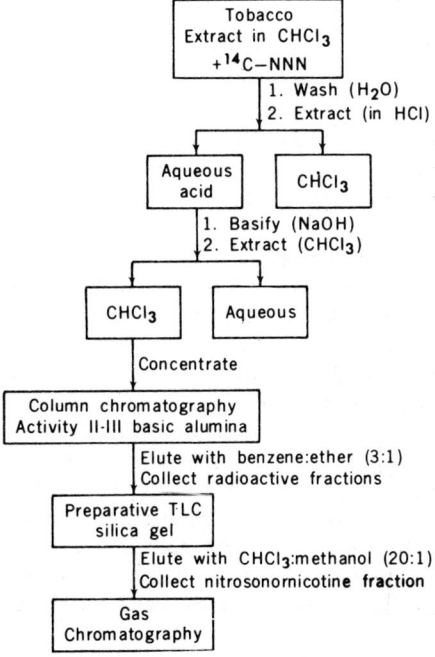

TABLE 1.14

Retention Data for Tobacco Alkaloids[a]

Alkaloid	RRT	RT (min)
Nicotine	0.80	4.14
Quinaldine (IS)	1.00	5.18
Nornicotine	1.95	10.10
Anabasine	2.19	11.33
Myosmine	2.19	11.33
Nicotyrine	3.25	16.80
Anatabine	3.52	18.20

[a] Adapted from Burns and Collin [33].

Tobacco (50 to 60 g) from various commercially available products (1972 to 1973; stored in a cold room) was ground to a powder and extracted in a Soxhlet apparatus with $CHCl_3$ for 16 hr. N'-Nitroso[2'-^{14}C]nornicotine (0.56 µg, 14.1 mCi/mmole) was added, and the extract was analyzed (see above). The basic fractions were chromatographed on basic alumina, and the radioactive fractions were collected, combined, and chromatographed by preparative thin-layer chromatography on silica gel. The band corresponding in R_f to NNN was eluted and examined by gas chromatography. The peak corresponding in retention time to NNN [about 21 min with no GC details provided] was collected and analyzed by mass spectrometry; it was identical with that of synthetic NNN.

In a subsequent investigation, Hecht et al. [40] described a chemical analytical method to determine NNN and N'-nitrosoanabasine (NAB) in tobacco. NNN was found in the unburned tobacco of commercial products at concentrations between 0.3 and 88.6 µg/g. The highest levels were observed in highly fermented snuff (29.1 µg/g) and fine-cut chewing tobacco (88.6 µg/g). NAB was not detected (<0.5 ng/g) in any tobacco examined. Two new tobacco components, N'-carbomethoxynornicotine (CNN) and N'-carbomethoxyanabasine (CAB), were found. Possible origins of NNN were discussed, especially in relation to concentrations of nitrite, nitrate, and alkaloids, and in relation to pH and curing. Biological implications, including the possible function of NNN (the first organic carcinogen isolated from unburned tobacco) as a causative factor in cancer of the oral cavity in tobacco chewers and betel quid chewers were covered in detail.

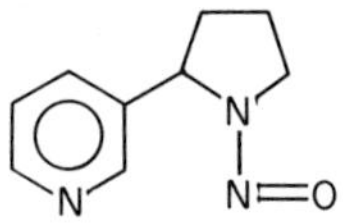

Chemical Composition of Tobacco Leaf

N' – Nitrosonornicotine
(NNN)

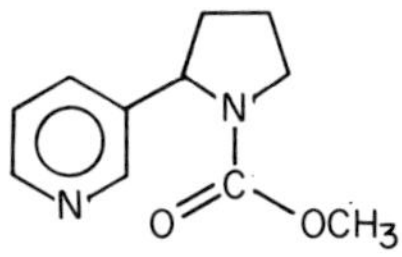

N' – Carbomethoxynornicotine
(CNN)

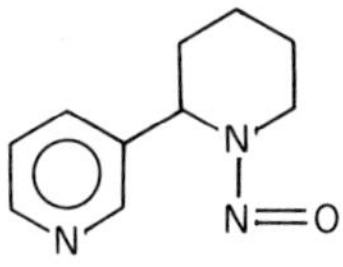

N' – Nitrosoanabasine
(NAB)

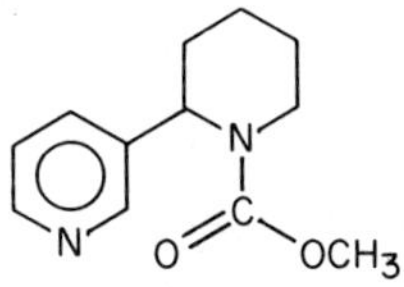

N' – Carbomethoxyanabasine
(CAB)

Following prescribed extraction procedures for (1) N'-carbomethoxynornicotine (CNN) and NNN and (2) N'-nitrosoanabasine (NAB) and N'-carbomethoxyanabasine (CAB), these compounds were resolved and analyzed with a Hewlett-Packard model 5711 gas chromatograph equipped with a flame ionization detector and two different column systems: column A, 5 ft by 1/8 in., packed with 10% Carbowax 20M-TPA on Gas Chrom Q (mesh size not specified); column B, 12 ft by 1/8 in., packed with 10% UCW-98 on Gas Chrom Q. Using a helium carrier-gas flow rate of 60 ml/min and injector, column, and detector temperatures of 300, 200, and 300°C, respectively, the absolute and relative retention times of NNN, CNN, NAB, and CAB on both columns are given below:

	Column A		Column B	
	RT (min)	RRT	RT (min)	RRT
NNN	24.00	1.00	20.00	1.00
CNN	13.90	0.58	20.00	1.00
NAB	24.00	1.00	25.40	1.27
CAB	15.60	0.65	28.00	1.40

In addition to retention time data, the presence of NNN, CNN, and CAB was demonstrated by comparison of their electron impact mass spectra to reference spectra, which are shown in Figure 1.9.

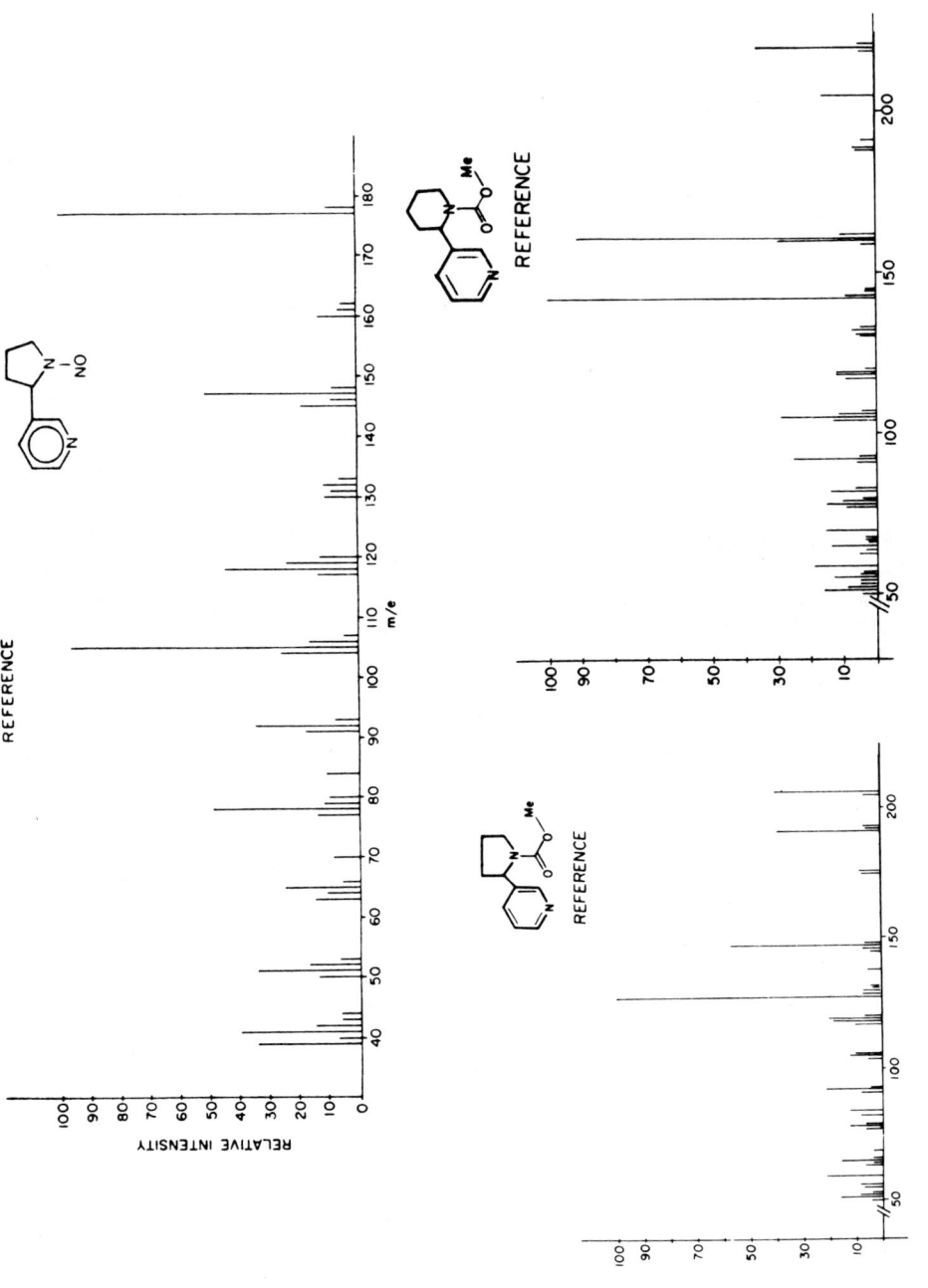

Figure 1.9. Reference mass spectra of NNN, CNN, and CAB (electron impact). Adapted from Hecht et al. [40].

In 1977, Munson and Abdine [41] determined NNN in tobacco products, quantitation being accomplished by GC-MS in the chemical ionization mode with selected ion recording of NNN and the deuterated internal standard, N'-nitrosonornicotine-d_4 (NNN-d_4).

The analysis of NNN by Munson and Abdine was carried out as noted below:*

1. Extraction of Tobacco Samples

Two-gram tobacco samples were extracted overnight with 45 ml of a pH 4.5 citric acid buffer containing ascorbic acid. The samples were filtered and washed with 15 ml of citric acid buffer. One milliliter of the internal standard solution (80 μg/ml) was added to the combined extracts which were adjusted to pH 6.0 and extracted with 5-50 ml portions of methylene chloride. The organic layers were combined and reduced in vacuo to about 50 ml. This residue was then extracted with 5-25 ml portions of 0.1 N HCl. The aqueous layer was adjusted to pH 6 (10 ml of 0.5 M sodium phosphate added for buffering) and extracted with 5-25 ml portions of chloroform. The chloroform layers were combined and evaporated to dryness. The residue was redissolved in a 10-ml tapered, glass centrifuge tube. The residue was then dissolved in 200 μl of methanol and 2 to 5 μl were injected into the GC-MS. Quantification was then performed from peak-height measurements.

2. Analytical Instrumentation

A Finnigan 3200F GC-MS instrument equipped with a programmed multiple-ion detection system (PROMIN) was used for quantitation. A 2-m by 2-mm-i.d. glass column packed with 3% OV-17 on 80-100 mesh Gas Chrom Q was used. Methane was used as the carrier gas and reagent gas. Column temperature was maintained at 180°C and the carrier-gas flow rate was adjusted to maintain a reagent gas pressure of 450 μm in the analyzer. With the ionization potential set at 120 eV, the PROMIN system was adjusted to record m/e 178 (NNN) and m/e 182 (NNN-d_4).

A Varian 2700 gas chromatograph was used in the study of extraction efficiency of NNN as a function of pH. A 6-ft by 2-mm-i.d. glass column packed with 3% OV-17 on 80-100 mesh Gas Chrom Q was used. A flow rate of 40 ml/min (nitrogen) and a column

---

*From Munson and Abdine [41], reproduced in part with permission of Marcel Dekker, Inc.

temperature of 220°C yielded a retention time of 11.0 min (flame ionization detection).

3. Results and Discussion

Chemical ionization mass spectra of NNN (Fig. 1.10) and the deuterated analog (Fig. 1.11) show major peaks corresponding to the M+1 ion (m/e 178 and m/e 182, respectively). These peaks were used for selected ion recording for all samples. Figure 1.12 shows a typical calibration curve in which the peak height ratio is plotted as a function of the weight ratio (NNN/NNN-d4). The slope of 1.03 indicates that the deuterated internal standard has essentially the same chromatographic and spectral characteristics as N-nitrosonornicotine.

Figure 1.13 shows the effect of pH on extraction efficiency for NNN. At pH values greater than 5 the efficiency approaches 100%. Therefore, the initial extraction of NNN from the aqueous extract of the tobacco was carried out at pH 6. While this allowed the complete extraction of NNN, it inhibited the extraction of some of the very basic compounds present in tobacco.

Figure 1.14 illustrates a typical mass chromatogram of a tobacco sample. The two large peaks on either side of the NNN peak are unknown substances that are carried through the extraction procedures.

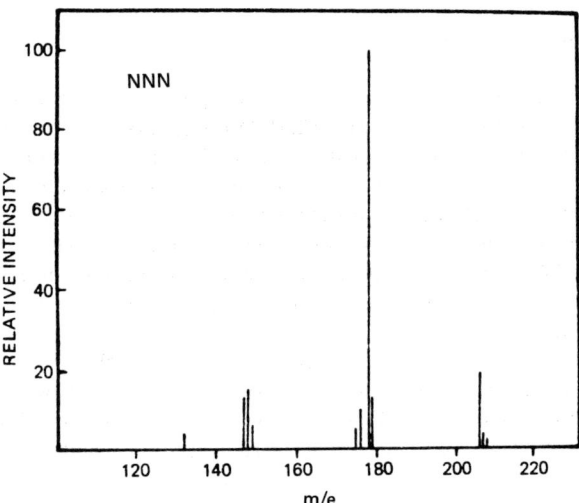

Figure 1.10. Chemical ionization mass spectrum of N-nitrosonornicotine. Reagent gas is methane. From Munson and Abdine [41], courtesy of Marcel Dekker, Inc.

Chemical Composition of Tobacco Leaf

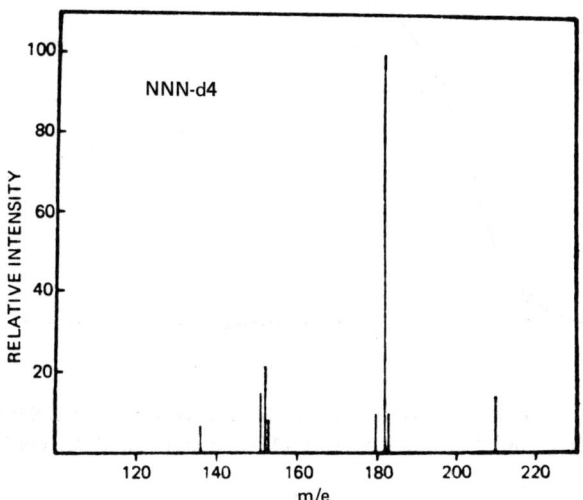

Figure 1.11. Chemical ionization mass spectrum of N-nitrosonornicotine-d_4. Reagent gas is methane. From Munson and Abdine [41], courtesy of Marcel Dekker, Inc.

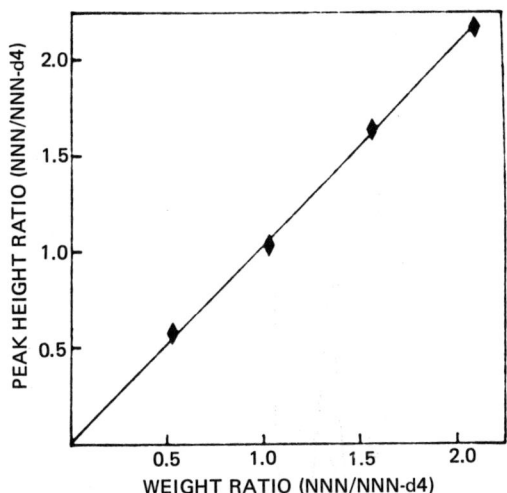

Figure 1.12. Typical calibration curve for N-nitrosonornicotine. NNN monitored at m/e 178, NNN-d_4 at m/e 182. From Munson and Abdine [41], courtesy of Marcel Dekker, Inc.

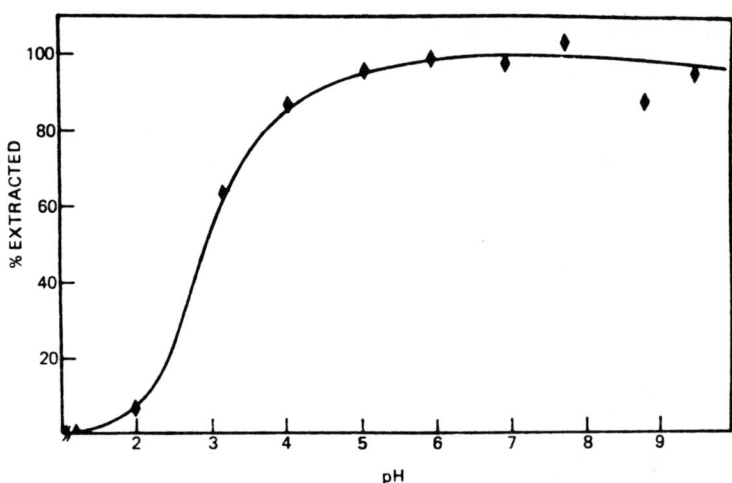

Figure 1.13. Effect of pH on the extraction efficiency of NNN from aqueous solutions. From Munson and Abdine [41], courtesy of Marcel Dekker, Inc.

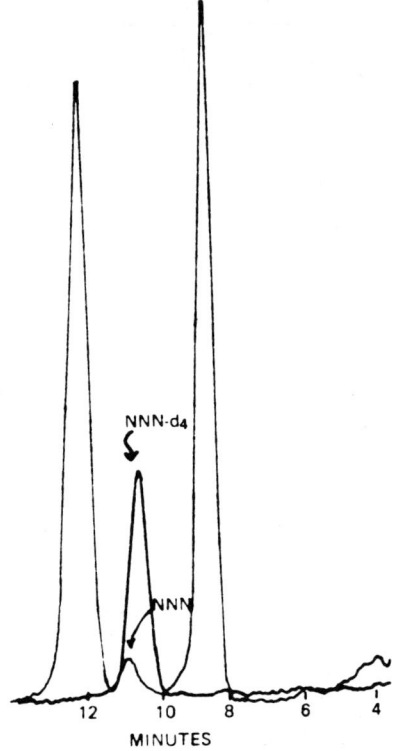

Figure 1.14. Typical mass chromatogram of tobacco sample monitored at m/e 178 (NNN) and m/e 182 (NNN-d4). From Munson and Abdine [41], courtesy of Marcel Dekker, Inc.

The apparent molecular weight of these compounds are both 176. Since they are present in very high concentrations, compared to NNN, satellite m/e peaks (due to the normal isotope distribution) give a significant response at m/e 178. At the concentration levels encountered in this study, the peaks did not cause interference.

The precision and accuracy of this method are good. A quadruplicate determination of tobacco obtained from research cigarettes gave a mean value of 0.88 ppm with a standard deviation of 0.087 ppm. Spiked tobacco samples were analyzed to determine the accuracy of the method. The results are shown in Table 1.15.

A variety of tobacco products have been analyzed for NNN by this method. In each case, two gram samples were analyzed. The results obtained to date are shown in Table 1.16.

TABLE 1.15

Recovery of NNN from Spiked Tobacco Samples[a]

NNN added (ppm)	NNN found (ppm)
5.1	5.7
10.2	10.6
10.2	10.7
15.3	14.6

[a] From Munson and Abdine [41], courtesy of Marcel Dekker, Inc.

TABLE 1.16

Results of Analysis of Tobacco Products[a]

Cigarettes			Chewing tobacco		
1.	Foreign	0.8 ppm	1.	Domestic	1.2 ppm
2.	Foreign	0.9 ppm	2.	Domestic	26.4 ppm
3.	Foreign	3.4 ppm	3.	Domestic	1.0 ppm
4.	Domestic	2.8 ppm	4.	Domestic	1.2 ppm
5.	Research	0.9 ppm	5.	Domestic	5.3 ppm
			6.	Domestic	1.3 ppm
Cigars			Pipe tobacco		
1.	Foreign	2.7 ppm	1.	Domestic	1.6 ppm
2.	Domestic	1.7 ppm	2.	Domestic	3.1 ppm
			Snuff and smokeless tobacco		
			1.	Domestic	9.3 ppm
			2.	Domestic	3.2 ppm

[a] From Munson and Abdine [41], courtesy of Marcel Dekker, Inc.

C. Hydrocarbons

Over the past several decades, studies have been performed to elucidate the composition of the paraffin wax fraction extracted from tobacco [42-46] using both GC and MS techniques. In 1959, Carruthers and Johnstone [42] reported their results derived from GC and MS studies of the paraffin wax composition of green tobacco leaf, fermented tobacco, and cigarette smoke. They reported the presence of a homologous series of normal paraffins (C_{25}-C_{33}) with the compounds containing odd numbers of carbon atoms being predominant for both the iso- and n-paraffins as indicated by the mass spectrometric data shown in Table 1.17 for green leaf and fermented tobacco as well as GC data for n-paraffins in both species. Whereas no details were given for the MS analysis of the paraffin waxes, the GC analysis was performed with a 3-ft column packed with silicone grease E 301 (percent not specified) coated on 52-85 mesh Silocel and maintained isothermally at 288°C with a nitrogen carrier-gas flow rate of 1.6 liters/hr.

For the preparation of the materials for analysis, neutral extracts of the tobacco were chromatographed on alumina and the initial waxy fractions eluted with light petroleum (b.p. = 40 to 60°C) were treated with urea in warm methanol. The resulting adducts were washed with light petroleum, decomposed with water, and the recovered paraffins crystallized once from benzene-ethanol.

Eglinton et al. [43] in 1962 examined the surface wax of leaves of certain species of the genera Monanthes, Greenovia, Aichryson, and Aeonium of the subfamily Sempervivoideae (Crassulaceae) by gas chromatography and showed that all contained also a significant amount of the homologous iso-paraffins, those with odd numbers of carbon atoms predominating. In some of the materials examined, the iso-paraffins were in greater amount than the normal series.

The surface wax from each species was removed from the leaves (about 100 g) by dipping the latter briefly into chloroform. The hydrocarbon fraction was then isolated by saponification of the crude wax followed by brief chromatography of the neutral fraction over alumina in light petroleum. The initial eluate contained the hydrocarbon fraction and was examined by gas chromatography.

The separation of the hydrocarbons derived from the surface waxes was performed with a Pye gas chromatograph equipped with an argon ionization detector and a column (dimensions not specified) packed with 0.5% Apiezon L on 80-100 mesh Celite. Using the specified GC conditions (column temperature, 237°C; argon carrier-gas flow rate, 45 ml/min; detector voltage, 1750 V), the retention times of i-C_{29}, n-C_{29}, i-C_{30}, n-C_{30}, i-C_{31}, n-C_{31}, i-C_{32}, n-C_{32}, i-C_{33}, i-C_{35}, and n-C_{35} relative to n-C_{33} (RRT = 1.00, RT = 34.4 min) were 0.24, 0.27, 0.33, 0.37, 0.47, 0.52, 0.65, 0.73, 0.89, 1.62, and 1.84, respectively.

TABLE 1.17

MS and GC Analyses of Tobacco Paraffin Waxes[a]

Carbon no.	MS analysis				GC analysis	
	Green leaf		Fermented tobacco		Green leaf	Fermented tobacco
	n-Alkanes	iso-Alkanes	n-Alkanes	iso-Alkanes	n-Alkanes	n-Alkanes
25	0.9	0.0	0.7	0.0	0.5	
26	0.6	0.0	1.2	0.0	0.3	0.3
27	3.0	0.9	5.9	0.8	7.5	4.4
28	0.1	0.0	0.9	0.0	0.6	1.0
29	6.6	15.9	6.3	11.0	8.8[b]	9.2[b]
30	0.9	2.5	0.6	1.6	3.9	7.1
31	24.1	24.4	26.6	20.4	47.0[b]	46.5[b]
32	3.9	2.4	5.1	2.0	12.5	16.0
33	10.8	3.3	13.1	3.9	18.9[b]	15.5[b]
Total	50.9	49.4	60.4	39.7	100.0	100.0

[a] Adapted from Carruthers and Johnstone [42].
[b] These peaks were composed of two unresolved peaks. The results given include both peaks. The impurity is probably a very slightly branched paraffin.

The main findings concerning the composition of the hydrocarbon fractions were summarized as follows:

1. Alkanes of carbon number less than C_{25} and more than C_{35} were not present to any appreciable extent.
2. The content of odd carbon-number alkanes is greater than that of even carbon-number alkanes by a factor of more than ten.
3. Some species contain quite high proportions of iso-alkanes, which occasionally even outweigh the n-alkane content.
4. The hydrocarbon pattern of the leaf wax of a given species was found to be substantially independent of season, age, or station of the individual.

In Figure 1.15 the hydrocarbon constituents of some of the leaf waxes studied are shown in histogram form, a single diagram for each authenticated species. As noted by Eglinton et al., the botanical classification is indicated, and within this the species are arranged such that, in general, the branched-chain isomer content increases from left to right and from top to bottom; the proportion of C_{31} to C_{33} seems to increase in a similar progression. They also further noted that even this limited investigation on a restricted area has shown at least some glimmerings of a taxonomic relevance in the leaf-wax hydrocarbon pattern.

Shortly thereafter, Mold et al. [44] established the presence of homologous series of normal, iso (2-methyl), and anteiso (3-methyl) paraffin hydrocarbons in several types of tobacco. The homologs with odd numbers of carbon atoms predominated for the normal and iso series, the $C_{31}H_{64}$ being the most abundant component of these series. For the anteiso series, the homologs with even numbers of carbon atoms are present in major amounts, with the $C_{32}H_{66}$ compound present in largest amount.

Following specified procedures for the extraction of paraffin hydrocarbons from tobacco and the separation of branched from normal paraffins, the GC resolution of these homologs were performed with a Micro-Tek GC-2500 gas chromatograph equipped with dual thermal conductivity detectors and 3-ft by 0.25-in.-o.d. copper columns packed with 3% SE-30 coated on 80-100 mesh Gas Chrom P. With the injector, column, and detector temperatures held at 300, 253, and 300°C, respectively, and the helium carrier-gas flow rate maintained at 46 ml/min, the retention times of the normal paraffins in a commercial blend of Bright, Burley, Turkish, and Maryland tobaccos were as follows: C_{25}, 1.30 (in min); C_{26}, 1.74; C_{27}, 2.17; C_{28}, 2.86; C_{29}, 3.65; C_{30}, 4.78; C_{31}, 6.52; C_{32}, 8.08; C_{33}, 10.42. The branched paraffin fraction from this commercial blend yielded the following retention data: C_{28}, 2.78; C_{29}, 3.48; C_{30}, 4.70; C_{31}, 6.08; C_{32}, 8.25; C_{33}, 10.10; C_{34}, 12.70 (in min).

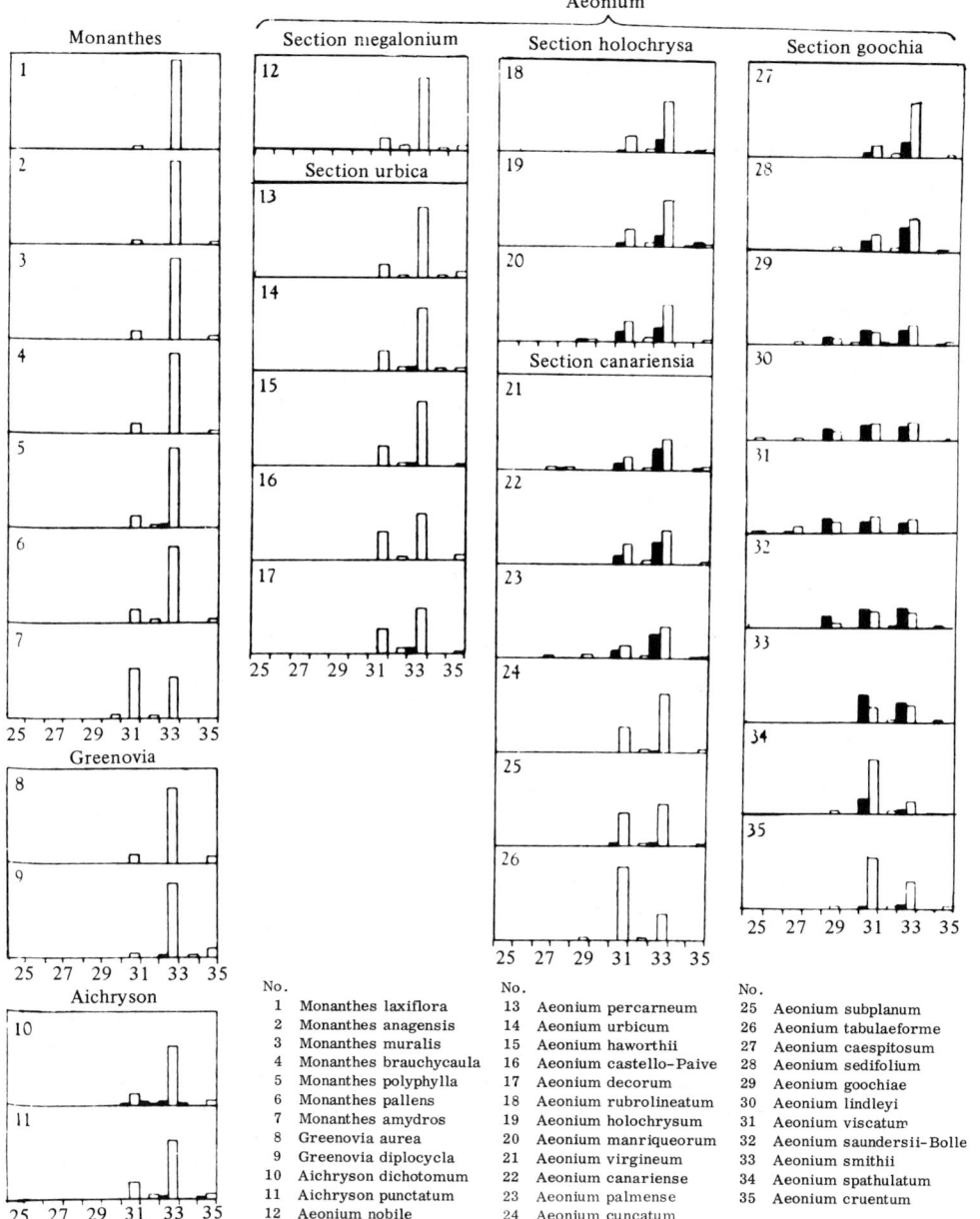

Figure 1.15. Distribution, in mole percentage, of n- and iso-alkanes C_{25}-C_{35} in the hydrocarbon fraction of the surface waxes from the leaves of individual species of the genera <u>Monanthes</u>, <u>Greenovia</u>, <u>Aichryson</u>, and <u>Aeonium</u>. Alkanes present at less than 2 mole % have been omitted.

From Eglinton et al. [43], courtesy of <u>Nature</u>.

Mass spectra of GC fractions collected in traps cooled in liquid air were obtained with the model 14-101 Bendix TOF mass spectrometer, equipped with an S-14-105 ion source. Spectra of synthetic reference compounds (isomeric tetratriacontanes) and of tobacco wax paraffins are given in Tables 1.18 and 1.19, respectively, whereas the relative amounts of individual hydrocarbons by GC are listed in Table 1.20.

Johnston and Jones [45] in 1968 also developed a GC method for the separation and determination of normal hydrocarbons in tobacco leaves. For this study, Johnston and Jones used an F & M model 300 gas chromatograph equipped with a flame ionization detector and 2-ft by 0.25-in. copper columns packed with the following liquid stationary phases and solid substrates: (1) 1% Apiezon L on 60-80 mesh Diatoport S; (2) 5% SE-30 on 30-60 mesh, acid-washed Chromosorb W; (3) 5% XE-60 on 60-80 mesh Diatoport S. With the injection and detector temperatures maintained at 230 and 210°C, respectively, and the helium carrier-gas glow rate set 75 ml/min, all runs were temperature programmed from 75 to 275°C at 9°C/min.

As noted by these investigators:

Identification of the n-paraffins was made by the method of Lewis, Patton, and Kaye [47] and consisted of plotting the retention times of a known series of hydrocarbons obtained on three different columns (under the same operating conditions) versus one another. The straight lines thus obtained were calculated by the method of least squares (Fig. 1.16). The same procedure was employed for the unknowns and the results are shown in Table 1.21.

To determine the qualitative utility of the procedure, a sample of the "gum" or resinous material which is found on a tobacco leaf was obtained. The sample was extracted with ether and the ether evaporated. Using the procedure described [see original text for details], except that urea adduction was accomplished in 30 min, the PTGC analysis indicated a homologous series of C_{15}-C_{35} hydrocarbons (Fig. 1.17).

Using the urea adduct technique followed by GC analysis, the recovery of hydrocarbons C_{12}-C_{24} is given in Table 1.22, whereas the percent composition determined with the Apiezon L and SE-30 columns by the usual normalization techniques is noted in Table 1.23, where it is seen that there was reasonable agreement between the data collected on both columns and, of the 19 hydrocarbons analyzed, only C_{22} and C_{28} differed by more than 1% absolute.

In their GC procedure for the quantitative determination of tobacco lipids, Ellington et al. [46] analyzed neophytadiene and hydrocarbon waxes on a Varian 2800 gas chromatograph equipped with a 9-ft by 1/8-in. stainless steel column packed with 3% Dexsil 300 coated on 100-120 mesh, acid-washed Chromosorb W. The hydrocarbons in the tobacco extract were separated

TABLE 1.18

Relative Ion Intensities for the Mass Spectra of Authentic Isomeric Tetratriacontanes[a,b]

Ion carbon number	n-Tetratriacontane (1-methyltritriacontane)	2-Methyltritriacontane	3-Methyltritriacontane	4-Methyltritriacontane	5-Methyltritriacontane	5% 2-Methyltritriacontane, 95% 3-Methyltritriacontane	5% 3-Methyltritriacontane, 95% 2-Methyltritriacontane	10% 3-Methyltritriacontane, 90% n-Tetratriacontane
25	4	4	4	2	4	3	8	6
26	4	5	3	2	4	6	5	5
27	4	4	2	2	3	4	4	6
28	4	4	2	2	21	2	5	5
29	4	4	2	18	6	3	4	4
30	3	5	13	2	100	12	5	5
31	3	38	2	100	8	12	34	2
32	2	5	100	10	8	100	15	25
33	0.6	20	6	8	8	14	15	1
34	100	100	26	30	21	52	100	100

[a] From Mold et al. [44], courtesy of Biochemistry.
[b] The intensities are calculated relative to the highest peak in the portion of the spectrum presented. Only the heavy fragment portions of the mass spectra are tabulated. The relative ion intensities are presented as functions of the carbon number of the ion. The peak height used is the highest for the unresolved cluster of peaks at that carbon number.

TABLE 1.19

Mass Spectra of Tobacco Wax Paraffins[a]

Ion carbon number	Relative ion intensities[b]								
	Branched isomers						Normal isomers		
	C_{29}	C_{30}	C_{31}	C_{32}	C_{33}	C_{34}	C_{28}	C_{30}	C_{31}
25	7	4	5	3	4	7	6	7	7
26	42	11	5	2	4	6	5	18	7
27	10	4	8	2	4	4	4	7	6
28	26	100	45	13	4	4	100	38	6
29	100	7	20	6	5	4		2	5
30		33	47	100	35	13		100	2
31			100	8	28	6			100
32				38	22	100			
33					100	8			
34						33			

[a] From Mold et al. [44], courtesy of <u>Biochemistry</u>.
[b] See footnote b, Table 1.18.

and quantitatively determined using docosane as internal standard with the following GC operating conditions: injector temperature, 290°C; column temperature, programmed from 150 to 330°C at 4°C/min; detector temperature, 330°C; helium carrier-gas flow rate, 50 ml/min.

Sample preparation for GC analysis consisted of the following:

The tobacco samples were equilibrated 2 days at ambient conditions, then ground to pass a 32-mesh screen. Moisture was determined by oven drying for 3 hr at 95°C. Seven to nine grams of ground tobacco of known moisture content were extracted with 300 ml of hexane in a Soxhlet apparatus under nitrogen. The hexane was degassed prior to use. The extraction was continued for 18 hr with rapid siphoning (10 to 12 times/hr) of the hexane. The solution was then cooled and solvent removed in vacuo. Redistilled benzene was added to the residue several times and evaporated in vacuo for azeotropic removal of any water. The solvent-free extract was dissolved in hexane and transferred quantitatively to a 10-ml volumetric flask.

TABLE 1.20

Relative Percent Composition of the Tobacco Hydrocarbons on the Basis of Gas-Liquid Chromatographic Information[a],[b],[c]

Paraffin carbon numbers	Commercial blend tobacco			Bright tobacco			Burley tobacco			Turkish tobacco		
	n	i	a	n	i	a	n	i	a	n	i	a
25	1.71			2.04			1.05			1.37		
26	0.83			1.02			0.49			0.69		
27	7.73			5.71			4.78			8.60		
28	0.89		0.13	1.38		0.43	1.05		0.36	1.78		0.18
29	6.72	1.24		5.87	3.06		5.43	2.54		7.93	1.82	
30	3.16		5.65	3.06		6.71	2.92		6.75	5.46		5.31
31	26.3	10.92		24.49	14.26		27.51	12.6		23.24	6.69	
32	4.88		13.02	4.24		11.30	5.60		11.73	7.39		8.94
33	10.77	5.62		7.19	6.35		8.09	6.45		12.85	4.89	
34			1.15			2.88			2.58			2.34

[a] From Mold et al. [44], courtesy of Biochemistry.
[b] n, normal; i, iso; a, anteiso.
[c] Mass spectrometric evaluation of the individual branched compounds for the wax from the commercial blend of tobaccos indicated the presence of the anteiso isomers to the extent of 2 to 3% of the C_{29}, 5 to 7% of the C_{31}, and 10 to 12% of the C_{33} homologs. Similarly, the iso compounds were present to the extent of 2 to 3% of the C_{32} and 6 to 7% of the C_{34} homologs. Small additional amounts of each branched compound would be expected in the corresponding normal isomer. The presence of the branched C_{35} homologs to the extent of 0.16% in the paraffins from the commercial blend of tobacco and 0.42%, 0.28%, and 0.10% for the Bright, Burley, and Turkish tobaccos was demonstrated in a study carried out subsequent to submitting this manuscript.

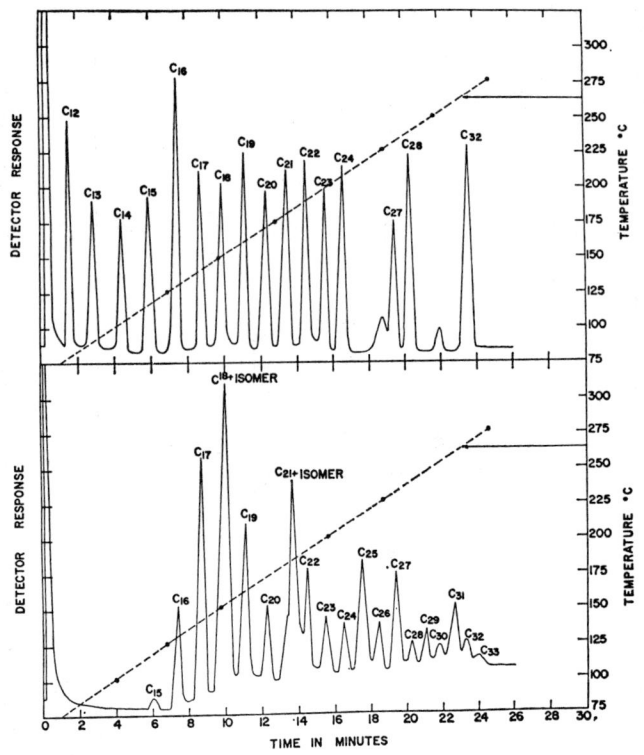

Figure 1.16. Chromatograms of normal hydrocarbons on a 2-ft by 1/4-in. column of 1% Apiezon L on Diatoport S. Top: Standard hydrocarbon mixture. Bottom: From tobacco aroma. From Johnston and Jones [45], courtesy of Analytical Chemistry.

For the separation of the sterols and hydrocarbons, a 9-mm by 150-mm column was slurry-packed (in hexane) with 2.5 g of Unisil 100/120 mesh silicic acid which had been heated at 120°C for 1 hr. A 1-ml aliquot of the above hexane extract was placed on the column, which was then sequentially eluted with 50 ml of hexane and 250 ml of 10% methanol in ether to separate the hydrocarbons and sterols, respectively.

GC analysis of the hydrocarbon fraction with the Dexsil column resolved the neophytadiene and the straight, iso-, and anteiso-isomers of the hydrocarbons as illustrated in Figure 1.18, whereas quantitative data are given

TABLE 1.21

Comparison of Calculated Equations[a]

Columns[b]	Sample	Equation[c]	r[d]	s[e]
XE-60 vs. AL	Standard	y = 1.27x - 0.87	0.999	0.22
	Unknown	y = 1.26x - 1.16	0.999	0.13
SE-30 vs. AL	Standard	y = 1.00x - 0.68	0.999	0.04
	Unknown	y = 1.09x - 0.01	0.990	0.08
XE-60 vs. SE-30	Standard	y = 1.15x - 0.17	0.999	0.23
	Unknown	y = 1.13x - 0.02	0.999	0.13

[a] From Johnston and Jones [45], courtesy of <u>Analytical Chemistry</u>.
[b] Columns: XE-60 = 5% XE-60 on Diatoport S; AL = 1% Apiezon L on Diatoport S; SE-30 = 5% silicone rubber on Chromosorb W-AW.
[c] Least-squares equation, where y = retention time in minutes of first named column and x = retention time in minutes of second named column.
[d] r = correlation coefficient.
[e] s = standard deviation of x from calculated line.

in Table 1.24. As noted, a simplified group analysis of neophytadiene and the C_{27}-C_{34} hydrocarbons can be achieved with a 2-ft 5% Dexsil column. With this column, which is also used for fatty acids, resolution is sacrificed for shorter analysis time and easier operation.

D. Alcohols of Tobacco

With regard to the analysis of tobacco alcohols, very few have been performed by gas chromatographic techniques [48-50].

Carruthers and Johnstone [48] in 1960 obtained additional evidence by mass spectrometry in support of the structure for solanesol shown below:

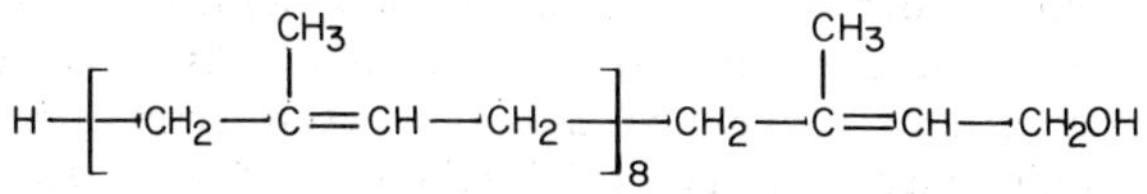

$$H-\left[-CH_2-\underset{\underset{CH_3}{|}}{C}=CH-CH_2-\right]_8 CH_2-\underset{\underset{CH_3}{|}}{C}=CH-CH_2OH$$

Solanesol

Following hydrogenation of solanesol, the mass spectrum of the saturated hydrocarbon exhibited a parent peak at m/e 632 and a major fragment

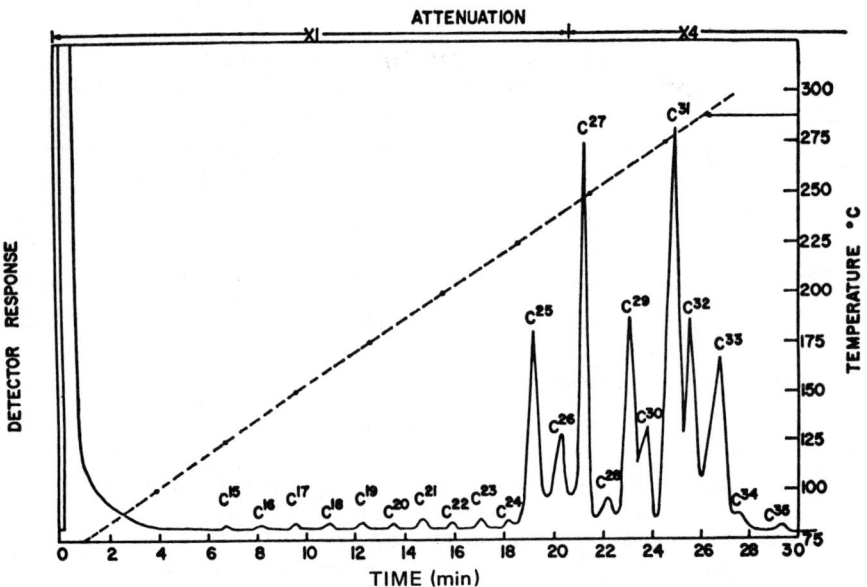

Figure 1.17. Chromatogram of normal hydrocarbons from tobacco gum on a 2-ft by 1/4-in. column of 1% Apiezon L on Diatoport S. From Johnston and Jones [45], courtesy of Analytical Chemistry.

ion at m/e 617, due to the loss of a methyl group. Further chromatography on grade I alumina of crude solanesol fractions from green tobacco extracts and elution with ether afforded, immediately following solanesol, a long-chain saturated primary alcohol. Mass spectrometry indicated the molecular weight to be 326, and the alcohol was identified as n-docosanol by comparison with an authentic specimen prepared from methyl behenate by reduction with lithium aluminum hydride.

Because solanesol, a trisesquiterpenoid alcohol in tobacco leaf, had been shown to be an important precursor of the tumorigenic polynuclear aromatic hydrocarbons of smoke, Severson et al. [49] developed procedures based on high-temperature gas chromatography for the analyses of free and total solanesol. The alcohol, as its trimethylsilyl derivative, was separated and quantitated on a short Dexsil 300 GC column, with 1,3-dimyristin as an internal standard. The free alcohol was determined by direct derivatization of ground tobacco and its hexane extract with N,O-bis(trimethylsilyl)acetamide-dimethylformamide reagents. For total solanesol, the ground tobacco or its hexane extract was saponified with ethanolic KOH. Acidification and hexane extraction yielded samples suitable for silylation and gas chromatographic analysis.

TABLE 1.22

Hydrocarbon Recovery from Urea Adducts[a]

n[b]	Standard[c]	Adduct[d]	Recovery (%)
12	3.27 ± 0.06	1.62 ± 0.23	49.54
13	3.01 ± 0.11	2.34 ± 0.08	77.23
14	3.23 ± 0.05	2.77 ± 0.19	85.23
15	3.75 ± 0.11	3.31 ± 0.24	88.03
16	3.67 ± 0.08	3.34 ± 0.18	91.01
17	3.58 ± 0.25	3.38 ± 0.31	94.41
18	4.27 ± 0.24	3.85 ± 0.20	89.74
19	4.11 ± 0.21	3.74 ± 0.26	91.00
20	4.00 ± 0.13	3.32 ± 0.17	83.00
21	4.24 ± 0.11	3.66 ± 0.23	86.32
22	4.04 ± 0.08	3.50 ± 0.15	86.63
23	3.75 ± 0.04	3.32 ± 0.11	88.53
24	4.34 ± 0.24	3.92 ± 0.33	90.32

[a] From Johnston and Jones [45], courtesy of <u>Analytical Chemistry</u>.
[b] n of C_nH_{2n+2}.
[c] Area ± mean deviation.
[d] Area ± mean deviation.

In this analytical investigation, Severson et al. performed their GC analyses with a Hewlett-Packard model 5750 gas chromatograph equipped with a 1.5-ft by 1/8-in. stainless steel column packed with 5% Dexsil 300 coated on 100-120 mesh, acid-washed Chromosorb W and flame ionization detection. Maintaining the specified GC conditions (injector temperature, 300°C; column temperature, held isothermally at 210°C for 4 min, then programmed from 210 to 330°C at 6°C/min, and then maintained at 330°C for an additional 6 min; detector temperature, 350°C; helium carrier-gas flow rate, 50 ml/min), they noted that the retention times of dimyristin-TMSi and solanesol-TMSi were approximately 15.80 and 24.96 min, respectively, with three minor peaks indicated in various chromatograms having retention times slightly less than that of solanesol, the retention times of these compounds being peak 1, 22.40; peak 2, 23.10; and peak 3, 23.80.

TABLE 1.23
Areas and Percent Composition for C_nH_{2n+2}[a]

n	1% Apiezon L Area[b]	% Comp.[c]	5% Silicone rubber Area[b]	% Comp.[c]	Average % comp.[d]
15	0.80 ± 0.05	0.51	0.07 ± 0.03	0.57	0.54
16	6.55 ± 0.04	4.19	7.10 ± 0.10	4.17	4.18
17	16.18 ± 0.36	10.37	16.90 ± 0.05	9.94	10.15
18[e]	27.55 ± 0.31	17.70	28.75 ± 0.08	16.91	17.30
19	12.72 ± 0.26	8.15	12.00 ± 0.15	7.06	7.61
20	5.92 ± 0.06	3.79	6.70 ± 0.18	3.94	3.87
21[e]	22.52 ± 0.20	14.44	25.95 ± 0.04	15.26	14.85
22	9.45 ± 0.14	6.06	8.18 ± 0.24	4.81	5.44
23	5.05 ± 0.04	3.24	4.48 ± 0.09	2.63	2.94
24	4.50 ± 0.02	2.88	4.35 ± 0.10	2.56	2.72
25	9.75 ± 0.08	6.25	9.82 ± 0.09	5.77	6.01
26	4.38 ± 0.09	2.81	4.88 ± 0.14	2.87	2.84
27	8.45 ± 0.04	5.42	9.40 ± 0.25	5.53	5.48
28	3.02 ± 0.14	1.94	5.68 ± 0.11	3.34	2.64
29	3.90 ± 0.05	2.50	5.60 ± 0.14	3.29	2.90
30	3.15 ± 0.05	2.02	4.35 ± 0.08	2.56	2.29
31	7.15 ± 0.11	4.58	8.52 ± 0.14	5.01	4.80
32	3.00 ± 0.08	1.92	3.70 ± 0.08	2.18	2.05
33	2.02 ± 0.02	1.29	2.75 ± 0.06	1.62	1.46

[a] From Johnston and Jones [45], courtesy of <u>Analytical Chemistry</u>.
[b] Average of four determinations ± mean deviation.
[c] Calculated from normalized areas.
[d] Average of two % compositions.
[e] Plus isomer.

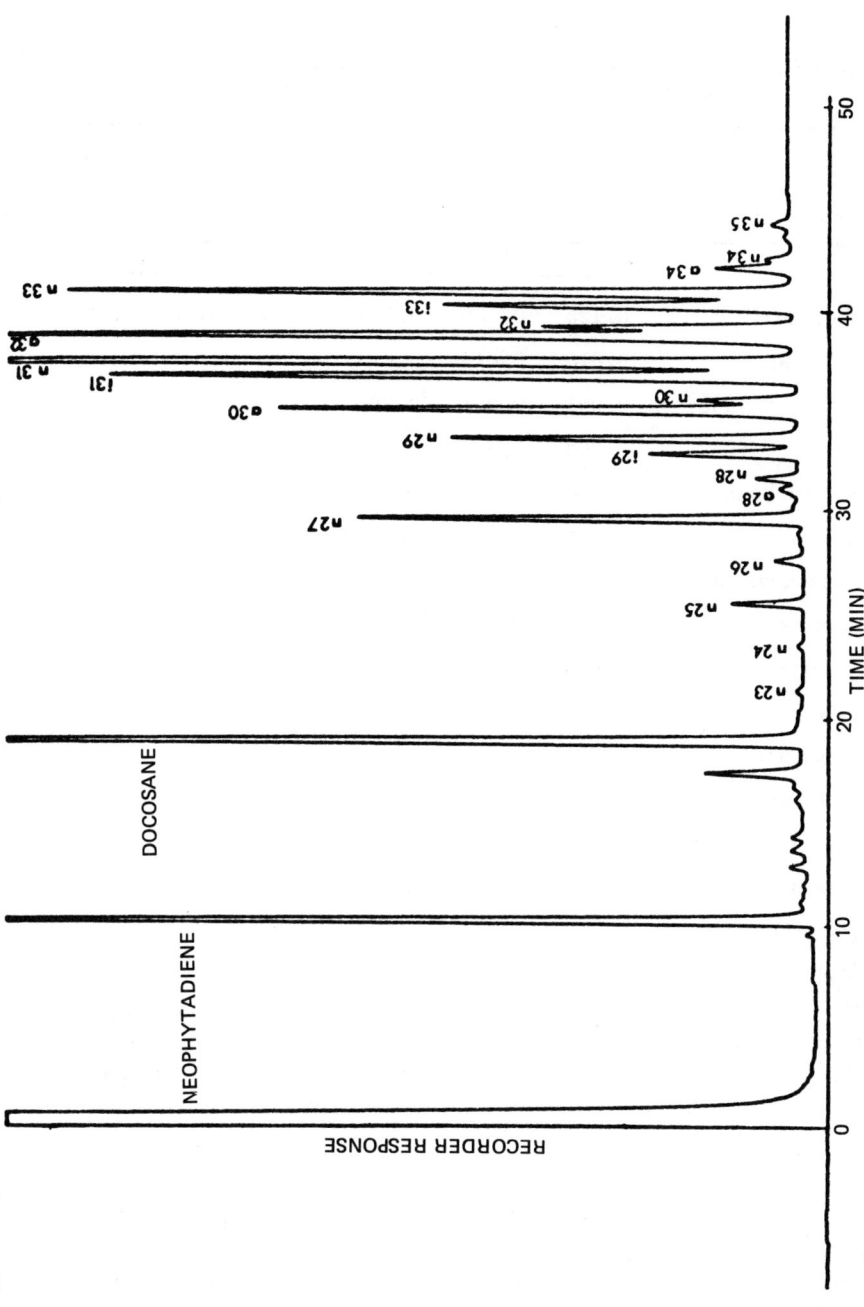

Figure 1.18. Gas chromatogram of neophytadiene and the hydrocarbon waxes (n = normal, i = iso, a = anteiso). Conditions: 9-ft by 1/8-in. stainless steel column, 3% Dexsil 300 GC, He flow 50 ml/min, temperature programmed from 150 to 330°C at 4°C/min. From Ellington et al. [46], reproduced from the Journal of Chromatographic Science, by permission of Preston Publications, Inc.

TABLE 1.24

Neophytadiene and C_{27}-C_{34} Hydrocarbons in Replicate Samples on 3% and 5% Dexsil 300 GC Columns[a]

Hydrocarbon	3% Dexsil[b]			5% Dexsil[b]		
	1	2	3	1	2	3
	---------------- % dry leaf ----------------					
Neophytadiene	0.123	0.130	0.118	0.13	0.13	0.134
n-Heptacosane	0.015	0.014	0.015			
3-Methylheptacosane	0.001	0.001	0.006			
n-Octacosane	0.002	0.002	0.002			
2-Methyloctacosane	0.004	0.004	0.004			
n-Nonacosane	0.012	0.011	0.012			
3-Methylnonacosane	0.013	0.012	0.013			
n-Triacontane	0.005	0.004	0.005			
2-Methyltriacontane	0.026	0.025	0.027			
n-Hentriacontane	0.039	0.037	0.041			
3-Methylhentriacontane	0.026	0.025	0.027			
n-Dotriacontane	0.008	0.008	0.009			
2-Methyldotriacontane	0.013	0.012	0.014			
n-Tritriacontane	0.018	0.017	0.019			
3-Methyltritriacontane	0.002	0.002	0.002			
n-Tetratriacontane	0.001	0.001	0.001			
Total C_{27}-C_{34}	0.185	0.175	0.196	0.216	0.203	0.213

[a] From Ellington et al. [46], reproduced from the <u>Journal of Chromatographic Science</u>, by permission of Preston Publications, Inc.
[b] Same set of replicates run on both columns.

Peaks 1 and 3 were subsequently identified as solanesene(s) and peak 2 as bombiprenone, the latter compound postulated to be formed during the biochemical breakdown of plastoquinone-A in the ripe leaf. On the other hand, solanesene(s) (C) and bombiprenone (B) can also result from thermal conversion of solanesol (A) as shown below:

$$\text{(A)} \quad H\text{-}[\text{CH}_2\text{-}C(\text{CH}_3)\text{=CH-CH}_2]_8\text{-CH}_2\text{-}C(\text{CH}_3)\text{=CH-CH}_2\text{-OH}$$

$$\xrightarrow{[O]} \text{(B)} \quad H\text{-}[\text{CH}_2\text{-}C(\text{CH}_3)\text{=CH-CH}_2]_8\text{-CH}_2\text{-}C(\text{=O})\text{-CH}_3$$

$$\xrightarrow{-H_2O, \, \Delta} \text{(C)} \quad H\text{-}[\text{CH}_2\text{-}C(\text{CH}_3)\text{=CH-CH}_2]_8\text{-CH}_2\text{-}C(\text{=CH}_2)\text{-CH=CH}_2$$

or

$$H\text{-}[\text{CH}_2\text{-}C(\text{CH}_3)\text{=CH-CH}_2]_8\text{-CH=}C(\text{CH}_3)\text{-CH=CH}_2$$

Applying their procedures for the determination of free and total solanesol in Eastern Carolina flue-cured tobacco leaf, comparable data were obtained as shown in Table 1.25. Commenting on their results, Severson et al. noted that:

> Both procedures for the analysis of the free isoprenoids yielded essentially identical values. However, the direct silylation procedure yielded data with a much larger deviation. Comparison of the two methods for total solanesol showed that the average value was about 4% higher by the direct hydrolysis method. However, levels for solanesenes and bombiprenone were the same for both methods. The direct hydrolysis procedure, with only a 1.9% relative standard deviation from the mean, appeared to be the better method for determining total solanesol. By difference, about 14 to 19% of the C_{45} isoprenoids were bound by base-hydrolyzable linkages.

The total solanesol values derived from a series of tobacco samples by both KOH hydrolysis methods are compared in Table 1.26, where it is noted that the direct KOH hydrolysis of ground tobacco gave a higher value for total solanesol than the KOH hydrolysis of the hexane extract.

In 1978, Hecht et al. [50] developed an analytical method for the determination of 2-(p-hydroxyphenyl)-ethanol (PHPE) and m-hydroxybenzyl alcohol (MHBA) in tobacco and tobacco smoke, because both compounds were

TABLE 1.25

Comparison of Methods for Free and Total Solanesol in Tobacco Leaf[a]

		Method	Percent weight of dry leaf	
			Solanesol	Solanesenes + bombiprenone
Free	1.	Hexane extraction	2.60 ± 0.11	0.17 ± 0.08[b]
	2.	BSA-DMF extraction	2.57 ± 0.28	0.14 ± 0.04[b]
Total	1.	KOH hydrolysis of hexane extract	2.98 ± 0.05	0.24 ± 0.01[b]
	2.	KOH extraction	3.11 ± 0.06	0.25 ± 0.03[b]

[a] Adapted from Severson et al. [49].
[b] Calculated assuming a detector response identical to that for solanesol-TMSi.

TABLE 1.26

Total Solanesol Results

Tobacco type	Percent weight of dry leaf	
	Method A[b]	Method B[b]
Flue cured	2.99	3.11
Burley I	2.07	2.14
Maryland	2.04	2.09
Turkish	0.87	1.20
Burley II	0.88	0.81
Cigar filler	0.78	0.98

[a] Adapted from Severson et al. [49].
[b] Method A: KOH hydrolysis of hexane extract; method B: KOH extraction.

major components of subfractions of the tumor-promoting fractions of cigarette smoke condensate (CSC).

Sample preparation for tobacco analysis consisted of the following procedure:

> Tobacco from 50 to 100 cigarettes or from 10 cigars was ground in a blender. An aliquot was removed for water determination and the remainder was extracted overnight with 800 ml of water. The resulting mixture was filtered through a 5-μm nylon filter cup. The insolubles were washed with 400 ml of H_2O, and PHBA, p-hydroxybenzyl alcohol (internal standard, 5 μg/cigarette), was added to the extracts. The extract was brought to 1 N in NaOH by addition of 10 N NaOH. This basic solution was partitioned by adjusting the pH to 6.1 with 3 N HCl and then extracting three times with equal volumes of preequilibrated $CHCl_3$. The $CHCl_3$ phases were combined and washed with 50 ml of H_2O (pH 6). The combined pH 6.1 solutions were then extracted three times with equal volumes of ethyl acetate. The combined ethyl acetate layers were dried (Na_2SO_4), filtered, and concentrated under reduced pressure at 30°C to near dryness. This final residue from the ethyl acetate extracts was triturated with small portions of ethyl acetate and brought to a total volume of 4.0 ml with ethyl acetate. A 1.0-ml aliquot was removed and applied to two silica gel plates.

The hydroxyphenyl alcohol fractions (R_f = 0.4) were scraped from the plates, extracted from the silica with ethyl acetate, filtered, and carefully concentrated to a volume of 1.0 ml. From this EA solution a 50-μl aliquot was transferred to a 100-μl Reacti vial and brought to dryness on a rotary evaporator at 20°C. Following the addition of 50 μl of silylating reagent [N,O-bis-(trimethylsilyl)trifluoroacetamide containing 1% trimethylchlorosilane], the vial was heated to 80°C for 20 min in an oil bath. From this final solution, 5-μl aliquots were withdrawn and injected into the gas chromatograph.

GC separations of the TMSi derivatives were performed with a Hewlett-Packard model 5700A gas chromatograph equipped with a glass-lined injection port, a flame ionization detector, and a 6-ft by 1/8-in. column packed with 10% UCW-98 coated on 80-100 mesh WHP 7620. Using a helium carrier-gas flow rate of 50 ml/min and a column temperature program (held at 100°C isothermally for 8 min, then programmed to 240°C at 4°C/min), retention data for PHPE, MHBA, PHBA, OHBA (o-hydroxybenzyl alcohol), OHPE (o-hydroxyphenyl ethanol), and MHPE (m-hydroxyphenyl ethanol) are listed in Table 1.27. Based on calibration curves prepared by plotting peak height versus concentration of TMSi standard, the average recovery for PHBA using this procedure was 70%.

TABLE 1.27

Retention Time Data for TMSi Derivatives of Hydroxybenzyl
Alcohols and Hydroxyphenyl Ethanols[a]

Compound	RRT	RT (min)
OHBA	0.91	21.50
MHBA	0.96	22.60
PHBA (I.S.)	1.00	23.60
OHPE	1.00	23.60
MHPE	1.05	24.80
PHPE	1.09	25.70

[a] Adapted from Hecht et al. [50].

To obtain further information on the origins of these compounds, unburned tobacco was analyzed. As reported by these investigators, gas chromatograms of the hydroxyphenyl alcohol fraction were similar to those observed for smoke samples. Of the six hydroxyphenyl alcohols considered, only PHPE was detected in tobacco (detection limit = 0.05 µg/g of tobacco), as shown in Table 1.28. However, by extraction of the tobacco with the hexane-ethanol azeotrope, PHPE levels in cigarette tobacco could be reduced, indicating its presence in the wax layer of tobacco leaf.

E. Sterols

In addition to determining lipids in tobacco, Ellington et al. [46] in 1977 determined sterols (free and total) by gas chromatography. The free sterols [removed from tobacco via a Soxhlet extraction (18 hr, hexane) and separated from tobacco hydrocarbons by silicic acid column chromatography; both separations are described in Section I.C] were converted to their respective TMSi derivatives with a mixture of BSA/DMF (50 µl each) which was then heated in a sealed tube for 15 min at 76°C. An aliquot of about 2 to 3 µl was withdrawn for GC analysis. However, prior to silylation, hexacosanol was added as internal standard.

For the determination of total sterol content, the sample was prepared as follows:

> An aliquot (1 to 2 ml) of the hexane extract or 2 g of ground tobacco was transferred to a 300-ml saponification flask containing 40 ml of 2 N KOH in 95% ethanol and fitted with a reflux condenser. The sample

TABLE 1.28

PHPE in Tobacco[a]

Product	PHPE (μg/g dry weight)
Cigarette A	15.00
Burley cigarette	0.96
Bright cigarette	12.00
Bright cigarette—(homogenized leaf cured)	10.00
Standard experimental-blend cigarette	13.00
Standard experimental-blend cigarette, extracted	9.60
Cigar	2.10

[a] From Hecht et al. [50], reproduced from the Journal of Analytical Toxicology, by permission of Preston Publications, Inc.

was refluxed for 2 hr under nitrogen. After cooling, 5 ml of a saturated KCl solution was added and the pH of the solution adjusted to 2 with concentrated HCl. The solution was shaken with 10 ml of hexane and, if necessary, water was added until a clear meniscus was obtained. The mixture was transferred to a 125-ml separatory funnel and extracted with 10-ml portions of hexane until two successive hexane layers were colorless (5 to 6 times). The combined hexane layers were diluted to 100 ml and stored at 4°C in tightly closed brown bottles. Aliquots of these stock solutions were removed and applied to the silicic acid column to separate again the hydrocarbons from the sterols. Total sterols were derivatized by the procedure discussed above except that an aliquot of the SA eluate equivalent to 1 ml of hydrolyzate was used and again hexacosanol added as internal standard.

The silyl derivatives of the sterols (cholesterol, campesterol, stigmasterol, and β-sitosterol) were analyzed with a Hewlett-Packard model 7610A gas chromatograph equipped with flame ionization detection and a 1.8-m by 2-mm U-shaped glass column packed with 3% SP-2250 coated on 80-100 mesh, acid-washed, DMCS-treated Supelcon. Using carefully specified GC conditions (injector temperature, 275°C; column temperature, held isothermally at 200°C for 10 min, then programmed to 240°C at 2°C/min, held 10 min at 240°C, increased to 330°C at 15°C/min and held at 330°C for 15 min; detector temperature, 350°C; helium carrier-gas flow rate, 50 ml/min), the retention times of the TMSi derivatives of hexacosanol, cholesterol, campesterol, stigmasterol, and β-sitosterol were approximately 7.90, 19.30, 23.30, 24.40, and 26.50 min, respectively.

Quantitative results for these tobacco sterols are given in Table 1.29, where it is noted that extension of the alkaline hydrolysis time to 18 hr did not affect the sterol levels. Free sterols in the hexane extracts are listed in Table 1.30. These values are based on the results of three replicates.

In 1976, Novotny, Lee, Low, and Maskarinec [51] developed a high-resolution GC-MS procedure for the determination of components of the sterol fraction of tobacco and marijuana based on the separation of their TMSi derivatives with GC glass capillary columns. In their investigation, ten phytosterols in tobacco and five in marijuana were identified by comparisons of their retention with authentic compounds on three different stationary phases and through mass spectral data.

TABLE 1.29

Total Sterols in Flue-Cured Tobacco[a]

Sample	Hydrolysis time (hr)	Cholesterol	Campesterol	Stigmasterol	β-Sitosterol
			(% dry leaf)		
1	2	0.025	0.054	0.072	0.088
2	18	0.028	0.052	0.071	0.082
3	18	0.030	0.053	0.075	0.088

[a] From Ellington et al. [46], reproduced from the Journal of Chromatographic Science, by permission of Preston Publications, Inc.

TABLE 1.30

Free Sterols in Hexane Extract of Flue-Cured Tobacco[a]

Sample	Cholesterol	Campesterol	Stigmasterol	β-Sitosterol
		(% dry leaf)		
1	0.005	0.013	0.025	0.018
2	0.007	0.012	0.023	0.020
3	0.008	0.015	0.024	0.020

[a] From Ellington et al. [46], reproduced from the Journal of Chromatographic Science, by permission of Preston Publications, Inc.

Preparation of tobacco and marijuana samples for sterol analysis consisted of the following:

Fifteen-gram samples each of moisture-free tobacco and Mexican marijuana were homogenized, extracted, and fractionated according to the procedure of Keller, Bush, and Grunwald [52]. Since this method involves a hydrolysis step, total sterols (both in free and conjugated forms) were determined. The digitonin fractionation yielded sterol mixtures for both tobacco and marijuana, that were further converted to TMSi derivatives according to the method of Chambaz and Horning [53] prior to GC analysis.

For the determination of retention data and the recording of chromatograms, a Perkin-Elmer model 990 gas chromatograph equipped with modified injector and detector systems and 11 to 15-m by 0.25-mm-i.d. glass columns treated according to the methods established in their laboratory [54] and statically coated [55] with SE-30, SE-52, and Poly I-110 stationary phase was used. With the P-E 990, samples were introduced by either a splitless injection [56] or a precolumn technique [57].

To obtain spectral data for identification purposes, Novotny et al. employed a Hewlett-Packard model 5980A integrated GC-MS (dodecapole) instrument; the GC unit was provided with a modified sampling system (to use a precolumn) and the capillary column was interfaced to the MS via a single-stage, all-glass jet separator. To obtain EI mass spectra, an electron energy of 70 eV was used.

Measured under temperature programming conditions (1°C/min, from 220 to 260°C), retention data for several standard TMSi sterols on the three stationary phases are shown in Table 1.31. Based on retention indices on all three columns and mass spectra comparable to previously published data [58-62], individual sterols were assigned their structures.

In Figure 1.19, an effective resolution of ten tobacco sterols as their TMSi derivatives on the SE-52 glass capillary column is shown; assignment of structures to peaks 8, 9, and 10 was tentatively made as (I) 4-methyl,5α-ergosta-7,24(28)-dien-3β-ol, (II) 5α-stigmasta-7,24(28)-dien-3β-ol, and (III) 4-methyl,5α-stigmasta-7,24(28)-dien-3β-ol, respectively, based on MS interpretation of the TMSi structure shown below.

TABLE 1.31

Retention Data for Several Standard Trimethylsilyl Sterols[a]

Sterol	Retention index[b]		
	SE-30	SE-52	Poly I-110
5-Cholesten-3β-ol (cholesterol)	3092	3122	3214
5α-Cholest-7-en-3β-ol	3121	3149	3307
5-Ergosten-3β-ol (campesterol)	3187	3217	3552
5,21-Stigmastadien-3β-ol (stigmasterol)	3227	3260	3348
5,24(28)-Stigmastadien-3β-ol (fucosterol)	3263	3297	3438
5-Stigmasten-3β-ol (β-sitosterol)	3294	3331	3427
5A-Stigmast-7-en-3β-ol	3312	3349	3530

[a] From Novotny et al. [51], courtesy of Steroids.
[b] Measured under temperature programming conditions: 1°C/min, from 220 to 260°C.

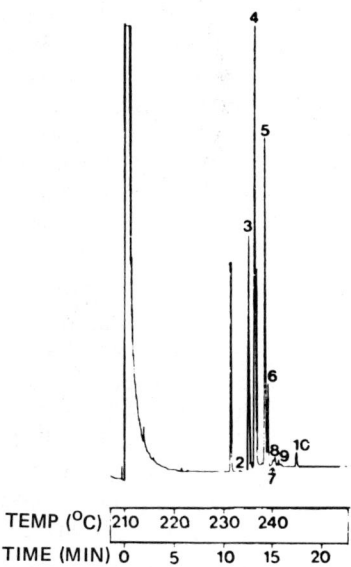

Figure 1.19 Chromatogram of the tobacco sterol fraction (as TMS* ethers) obtained with an SE-52 capillary column. Conditions: 11 m by 0.26 mm, i.d., glass column; splitless injection procedure. Key: (1) cholesterol; (2) 5α-cholest-7-en-3β-ol; (3) campesterol; (4) stigmasterol; (5) β-sitosterol; (6) 5,24(28)-stigmastadien-3β-ol (fucosterol); (7) 7-stigmasten-3β-ol; (8) 4-methyl,5α-ergosta-7,24(28)-dien-3β-ol; (9) 5α-stigmasta-7,24(28)-dien-3β-ol; (10) 4-methyl,5α-stigmasta-7,24(28)-dien-3β-ol. (*TMS = trimethylsilyl.) From Novotny et al. [51], courtesy of Steroids.

F. Fatty Acids and Esters

As early as 1959, methyl esters of acids obtained from the flue-cured tobacco of a variety of British cigarettes were analyzed by GC and MS [42]. The acids were extracted from tobacco with chloroform, and the alkali-soluble fraction was reacted with ethereal diazomethane, which yielded the methyl esters. MS results of a complex mixture containing methyl caprate, laurate, myristate, palmitate, margarate, arachidonate, linolenate, linoleate, oleate, stearate, nonadecylate, arachidate, and behenate showed that methyl palmitate was the major component, accompanied by some stearate and smaller amounts of the other higher and lower homologs. A considerable amount of C_{18}-unsaturated esters was present, with methyl linolenate predominating. As noted by Carruthers and Johnstone, an essentially similar result was obtained in the GC analysis.

In 1963, Schmeltz and co-workers [63,64] reported their results for the GC analysis of steam-volatile acids of various tobaccos. Using methods of steam distillation and isolation of acidic substances that had been described previously [65], the resulting concentrate of acidic constituents were chromatographed with either of two gas chromatographs: (1) an F&M model 500 with thermal conductivity detection or (2) a Perkin-Elmer model 800 with dual columns and flame ionization detectors. The column systems and operating conditions used with each gas chromatograph are given below:

GC unit	Operating parameters
F&M 500	System I: 2-ft by 0.25-in. column packed with 25% Tween 80/2% H_3PO_4 coated on 60-80 mesh Gas Chrom P; injector temperature, 250°C; column temperature, 110°C; detector temperature, 290°C; helium carrier-gas flow rate, 60 ml/min
F&M 500	System II: 2-ft by 0.25-in. column packed with 25% DEGA/2% H_3PO_4 coated on 60-80 mesh Gas Chrom P; injector temperature, 260°C; column temperature, programmed from 100 to 211°C at 4°C/min; detector temperature, 290°C; helium carrier-gas flow rate, 60 ml/min
P-E 800	System III: 5-ft by 0.13-in. column packed with 25% Tween 80/2% H_3PO_4 coated on 60-80 mesh Gas Chrom P; injector temperature, 170°C; column temperature, 110°C; detector temperature, 290°C; helium carrier-gas flow rate, 40 ml/min
P-E 800	System IV: 6-ft by 0.13-in. column packed with 25% DEGA/2% H_3PO_4 coated on 60-80 mesh Gas Chrom P; injector temperature, 170°C; column temperature, 125°C; detector temperature, 290°C; helium carrier-gas flow rate, 40 ml/min

Schmeltz et al. summarized their findings as follows [63]:

Comparisons of the four major types of cigarette tobaccos (bright, burley, Maryland, and Turkish) revealed major differences in the amounts and distribution of lower fatty acids. In the various tobaccos, formic, acetic, propionic, isobutyric, n-butyric, isovaleric, n-valeric, β-methylvaleric, isocaproic, n-caproic, n-heptylic, and n-caprylic were found. Except for n-butyric, n-heptylic, and isocaproic, the occurrence of these acids in tobacco leaf has been previously known; n-butyric, n-heptylic, and n-caprylic have been reported in tobacco smoke [66,67]. In general, burley and Maryland tobaccos contained relatively small amounts of the acids, with Maryland showing more than burley. Turkish (Samsun) and bright tobaccos contained relatively large amounts of the acids with distinct differences between the two types (see Figs. 1.20, 1.21, and 1.22). Turkish (Samsun) showed much larger amounts (more than fourfold) of β-methylvaleric and smaller amounts (less than one-half) of n-valeric acid compared to most samples of bright. An unidentified peak (No. 9 in Fig. 1.20) eluting between isocaproic and caproic was present in bright and absent in Turkish (Samsun). Turkish (Smyrna) was similar to Turkish (Samsun) except that smaller amounts of all acids were apparent.

Differences were also observed between "aromatic" and "aroma-deficient" grades of bright tobacco (Figs. 1.20 and 1.21), although

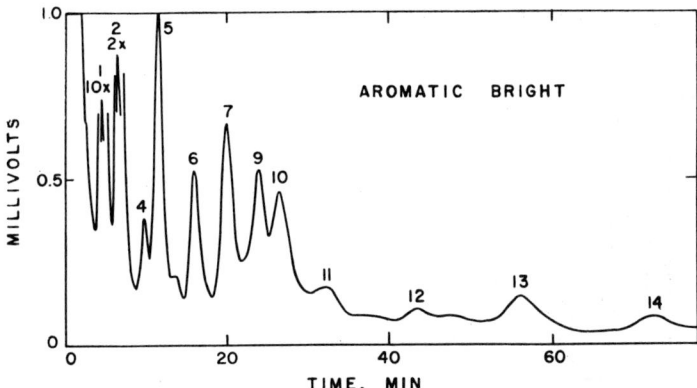

Figure 1.20. Separation of volatile acids from bright tobacco (aromatic grade). Chromatographic conditions: dual columns (5 ft by 0.125 in.) containing 25% Tween 80-2% phosphoric acid on Gas Chrom P; column temperature, 110°C; carrier-gas (helium) flow, 40 ml/min; dual flame ionization detectors (hydrogen pressure, 20 psi; air pressure, 38 psi). Peak identities (all acids): 1, acetic; 2, propionic and formic; 4, n-butyric; 5, isovaleric; 6, n-valeric; 7, β-methylvaleric; 10, n-caproic; 12, n-heptylic; 14, n-caprylic. Peaks 9, 11, and 13 are unidentified. From Schmeltz et al. [63], reproduced from the Journal of Chromatographic Science, by permission of Preston Publications, Inc.

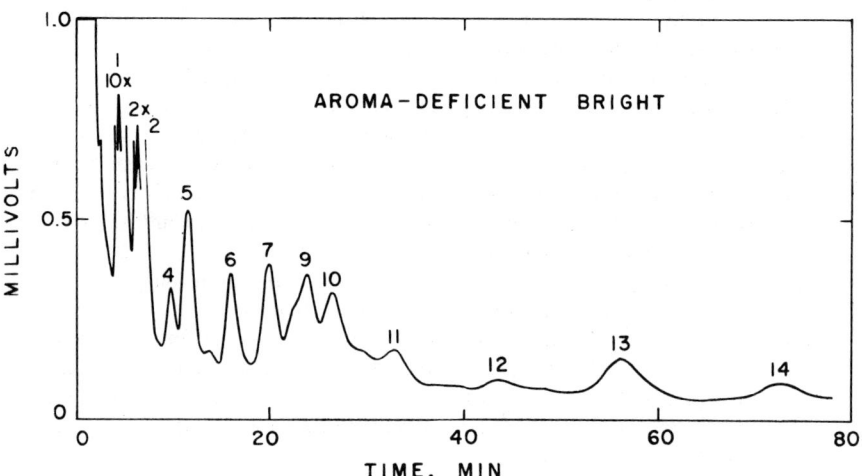

Figure 1.21. Separation of volatile acids from bright tobacco (aroma-deficient grade). See Figs. 1.20 and 1.22 for conditions and peak identities. From Schmeltz et al. [63], reproduced from the Journal of Chromatographic Science, by permission of Preston Publications, Inc.

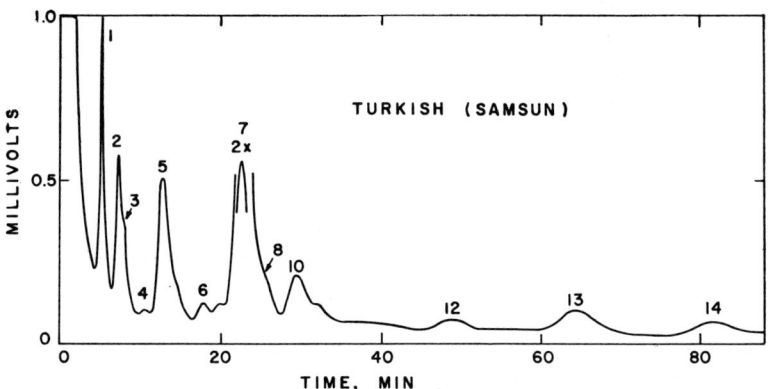

Figure 1.22. Separation of volatile acids from Turkish (Samsun) tobacco. See Fig. 1.20 for chromatographic conditions and peak identities. Peak 3, isobutyric acid; peak 8, isocaproic acid. From Schmeltz et al. [63], reproduced from the Journal of Chromatographic Science, by permission of Preston Publications, Inc.

the variations were slighter than in the above comparison. In general, aromatic grades showed larger amounts of isobutyric, n-valeric, isovaleric, and β-methylvaleric acids than aroma-deficient tobaccos but the differences in the case of the last three acids were small and possibly of questionable significance. Aroma-deficient tobaccos also gave a component which eluted as an inflection on unidentified peak 9 and was absent in the aromatic samples.

In 1966, Mold, Means, and Ruth [68] showed that the principal free and combined higher fatty acids of flue-cured (bright) tobacco contained 16 or 18 carbon atoms. Seventy-six percent of the total acids were found to be relatively nonpolar saturated or olefinic types, with palmitic acid comprising 20.7% [stearic acid (2.1%), oleic acid (3.5%), linoleic acid (5.8%), and linolenic acid (26.2%)]. Ninety percent of the total acids were shown to have unbranched structures of C_{10} to C_{34} carbon atoms. Only 4.1% were homologous monomethyl- or cyclohexyl-substituted compounds, and 5.9% had more complex branched structures. The saturated hydrocarbons derived from the simply branched acids proved to be homologs of 2-methyl and 3-methyl isomers with 15 to 26 carbon atoms and 1-cyclohexyl isomers with 22 to 25 carbon atoms.

Their approach to this analytical problem was to convert the total fatty acids from tobaccos into their corresponding saturated hydrocarbons of the same skeletal arrangement. By this procedure, evidence regarding the nature of the branching could be obtained without complications due to the presence of unsaturated or hydroxy acids. The procedure used for this conversion involved the catalytic reduction of the unsaturated methyl esters with hydrogen over Pt, reduction of the saturated esters to alcohol with $LiAlH_4$, conversion of these alcohols to iodides, and subsequent reduction of the iodides with $LiAlH_4$ to yield the saturated hydrocarbons. Following the removal of normal hydrocarbons by inclusion into 5A molecular sieves and the separation of the branched isomers into two groups by urea adduction, these fractions were subjected to GC analysis in order to quantitatively determine individual homologs.

These hydrocarbon fractions were resolved using a Micro-Tek 2500-DPFF gas chromatograph equipped with dual flame ionization detectors and 6-ft by 3/16-in.-i.d. stainless steel columns packed with 6% SE-30 coated on 80-100 mesh Gas Chrom P. To obtain chromatograms A, B, and C illustrated in Figure 1.23, the following GC conditions were used: injector temperature, 260°C; column temperature, initially 100°C, then programmed to 280°C at 4°C/min; detector temperature, 260°C; helium carrier-gas flow rate, 60 ml/min.

As noted by Mold et al., structural assignments were based on a comparison of GC retention data, infrared and mass spectra with those for compounds of known, similar structures. The very intense peaks representing ion fragments of m/e 82 and 83, coupled with the absence of any

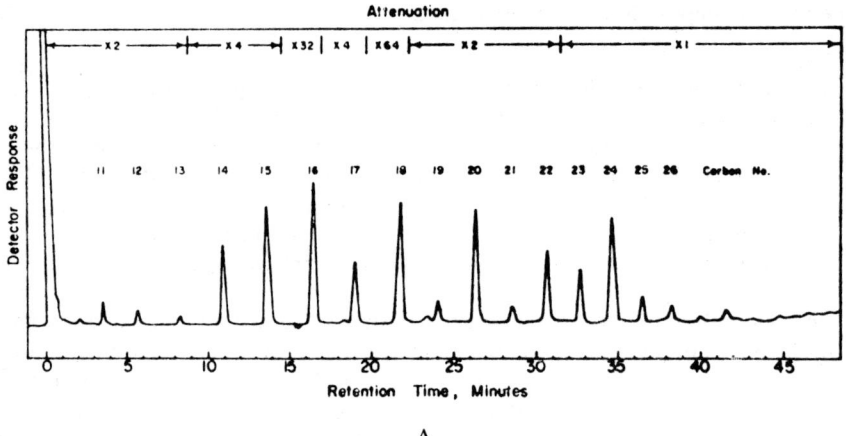

A

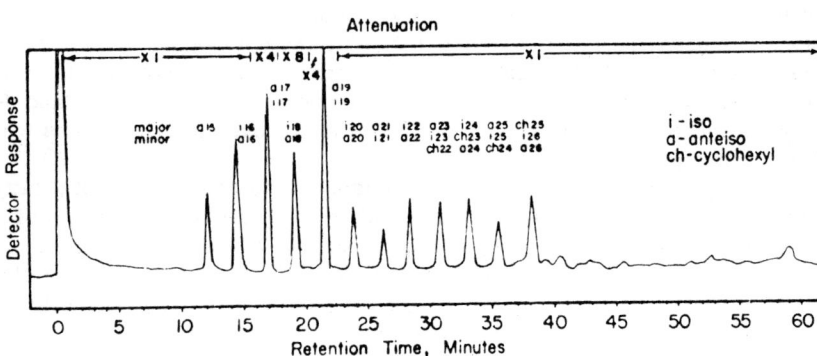

B

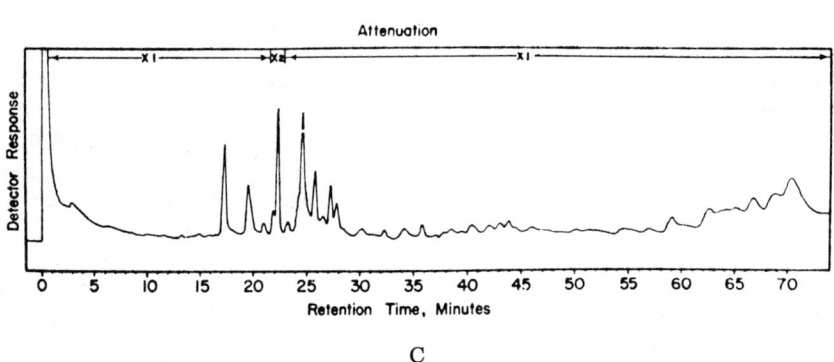

C

Figure 1.23. GC chromatograms of (A) saturated normal hydrocarbons, (B) simply branched saturated hydrocarbons, and (C) more highly branched saturated hydrocarbons derived from the higher fatty acids of bright tobacco. Adapted from Mold et al. [68].

significant increase in the intensity of the m/e 68 and 69 peaks over those noted for the spectrum of the reference cyclohexyl alkane, demonstrated that the cycloalkyl alkanes derived from tobacco acids were cyclohexyl compounds with no observable quantity of cyclopentyl alkanes present. Furthermore, approximate compositions for the mixed fractions were calculated using the intensities for the $(M-43)^+$ peak as a measure of the 2-methyl compounds, the $(M-29)^+$ peak as a measure of the 3-methyl compounds, and the M^+ peak as a measure of the cyclohexyl compounds.

In addition to the GC and MS analyses of tobacco fatty acids performed by other investigators [69,70], Ellington et al. [71] developed a GC-MS procedure in which the acids obtained by saponification of a hexane-soluble fraction of flue-cured tobacco were converted to their methyl esters, purified by TLC, and identified by GC-MS. For acids of chain lengths C_{12}-C_{34}, both qualitative and quantitative data were reported.

For the quantitative analysis of the methyl esters prepared with ethereal diazomethane, Ellington et al. used a Hewlett-Packard 7610A gas chromatograph equipped with a flame ionization detector, an Infotronics automatic digital integrator, and 1.8-m by 2-mm-i.d. U-shaped columns packed with 10% SILAR-10C on 100-120 mesh Gas Chrom Q. With methyl undecanoate added as internal standard, these esters were resolved with the following GC operational conditions: injector temperature, 275°C; column temperature, programmed from 100 to 250°C at 2°C/min; helium carrier-gas flow rate, 40 ml/min.

On the other hand, GC-MS studies were performed with a Varian 1400 gas chromatograph interfaced via a jet separator to a DuPont 21-492 mass spectrometer equipped with a Datagraph 5-134 recording oscillograph, and a 10-ft by 1/8-in. stainless steel column packed with SILAR-10C which was operated under the same conditions as given above. The effluent from the column was split 1:1, with one-half going to the FID of the chromatograph and the other half to the source area of the MS unit.

With the HP chromatograph, Figure 1.24 shows the elution pattern of the fatty-acid methyl esters. Retention times and those relative to the internal standard are listed in Table 1.32.

With regard to mass spectral studies, Ellington et al. noted that:

> In mass spectral studies, positive identification of each component was based on the presence of a large ion at m/e 74, due to the classical McLafferty rearrangement of methyl esters, and molecular (M^+) and fragment ions consistent with those reported for methyl esters. Both reference data and, in some cases, known standards were used for the identifications. The m/e 74 peak was the base or 100% peak in all spectra. The iso and anteiso branched esters were differentiated by the ratios of the high-mass branched segments: M-15, M-29, M-31, M-43, and M-57 [72,73]. The fragments indicated loss of the methyl, ethyl, isopropyl, and isobutyl groups from the molecular ions.

TABLE 1.32

Retention Time Data for Fatty Acid Esters[a]

Fatty acid[b]	RT (min)	RRT[c]
11	8.20	1.00
12	10.35	1.26
13	13.00	1.58
14	16.10	1.96
15[d] (A)	18.20	2.22
15[e] (I)	18.72	2.28
15	19.15	2.33
15:1	20.00	2.44
16	22.00	2.68
17[e] (I)	23.80	2.90
17	24.80	3.02
18[e] (I)	26.20	3.19
18	27.60	3.36
18:1	29.10	3.54
19	30.80	3.76
18:2	31.55	3.84
20	32.60	3.97
18:3	34.30	4.18
21	35.00	4.26
22	37.20	4.53
23	39.40	4.80
24	41.70	5.07
25	43.80	5.33
26	45.80	5.57
27	47.80	5.82
28	49.80	6.06
29	51.70	6.29

TABLE 1.32 (continued)

Fatty acid[b]	RT (min)	RRT[c]
30	53.50	6.51
31	55.30	6.73
32	57.10	6.95
33	58.90	7.16
34	60.50	7.38

[a] Adapted from Ellington et al. [71].
[b] Numbers refer to carbon number of fatty acid.
[c] Retention times, relative to C_{11} = 1.00.
[d] Anteiso branched.
[e] Iso branched.

Whereas Ellington et al. used SILAR-10C for resolving fatty acids, Myher, Marai, and Kuksis [74] employed in 1974 a similar but less polar liquid stationary phase (SILAR-5CP) for the GC-MS analysis of standard mixtures of fatty acid methyl esters, the maximum chain length studied being 24 carbon atoms and the separations dependent on the degree of unsaturation as well as molecular weight. In the Myher et al. investigation, the methyl esters and dialkylacetals were separated with an F&M model 402 gas chromatograph equipped with a hydrogen flame ionization detector and 180-cm by 3-mm-i.d. glass columns packed with 3% SILAR-5CP on 100-120 mesh Gas Chrom Q or 10% EGSS-X on 100-120 mesh Gas Chrom P. Using the specified GC conditions (injector temperature, 225°C; helium carrier-gas flow rate, 40 ml/min), Table 1.33 lists retention time data obtained from both columns. One will note that SILAR-5CP data compares favorably with that obtained with EGSS-X, possibly the most effective liquid stationary phase for the resolution of fatty acid methyl esters at the present time.

Compound identifications were made with a Varian 2700 gas chromatograph interfaced via a Watson-Biemann dual-stage helium separator to a Varian MAT CH-5 single-focusing mass spectrometer and equipped with a 180-cm by 2-mm-i.d. stainless steel column which contained 30 cm of a 3% OV-1 packing at the outlet end and 150 cm of 3% SILAR-5CP on Gas Chrom P in the rest of the column. The operating conditions specified for the integrated GC-MS instrument were injector temperature, 225°C; column temperature, run isothermally at 180°C or programmed from 180 to 220°C (heating rate not specified); transfer-line temperature, 275°C; separator temperature, 275°C; ion source temperature, 270°C; helium carrier-gas flow rate, 10 ml/min; ionization voltage, 70 eV; accelerating voltage, 3000 V; electron emission energy, 100 µA; scan rate, 4 sec/decade at a resolution of 800-1000.

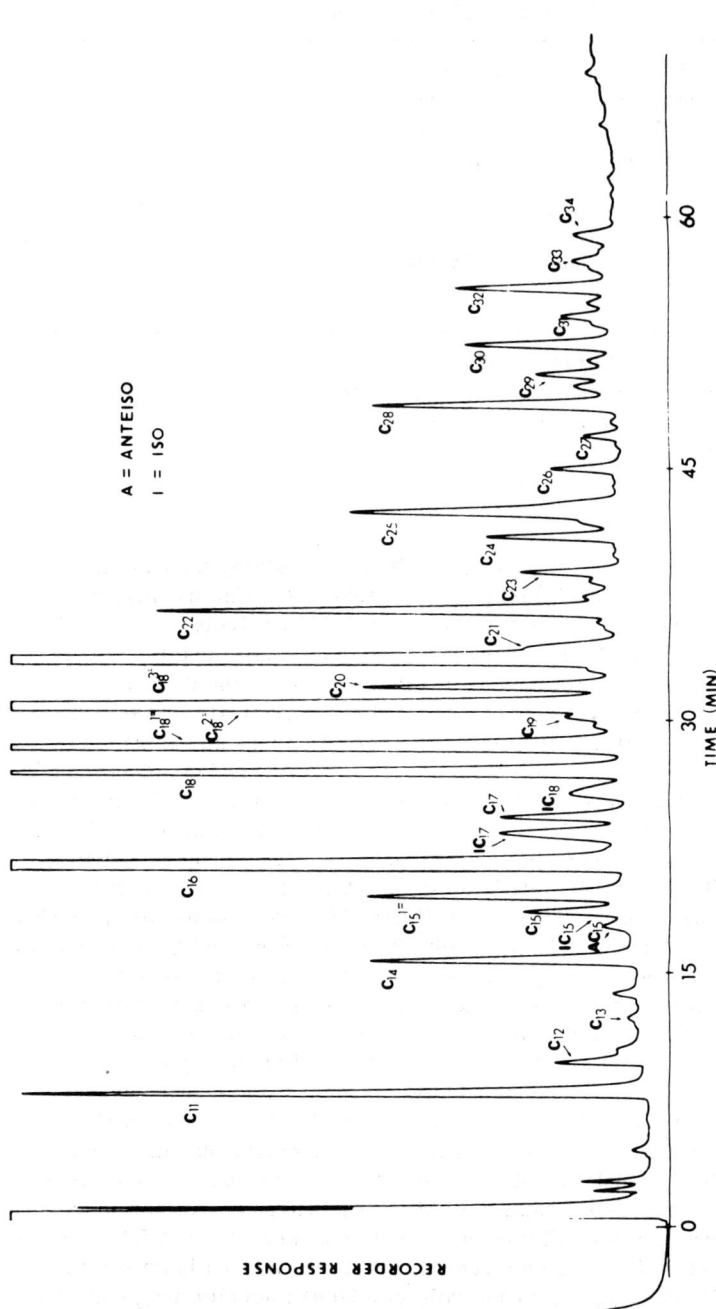

Figure 1.24. GLC chromatography of fatty-acid methyl esters of tobacco. Conditions: 1.8-m by 2-mm glass "U" column, 10% SILAR-10C, He flow 40 ml/min, temperature programmed 100 to 250°C at 2°C/min. Numbers refer to the chain length of the acid portion of the ester. From Ellington et al. [71], reproduced from the Journal of Chromatographic Science, by permission of Preston Publications, Inc.

TABLE 1.33

Equivalent Chain Lengths of Fatty Acid Methyl Esters and Dimethylacetals[a]

Fatty acids and dimethylacetals	Retention time (min)	
	3% SILAR-5CP	10% EGSS-X
14:0 DMA	13.65	13.21
14:0	14.00	14.00
16:0 DMA	15.66	15.20
16:0	16.00	16.00
16:1ω7	16.31	16.72
18:0 DMA	17.66	17.20
18:0	18.00	18.00
18:1ω9 (cis)	18.25	18.64
18:1ω9 (trans)	18.25	18.64
18:2ω6	18.73	19.42
18:3ω3	19.28	20.62
18:4ω3	19.63	
20:0	20.00	20.00
20:1ω9	20.23	20.65
20:4ω6	21.15	22.68
20:5ω3	21.80	23.81
22:0	22.00	22.00
22:1	22.23	22.55
22:5ω6	23.38	25.23
22:5ω3	23.81	25.79
22:6ω3	23.96	26.40
24:0	24.00	24.00

[a] Adapted from Myher et al. [74].

For the separation of tobacco lipids, Ellington et al. [46] prepared samples for GC analysis following a hexane extraction of the tobacco (see Sec. I.C for details) using the following procedure:

An aliquot (1 to 2 ml) of the hexane extract or 2 g of ground tobacco was transferred to a 300-ml saponification flask containing 40 ml of 2 N KOH in 95% ethanol and fitted with a reflux condenser. The sample was refluxed for 2 hr under nitrogen and, after cooling, 5 ml of a saturated KCl solution was added and the pH of the solution adjusted to 2 with concentrated HCl. The solution was shaken with 10 ml of hexane and, if necessary, water was added until a clear meniscus was obtained. The mixture was transferred to a 125-ml separatory funnel and extracted with 10-ml portions of hexane until two successive hexane layers were colorless (5 to 6 times). The combined hexane layers were diluted to 100 ml and stored in tightly closed brown bottles. For fatty acid analysis, a 1-ml aliquot (equivalent to about 20 mg of tobacco) was withdrawn; to this extract hydrolyzate, undecanoic

TABLE 1.34

Total Fatty Acids in Flue-Cured Tobacco and Hexane Extract After Alkaline Hydrolysis of Replicate Samples[a]

	Myristic acid	Palmitic acid	Oleic + linoleic + linolenic acids	Stearic acid
Tobacco				
1	0.034	0.223	0.413	0.168
2	0.037	0.232	0.416	0.177
3	0.032	0.222	0.416	0.175
Hexane extract				
1	0.032 (tr)[b]	0.175 (0.04)	0.303 (0.12)	0.112 (0.007)
2	0.030	0.185	0.346	0.128
3	0.030	0.173	0.315	0.113

[a] From Ellington et al. [46], reproduced from the Journal of Chromatographic Science, by permission of Preston Publications, Inc.
[b] Numbers in parentheses refer to fatty acids in residue (marc) after extraction of sample 1.

acid was added as internal standard. After solvent evaporation in a stream of nitrogen, 50 µl each of N,O-bis(trimethylsilyl)acetamide/dimethylformamide were added to the residue and then this residue-reagent mixture was heated for 15 min at 76°C in a sealed tube. For GC analysis, 2 to 3 µl were withdrawn and injected into the GC unit.

The silylated fatty acids were analyzed with a Hewlett-Packard 7610A gas chromatograph equipped with a 2-ft by 1/8-in. stainless steel column packed with 5% Dexsil 300 coated on 100-120 mesh, acid-washed Chromosorb W. To obtain the GC chromatogram for the hydrolyzate of the flue-cured tobacco after treatment with BSA/DMF as shown in Figure 1.25, the following GC conditions were employed: injector temperature, 275°C; column temperature, 90°C for 4 min, 90 to 210°C at 4°C/min, held at 210°C for 10 min, then 210 to 330°C at 15°C/min and held at 330°C; FID detector temperature, 350°C; helium carrier-gas flow rate, 50 ml/min.

Assays of total fatty acids in replicate hydrolyzates of ground tobacco, hexane extracts, and in extracted tobacco (marc) were compared as shown in Table 1.34.

As a further check on the completeness of the hydrolysis, known amounts of cholesteryl acetate, cholesteryl palmitate, tripalmitin, and methyl palmitate were added to separate samples of ground tobacco prior to hydrolysis. The data in Table 1.35 indicate essentially complete hydrolysis and good recovery of the esters.

TABLE 1.35

Results of Standard Additions to Flue-Cured Tobacco Prior to Hydrolysis[a]

Ester added	Palmitic acid			Cholesterol		
	Added[b]	Found[c]	Percent recovery	Added[b]	Found[c]	Percent recovery
Flue-cured blank	0	2.09			0.26	
Cholesteryl acetate				4.50	4.70	104.0
Cholesteryl palmitate	1.23	1.16	94.3	1.85	1.60	86.5
Methyl palmitate	2.37	2.36	99.0			
Tripalmitin	0.95	0.93	98.0			

[a] From Ellington et al. [46], reproduced from the Journal of Chromatographic Science, by permission of Preston Publications, Inc.
[b] Milligrams of component added as the ester.
[c] Milligrams found in excess of blank.

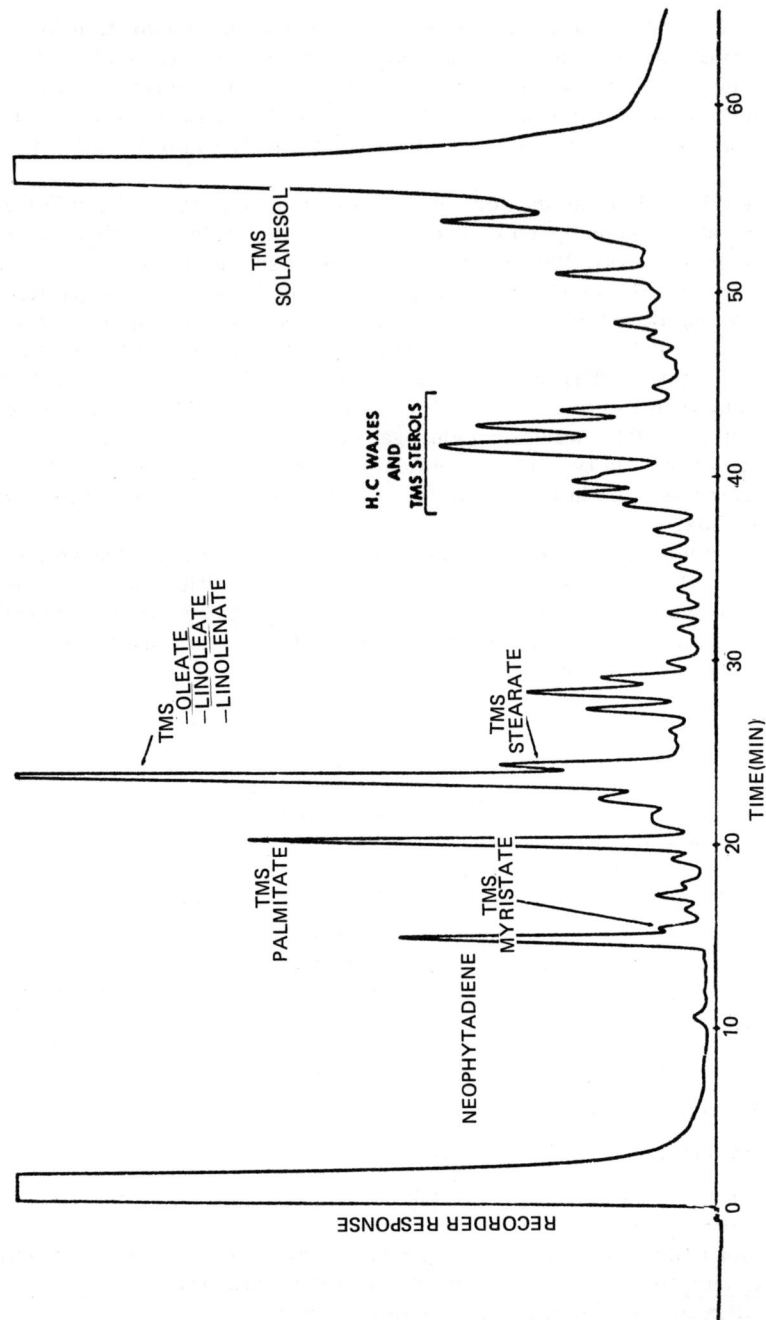

Figure 1.25. Gas chromatogram of the TMS-derivatized hydrolyzate of flue-cured tobacco. From Ellington et al. [46], reproduced from the Journal of Chromatographic Science, by permission of Preston Publications, Inc.

G. Ammonia [75] and Hydrazine [76]

In 1969, Ayers [75] analyzed ammonia in tobacco by gas-liquid chromatography, developing a method in which sample preparation consisted of the following:

> The tobacco specimen was weighed accurately and transferred to the outer chamber of the Conway-type diffusion cell. For ammonia contents <250 ppm, 250 to 1500 ppm, or >1500 ppm, the sample size specified was approximately 200, 50, and 25 mg, respectively. The 0.05 M H_2SO_4 absorbent solution (3 ml) was placed by pipette into the inner chamber of the diffusion cell. Then 3 ml of buffer solution [potassium dihydrogen phosphate (10.2 g) and borax (4.8 g) in ammonia-free water (1 liter)], followed by 1.5 ml of an alkaline solution [borax (5 g) in 0.5 M sodium hydroxide (100 ml)] were added to the outer cell chamber, which was immediately covered with the lid. Following an incubation period of 3 hr at 27.5°C, the cell was opened and an aliquot of the inner chamber's solution was withdrawn and injected into the gas chromatograph.

The GC separation was performed with a Pye model 104 gas chromatograph, equipped with a thermal conductivity detector. The operating conditions employed and specific columns used were as follows:

Column:	1.2-m by 3-mm stainless steel packed with 5% polyethyleneimine coated on base-washed Phasepak Q
Column temperature (°C):	45
Injector temperature (°C):	Room temperature
Carrier-gas (helium) flow rate (ml/min):	20
Precolumn:	25-mm glass tubing fitted with a S 13 ball-and-socket joints and packed with anhydrous barium hydroxide
Sample size (μl):	5.0
Ammonia retention time (min):	1.0

Applied to the determination of ammonia in tobacco, its concentration ranged from 0.17 to 0.37%, with the coefficient obtained from a series of 12 separate assays on one sample found to be ±2%.

For the determination of hydrazine in tobacco, sample preparation for GC analysis was as follows:

Tobacco from 20 cigarettes was ground in a blender and mixed with 1.2 g of pentafluorobenzaldehyde in 300 ml of phosphate buffer and 50 ml of methanol. The mixture was stirred overnight, filtered, and the filtrate was saturated with NaCl and extracted with ether (6 × 50 ml); the ether layer was extracted with 20% sodium bisulfite (4 × 50 ml) to remove excess PFB reagent. The bisulfite solution was saturated with NaCl and back-extracted with ether (2 × 50 ml). The combined ether solutions were dried (Na_2SO_4), concentrated, and chromatographed consecutively on two silica gel plates (2 mm) with benzene as developing solvent. The band corresponding to the reference compound was scraped off the two plates, extracted with ether, and purified further by chromatography on one aluminum oxide plate (1 mm) with hexane/benzene (1:1) as developing solvent. The band corresponding to DFBA (decafluorobenzaldehyde azine) was again scraped off the plate and extracted with ether. DFBA was then determined by injection of an aliquot into the gas chromatograph.

For quantitative analysis, a Varian Aerograph model 1200 gas chromatograph equipped with an electron-capture detector and a 1.8-m by 3-mm column packed with 11% OV-17 plus QF-1 (mixed phase) on 80-100 mesh Gas Chrom Q. With the nitrogen carrier-gas flow rate and the column temperature maintained at 40 ml/min and 145°C, respectively, the retention time for DFBA was nearly 18 min. With this arrangement, the lower limit of detection of hydrazine as DFBA was 0.1 ng.

On the other hand, for the separation of benzalazine, decafluorobenzaldehyde azine (DFBA), and pentafluorobenzaldehyde azine, a 1.3-m by 3-mm column packed with the same packing noted above was used with a Perkin-Elmer model 800 gas chromatograph equipped with dual flame ionization detectors and a 4:1 splitter for the isolation of unknowns. At a column temperature of 180°C and the helium carrier-gas flow set at 40 ml/min, the retention times of DFBA, pentafluorobenzaldehyde azine, and benzalazine were 2.7, 4.4, and 9.0 min, respectively.

II. CHEMICAL COMPOSITION OF TOBACCO SMOKE

Like the chemical composition of tobacco itself, that of tobacco smoke is also extremely complex, as indicated in Table 1.36 by the various compounds that have been identified in its smoke. Since this 1968 tabulation, the identification of literally thousands of other complex molecular structures have been achieved as the result of technological advances in GC capillary column technology and interactive GC-MS-COMP systems [77,78].

With regard to carcinogenic compounds isolated in smoke condensates, benzo(a)pyrene is the most widely and ubiquitously distributed hydrocarbon

TABLE 1.36

Chemical Composition of Tobacco Smoke[a]

1. Alkanes
   Normal C_1-C_9, C_{12}-C_{36}
   Iso C_4-C_6, C_{27}-C_{33}
   Anteiso C_6
   Cyclic C_5, C_6

2. Isoprenoid hydrocarbons
   2,4-Dimethyl-4-vinylcyclohexene
   Dipentene
   Farnesene
   Isoprene
   1-Methyl-4-isopropyl-1-cyclohexene
   Neophytadiene
   Neophytadiene dimers
   Norphytene
   β-Pinene
   Solanesenes
   Squalene

3. Alkenes and alkynes
   Acetylene
   Allene
   1,2-Butadiene
   1,3-Butadiene
   1-Butene
   2-Butene
   3-Buten-1-yne
   1-Butyne
   Cyclohexene
   1,3-Cyclopentadiene
   Cyclopentene
   1-Decene
   2,3-Dimethyl-1-butene
   3,3-Dimethyl-1-butene
   Ethylene
   1-Hexene
   2-Hexene
   Methylacetylene
   2-Methyl-1-butene
   2-Methyl-2-butene
   3-Methyl-1-butene
   1-Methyl-1-cyclopentene
   3-Methyl-1-cyclopentene

(continued)

TABLE 1.36 (continued)

4-Methyl-1-cyclopentene
2-Methyl-1-pentene
2-Methyl-2-pentene
3-Methyl-1-pentene
4-Methyl-1-pentene
4-Methyl-2-pentene
2-Methylpropene
1,2-Pentadiene
1,3-Pentadiene
1,4-Pentadiene
1-Pentene
2-Pentene
Propene

4. Aromatic hydrocarbons
Acenaphthene
Acenaphthylene
Alkylbenzo(a)pyrene
Alkylchrysene
Alkylfluoranthene
Alkylpyrene
Anthracene
Azulene
Benz(a)anthracene
Benzene
Benzo(b)fluoranthene
Benzo(g,h,i)fluoranthene
Benzo(j)fluoranthene
Benzo(k)fluoranthene
Benzo(m,n,o)fluoranthene
5H-Benzo(a)fluorene
11H-Benzo(a)fluorene
Benzo(b)fluorene
7H-Benzo(c)fluorene
11H-Benzo(b)fluorene
Benzo(a)naphthacene
Benzo(g,h,i)perylene
Benzo(c)phenanthrene
Benzo(a)pyrene
Benzo(e)pyrene
Biphenyl
Chrysene
Coronene
Dibenz(a,h)anthracene

TABLE 1.36 (continued)

Dibenzo(a,i)fluorene
Dibenzo(a,c)naphthacene
Dibenzo(a,j)naphthacene
Dibenzo(b,h)phenanthrene
Dibenzo(a,h)pyrene
Dibenzo(a,i)pyrene
Dibenzo(a,l)pyrene
Dibenzo(cd,jk)pyrene
9,10-Dihydroanthracene
5,6-Dihydro-8H-benzo(a)cyclopent(h)-
  anthracene
10,11-Dihydro-9H-benzo(a)cyclopent(i)-
  anthracene
3,4-Dihydrobenzo(a)pyrene
16,17-Dihydro-15H-cyclopent(a)-
  phenanthrene
9,10-Dimethylbenz(a)anthracene
Dimethylchrysene
Dimethylfluoranthene
1,6-Dimethylnaphthalene
1,8-Dimethylnaphthalene
2,6-Dimethylnaphthalene
2,7-Dimethylnaphthalene
2,5-Dimethylphenanthrene
Ethylbenzene
Ethyltoluenes (o-, m-, p-)
Fluoranthene
Fluorene
Indene
Indeno(1,2,3-cd)fluoranthene
Indeno(1,2,3-cd)pyrene
Ionene
4-Isopropenyltoluene
Isopropylbenzene
4-Isopropyltoluene
2-Methylanthracene
9-Methylanthracene
3-Methylbenz(a)anthracene
5-Methylbenz(a)anthracene
11-Methyl-11H-benzo(a)fluorene
Methylbenzo(a)pyrene
Methylchrysene
8-Methylfluoranthene
1-Methylfluorene

(continued)

TABLE 1.36 (continued)

9-Methylfluorene
1-Methylnaphthalene
2-Methylnaphthalene
1-Methylphenanthrene
9-Methylphenanthrene
1-Methylpyrene
2-Methylpyrene
4-Methylpyrene
Methylstyrenes (o-, m-)
Naphthacene
Naphthalene
11H-Naphtho(2,1-a)fluorene
Naphtho(2,3-a)pyrene
Perylene
Phenanthrene
Phenylacetylene
Pyrene
Styrene
Tribenz(a,c,h)anthracene
1,2,3-Trimethylbenzene
1,2,4-Trimethylbenzene
1,3,5-Trimethylbenzene
1,3,6-Trimethylbenzene
Xylenes (o-, m-, p-)

5. Sterols
　Campesterol
　β-Sitosterol
　Stigmasterol

6. Diterpenes
　12α-Hydroxy-13-epimanoyl oxide
　Levantenolide (α-, β-)
　α-5,8-Oxido-3,9,13-duvatrien-1-ol
　α-5,8-Oxido-3,9(17),13-duvatrien-1-ol
　Phytol

7. Triterpenes
　β-Amyrin

8. Trisesquiterpene
　Solanesol

9. Related isoprenoids
　Farnesylacetone
　Hexahydrofarnesylacetone
　Solanone

TABLE 1.36 (continued)

---

10. Alcohols
    a. Aliphatic
       Butyl alcohol
       sec-Butyl alcohol
       1-Docosanol
       1-Eicosanol
       Ethyl alcohol
       1-Heneicosanol
       1-Heptadecanol
       Isobutyl alcohol
       Methanol
       1-Nonadecanol
       1-Octadecanol
       Propyl alcohol
       1-Tetracosanol
       1-Tricosanol
    b. Aromatic
       Benzyl alcohol
       β-Phenethyl alcohol
    c. Polyols
       Diethylene glycol
       Ethylene glycol
       Glycerol
       Propylene glycol
       Triethylene glycol
    d. Cyclic
       Furfuryl alcohol
       Menthol

11. Esters
    β-Amyrenyl esters
    Benzyl acetate
    Benzyl benzoate
    Benzyl cinnamate
    Butyl acetate
    Ethyl acetate
    Ethyl butyrate
    Ethyl caproate
    Ethyl formate
    Ethyl isovalerate
    Ethyl β-methylvalerate
    Ethyl propionate
    Glyceryl triacetate
    Hentriacontanyl hentriacontanoate

(continued)

TABLE 1.36 (continued)

    Isopropyl formate
    Methyl acetate
    Methyl acrylate
    Methyl formate
    Methyl isocyanate
    Methyl nitrite
    Methyl propionate
    Methyl thionitrite
    Vinyl acetate

12. Aldehydes
    Acetaldehyde
    Acrolein
    Benzaldehyde
    Caproaldehyde
    Crotonaldehyde
    Formaldehyde
    Furfural
    Glyoxal
    5-Hydroxymethylfurfural
    Isobutyraldehyde
    Isovaleraldehyde
    Methacrolein
    2-Methylbutyraldehyde
    5-Methylfurfural
    Methylglyoxal
    2-Methyl-4-pentenal
    2-Methylvaleraldehyde
    Pivaldehyde
    Propionaldehyde
    Valeraldehyde

13. Ketones
    Acetone
    2-Acetylfuran
    2,3-Butadione
    2-Butanone
    Butenone
    Cyclopentanone
    2,4-Dimethyl-3-pentanone
    4-Heptanone
    2-Hexanone
    3-Hexanone
    3-Methyl-2-butanone

TABLE 1.36 (continued)

    3-Methyl-3-buten-2-one
    Methyl naphthyl ketone
    2-Methyl-3-pentanone
    3-Methyl-2-pentanone
    4-Methyl-2-pentanone
    Palmitone
    2,3-Pentadione
    2-Pentanone
    3-Pentanone
    4-Penten-2-one
    4-Penten-3-one
    Reductic acid

14. Quinones
    2,3,6-Trimethyl-1,4-naphthoquinone

15. Nitriles
    Acetonitrile
    Acrylonitrile
    Butyronitrile
    Capronitrile
    Cinnamonitrile
    Crotononitrile
    Cyanogen
    Hydrogen cyanide
    Isobutyronitrile
    Isocapronitrile
    Isovaleronitrile
    Methacrylonitrile
    3-Phenylpropionitrile
    Propionitrile
    Valeronitrile

16. Cyclic ethers
    2,5-Dimethylfuran
    Furan
    Methylfuran
    Tetrahydrofuran
    Tetrahydropyran

17. Sulfur compounds
    Carbon disulfide
    Carbonyl sulfide
    Dimethyl disulfide
    Dimethyl sulfide

(continued)

TABLE 1.36 (continued)

    Hydrogen sulfide
    Methanethiol
    Methyl thionitrite
    Thiocyanic acid
    Thiocyanogen
    Thiophene

18. Acids
    Acetic
    Adipic
    Arachidic
    Benzoic
    Butyric
    $C_{10}H_{12}O_2$
    $C_{12}H_{12}O_5$
    Caproic
    Cerotic
    Decanoic
    Formic
    Furoic
    Glutaric
    Glycolic
    Glyoxylic
    Heptanoic
    Isobutyric
    Isocaproic
    Isovaleric
    α-Ketoglutaric
    Lactic
    Lauric
    Levulinic
    Linoleic
    Linolenic
    Malic
    Malonic
    β-Methylvaleric
    Myristic
    Nonanoic
    Octanoic
    Oleic
    Oxalic
    Palmitic
    Palmitoleic
    Phenylacetic

TABLE 1.36 (continued)

    α-Phenylpropionic
    Phthalic
    Propionic
    Pyruvic
    Sorbic
    Stearic
    Succinic
    Toluic acids (m-, p-)
    Valeric

19. Phenols and related compounds
    Anisole
    Caffeic acid
    Catechol
    Chlorogenic acid
    p-Coumaric acid
    m-Cresol
    Cresols (o-, p-)
    2,6-Dimethoxyphenol
    Esculetin
    2-Ethylphenol
    3-Ethylphenol
    4-Ethylphenol
    Eugenol
    Ferulic acid
    Guaiacol
    Hydroquinone
    Hydroxyacetophenone (o-, m-, p-)
    Hydroxybenzaldehyde (m-, p-)
    3-Hydroxybenzoic acid
    4-Hydroxybenzoic acid
    2-Hydroxyphenylacetic acid
    3-Hydroxyphenylacetic acid
    4-Hydroxyphenylacetic acid
    3-Hydroxyphenylpropionic acid
    4-Hydroxyphenylpropionic acid
    Isoeugenol
    Isovanillic acid
    3-Methoxyphenol
    4-Methoxyphenol
    1-Naphthol
    2-Naphthol
    Neochlorogenic acid
    Phenol

(continued)

TABLE 1.36 (continued)

Protocatechuic acid
Protocatechuic aldehyde
Quinic acid γ-lactone
Resorcinol
Salicylaldehyde
Salicylic acid
Scopoletin
Sinapic acid
Syringaldehyde
Syringic acid
Thymol
2,3,5-Trimethylphenol
2,4,6-Trimethylphenol
Vanillic acid
Vanillin
2,3-Xylenol
2,4-Xylenol
2,5-Xylenol
2,6-Xylenol
3,4-Xylenol
3,5-Xylenol

20. Alkaloids and other bases
Allylamine
Ammonia
Amylamine
sec-Amylamine
Anabasine
Anatabine
Aniline
m-Anisidine
Benzimidazole
Butylamine
sec-Butylamine
Carbazole
Collidine
Cotinine
3,4-Dehydropiperidine
Dibenz(a,h)acridine
Dibenz(a,j)acridine
7H-Dibenzo(c,g)carbazole
Diethylamine
Dihydrometanicotine
9,9-Dimethylacridan

TABLE 1.36 (continued)

Dimethylamine
2,3-Dimethylaniline
2,4-Dimethylaniline
2,5-Dimethylaniline
2,6-Dimethylaniline
3,5-Dimethylaniline
2,6-Dimethylpyrazine
2,3-Dimethylpyridine
2,4-Dimethylpyridine
2,5-Dimethylpyridine
2,6-Dimethylpyridine
3,4-Dimethylpyridine
3,5-Dimethylpyridine
Diphenylamine
Dipropylamine
Ethylamine
2-Ethylaniline
4-Ethylaniline
3-Ethylindole
3-Ethylpyridine
Hexylamine
Harmane
Indole
Isoamylamine
Isobutylamine
N-Isobutylbutylamine
Isonicotein (2,3'-bipyridyl)
Isopropylamine
N-Isopropylpropylamine
Isoquinoline
Metanicotine
Methylamine
3-Methylaminopyridine
N-Methylaniline
2-Methylbutylamine
N-Methylbutylamine
Methylcarbazole
Methylethylamine
3-Methylindole
N-Methylmyosmine
N-Methyl-2-phenethylamine
2-Methylpyrazine
2-Methylpyridine

(continued)

TABLE 1.36 (continued)

    3-Methylpyridine
    4-Methylpyridine
    N-Methylpyrrole
    2-Methylpyrrolidine
    Myosmine
    1-Naphthylamine
    Nicotelline
    Nicotinamide
    Nicotine
    Nicotinic acid
    Nicotinonitrile
    Nicotyrine
    Norharmane
    Nornicotine
    Nornicotyrine
    1,8,9-Perinaphthoxanthene
    2-Phenethylamine
    N-Phenyl-4-isopropylphenylamine
    N-Phenyl-2-naphthylamine
    Piperidine
    Propylamine
    Pyrazine
    Pyridine
    Pyridine-3-aldehyde
    3-Pyridinol
    Pyrido(2,3-b)indole
    3-Pyridyl ethyl ketone
    3-Pyridyl methyl ketone
    3-Pyridyl propyl ketone
    Pyrocoll
    Pyrrole
    Pyrrolidine
    3-Pyrroline
    Pyrrolo(2,3-b)pyridine
    Quinoline
    Quinoxaline
    Toluidines (o-, m-, p-)
    Trimethylamine
    2,4,6-Trimethylaniline
    3-Vinylpyridine

21. Volatile bases and alkaloids in brown pigments
    Cotinine
    Dihydrometanicotine

TABLE 1.36 (continued)

Metanicotine
2-Methylpyridine
3-Methylpyridine
N-Methylpyrrolidine
Nicotine
Nornicotine
Pyridine
Pyrrole
3-Vinylpyridine

22. Amino acids and related compounds
    α-Alanine
    β-Alanine
    γ-Aminobutyric acid
    Aspartic acid
    Glutamic acid
    Glutamine
    Glycine
    Leucine
    Ornithine
    Phenylalanine
    Proline
    Serine
    Threonine
    Valine

23. Miscellaneous components
    Carbon dioxide
    Carbon monoxide
    Methyl chloride
    Methyl isocyanate
    Nitric oxide
    Nitrous oxide

---

[a] Adapted from Stedman [2].

carcinogen in the environment. In his review of the methods of testing chemicals for carcinogenic activity, the guidelines for evaluating the significance of the test results, the major classes of chemical carcinogens, and the types of interaction that occur when several carcinogens act in combination, Arcos [79] noted that:

While hydrocarbons have a predominantly local effect at the site of application, such as the mouse skin tumor shown, oral administration of hydrocarbons produces tumors in distant tissue targets, mainly malignant mammary and lung tumors, and leukemia.

The reason why hydrocarbon carcinogens, and especially benzo(a)-pyrene, are so ubiquitous is that they readily arise from any kind of organic material by pyrolysis. Thus, benzo(a)pyrene has been identified in food materials and in consumer and industrial products, the processing of which involves high temperatures. These include, for example, lubricating machine oils, medicinal paraffin oil used as a laxative, cigarette smoke, the exhaust fumes of internal combustion engines, charcoal-broiled steak, roasted coffee and chicory, smoked sausage, smoked fish, charred dough, caramellized sugar, and French fries, to name but a few. The pollution of lakes and rivers with polycyclic hydrocarbons originates mainly from the untreated effluents of industrial plants and from the lubricating oils and exhaust fumes of ships. Some of these pollutants eventually find their way into the water table and, hence, into the drinking water.

Below is summarized the chemical mechanism of the formation of benzo(a)pyrene from organic materials during pyrolysis. A common pyrolytic decomposition product of all organic compounds is acetylene. However, molecules of acetylene in turn undergo gradual recombination at high temperatures to yield larger and larger structures, ultimately yielding polycyclic hydrocarbons. These compounds are relatively stable at high temperatures and, hence, accumulate as pyrolysis progresses.

A. Tobacco Alkaloids in Smoke and Biological Fluids

With regard to the specific analysis of nicotine and, to a lesser degree, tobacco alkaloids in smoke and biological fluids, most of the GC and/or GC-MS investigations have dealt with the determination of nicotine in biological fluids [23,91-112] rather than in tobacco smoke [8,11,18,23,80-90].

1. Tobacco Smoke

Although it had been known for years that other alkaloids accompany nicotine in tobacco smoke, definitive studies were lacking on the alkaloidal content of tobacco smoke. Therefore, Quin [11], anticipating that the tobacco alkaloids, generally boiling in the range 200 to 300°C, might be chromatographed, found that good separation of a majority of the alkaloids studied could be achieved at 190°C on 1-m columns containing several glycols as liquid stationary phases as noted in Table 1.37. In this preliminary work on cigarette smoke, Quin found it necessary to perform the gas chromatographic separation of the gross alkaloid extract mixture under three different sets of conditions to provide adequate overall resolution and to overcome the interference of nicotine, which accounts for 90% or more of the alkaloidal fraction of cigarette smoke under consideration, in the elution of neighboring peaks. At 140 to 150°C, several peaks noted prior to nicotine were due to substances boiling above about 170°C and, at a column temperature of 190°C, six peaks were readily detected after nicotine had been eluted. Typical conditions employed for these separations are summarized in Table 1.38.

In a subsequent paper, Quin [80] conclusively identified in burley tobacco cigarette smoke, by a combination of gas and paper chromatography and ultraviolet spectroscopy, myosmine, nornicotine, anabasine, anatabine, 2,3'-dipyridyl, and cotinine. By comparing the area of a chromatographic peak with the area from a known sample, the following values of micrograms per cigarette smoked were obtained: myosmine-nicotine, 88; anabasine, 11; anatabine, 14; 2,3'-dipyridyl, 7. The experimental data for these identifications are summarized in Table 1.39.

Concerned with the pyrolysis of nicotine in air streams, Kobashi et al. [82] examined the pyrolytic products using two different columns with a Perkin-Elmer model 188 gas chromatograph:

	Column I	Column II
Column length (m)	4	2
Stationary phase	Carbowax 1500	Carbowax 6000
Liquid phase concentration	Not specified	
Solid support	Not specified	
Helium carrier-gas flow rate (ml/min)	94	40
Column temperature (°C)	140	210

Using both columns, Kobashi et al. were able to separate the following compounds: myosmine, nornicotine, 2,2'-dipyridyl, isoquinoline, 3-vinyl pyridine, 3-cyanopyridine, 3-ethyl pyridine, 3-picoline, 3-pyridyl methyl ketone, pyridine, and pyrrole.

Jacin et al. [18] determined nicotine in particulate matter of smoke by gas chromatography; sample preparation for subsequent GC analysis consisted of the following:

> Weight-matched commercial cigarettes (70-mm length) were smoked on a smoking machine to a 23-mm butt length through a Cambridge filter. A 35-ml puff of 2-sec duration was taken once every minute. Five cigarettes were smoked through each Cambridge filter. The weights of the cigarettes, before smoking, and the weights of the butts were determined, thus allowing one to calculate the percent nicotine on the basis of tobacco consumed. After smoking was completed, the Cambridge filter was removed, and placed in a screw-capped vial. A 10-ml volume of benzene-chloroform (9:1) was added and the mixture was shaken on the wrist-action type shaker for 20 min. An aliquot of the supernatant liquid was subsequently injected into the gas chromatograph for nicotine determination, using the instrument and GC conditions described in Section I.A.

Levins and Ikeda [85] presented two variations of a simple method for obtaining volatiles from solid samples for GC analysis. For thermally stable samples, a glass tube partially filled with the sample is inserted into the injection port, where it is swept with carrier gas. The injection port and column are simultaneously temperature programmed. For thermally unstable samples, a relatively low-temperature presweep is used to elute the volatiles onto the column, which is held at room temperature; the sample is then removed, and the column temperature programmed in the normal manner.

TABLE 1.37

Gas Chromatography of Individual Tobacco Alkaloids[a]

Liquid phase	Retention time (min) for		
	Polypropylene glycol[b]	Polybutylene glycol[c]	Polyethylene glycol[d]
3-Pyridyl methyl ketone	4.3	3.1	4.3
3-Pyridyl ethyl ketone	6.1	5.0	5.3
3-Pyridyl n-propyl ketone	8.1	7.0	6.6
Nicotine	8.6	8.2	5.2
Nornicotine	16.1	14.3	12.3
Myosmine	16.4	14.7	13.4
Anabasine	19.4	18.1	13.8
Nicotyrine	21.0	18.3	19.4
Metanicotine	23.5	20.9	16.5
Anatabine	25.2	22.5	21.1
2,3'-Dipyridyl	31	26	29
N-Methyl nicotinamide	42	30	64
Nornicotyrine	73	55	101
Cotinine	79	63	85

[a] From Quin [11], courtesy of the Journal of Organic Chemistry.
[b] Molecular weight, 1025. Conditions: temp., 190°C; helium flow rate, 45 ml/min.
[c] Molecular weight, 1500. Conditions: temp., 180°C; helium flow rate, 50 ml/min.
[d] Molecular weight, 20,000. Conditions: temp., 190°C; helium flow rate, 48 ml/min.

Using a modified F&M model 810 gas chromatograph equipped with dual flame ionization detectors, gas chromatograms were obtained using the following GC conditions:

Column: 12-ft by 1/4-in.-o.d. copper tubing filled with 10% Carbowax 20M on 60-80 mesh Gas Chrom Q

Column temp. (°C):	80 to 230°C at 4°C/min, then held at 230°C
Injector temp. (°C):	Variable, depending on the sample
Detector temp. (°C):	290
Helium carrier-gas flow rate (ml/min):	123
Low-temp. presweep:	Usually 30 min, with the injection port at a selected temperature and the column at room temperature
Glass inserts:	6-mm-o.d. Pyrex cut so that one end protrudes about 0.5-mm beyond the lip of the injection port when fully inserted

Applied to the analysis of volatile, thermally stable particulate matter of cigarettes in order to study differences in the volatile fraction of the particulate matter from tobacco smoke [113], Figure 1.26 is a typical chromatogram obtained under the prescribed operating conditions. With such an analytical system, the reproducibility one can expect for the major components of filter cigarettes can be seen in Table 1.40.

TABLE 1.38

Fractionation of Cigarette Smoke Alkaloids by Gas Chromatography[a]

Separation performed	Prenicotine	Immediate post-nicotine	Late post-nicotine
Column[b]	1 m × 6 mm	1 m × 10 mm	1 m × 6 mm
Temp. (°C)	145	190	190
Helium flow rate (ml/min)	47	75	60
Retention time of nicotine (min)	24[c]	18	5.5
Retention time of other peaks	4.9, 6.2, 7.9, 11.3, 4.4, 17.0, 20	23, 28, 33, 39, 51, 60	27, 31, 49

[a] From Quin [11], courtesy of the Journal of Organic Chemistry.
[b] Packed with polypropylene glycol, MW 1025, on firebrick, 1:4.
[c] On separately run sample.

TABLE 1.39

Identification of Tobacco Smoke Alkaloids[a]

	Retention time (min)	
	I[b]	II[c]
Fraction 10	13.5	13.9
Nornicotine	13.3	13.2
Myosmine	13.5	14.0
Fraction 11	17.4	11.8, 13.6
Anabasine	17.4	13.6
Fraction 12	23.2	20.9
Anatabine	23.0	20.8
Fraction 13	20.0	23.2
2,3'-Dipyridyl	20.1	23.0
Fraction 16	49	52
Cotinine	49	51

[a] From Quin [80], courtesy of the Journal of Organic Chemistry.
[b] On polypropylene glycol.
[c] On polyethylene glycol.

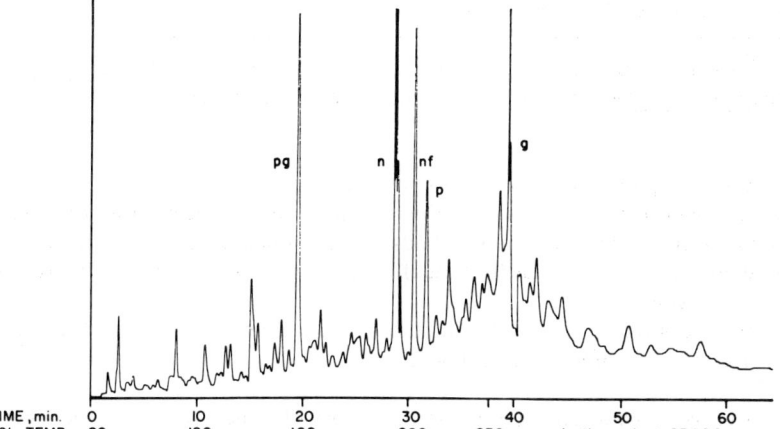

Figure 1.26. Volatile particulate matter from smoke of a nonfilter cigarette. 1.8-mm strip cut from Cambridge filter pad. Injection port: 120 to 220°C in 14 min. Column: 80 to 230°C (hold) at 4.0°C/min. Peaks: pg, propylene glycol; n, nicotine, nf, neophytadiene; p, phenol; g, glycerine. From Levins and Ikeda [85], reproduced from the Journal of Chromatographic Science, by permission of Preston Publications, Inc.

TABLE 1.40

Major Components of Volatile Particulate Matter (Filter Cigarettes)[a]

Peak	Attenuation	Average peak height	Standard deviation	RSD at 2 $(2\sigma)/\sigma$
Propylene glycol	×8	116	±24	23%
Nicotine	×32	122	± 9	7%
Neophytadiene	×4	134	± 9	7%
Phenol	×4	44	± 8	18%
Triacetin	×8	140	±24	17%
Glycerine	×8	188	±55	29%
Triethylene glycol	×4	73	±11	15%
Weight particulate matter/ 4 cigarettes		0.0758 g ± 0.0080 g		11%
Time for injection port program (120 to 220°C)		13.7 min ± 0.4 min		3%
Weight unconditioned cigarettes		1.058 g ± 0.040 g		
n = 8 samples (4 cigarettes each)				

[a] From Levins and Ikeda [85], reproduced from the Journal of Chromatographic Science, by permission of Preston Publications, Inc.

For comparative studies, Figure 1.27 shows a scan of the volatiles obtained from the charcoal granules taken from the filter of a commercial carbon-filter cigarette (unsmoked). It demonstrates that volatiles trapped on charcoal may be readily transferred to the GC column with little sample manipulation.

With regard to thermally unstable components, Levins and Ikeda demonstrated this technique as follows:

> Figure 1.28 shows the chromatograms obtained from the tobacco taken from three different major brands of cigarettes. Some of these peaks have been related to the known constituents of various tobacco flavoring formulations. By using a 1:1 split at the exit of the column, we have been able to subjectively evaluate the odor of the eluants. Some of the more odorous compounds have been trapped in glass capillaries and have been identified by mass spectroscopy.

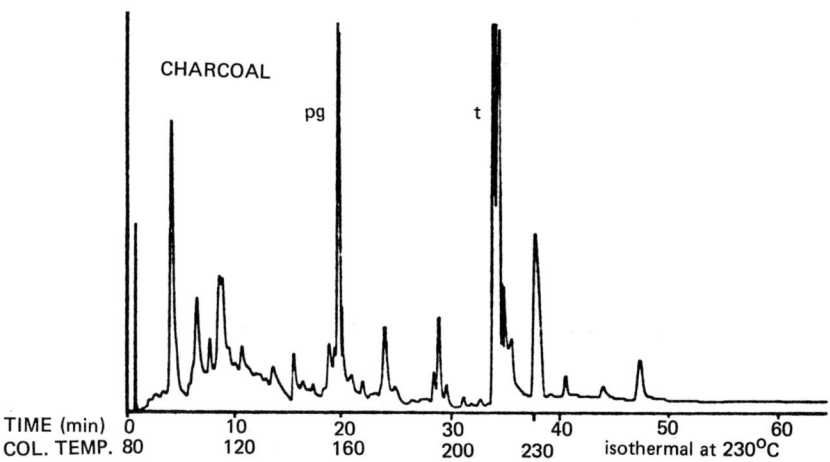

Figure 1.27. Volatiles from cigarette filter charcoal, 8.0-mg sample. Same conditions as Figure 1.26. Peaks: pg, propylene glycol; t, triacetin. From Levins and Ikeda [85], reproduced from the Journal of Chromatographic Science, by permission of Preston Publications, Inc.

For example, the large peak occurring in Figure 1.28a at 21 min represents the menthol from a fully mentholated cigarette (2200 ppm menthol). The tobacco used to obtain Figure 1.28b contained practically no menthol. The tobacco used to obtain Figure 1.28c was from a cigarette which is not considered to be mentholated; it contained approximately 200 ppm menthol. The taller of the four peaks clustered around 39 min in Figure 1.28b represents myristicin (5-allyl-2,3-methylenedioxy phenyl methyl ether), a major constituent of nutmeg. It was trapped in a glass capillary and identified by mass spectroscopy, and it is estimated to be present at approximately 100 ppm in the tobacco.

Cano et al. [23] determined the nicotine concentration in both tobacco smoke and urine specimens. In their procedure, the nicotine extracted from either urine or cigarette smoke condensate was analyzed with an Aerograph model 204 gas chromatograph equipped with a flame ionization detector and a 1.5-m by 1/8-in. o.d. column packed with 3.2% UCON 50 HB 2000 coated on Gas Chrom Q (mesh size not specified) which had been previously treated with a methanolic solution of 6% KOH. With the injector, column, and detector temperatures maintained at 175, 120, adn 135° C, respectively, and the nitrogen carrier gas adjusted to 18 ml/min, the retention time of nicotine was about 11 min. Using aqueous and methanolic solutions of known concentrations of tobacco smoke condensate obtained from SEITA, the recovery of nicotine was 90 ± 3% based on peak area measurements.

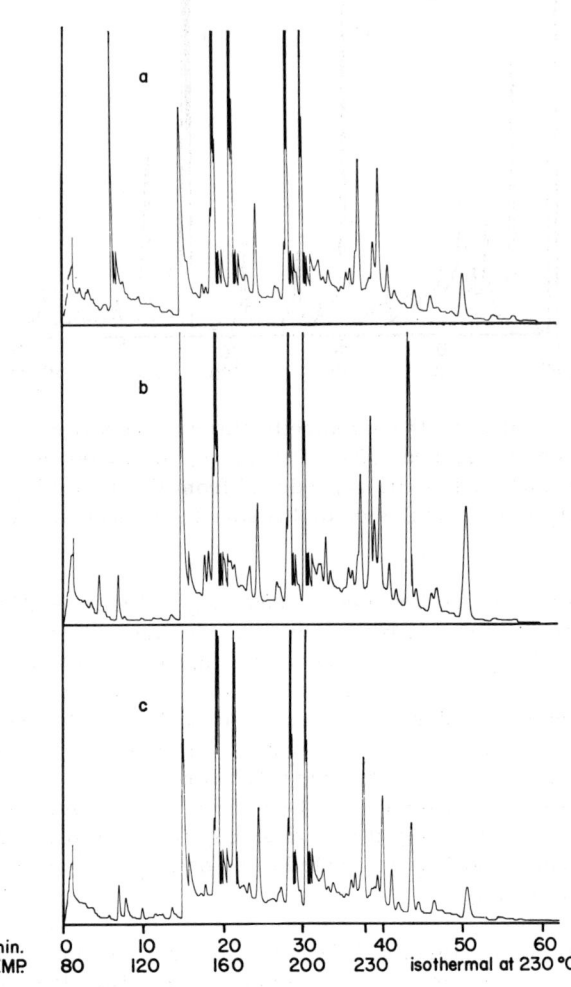

Figure 1.28. Tobacco volatiles from commercial cigarettes. Brands a, b, and c. 150-mg samples, milled to 20 mesh. Injection port: 30 min at 80° C (presweep). Column: room temperature during presweep, then 80 to 230° C (hold) at 4.0° C/min. From Levins and Ikeda [85], reproduced from the Journal of Chromatographic Science, by permission of Preston Publications, Inc.

In 1974, Guerin et al. [87] demonstrated GC component profiling to be an effective means of increasing the number of cigarette smoke components which can be surveyed for biological significance. Smoke particulate matter was routinely profiled on standard columns following trimethylsilylation, whereas the gas phase of smokes were profiled directly. In their investigation, the use of sulfur- and nitrogen-selective detectors permitted the convenient acquisition of data previously requiring prohibitive effort. Furthermore, the reproducibility of the profiles and high correlations between profile peak areas and independently determined quantitative results for the corresponding constituents demonstrated the applicability of the method to quantitative studies.

Following detailed procedures for the preparation of gas phase and TPM samples for subsequent GC analysis, the separations were performed with several GC instruments as indicated below:

1. Total particulate matter

Instrument:	Tracor model 220 gas chromatograph equipped with dual flame ionization detectors and 6-ft by 1/4-in. o.d. glass U-shaped columns packed with 4% OV-101 on 80-100 mesh Chromosorb G (HP)
Injector temp. (°C):	250
Column temp. (°C):	Held isothermally for 10 min after sample injection at 70°C, then programmed to 250°C at 5°C/min
Detector temp. (°C):	250
Nitrogen carrier-gas flow rate (ml/min):	60

2. Gas-phase profiling

Instrument:	Tracor model 150 cryothermal gas chromatograph equipped with flame ionization detection and 3-ft by 1/4-in. copper columns packed with 20% Carbowax 20M on 30-60 mesh, acid-washed Firebrick in series with 6-ft by 1/4-in. β,β'-oxydipropionitrile on the same support.
Injector temp. (°C):	100
Column temp. (°C):	Maintained at -75°C until the cigarette was consumed and all puffs, including

	a clearing puff, were collected; then increased to +20°C at 25°C/min, held at +20°C for 20 min, raised to +35°C and held at this temperature for 16 min, then raised to +90°C and held for the duration of the run
Detector temp. (°C):	100
Helium carrier-gas flow rate (ml/min):	90

3. Sulfur profiling

Instrument:	Tracor model 220 gas chromatograph equipped with a flame photometric detector (FPD) (sulfur-selective) and a 18-ft by 1/4-in.-o.d. glass column packed with 20% FFAP on 60-80 mesh, acid-washed Chromosorb W
Injector temp. (°C):	200
Column temp. (°C):	Held for 6 min at RT (about 25°C), then programmed to 225°C at 5°C/min
Detector temp. (°C):	200
Flow rates (ml/min):	Hydrogen, 150; air, 40; oxygen, 20

4. Nitrogen profiling

Instrument:	Tracor model 220 gas chromatograph equipped with a Coulson electrolytic conductivity detector (nitrogen-selective), a 12-ft by 1/4-in. glass column packed with 28% Pennwalt 223 plus 4% KOH on 80-100 mesh Gas Chrom R, and a reduction furnace of the detector operated at 800°C with a hydrogen flow of 50 ml/min and helium flush of 45 ml/min
Column temp. (°C):	55°C for 12 min, then programmed to 200°C at 5°C/min
Injector temp. (°C):	225
Detector temp. (°C):	225
Helium carrier-gas flow rate (ml/min):	45

Using these varied techniques, the type of information that could be derived by Guerin et al. was as follows:

1. Total particulate matter (TPM): In Figure 1.29 the peaks located by cochromatography and located by GC-MS corresponded to phenol, o-cresol, m-cresol, p-cresol, glycerol, nicotine, palmitic acid, oleic plus linoleic plus linolenic acids, and stearic acid. As noted, other smoke components eluting under these GC conditions included propylene glycol (with peak 1), indole (with peak 8), hydroquinone (with peak 9), catechol (with peak 7), and menthol (following peak 7).

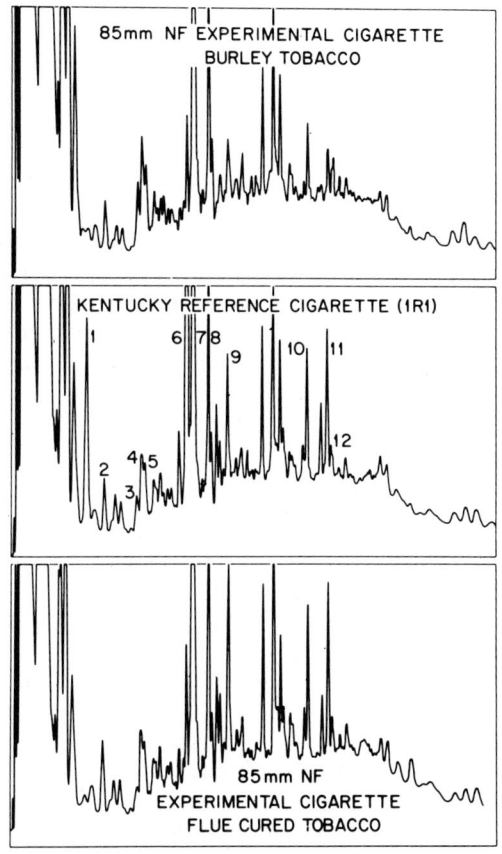

Figure 1.29. Profiles of silylated total particulate matter (TPM). Peak 2, phenol; 3, o-cresol; 4, m-cresol; 5, p-cresol; 6, glycerol; 7, nicotine; 10, palmitic acid; 11, oleic, linoleic, and linolenic acids; 12, stearic acid. From Guerin et al. [87], reproduced from the Journal of Chromatographic Science, by permission of Preston Publications, Inc.

On the other hand, Figure 1.30 illustrates the effect of cellulose acetate and charcoal filtration on the TPM composition.

2. Gas-phase profiling: A typical chromatogram which illustrates a gas-phase profile generated using the cryothermal GC procedure is shown in Figure 1.31, in which components in that portion of the smoke which passes through the standard TPM filter have been identified by cochromatography only.

3. Sulfur and nitrogen profiling: With the use of selective detectors, profiles become even more informative to the analyst, enabling them to observe differences as shown in Figures 1.32 and 1.33, which are sulfur- and nitrogen-specific profiles of the gas phase of cigarette smokes, respectively.

2. Biological Fluids

a. Urine

In 1965, since the amount of nicotine and its metabolites in the body fluids of smokers might be correlated with nicotine exposure and heart disease, McNiven et al. [91] described a GC method for the determination of nicotine in smokers' urine. Previous work by Corcoran, Halmer, and Page [114] and by Bowman, Turnbull, and McKennis [115] showed respectively that nicotine is present in smokers' urine and that cotinine is the main metabolite of nicotine in human urine.

Sample preparation prior to GC analysis consisted of the following:

Five-milliliter portions of 24-hr urine collections (smokers and nonsmokers) were adjusted to pH 1 with 6 M sulfuric acid and exhaustively extracted with methylene chloride. This removed neutral, phenolic, and other acidic materials. The aqueous layer was adjusted to pH 11 with 50% aqueous NaOH. The liberated nitrogen bases were then extracted immediately with methylene chloride and the aqueous layer discarded. The extract was then acidified with HCl and evaporated to dryness with nitrogen. The residue was then made alkaline with 3 µl of 2 N aqueous NaOH to liberate the nitrogen bases and diluted with 200 µl of acetonitrile. Aliquots of this solution (2 to 3 µl) were injected into the gas chromatograph.

A Glowall Chromalab gas chromatograph equipped with a strontium-90 argon detector and a 12-ft by 3.4-mm-i.d. glass column packed with 14.5% SE-30 on 100-110 mesh Anakrom ABS was used. With the argon flow rates, column temperature, and detector voltage maintained at 10 to 20 ml/min, 200° C, and 1350 V, respectively, an extract of a cigarette smoker's urine gave the chromatogram shown in Figure 1.34, which is compared with the chromatogram of a nonsmoker. Since no interferences due to other compounds

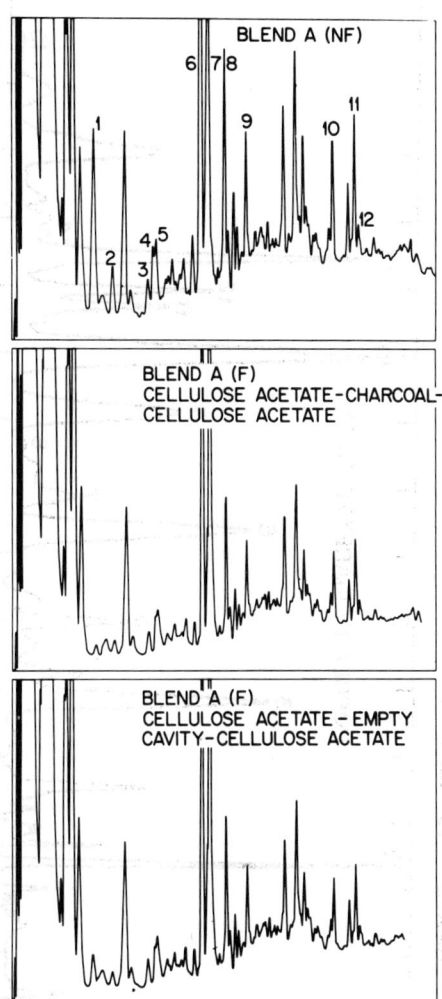

Figure 1.30. Effect of filtration on TPM composition. From Guerin et al. [87], reproduced from the Journal of Chromatographic Science, by permission of Preston Publications, Inc.

116   Natural, Pyrolytic, and Carcinogenic Products of Tobacco

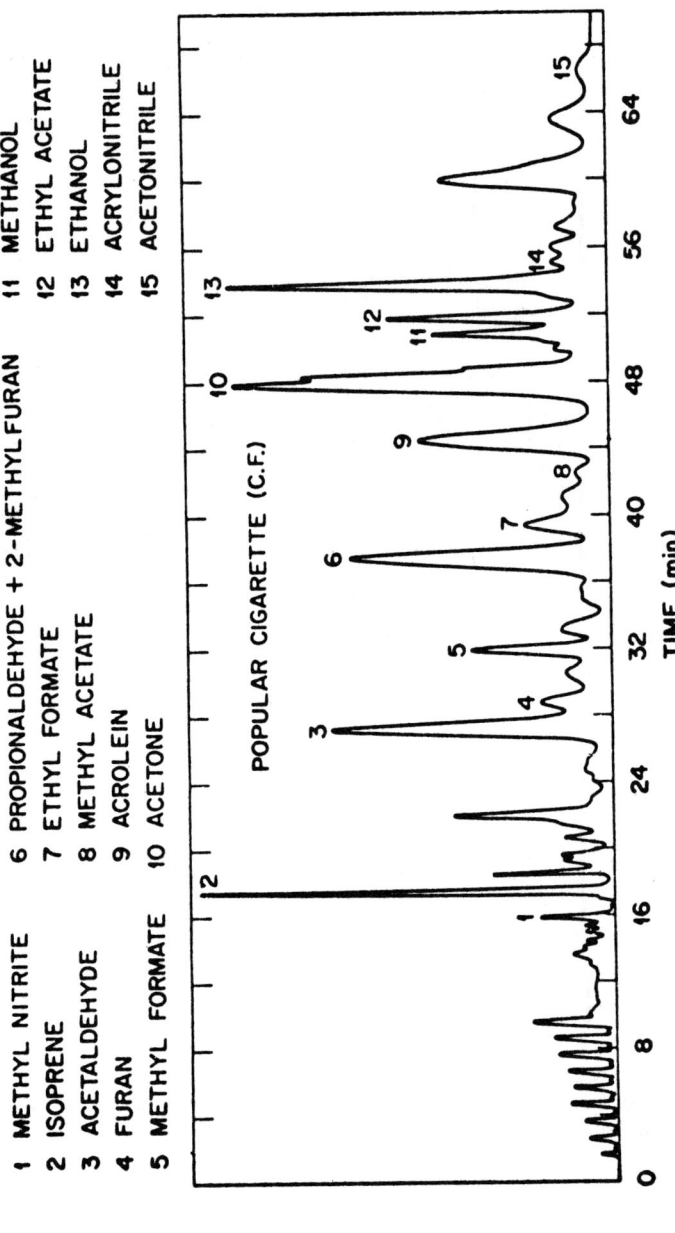

Figure 1.31. Profile of the gas phase of cigarette smoke. From Guerin et al. [87], reproduced from the Journal of Chromatographic Science, by permission of Preston Publications, Inc.

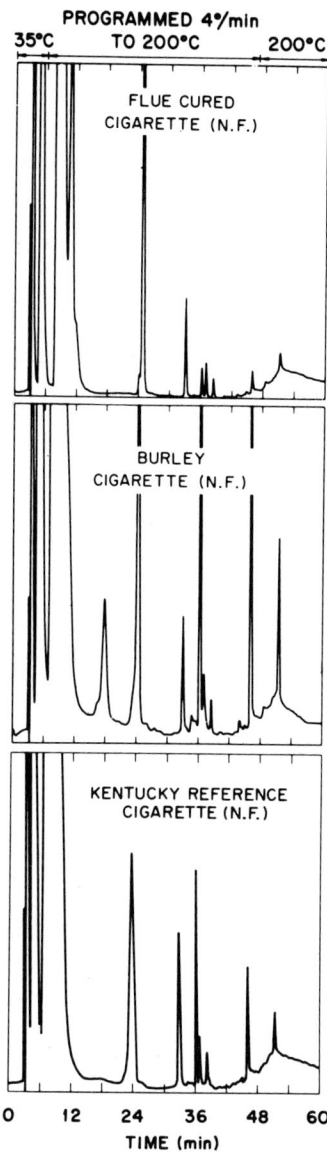

Figure 1.32. Sulfur-specific profiles of the gas phase of cigarette smokes. From Guerin et al. [87], reproduced from the Journal of Chromatographic Science, by permission of Preston Publications, Inc.

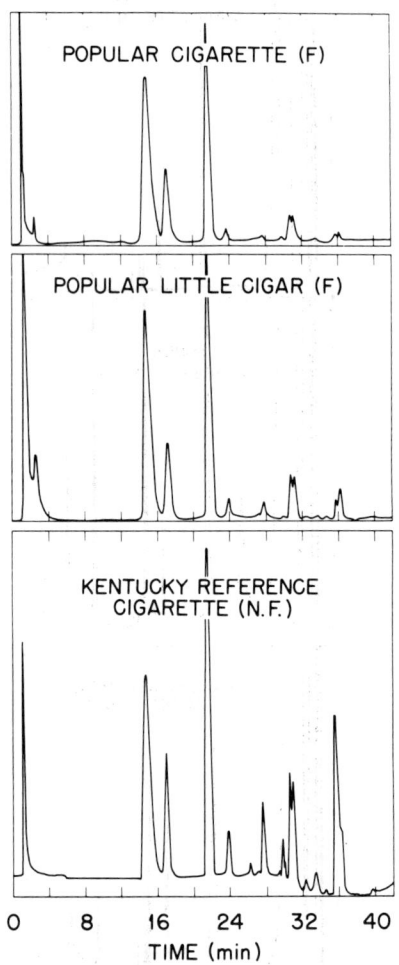

Figure 1.33. Nitrogen-specific profiles of the gas phase of cigarette smokes. From Guerin et al. [87], reproduced from the Journal of Chromatographic Science, by permission of Preston Publications, Inc.

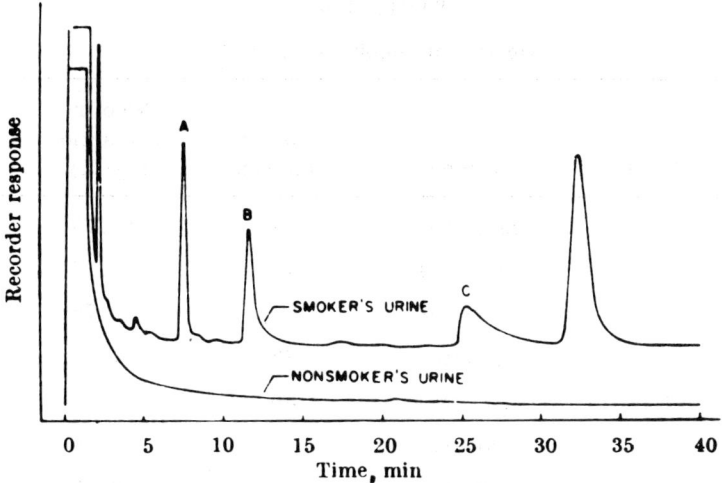

Figure 1.34. Gas chromatography of extracts from smoker's and nonsmoker's urine. A, nicotine; B, added standard 3-methyl 3-phenyl piperidine; C, cotinine. From McNiven et al. [91], courtesy of Nature.

in urine are observed, this proposed extraction and gas chromatographic method is specific for nicotine and cotinine. McNiven et al. reported overall recoveries of 80 to 92% for the analysis of nonsmokers' urine to which had been added known amounts of nicotine, and some of their preliminary results by individuals, tobacco forms, and amount of tobacco consumed are shown in Table 1.41.

Beckett and co-workers [92,93] studied the significance of smoking in investigations of urinary excretion rates of amines in man [92] and described a GC method for the determination of nicotine and its metabolite, cotinine, in urine [93].

In the determination of amphetamine by gas chromatography [116], Beckett et al. [92] observed an additional peak in urine extracts of some but not all subjects; only subjects who smoked gave this latter peak. They established that the additional component, with a retention time of about 26 min (amphetamine eluted in 9 min under the same GC conditions), was nicotine (confirmed by GC, thin-layer chromatography, and the infrared spectrum of the isolated base) and that the sum of the amphetamine and nicotine concentrations in urine, both determined by GC, yielded values comparable to the "apparent amphetamine" content determined by the methyl orange complexing method.

They also noted that, using the GC method for the determination of nicotine, the rate of nicotine excretion in urine is influenced by the pH of the urine. Subjects smoking 20 cigarettes per day at regular intervals excreted

TABLE 1.41

Nicotine in Smokers' Urine[a]

Subject	Tobacco	No. or amount	Nicotine extraction (mg/day)
1	Cigarettes	4	<0.10
2	Cigarettes	10	0.27
3	Cigarettes	25	0.74
4	Cigarettes	27	1.37
3	Cigarettes	30	10.50
3	Cigarettes	31	5.46
5	Cigarettes	31	9.06
6	Cigarettes	43	5.81
7	Cigars	20 g	<0.10
8	Cigars	55 g	8.50
9	Pipe	20 g	0.11

[a] From McNiven et al. [91], courtesy of Nature.

0.1 to 4.3 µg of nicotine per minute under normal conditions, 0.4 to 13.0 µg/min when the urine was acidic after oral administration of NH_4Cl, and less than 0.1 µg/min when the urine was alkaline after the oral administration of sodium bicarbonate.

With regard to the determination of nicotine and cotinine using chlorphentermine and lidocaine as internal standards, respectively, Beckett and Triggs [93] employed two extraction procedures as described below:

1. Nicotine: Five milliliters of urine and 0.1 ml of 5 N HCl were added to 1 ml of internal standard solution (chlorphentermine·HCl, 5 µg/ml of distilled water) in a glass-stoppered centrifuge tube. After extracting the solution with 3 × 2.5 ml of diethyl ether and centrifuging to break up any emulsion, the ether extracts were discarded and the urine made alkaline with 0.5 ml of 5 N NaOH. Again, the basic urine solution was extracted with 3 × 2.5 ml of ether; these ether extracts were then combined in a Quickfit test tube, finely tapered at the base, and evaporated to approximately 50 µl at 42°C. From this concentrated extract solution, a 2- to 3-µl aliquot was withdrawn and injected onto the GC column.

2. Cotinine: For cotinine, a similar procedure was used except that (a) there was no acid solvent extraction, (b) NH₄OH was employed rather than 5 N NaOH, (c) freshly distilled methylene dichloride was the extraction solvent, and (d) lidocaine (lignocaine) was added as internal standard (5 µg/ml of distilled water).

These components were resolved and quantitatively determined with a Perkin-Elmer F-11 gas chromatograph equipped with a flame ionization detector and a 1-m by 1/8-in.-o.d. stainless steel column packed with 5% KOH:2% Carbowax 20M coated on 80-100 mesh Diatoport S. With the injector temperature set at 250°C and the nitrogen carrier gas adjusted to 33 ml/min, the retention times of nicotine and chlorphentermine at a column temperature of 135°C were about 3.90 and 4.47 min, respectively, whereas at a column temperature of 209°C, lidocaine and cotinine were eluted in 3.93 and 5.85 min, respectively.

Using peak height measurements and previously constructed calibration curves, the reported absolute recoveries of these compounds and their standards from urine were as follows: (1) nicotine, 90 to 95%; (2) chlorphentermine, 90 to 95%; (3) cotinine, 85 to 90%; and (4) lidocaine, 95 to 100%. A relative recovery of 95 to 100% for nicotine and cotinine was obtained with respect to their internal standards, which are added to the urine at the start of the assay procedure.

In 1967, McKennis et al. [94] noted that in all attempts to describe or propose a sequence of events responsible for the rapid termination of physiological responses to nicotine, a knowledge of the transportation and distribution of submicrogram quantities of nicotine becomes extremely desirable. They continue by stating that: "These needs have been previously met, in part by isotopic procedures, and in part by GC methods. To increase the effectiveness of the latter, a number of perfluoro derivatives of nicotine—suitable for determination with an electron capture detector—have been synthesized. The most promising of these, pentafluoropropionylmetanicotine, was prepared by reacting nicotine with pentafluoropropionic anhydride in the presence of triethylamine. The product, which was identified as a picrate (m.p. 129 to 131°C), gave satisfactory ECD signals. Comparative analyses of the urine of a male smoker by the direct chromatographic methods, modified from procedures of McNiven and others, and the pentafluoro method indicated a satisfactory agreement. Urinary excretions of 2.4 to 3.3 mg of nicotine per day were noted. Determinations by the PFP method on a Carbowax 20M column permit the determination of as little as 1×10^{-9} g of nicotine without apparent interference from other urinary constituents or side products which form during the course of the chemical manipulations."

Because nicotine is commonly observed in smokers' urine, Horning et al. [95,98] used this circumstance as a means of developing atmospheric pressure ionization (API) mass spectrometric techniques for the detection of volatile components of urine which behave as bases in the gas phase. When these methods were applied in an examination of urinary bases of nonsmokers who shared the laboratory areas with smokers, nicotine was found to be present to the extent of about 5% of that observed for smokers. Two routes of nicotine transfer seemed possible: water and air. The water supply was found to be nicotine-free. An air analysis method was devised; this showed that nicotine was present in the laboratory air, so that the probable route of transfer for nonsmokers is through room air.

With regard to their API-MS studies, Horning et al. [98] clearly substantiated their conclusions via several experiments discussed below.*

Figure 1.35 shows the time course of events when a small sample of nicotine is injected in benzene solution. Nicotine is a gas-phase base, and it also has an ionization potential lower than that of benzene. Two ions are formed: M^+ by charge transfer and MH^+ by proton transfer. The reagent ions disappear immediately, and nicotine is ionized. As the sample is swept from the reaction chamber by the carrier gas, the concentration of nicotine decreases, and the reagent ions reappear. When the nicotine concentration falls below the detectable limit, some benzene is still present, and the reagent ions will persist until the solvent-reagent (benzene) is swept completely from the chamber.

Examples of analyses of biologic samples are shown in Figures 1.36 and 1.37. Figure 1.36 shows an ion profile of the organic bases in a urinary extract of a patient receiving the drug Ethambutol. Figure 1.37 shows an ion profile of the organic bases in urinary extracts from a smoker and a nonsmoker. Nicotine is a major organic base in the urine of smokers.

In 1975, Dumas et al. [97] developed a GC method for the analysis of both nicotine and cotinine in urine extracts, a procedure in which an Aerograph model 1440 gas chromatograph equipped with a flame ionization detector and a 1-m by 1/8-in.-o.d. column packed with 6% KOH:3.2% UCON 50 HB 2000 coated on 100-120 mesh Gas Chrom Q was employed. With the nitrogen carrier-gas flow rate, injector temperature, and detector temperature maintained at 30 ml/min, 215° C, and 220° C, respectively, the column temperature was variable: 120° C for nicotine and 170° C for cotinine. Under these conditions, the retention time of nicotine was 6.30 min (120° C), whereas cotinine was eluted at 170° C in 15.60 min. The reliability of the extraction procedure was examined using several known solutions containing 2 µg/ml and 1 µg/ml of nicotine and cotinine, respectively. Their data showed that the recovery of nicotine was 90 ± 4%, whereas that for cotinine

---

*Reproduced in part with permission of the National Institute of General Medical Sciences (HEW).

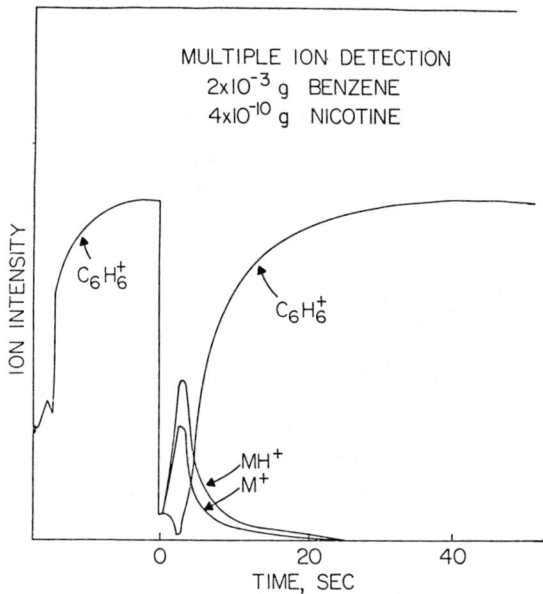

Figure 1.35. Time course of change in concentration of ions showing effect of injection of a sample of nicotine in benzene. The source was first filled with benzene vapor by injection of benzene alone. When the sample was injected, nicotine was ionized to form MH^+ and M^+ ions and all ions derived from benzene disappeared. As the nicotine sample was swept from the source, the benzene ions reappeared. Continued sweeping with carrier gas resulted in a nitrogen profile alone. From Horning et al. [98], courtesy of the National Institute of General Medical Sciences (HEW).

was 95 ± 4%. Applied to the urinary excretion of these two compounds, their procedure showed that the excretion of nicotine was rapid and slow for cotinine and that their data exhibited a correlation between the urinary flow and the eliminated amounts of both these substances.

Using the GC procedure of Beckett and Triggs [93], Goldfarb et al. [99] "carried out experiments to examine the effects of nicotine and 'tar' on the extent of and subjective reactions to cigarette smoking. It was confirmed that smokers rate commercial, low-nicotine cigarettes as less 'strong' and less 'satisfying' than their usual brands. Since such cigarettes deliver reduced amounts of tar as well as of nicotine, an experiment to distinguish between the two was carried out with special cigarettes. Ratings of 'strength' were directly related to nicotine but were not affected by tar. The number of cigarettes smoked fell slightly as their estimated delivery of nicotine increased, but tar had no effect on this index. The urinary excretion of nicotine was correlated with the rated yields of nicotine for the different

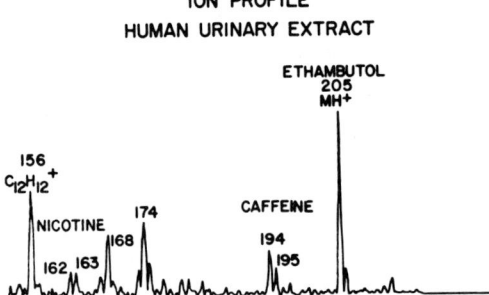

Figure 1.36. Ion profile showing urinary bases for a patient who was being treated with ethambutol. Nicotine and caffeine are also present in ionized form. From Horning et al. [98], courtesy of the National Institute of General Medical Sciences (HEW).

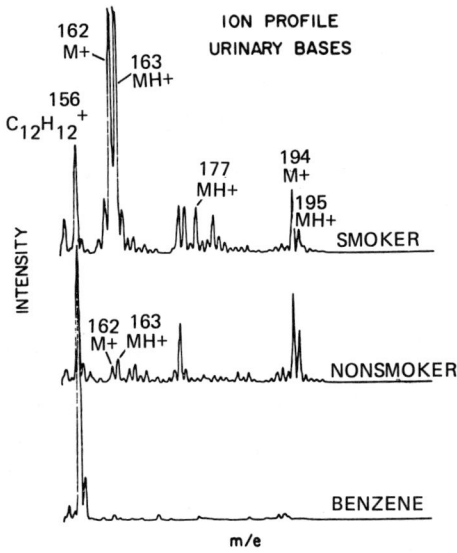

Figure 1.37. Ion profiles showing the presence of nicotine in the urine of a smoker. Nonsmokers were observed to excrete about 5% of the amount of nicotine excreted by smokers. The route of transfer is by air. From Horning et al. [98], courtesy of the National Institute of General Medical Sciences (HEW).

cigarettes, but there was also evidence that subjects tended to adjust their manner of smoking so as to titrate their doses of nicotine. The results are interpreted as indicating a role for nicotine, but not for tar, in the maintenance of cigarette smoking behavior, and as support for the view that less harmful cigarettes should have a high yield of nicotine to tar."

In a subsequent investigation, Gritz et al. [100] studied the titration, the self-regulation of nicotine intake, with full-length and half-length cigarettes in 12 smokers by GC assays of urinary nicotine levels.

The GC assay for urinary nicotine was performed in the following manner:

Replicate urine samples (5.0 ml) were placed in centrifuge tubes, and 100 µl of dimethylphentermine hydrochloric acid (DMP) (20 µg/100 µl) was added as internal standard. After 20% NaOH (0.5 ml) and benzene (6.0 ml) were added, the tubes were shaken and centrifuged to extract the nicotine and internal standard into the benzene layer. The nicotine was reextracted into an aqueous solution by transferring 5 ml of the benzene phase into smaller tubes and adding 1 N HCl (0.3 ml). Twenty percent NaOH (0.2 ml) was added to the benzene solutions, and the samples were extracted into dichloromethane (50 µl). The concentrated nicotine extracts were assayed by gas chromatography. A 6-ft by 2-mm i.d. silanized glass column packed with 1% OV-101 and 1% succinamide polymer on Gas Chrom Q was used in a Varian 2100 gas chromatograph equipped with a flame ionization detector. The concentrations of nicotine were calculated from the height ratio of nicotine and the internal standard using a curve of known quantities of nicotine in 0.9% saline.

Using the above GC method, results demonstrated that excretion of urinary nicotine in the proximal condition (half cigarette closer to the filter) did not differ significantly from the whole cigarette condition; however, less nicotine was excreted in the distal condition (half cigarette farther from the filter) because of a rod filtration effect. Subjects extracted proportionately more nicotine from the half than from the whole cigarettes: titration was approximately the same in both half-cigarette conditions. On scales of strength and satisfaction, full-length cigarettes were given the highest rating, followed by proximal and then distal cigarettes.

In 1977, Veal [101] presented a GC procedure for the analysis of nanogram levels of nicotine in urine and blood. Using a modified Hewlett-Packard model 5830 gas chromatograph equipped with flame ionization detection and glass capillary columns, the assay technique involved extraction of the biological sample with methylene chloride, concentration of the organic extract, and subsequent GC analysis. The columns utilized were 30 m in length and coated with 95% SE-30 and 5% BTPPC. The columns had an N_{eff} greater than 35,000, with a partition ratio of 5.5 based on the nicotine peak. Using this procedure, nicotine concentrations as low as 5 ng/ml could be measured quantitatively. The analysis procedure achieved recoveries of 95% or better, and with the use of autoinjection techniques replication of 5% or better was achieved for routine sample analysis.

In 1977, Wolstenholme and Gerber [102] noted that, more recently, a very significant step forward has been taken toward relieving the GC-MS user of the concern for instrument optimization; this has been achieved in part via microprocessor control of the complete integrated GC-MS instrument. Discussing features of the latest instrumentation (automatic calibration, optimization, fault diagnosis, unattended operation, and a clean, rugged vacuum system that are characteristic of the Varian MAT 44 system), they illustrated its applicability to toxicological analysis as illustrated in Figure 1.38a, which shows the TIC chromatogram obtained from the ether extract of the urine from an overdose patient. Simultaneously, normalized spectra were displayed on the TV display, and Figure 1.38b shows two of the spectra obtained in this case. These compounds were quickly identified as nicotine (MW 162) and codeine (MW 299). Sample quantities used in this analysis were in the range of 10 to 100 ng injected onto the column. A 6-ft OV-17 packed column was used with a flow rate of 20 ml/min. The temperature was programmed from 70 to 290° C at 12° C/min. The rapid real-time presentation of the mass spectra is particularly important in such cases where speed of identification is of the essence.

b. Blood and Plasma

In 1968, Schievelbein and Grundke [103] described a GC procedure for the estimation of nicotine in blood and tissues. Capable of separating nicotine from its metabolites and of estimating it at concentrations of 0.01 μg/ml of blood, the analysis was performed with a Hewlett-Packard model 810 gas chromatograph equipped with a flame ionization detector and a 180-cm by 1/4-in. glass column packed with PEG 4000 on 60-100 mesh Celite and maintained isothermally at 190° C (no other conditions specified).

Shortly thereafter, Isaac and Rand [104] used a sensitive, specific GC method to assay nicotine in smoke and in blood after cigarette smoke was introduced into the respiratory tract of artificially ventilated, anesthetized dogs by a machine simulating the pattern of human smoking. Observations were made on the abstraction of nicotine from the smoke introduced into the respiratory tract, the appearance of nicotine in the blood, and on some physiological effects (intratracheal pressure, blood pressure, heart rate, patellar tendon reflex) of nicotine from the smoke. The blood levels of nicotine were determined with a Varian 1520 gas chromatograph equipped with dual flame ionization detectors and dual 5-ft by 3.2-mm-o.d. stainless steel columns packed with 12% KOH and 5% Carbowax 20M coated on 80-100 mesh Aeropak 30. The retention times for nicotine and chlorphentermine (internal standard) were 4.2 and 5.1 min, respectively, using the following GC conditions: injector temperature, 250° C; column temperature, 150° C; detector temperature, 275° C; nitrogen carrier-gas flow rate, 30 ml/min.

Chemical Composition of Tobacco Smoke

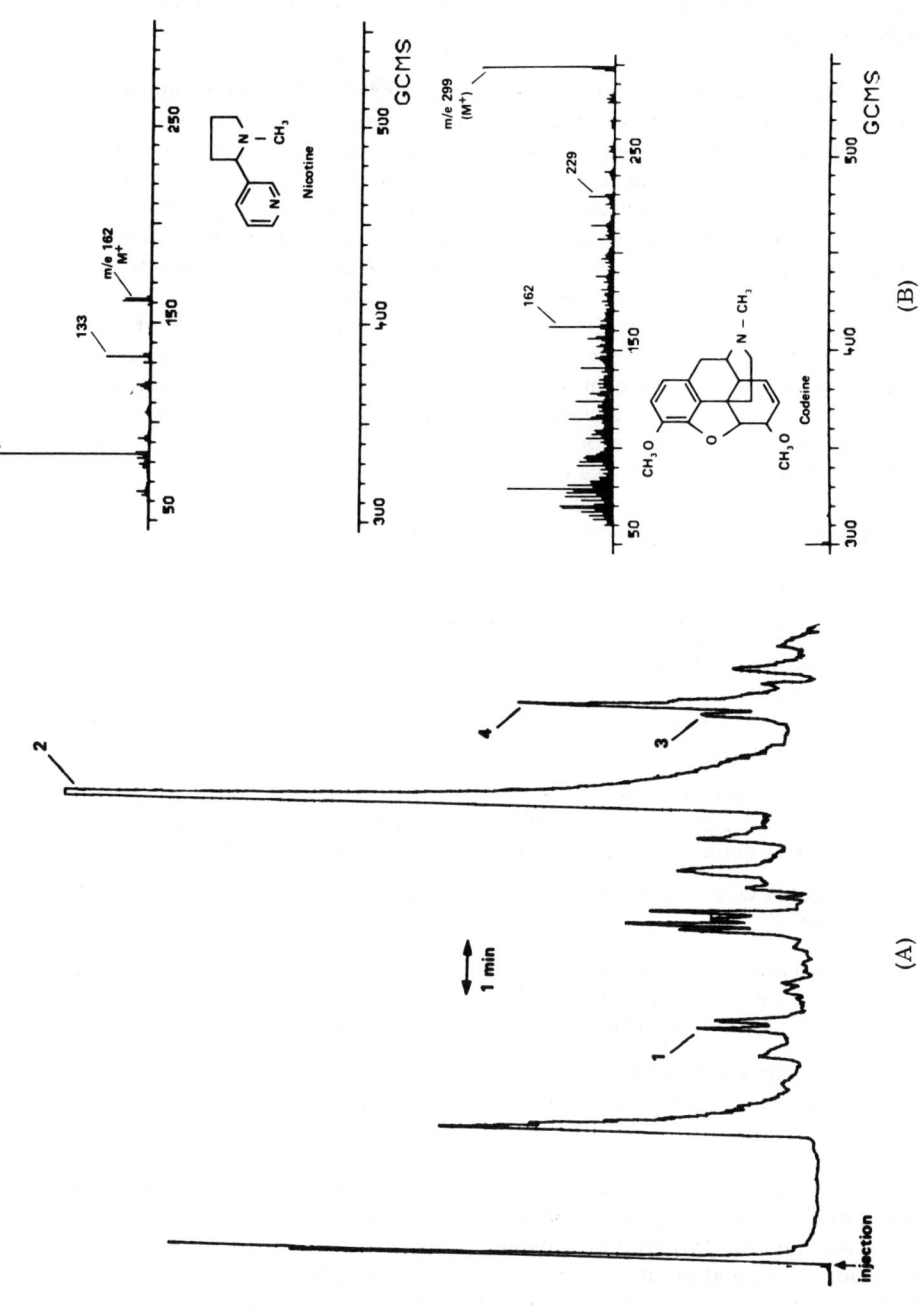

Figure 1.38. (a) Total ion chromatogram of urine extract; (b) spectra of two drugs in urine extract. Adapted from Wolstenholme and Gerber [102].

The extraction procedure for nicotine in blood involved the following preparative stages:

Samples of exactly 1 or 2 ml of heparinized whole blood were added to 3 ml of distilled water containing 1 µg of chlorphentermine in a 40-ml glass-stoppered centrifuge tube and 1 ml of 5 M NaOH was added. The mixture was extracted with one 5-ml and two 3-ml portions of diethyl ether. It was necessary to centrifuge for 15 min at 3000 rpm between extractions to break the emulsion. The ether extracts were combined in a 15-ml glass-stoppered tube, the end of which was drawn out to a capillary of 20 mm length and 2 mm internal diameter. The ether was evaporated by allowing the tubes to stand in a water bath at about 40°C. Care was taken to remove any aqueous phase which separated out before the concentration had proceeded to a marked degree. Evaporation was continued until only the capillary contained ether (about 50 µl). The tube was then stoppered and immediately placed in ice until the sample was assayed.

They summarized their findings as follows:

1. The mean amount of nicotine in a puff of smoke was 120 µg; about 50% was abstracted in the respiratory tract.
2. Nicotine appeared in brachial artery blood after administration of single puffs of smoke: the levels rose rapidly to 0.15 to 0.25 µg/ml after 7 to 10 sec and then fell markedly before the next smoke puff was administered.
3. After intravenous injections of 28 µg/ml of nicotine, peak levels of 0.9 to 1.3 µg/ml were attained in arterial blood in about 25 sec, but then fell rapidly, being undetectable 2 min after injection.
4. Nicotine was distributed between plasma and blood cells in the ratio 1.0:0.8.
5. The intensity and duration of heart rate, blood pressure, and tracheal pressure responses produced by administration of smoke or injection of nicotine ran parallel to the concentration of nicotine in the blood.
6. The depression of the knee jerk produced by nicotine persisted beyond the time when the blood level had dropped to undetectable concentrations.

Burrows et al. [105] developed a method for both the extraction and determination of submicrogram amounts in blood that involved steam distillation of the nicotine followed by solvent partition and column-chromatographic clean-up. The final solution, containing quinoline added as internal standard, was injected directly onto the GC column for analysis.

For this investigation, all GC analyses were performed with a Perkin-Elmer F-11 gas chromatograph equipped with a flame ionization detector and a 2-m by 4-mm-i.d. glass column packed with 8% Carbowax 20M and 2% KOH on acid-washed, HMDS-treated, 80-100 mesh Chromosorb W. Using optimized chromatographic conditions (column temperature, 150°C; nitrogen carrier-gas flow rate, 60 ml/min), the retention times for nicotine and quinoline (IS) were approximately 13.0 and 15.0 min, respectively.

With regard to the method's sensitivity, at a signal-to-noise ratio of 3:1, the minimum amount of nicotine that could be detected was 0.2 ng, this representing a practical detection limit for nicotine of 1 ng and, in a 10-ml sample of human blood, a limit of 0.0001 ppm when using this method.

On the other hand, recovery experiments conducted by adding known amounts of nicotine (25 to 500 ng) to 10-ml samples of human blood and taking these through the entire analytical procedure (11 experiments) yielded an average value of 56.9 ± 1.6%.

In 1972, Isaac and Rand [106] improved their GC method for the determination of nicotine in blood by using a newly developed alkali flame ionization detector (AFID) which provided a system sensitive to 1 ng/ml of nicotine in a 2.5-ml sample.

The preparation of the plasma sample consisted of mixing 2.5 ml of plasma with 3 ml of an aqueous solution containing 15 ng/ml of modaline as internal standard and 1 ml of 5 M NaOH. After extracting the mixture three times with diethyl ether, the combined ether extracts (11 ml) were placed in a 15-ml glass-stoppered tube, the end of which was formed into a capillary of about 75 μl capacity. Using a 42°C water bath, the volume was carefully reduced to approximately 20 μl. From this extract sample, a 2-μl aliquot was withdrawn for GC analysis.

Isaac and Rand employed for this study a Varian model 1400 gas chromatograph equipped with an alkali flame ionization detector and a 5-ft by 1/8-in. stainless steel column packed with 6.5% Carbowax 20M and 13% KOH coated on 80-100 mesh Varaport 30. The other GC operating parameters were column temperature, 146°C; injector and detector temperatures, 250°C; nitrogen carrier-gas flow rate, 30 ml/min.

With these conditions, the retention times for nicotine and modaline were 8.3 and 12.0 min, respectively, whereas the minimum detectable quantity of nicotine was nearly 0.2 ng.

In Figure 1.39 one notes that plasma nicotine levels increased rapidly during the smoking of the cigarette and thereafter gradually fell. Furthermore, the rate of decline of the plasma levels of nicotine varied between smokes, the plasma half-life being less than 30 min in all four subjects.

In 1975, Feyerabend et al. [107] also developed a GC method for the determination of nicotine in plasma using a GC unit equipped with a nitrogen detector. Nicotine was extracted from alkalinized plasma into diethyl ether. This was then concentrated by evaporation and, after an acid back-extraction and the addition of quinoline as internal standard, was reextracted into n-heptane before injection into the gas chromatograph.

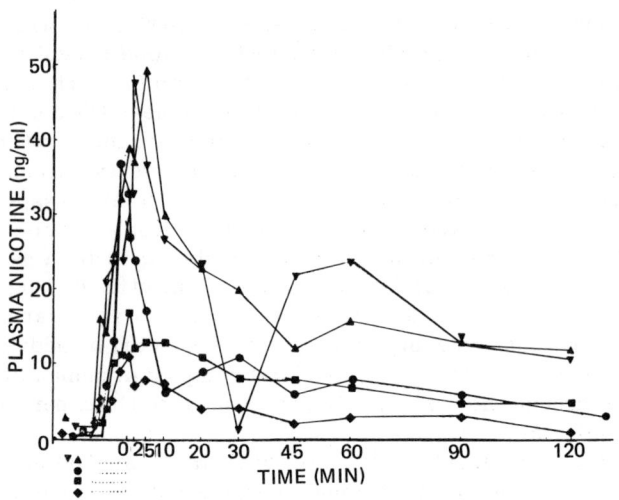

Figure 1.39. Effect of a single cigarette on plasma nicotine concentration in four subjects: P. H. (two experiments), (a) ▼ and (b) ▲; M. R., ●; J. G., ■; B. E., ♦. The number of puffs taken by each is shown beneath the abscissa, the curves being arranged as coincident on the last puff which is assigned to zero time on the abscissa. Blood samples were taken before smoking, 30 sec after every second puff, and at 0.5, 2, 5, 10, 20, 30, 45, 60, 90, and 120 min after the last puff. The presmoking nicotine levels are the means of assays on two samples taken 5 min apart. From Isaac and Rand [106], courtesy of Nature.

Quantitative data were obtained with a Hewlett-Packard model 5750 gas chromatograph equipped with a nitrogen detector (rubidium bromide crystal) and a 6-ft by 6-mm glass column packed with 10% Apiezon L and 10% KOH coated on 80-100 mesh Chromosorb W. To obtain retention times of 2.8 and 3.6 min for quinoline and nicotine, respectively, the instrument settings were injector temperature, 210°C; column temperature, 170°C; detector temperature, 400°C; helium carrier-gas flow rate, 60 ml/min; air flow rate, 300 ml/min; hydrogen flow rate, 26 ml/min. Using calibration curves prepared by peak height measurements over the 1- to 100-ng/ml range, the absolute recovery of nicotine achieved by adding an n-heptane solution containing 10 ng/ml of nicotine and 50 ng/ml of quinoline to plasma was about 75% of the known nicotine added.

Figure 1.40 shows typical venous plasma profiles where, in the inhaling smoker, nicotine levels rose steadily throughout the duration of smoking reaching a peak of about 30 ng/ml shortly after the cigarette had been discarded. The levels then fell over the next hour to a resting concentration of approximately 7 ng/ml whereas, by contrast, the noninhaling smoker

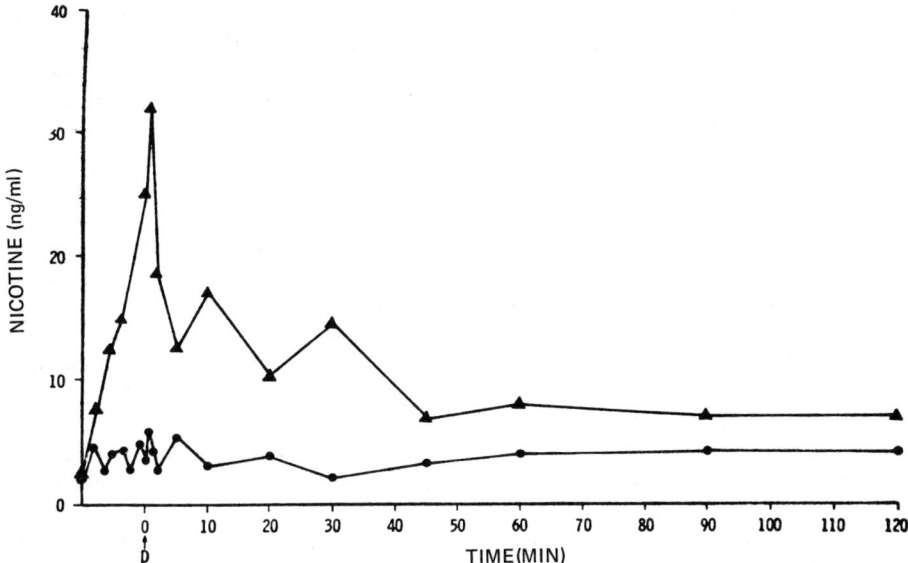

Figure 1.40. Plasma nicotine concentrations in an inhaling smoker (▲) and a noninhaling smoker (●) during and after smoking one cigarette which was discarded at the point D. From Feyerabend et al. [107], courtesy of the Journal of Pharmacy and Pharmacology.

maintained concentrations throughout of between 2 and 5 ng/ml. As noted, it was of interest that these "resting" levels were similar to those encountered in nonsmokers working in areas adjacent to smokers.

In 1975, Falkman et al. [108] modified the GC method developed by Burrows et al. [105] to improve the overall sensitivity of the method as well as decrease the level of interfering coextractives. The extraction technique employed was as follows:

Obtain blood samples from the subjects by venipuncture in the arm. Dilute 10 ml of heparinized whole blood with 10 ml of glass-distilled water, make the solution alkaline with 10 ml of alkaline salt solution (solution of 300 g of NaOH in 700 ml of water saturated with NaCl) and add 0.1 g of Antifoam A. Steam distill the mixture at a distillation rate of 5 ml/min, and collect the first 70 ml of distillate in a separating funnel containing 2 ml of 1 N H_2SO_4. Extract the acidified distillate with 10 ml of dichloromethane and discard the organic phase. Make the aqueous phase alkaline by adding 2 ml of 10 N NaOH solution and extract the mixture four times with 10 ml of dichloromethane. Extract the combined dichloromethane layers four times with 2 ml of

0.5 N H_2SO_4. Combine the aqueous layers in a 10-ml calibrated flask containing 2.5 g of NaCl and make the mixture alkaline with 1 ml of 10 N NaOH solution. Extract the nicotine with 100 µl of the benzene solution containing 5 µg/ml of quinoline as internal standard. Inject 6 µl of the benzene solution onto the GC column using a Hamilton syringe.

The GC separations were carried out with a gas chromatograph equipped with a flame ionization detector and a 2-m by 3-mm-i.d. glass column packed with 8% Carbowax 20M and 2% KOH coated on acid-washed, HMDS-treated, 80-100 mesh Chromosorb W. For the determination of nicotine, the optimum operating conditions were column temperature, 150°C; nitrogen carrier-gas flow rate, 27 ml/min. With these operating parameters, the retention times for nicotine and quinoline were about 6 and 8 min, respectively, whereas at a signal-to-noise ratio of 3:1, the detection limit for nicotine was 0.04 ng.

Using integrated GC-MS for compound identity, the various GC-MS parameters used to obtain mass spectra and a retention time of nearly 11 min for nicotine were 2.7-m by 2-mm-i.d. column packed with 5% Carbowax 20M and 2% KOH; injector temperature, 200°C; column temperature, 170°C; helium carrier-gas flow rate, 30 ml/min; ion source temperature, 270°C; inlet tube temperature, 230°C; electron energy, 20 or 70 eV; trap current, 50 µA; accelerating voltage, 3.0 kV. In their study, the mass numbers 162, 133, and 84 were used to confirm the presence of nicotine in the plasma extracts. As can be seen from the total ion chromatogram and the mass chromatograms in Figure 1.41, all the typical fragments were present in the presumed nicotine peak.

The average of the results of recovery experiments with control blood to which known amounts of nicotine were added was 95.8 ± 2.1%.

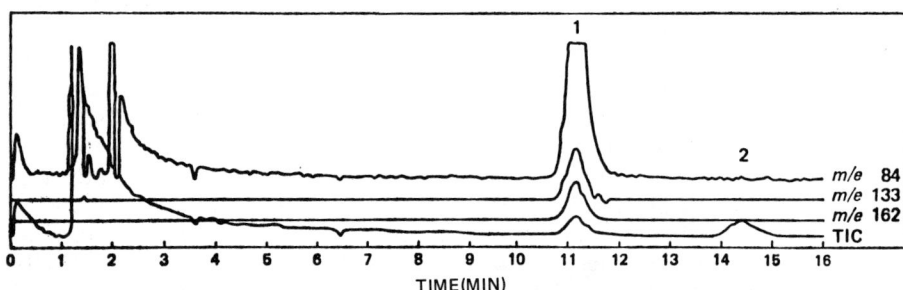

Figure 1.41. Total ion chromatogram (TIC) and mass fragmentograms for mass numbers 162, 133, and 84 from extract of blood from a smoker. 1, Nicotine; 2, quinoline. From Falkman et al. [108], courtesy of The Analyst.

In 1978, Hengen and Hengen [110] described a GC procedure for determining nicotine and cotinine in plasma in which nicotine is extracted from 1 ml of plasma with diethyl ether, back-extracted, and analyzed by GC with a nitrogen/phosphorus detector. Nicotine and its internal standard, modaline (2-methyl-3-piperidinopyrazine), had retention times of 1.9 and 2.9 min, respectively. On the other hand, cotinine was then extracted from the same plasma with dichloromethane and similarly analyzed. Cotinine and its internal standard, lidocaine, had retention times of 3.8 and 4.9 min, respectively. They noted that day-to-day reproducibilities (CV) within 14% for nicotine and within 6% for cotinine are attainable for the respective concentration ranges 1 to 100 µg/liter and 1 to 200 µg/liter. With no interference observed from nornicotine and related alkaloids, the sensitivity was such that less than 0.1 µg of nicotine and 0.1 µg of cotinine could be detected per liter.

For the determination of nicotine, a Hewlett-Packard model 5711 gas chromatograph equipped with a nitrogen flame ionization detector and a 1.8-m by 2-mm-i.d. glass column packed with 3% SP-2250 DB on 100-120 mesh Supelcoport was used; the other instrument settings were injector temperature, 200°C; column temperature, 155°C; detector temperature, 300°C; helium carrier-gas flow rate, 60 ml/min.

For the analysis of cotinine, they used a Perkin-Elmer model 3920 gas chromatograph equipped with a nitrogen-phosphorus detector and a 1.8-m by 2-mm-i.d. glass column packed as described above. This GC unit was operated as follows: injector temperature, 240°C; column temperature, 190°C; detector temperature, 250°C; helium carrier-gas flow rate, 30 ml/min.

With the methods described, the analytical recovery of nicotine was about 80 ± 6%, estimated by assaying an ether:heptane solution containing 10 µg of nicotine and 50 µg of modaline per liter added to plasma. Using a similar approach with lidocaine as internal standard, the recovery of cotinine added to plasma was 95 ± 5%.

Dow and Hall [111] described a method for the estimation of nicotine in plasma by selective ion monitoring by attaching a GC glass capillary column directly to the mass spectrometer. Nicotine was extracted from plasma by a modification of the method of Feyerabend et al. [107] which permitted the direct addition of the internal standard (quinoline) to plasma prior to extraction.

The GC-MS investigations were carried out with a Varian 1400 gas chromatograph equipped with a capillary injector (split ratio of 1:10) and a 20-m by 0.3-mm-i.d. glass capillary column coated with SP 1000. Via a 0.15-mm-i.d. glass capillary restriction, the GC was coupled to a V. G. Micromass 12B mass spectrometer. The various operating settings used with both analytical instruments were column temperature, 160°C; helium carrier-gas flow rate, 0.5 ml/min; interface temperature, 250°C; ion source temperature, 200°C; ionizing potential, 70 eV; accelerating voltage, 4.0 kV; ion source pressure, 10^{-5} torr.

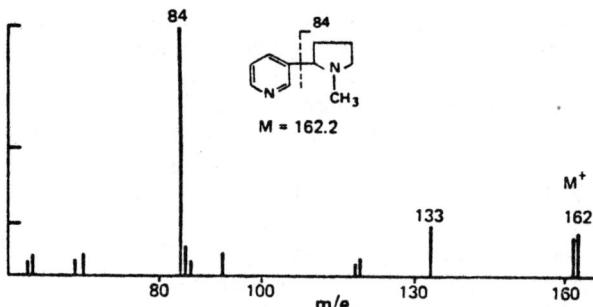

Nicotine mass spectrum

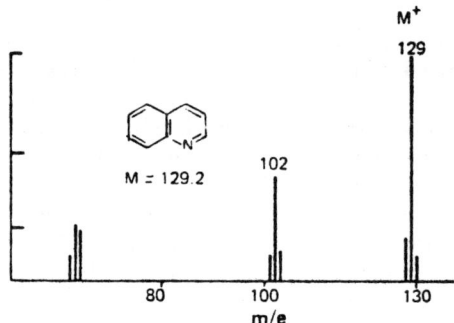

Quinoline mass spectrum

In this SID method, only the base peak of the nicotine spectrum (m/e 84) and the molecular ion of quinoline (m/e 129) were monitored; the accelerating voltage switched between 2.6 and 4.0 kV to bring these ions into focus.

With these conditions, the retention times of nicotine and quinoline were 2.40 and 3.00 min, respectively, whereas the lower limit of detection of nicotine was reported to be below 5 ng/ml. Using calibration curves constructed by plotting the ratio of the peak heights of the m/e 84 ion of nicotine and the m/e 129 ion of quinoline versus the concentration of nicotine (ng/ml), the recovery of nicotine added to plasma in quadruplet to give solutions containing 50 and 10 ng/ml was 103.5 ± 4.8% and 106.0 ± 8.0%, respectively, these values comparing favorably with data obtained by Feyerabend et al. [107] using the nitrogen flame ionization detector.

c. Breast Fluid

Using a combination of gas chromatography, mass spectrometry, and selected ion recording techniques, Petrakis et al. [112] identified nicotine and its major metabolite, cotinine, in the breast fluid of nonlactating women smokers.

Sample preparation of the sample for GC-MS analysis was performed in the following manner:

Breast fluid samples were obtained from nonlactating volunteers with a history of smoking, by a standardized nipple aspiration technique with a breast pump 15 min after a single cigarette had been smoked. Nonsmokers were used as controls. Each sample of breast fluid was placed in a 1-ml reaction vial and treated with 5.0 µl of (^{2}H)-nicotine standard solution (9.1 ng/µl) in methanol, 100 µl of 1 N NaOH solution, and 200 µl of dichloromethane (spectroscopic grade, purified by washing 100 ml of solvent two times with 50 ml of 1 N HCl). All aqueous reagents were made from doubly distilled deionized water purified by washing 100 ml of water twice with 40 ml of dichloromethane.

After being agitated on a Vortex mixer for 5 min, the vial was centrifuged for 10 min and the organic layer was removed to a second tube. This extraction was repeated with another 200-µl portion of dichloromethane; the combined dichloromethane extracts were then extracted once with 200 µl of 1 N HCl. The aqueous layer was removed, brought to a pH of at least 10 with 150 µl of aqueous 2 N NaOH solution, and the resulting solution extracted with two 300-µl portions of dichloromethane. The combined organic extracts were dried for 1 hr with 5 mg of anhydrous potassium carbonate, and the solution then decanted in portions to a 0.3-ml reaction vial and carefully reduced in volume to about 15 µl under a stream of nitrogen. Absolute ethanol (10 µl) was added, the total volume reduced to about 5 µl, and the entire sample analyzed on the GC-MS-SID system in a single run.

In this investigation, Petrakis et al. used an Infotronics model 2400 gas chromatograph equipped with a 2-m by 2-mm U-shaped glass column packed with 2% Carbowax 20M and 2% KOH (solid support not specified). The GC unit was interfaced to an AEI MS-12 mass spectrometer. The settings maintained for nicotine analysis were column temperature, 140°C for 1 min, then increased at 5°C/min; injector temperature, 175°C; helium carrier-gas flow rate, 25 ml/min; ion source temperature, 200°C; base accelerating voltage, 8 kV; multiplier voltage, 2.2 kV; trap current, 500 µA; electron energy, 70 eV; resolving power, 1200; m/e ions monitored, 84 for natural nicotine and 86 for deuterated nicotine. With these conditions, the retention times for both compounds were essentially the same, 4.5 min.

Using the procedure described above, the results for nicotine in the breast fluids from smokers and nonsmokers are given in Table 1.42. They further noted that these amounts were considerably greater than those reported in plasma (10 to 20 ng/ml). In addition, it was possible to identify cotinine in the same breast fluids by using (3,3-^{2}H)-cotinine as internal standard and carrier. Cotinine levels of 200 to 300 ng/ml were found in breast fluid, and the same concentrations of cotinine were also detected in plasma.

TABLE 1.42

Nicotine Data in Breast Fluids from Smokers and Nonsmokers[a]

Subject	Smoking history	Nicotine conc. (ng/ml)
A	Nonsmoker	0
B	Nonsmoker	0
C	Two packs/day	195
D	One pack/day	60
E	One pack/day	46
F	About one pack/day	59

[a] Adapted from Petrakis et al. [112].

B. Low-Boiling Volatiles (General)

As early as 1962, results of investigations have appeared in the literature for the GC and GC-MS analysis of low boiling and "semivolatile" components of cigarette smoke [117-129].

Johnstone et al. [117] used preparative-scale GC to isolate and identify the components of the methanol-volatile fractions from Russian (flue-cured), Argentinian ("black" fermented), and British (flue-cured) cigarettes. As noted, the results were broadly similar, each containing such compounds as benzene, 2,5-dimethylfuran, toluene, ethylbenzene, m-xylene, o-xylene, styrene, mesitylene, ψ-cumene, dipentene, p-isopropenyltoluene, as well as a mixture of unsaturated aliphatic compounds containing vinyl groups.

The preparation of the methanol-volatile fraction of cigarette smoke consisted of the following:

> Uncased British cigarettes containing a blend of flue-cured tobaccos were smoked mechanically with a 2-sec puff of 25 ml every 15 sec. The mainstream smoke from 25,000 cigarettes was collected at -70°C and the condensate dissolved in ether-methanol (1:1). The solvents were distilled off, and further quantities of methanol were repeatedly distilled from the residue until no more material codistilled. The combined distillates were diluted with a large volume of water and the ether layer was washed several times with water and dried (calcium chloride). Distillation of the ether gave a pale yellow liquid, boiling point approximately 60 to 200°C, which was filtered through a column

of 250 g of neutral alumina. Elution with petrol (b.p. 40 to 45° C) and removal of the solvent by distillation through a Vigreux column yielded 16.9 g of a mobile yellow liquid (M).

This petrol eluate (M) was chromatographed with a 6-ft by 1-in. column packed with 20% E-301 (silicone elastomer) coated on 30-60 mesh C-22 Firebrick and operated isothermally at 150° C with a nitrogen carrier-gas flow rate of 100 ml/min. Seven fractions were collected, each of which was examined in more detail by GC on a 15-ft by 1/4-in. column filled with 20% Apiezon M grease on 30-60 mesh Firebrick and operated at 95° C with an argon CG flow of 40 ml/min. In addition to the compounds listed above, 1,3,5-trimethylbenzene and 1,2,4-trimethylbenzene were identified as smoke components.

Grob [119] studied the separation of the overlap region of cigarette smoke gas and particulate phase (substances with boiling points between 100 and 200° C) by capillary columns, the gas-phase constituents boiling within the range of 20 to 100° C having been investigated and reported in 1962 [118]. Using 50-m by 0.02-in. columns coated with a methylated fraction of PEG 600 isolated by gel filtration on Sephadex G-25 and Carlo Erba gas chromatographs equipped with flame ionization detectors, adequate separations were achieved with a nitrogen carrier-gas flow rate of approximately 5 ml/min. As noted by Grob, the sample volume of 0.5 ml was split in the ratio of 20:1, with the column temperature normally programmed from 30 to 100° C at a program rate of 3 to 4° C/min. With this type of GC arrangement and the conditions specified, Grob was able to identify in both studies [118,119] the following compounds: acetone, methanol, acetonitrile, toluene, methyl ethyl ketone, acrolein, propionaldehyde, benzene, methyl formate, butenone, methylfuran, isovaleraldehyde, furan, dimethylfuran, crotonaldehyde, methyl propyl ketone, isobutyraldehyde, diethyl ketone, n-butyraldehyde, methyl acetate, methacrolein, n-valeraldehyde, methyl isopropyl ketone, pentadiene, acrylonitrile, tetrahydrofuran, isobutyronitrile, pivalaldehyde, methyl acrylate, methacrylonitrile, cyclohexane, ethyl acetate, ethanol, thiophene, tetrahydropyran, isopentenone, limonene, m-xylene, ethyl benzene, capronaldehyde, styrene, p-xylene, o-xylene, pseudocumene, isobutanol, n-butanol, anisole, n-butyronitrile, n-propanol, sec-butanol, β-pinene, hemimellitene, 3-ethyl toluene, 4-ethyl toluene, n-valeronitrile, m-methylstyrene, o-methylstyrene, n-butyl acetate, ethylcapronate, n-capronitrile, mesitylene, and 2-ethyl toluene.

The GC data also indicated that the partition of certain substances between the gas phase and particulate phase to be of great importance, the composition of the gas phase depending on a partition equilibrium that is easily influenced by temperature. In addition to the above, the gas phases from the first and seventh puff of a cigarette show similar differences to the gas phases filtered by a cold and a heated filter, and data was presented in which the smoke from a cigarette of a commercial brand with charcoal filter

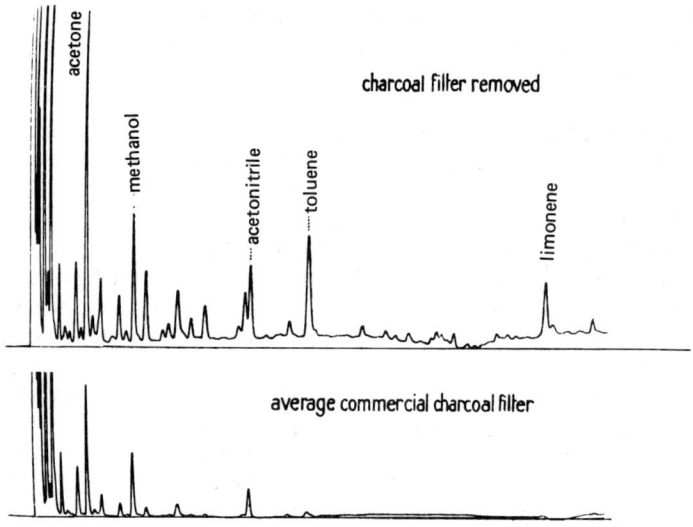

Figure 1.42. Effect of a commercial charcoal filter of average efficiency (column 50 m by 0.02 in.). From Grob [119], reproduced from the Journal of Chromatographic Science, by permission of Preston Publications, Inc.

was compared with that from a cigarette of the same brand, from which the filter tip was removed (Fig. 1.42).

Rushneck [120] showed that cryogenic chromatography provides superior resolution of widely boiling mixtures and an additional advantage when organic compounds of low concentration are contained in a gas, in that the sample is condensed in the initial portion of the column. Since the sample does not move until the column is heated, large or repetitive injections can be made to accumulate amounts of each trace component sufficient for detection.

Applied to the separation of cigarette smoke, Figure 1.43 shows an injection at 30°C of 30 ml of cigarette smoke directly into a 150-ft by 0.01-in.-i.d. SF-96 capillary column. Figure 1.44 shows a similar 30-ml smoke sample injected into the same column at -65°C. The advantages of the cryogenic injection are apparent.

In 1966, Vollmin et al. [122] used an integrated GC-MS instrument which permitted the comparison of FID and TIC (total ion current) chromatograms. FID studies were performed with a Carlo Erba model D gas chromatograph equipped with a 110-m by 0.42-mm-i.d. capillary glass column coated with polypropyleneglycol. Using a nitrogen carrier-gas flow rate of 5.6 ml/min, the column was temperature programmed: held isothermally for 8 min at 0°C, then programmed at 3°C/min to 95°C.

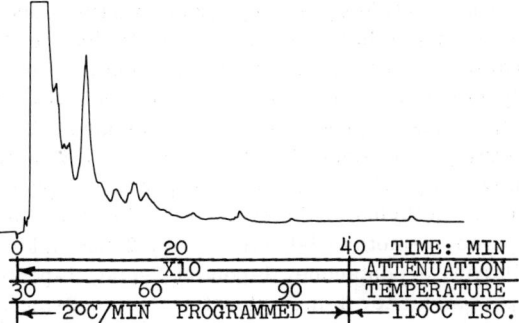

Figure 1.43. Ambient injection of cigarette smoke. From Rushneck [120], reproduced from the Journal of Chromatographic Science, by permission of Preston Publications, Inc.

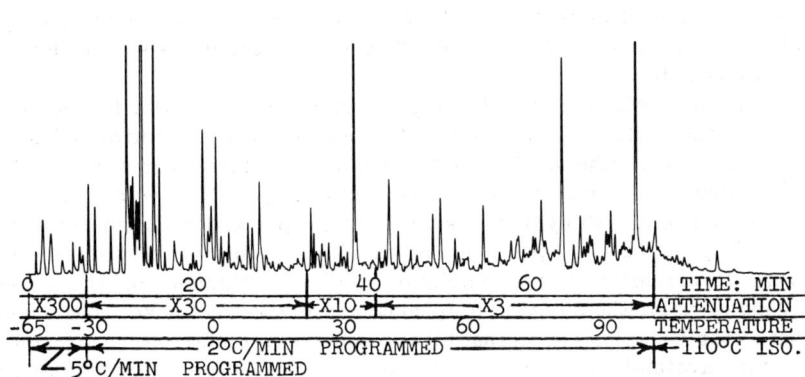

Figure 1.44. Cryogenic injection of cigarette smoke. From Rushneck [120], reproduced from the Journal of Chromatographic Science, by permission of Preston Publications, Inc.

On the other hand, the GC-MS unit (consisting of a Perkin-Elmer model 800 gas chromatograph interfaced with a Hitachi Perkin-Elmer RMU-6D mass spectrometer) also used a 110-m by 0.42-mm-i.d. capillary glass column coated with polypropyleneglycol and the following operating conditions: nitrogen carrier-gas flow rate, 4.0 ml/min; column temperature, held isothermally for 7 min at 28° C, then programmed to 100° C at 2.1° C/min; ionizing voltage, 70 eV; ion source temperature, 240° C.

Based on data obtained with both analytical systems, some of the smoke components identified were carbon monoxide, carbon dioxide, methylchloride, butane, butadiene, acetaldehyde, n-pentane, methylmercaptan, 1-pentene, ethyl chloride, diethyl ether, isoprene, 2-methylpentane, 1,3-pentadiene, propionaldehyde, 1,4-pentadiene, furan, acetone, n-hexane, propenaldehyde, 2,4-dimethylpentane, 3,3-dimethyl-1-butene, methyl acetate, methylpropionaldehyde, 2,2-dimethylpropionaldehyde, methanol, 2-methylpropenaldehyde, acetonitrile, butyraldehyde, 2-methylfuran, 3-methylhexane, 2-butanone, 1-buten-3-one, 2,3-butanedione, 2-propanol, propionitrile, 3-methylbutyraldehyde, 2-methylbutyraldehyde, 3-methylbutan-3-one, isobutyronitrile, n-heptane, 2-methyl-1-buten-3-one, 2-butenaldehyde, 2-pentanone, 2,5-dimethylfuran, 3-pentanone, 1-penten-3-one, 1-penten-4-one, butyronitrile, 2,3-pentanedione, 4-methylpentan-2-one, 2-methylpentan-3-one, 3-methylpentan-2-one, n-octane, toluene, 2,4-dimethylpentan-3-one, isovaleronitrile, N-methylpyrrole, hexan-3-one, hexan-2-one, n-nonane, cyclopentanone, valeronitrile, ethylbenzene, 3-methylcyclopentanone, 1,4-dimethylbenzene, 1,3-dimethylbenzene, heptan-4-one, 2-methylpenten-4-aldehyde, isocapronitrile, 1,2-dimethylbenzene, styrene, 1-decene, capronitrile, cyclohexanone, furfural, 4-ethyltoluene, 3-ethyltoluene, anisole, 1-methyl-4-isopropylcyclohex-1-ene, 1,3,5-trimethylbenzene, 2-ethyltoluene, limonene, 1,2,4-trimethylbenzene, pyrrole, 4-isopropyltoluene, and 2-acetylfuran.

Grob and Vollmin [125] also applied an approach similar to that described by Vollmin et al. [122] for the GC-MS analysis of the "semivolatiles" of cigarette smoke (the smoke fraction having a boiling-point range of about 180 to 350°C). The compounds indicated in Figure 1.45 were identified as shown in Table 1.43 by both gas chromatography and mass spectrometry.

In 1973, Grob [127] published an excellent review of the progress made in the development of high-resolution capillary columns, and the associated technology, for the analysis of tobacco and cigarette smoke. Grob noted that experience has shown that tobacco smoke is by far the most complex mixture available for analysis and, as such, is an ideal mixture for high-resolution capillary columns. To date some 6000 components have been detected, and this list is still growing.

Holzer et al. [128] evaluated exhaled tobacco smoke by GC-MS in 1976, determining for the intermediate volatility range the impact of cigarette smoking on the distribution of organic substances in ambient air. A simple sampling procedure was employed, involving gas-solid adsorption onto an organic polymer (Tenax GC and Carbopack BHT), followed by direct thermal elution onto a capillary column.

For these investigations, a Hewlett-Packard model 5830 and a Hewlett-Packard model 5720 gas chromatograph were used, each equipped with flame ionization detectors and modified to accept capillary columns and to accommodate the adsorbent tubes. With either GC unit, glass capillary columns of 25 to 55 m by 0.35 mm i.d. coated with OV-101 were employed which yielded average efficiencies, measured for n-decane at 100°C, of 2000 to 2600 effective plates per meter. Identifications were performed on a LKB 9000 GC-MS instrument as described previously [130].

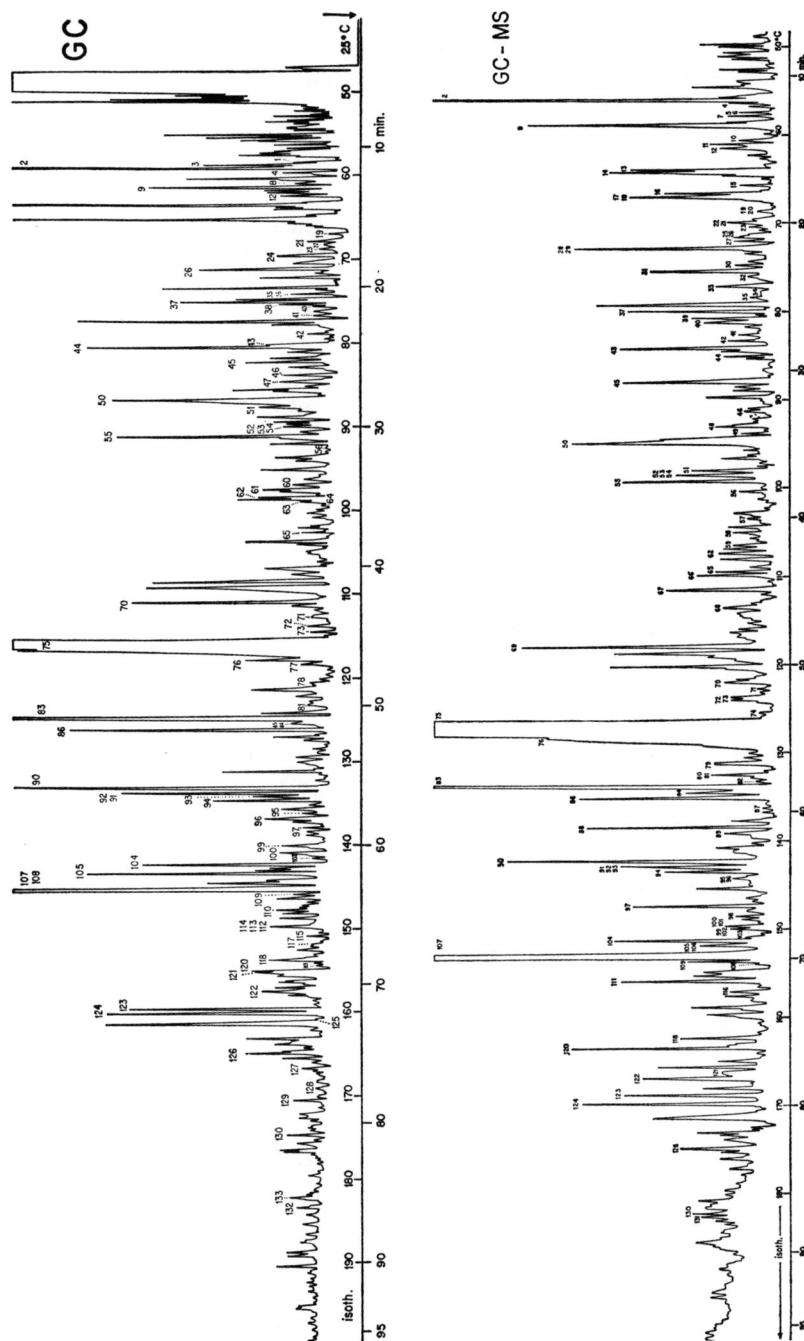

Figure 1.45. 4.0 µl of smoke solution injected with solvent bypassing without splitting on 1-m inlet capillary. Glass capillary column, 55 m/0.35 mm coated with Emulphor O. Detection: GC: FID, GC/MS: total ion monitor. From Grob and Vollmin [125], reproduced from the Journal of Chromatographic Science, by permission of Preston Publications, Inc.

TABLE 1.43

Identified Components in "Semivolatile" Fraction of Tobacco Smoke[a,b]

1	2,6-dimethylpyridine	(GC)[b]
2	limonene	(GC) (MS)[b]
3	pyridine	(GC)
4	methylpyrazine	(GC) (MS)
5	C_3-alkylbenzene	(MS)
6	n-undecane	(GC) (MS)
7	acetoin?	(MS)
8	anisole	(GC)
9	2,5-dimethylpyridine	(GC) (MS)
10	C_3-alkylbenzene	(MS)
11	C_4-alkylbenzene	(MS)
12	3-methylpyridine	(GC) (MS)
13	C_4-alkylbenzene	(MS)
14	quinone?	(MS)
15	C_3-alkylbenzene	(MS)
16	C_3'-alkylbenzene	(MS)
17	C_4-alkylbenzene	(MS)
18	C_3-alkylbenzene	(MS)
19	2,4-dimethylpyridine	(GC) (MS)
20	C_4'-alkylbenzene	(MS)
21	2,3-dimethylpyridine	(GC) (MS)
22	n-dodecane	(GC) (MS)
23	indan	(GC) (MS)
24	2-vinylpyridine	(GC)
25	C_4-alkylbenzene	(MS)
26	furfural	(GC) (MS)
27	C_4'-alkylbenzene	(MS)
28	C_4'-alkylbenzene	(MS)
29	C_5-alkylbenzene	(MS)
30	methylanisole	(MS)
31	dimethylpyrazole	(MS)
32	methylanisole	(MS)
33	C_5-alkylbenzene	(MS)
34	C_4'-alkylbenzene	(MS)
35	indene	(GC) (MS)
36	durene	(GC)
37	pyrrole	(GC) (MS)
38	benzofuran	(GC)
39	vinylpyridine	(MS)
40	2-acetylfuran	(GC) (MS)
41	benzaldehyde	(GC) (MS)

TABLE 1.43 (continued)

42	n-tridecane	(GC)	(MS)
43	tetralin	(GC)	(MS)
44	1-tridecene	(GC)	(MS)
45	5-methylfurfural	(GC)	(MS)
46	benzonitrile	(GC)	(MS)
47	dihydrobenzofuran	(GC)	
48	methylindene		(MS)
49	methylindene		(MS)
50	propylene glycol	(GC)	(MS)
51	acetylmethylfuran	(GC)	
52	acetophenone	(GC)	(MS)
53	n-tetradecane	(GC)	(MS)
54	p-tolualdehyde	(GC)	(MS)
55	furfuryl alcohol	(GC)	(MS)
56	1-tetradecene	(GC)	(MS)
57	$C_{15}H_{26}$		(MS)
58	$C_{16}H_{34}$		(MS)
59	$C_{15}H_{24}$		(MS)
60	menthol	(GC)	
61	3-cyanopyridine	(GC)	
62	naphthalene	(GC)	(MS)
63	propiophenone	(GC)	
64	aniline	(GC)	
65	n-pentadecane	(GC)	(MS)
66	$C_{15}H_{26}$		(MS)
67	$C_{15}H_{30}$		(MS)
68	methylacetophenone		(MS)
69	$C_{13}H_{22}O$		(MS)
70	2-methylnaphthalene	(GC)	(MS)
71	benzyl alcohol	(GC)	(MS)
72	1-methylnaphthalene	(GC)	(MS)
73	n-hexadecane	(GC)	(MS)
74	α-tolunitrile		(MS)
75	nicotine	(GC)	(MS)
76	quinoline	(GC)	(MS)
77	2,6-dimethylphenol	(GC)	(MS)
78	1-methylindole	(GC)	(MS)
79	dimethoxybenzene		(MS)
80	dimethylnaphthalene		(MS)
81	biphenyl	(GC)	(MS)
82	dimethylnaphthalene		(MS)
83	phenol	(GC)	(MS)
84	n-heptadecame	(GC)	(MS)

(continued)

TABLE 1.43 (continued)

85	1,6-dimethylnaphthalene	(GC) (MS)
86	o-cresol	(GC) (MS)
87	dimethylnaphthalene	(MS)
88	$C_{19}H_{38}$	(MS)
89	methylquinoline	(MS)
90	p-cresol	(GC) (MS)
91	m-cresol	(GC) (MS)
92	o-ethylphenol	(GC) (MS)
93	2,5-dimethylphenol	(GC) (MS)
94	2,4-dimethylphenol	(GC) (MS)
95	n-octadecane	(GC) (MS)
96	acenaphthene	(GC) (MS)
97	1,2-dimethylindole	(GC) (MS)
98	1-octadecene	(MS)
99	acenaphthylene	(GC) (MS)
100	3,5-dimethylphenol	(GC) (MS)
101	trimethylphenol	(MS)
102	$C_{20}H_{40}$	(MS)
103	myosmine	(GC) (MS)
104	3-ethylphenol	(GC) (MS)
105	4-ethylphenol	(GC) (MS)
106	trimethylphenol	(MS)
107	neophytadiene	(GC) (MS)
108	3,4-dimethylphenol	(GC) (MS)
109	n-nonadecane	(GC) (MS)
110	2,3,5-trimethylphenol	(GC)
111	$C_{20}H_{38}$	(MS)
112	pyrocatechol	(GC)
113	eugenol	(GC)
114	carbazole	(GC)
115	7-methylindole	(GC)
116	trimethylphenol	(MS)
117	fluorene	(GC)
118	nicotyrine	(GC) (MS)
119	o-methoxyphenol	(GC)
120	n-eicosane	(GC) (MS)
121	glycerol	(GC) (MS)
122	isoeugenol	(GC) (MS)
123	vinylphenol	(GC) (MS)
124	indole	(GC) (MS)
125	coumarin	(GC)
126	3-methylindole	(GC)
127	2-methylindole	(GC)

TABLE 1.43 (continued)

128	5-methylindole	(GC)
129	trans-stilbene	(GC)
130	n-docosane	(GC) (MS)
131	3-ethylindole	(MS)
132	anthracene	(GC)
133	phenanthrene	(GC)

[a] From Grob and Vollmin [125], reproduced from the Journal of Chromatographic Science, by permission of Preston Publications, Inc.
[b] (GC): gas chromatography; (MS): mass spectrometry.

Using the integrated GC-MS system and total ion current chromatograms of urban air and air contaminated with cigarette smoke (sample size of air sample, 3.5 liters; sample size of cigarette puff, 3 ml; glass capillary column, 38 m by 0.35 mm i.d., coated with OV-101; helium carrier-gas flow rate, 3 ml/min; column temperature, held isothermally for 8 min at 20°C, then programmed from 20 to 200°C at 2°C/min), the following volatiles were identified: 1-pentene, C_5H_{12}, acetaldehyde, n-pentane, acrolein, isoprene, C_6H_{12}, 2-methylbutane, furan, diethyl ether, dichloromethane, 2-methylpentane, 3-methylpentane, n-hexane, dimethylbutene, chloroform, ethyl acetate, 4-methyl-2-pentene, 2-methylcyclopentane, dichloroethylene, benzene, C_8H_{12}, C_6H_{10}, methylhexane, 1,5-hexadiene, C_7H_{16}, cyclohexene, 1,2-dimethylcyclopentane, C_7H_{14}, trichloroethylene, n-heptane, n-C_7H_{14}, C_2-alkylcyclopentane, 2-methyl-2-hexene, 2,4-dimethylhexane, C_8H_{16}, C_8H_{18}, 2-methylheptane, toluene, 2,5-dimethylhexane, 3-methylheptane, 1,1-dimethylcyclohexane, 2,5-dimethyl-1,2-hexene, 1,3-dimethylcyclohexane, tetrachloroethylene, C_2-cyclohexane, trimethylcyclopentane, C_3-cyclopentane, 1,1,3-trimethylcyclohexane, C_9H_{18}, C_9H_{16}, C_3-cyclohexane, ethylbenzene, o-xylene, m-xylene, p-xylene, 3-ethylheptane, diethylpentane, styrene, methyloctane, styrene, n-nonane, $C_{10}H_{20}$, $C_{10}H_{22}$, cumene, n-nonyne, C_3-benzene, α-pinene, β-pinene, 5-methyldecane, methylstyrene, $C_{11}H_{24}$, 1,3-dichlorobenzene, n-decane, C_4-benzene, limonene, $C_{11}H_{22}$, undecane, $C_{12}H_{26}$, methyldecane, 2-methyldecane, 1-methylindane, C_5-benzene, naphthalene, 1-methyl-(1,2,3,4-tetrahydronaphthalene), n-dodecane, phenylhexane, 2-methylnaphthalene, $C_{12}H_{24}$, 1-methylnaphthalene, tridecane, phenyloctane, dimethyl phthalate, and diethyl phthalate.

In 1978, Zeldes and Horton [129] developed a GC method in which the gas phase of cigarette smoke was also trapped and stored on Tenax GC for subsequent off-site analyses.

Using an ORNL single-port smoking machine and attached Tenax GC trap as illustrated in Figure 1.46, the GC analyses were performed with a Perkin-Elmer model 3920 gas chromatograph equipped with a flame ionization

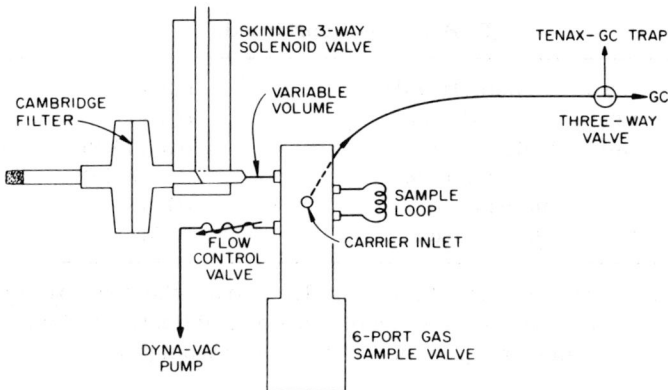

Figure 1.46. ORNL single-port smoking machine and attached Tenax-GC trap. From Zeldes and Horton [129], courtesy of <u>Analytical Chemistry</u>.

detector and a 15-ft by 2-mm-i.d. Pyrex column packed with 15% 3,3'-(trimethylenedioxy)dipropionitrile on 100-120 mesh Chromosorb W(HP). For the separation of isoprene, acetaldehyde, and acrolein as illustrated in Figure 1.47, the following GC conditions were used: injector port, 250°C; column temperature, -70 to +10°C at 8°C/min when sampling was complete, then held isothermally for 4 min before programming at 1°C/min to 70°C; detector temperature, 150°C; nitrogen carrier-gas flow rate, about 12 ml/min.

In addition to the above, optimum conditions were determined for adsorption and desorption of the gas phase, and the effects of aging on the trapped gases were studied. For example, the trapping efficiency relative to sample size was experimentally determined as indicated by the data listed in Table 1.44.

C. Hydrocarbons

In addition to general methods for the separation and identification of multicomponent mixtures containing compounds of different functionality (for example, the low-boiling volatile fraction of cigarette smoke), several studies have been concerned primarily with the hydrocarbon constituents [42,131-144].

In addition to determining the composition of a paraffin fraction from tobacco leaf, Carruthers and Johnstone [42] studied a similar fraction obtained from tobacco smoke. They reported the presence of a homologous series of n-paraffins (C_{24} to C_{34}), with the compounds containing odd numbers of carbon atoms being predominant also for both the iso- and n-paraffins

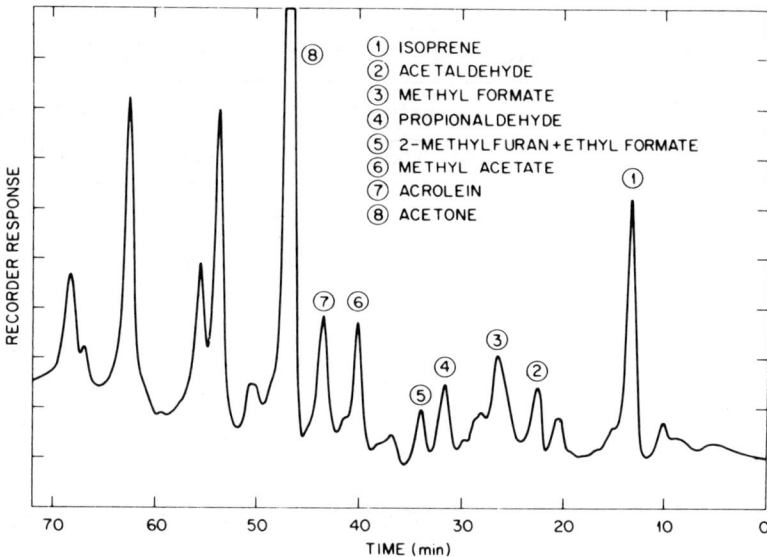

Figure 1.47. Chromatogram of the gas phase of a Kentucky Reference (1R1) cigarette (desorbed from Tenax-GC). Vol. of gas phase, 6.7 ml; trapping flow rate 10 ml/min; desorption time, 10 min at 250°C (trap, 5-1/4 in. by 5 mm i.d. packed with 60/80 mesh Tenax GC. From Zeldes and Horton [129], courtesy of Analytical Chemistry.

TABLE 1.44

Trapping Efficiency Relative to Sample Size[a,b]

	Isoprene, b.p. 34°C		Acetaldehyde, b.p. 21°C		Acrolein, b.p. 52.5°C	
Vol. of gas phase (ml)	4.2	6.7	4.2	6.7	4.2	6.7
Av μg (std run)	6.4	10.2	10.7	17.2	1.4	2.2
Av μg (trapped run)	6.3	8.0	8.6	10.1	1.4	1.9
Recovered (%)	98	78	80	59	100	86

[a] From Zeldes and Horton [129], courtesy of Analytical Chemistry.
[b] Trapping flow rate 10 ml/min, desorption time, 10 min.

TABLE 1.45
MS and GC Analysis of Paraffins in Tobacco Smoke[a]

Carbon no.	MS analysis			GC analysis
	n-Alkanes	iso-Alkanes	Total	n-Alkanes
24	0.0	0.0	0.0	0.1
25	0.0	0.0	0.0	0.6
26	0.5	0.0	0.5	0.4
27	5.2	0.8	6.0	6.3
28	0.5	0.0	0.5	1.1
29	5.2	15.3	20.5	7.4
30	1.0	1.5	2.5	3.8
31	25.7	20.2	45.9	43.3[b]
32	4.3	1.9	6.2	13.0
33	14.3	3.5	17.8	22.8[b]
34	0.0	0.0	0.0	1.1
Total	56.7	43.2	99.9	100.0

[a] Adapted from Carruthers and Johnstone [42].
[b] These peaks were composed of two unresolved peaks. The results given include both peaks. The impurity is probably a very slightly branched paraffin.

as indicated by the mass spectrometric data shown in Table 1.45 as well as results obtained by gas chromatography. Although no details were given for the MS analysis of the paraffin wax, the GC analysis was performed with a 3-ft column packed with silicone grease E 301 (percent not specified) coated on 52-85 mesh Silocel and maintained isothermally at 288° C with a nitrogen carrier-gas flow rate of 1.6 liters/hr.

Spears et al. [140] quantitatively determined the individual alkanes of cigarette smoke in a relatively short time through the use of gas chromatography, column chromatography, and isotope dilution techniques. After the concentrated pentane fraction derived from tobacco smoke was chromatographed on a column (50 by 2 cm) of aluminum oxide (Woelm, activity grade I, pH 7), the alkenes were removed by reacting these species preferentially with bromine-bromide reagent and the brominated products again chromatographed over alumina, and the normal and branched alkanes were separated with a column packed with 5A molecular sieves (isopentane used as eluting solvent), the resulting column effluent (about 150 ml of isopentane) was evaporated to dryness and diluted to 1.0 ml with toluene.

From this final toluene solution containing the alkanes, an aliquot was withdrawn and injected into a Perkin-Elmer model 154 equipped with a flame ionization detector, a linear temperature programmer, and a 2-m by 1/4-in.-o.d. stainless steel column packed with 2 in. of 80-100 mesh microbeads, 5 ft of 5% SE-30 coated on 80-100 mesh Gas Chrom Z, and the balance with 80-100 mesh microbeads. The operating conditions used for the analysis of the C_{12}-C_{33} alkanes (normal and branched) were injector temperature, 350°C; exit line temperature, 350°C; detector temperature, 265°C; column temperature, programmed from 80 to 325°C at 14°C/min; helium carrier-gas flow rate, 150 ml/min.

Using these conditions, the retention times and concentrations of the C_{12} to C_{33} alkanes are listed in Table 1.46, the quantitative data obtained with 85-mm nonfilter cigarettes composed of burley, flue-cured, Turkish, Maryland, and a commercial blend of tobaccos.

On the other hand, Caroff et al. [141] carried out the qualitative analysis of the C_1 to C_6 aliphatic hydrocarbons in the gas phase of cigarette smoke by means of two columns, one of hexadecane and the other of dimethylsulfolane; the column characteristics and GC operating conditions employed with the Aerograph Hy-Fi A-600B gas chromatograph equipped with flame ionization detection being as listed below:

	Column Characteristics	
Length	Column 1 6 m	Column 2 4 m 60
Nature and exterior diameter of column	Stainless steel: 3.17 mm (1/8 in.)	Copper: 3.17 mm (1/8 in.)
Nature of support and particle diameter	C-22 Firebrick: 0.25-0.35 mm	Acid-washed Sil-O-Cel: 0.21-0.25 mm
Nature and percentage[a] of the stationary phase	Hexadecane 20%	Dimethylsulfolane 28%

[a] The percentage of the stationary phase is such that weight of support plus weight of the stationary phase = 100 g.

	Operating Conditions	
	Column 1	Column 2
Column temperature	22.5°C	25°C
Carrier-gas flow rate	21 ml/min	8 ml/min
Pressure at column head	0.850 kg/cm^2	1.090 kg/cm^2
Pressure at column outlet	Atmospheric	Atmospheric
Hydrogen flow rate	22 ml/min	22 ml/min

TABLE 1.46

Retention Times and Quantitative Data for Alkanes of Cigarette Smoke[a]

Carbon no.	Retention time (min)	Burley[b]		Flue cured[c]		Turkish[d]		Maryland[e]		Blend[f]	
		Total	Branched	Total	Branched	Total	Branched	Total	Branched	Total	Branched
12	2.7	0.7		1.3	0.1			0.5			
13	3.3	0.7		0.8	0.1	1.7		3.7		1.2	
14	3.9	0.6		1.2	0.1	2.1		0.9		1.2	
15	4.6	0.9	0.1	2.2	0.2	2.5		1.3	0.2	2.1	0.2
16	5.3	0.6	0.2	1.0		1.5		0.4		2.2	
17	5.9	0.4		1.2		0.8		0.9		1.9	
18	6.5	0.7	0.3		0.2	1.8		0.8	0.1		
19	7.0	0.1		0.4		1.3		0.9			0.7
20	7.6	0.5	0.9	0.5		1.2		0.7	0.6	2.2	
21	8.1	0.6		0.6		1.5		1.3		1.9	
22	8.6	0.8	0.9	0.6	0.4	1.6		0.9	0.6	2.2	
23	9.1	0.4	0.9	0.5		0.9		0.7	0.6	0.9	

Chemical Composition of Tobacco Smoke

24	9.6	0.3	0.6	0.4		10.5		0.7	0.8	2.3	
25	10.0	0.3		0.7		3.1		0.5		1.2	1.5
26	10.4	0.3		0.3		1.5		3.2		1.1	
27	10.8	2.1		5.3	0.3	13.1	1.0	4.9	0.7	4.7	
28	11.2	0.3	0.5	0.5		2.3	0.1	0.4	0.3	0.9	
29	11.6	3.0	0.6	4.6	0.8	10.4	1.1	4.5	0.6	5.5	1.1
30	11.9	3.7	1.2	7.1	2.6	10.1	2.9	5.0	2.2	7.1	2.5
31	12.2	16.4	2.3	19.8	4.6	34.1	5.3	28.8	2.6	19.4	4.2
32	12.5	9.7	2.7	11.6	4.3	19.4	5.5	15.0	2.0	11.3	4.6
33	12.8	11.3	1.3	14.3	3.0	16.4	2.5	26.8	0.6	7.3	3.2
Total		54.4	12.5	74.9	16.7	137.8	18.4	99.8	11.9	76.6	18.0

[a] Adapted from Spears et al. [140].
[b] Tobacco burned per 100 cigs. = 80.0 g; total condensate per 100 cigs. = 1.44 g.
[c] Tobacco burned per 100 cigs. = 90.4 g; total condensate per 100 cigs. = 2.68 g.
[d] Tobacco burned per 100 cigs. = 110.0 g; total condensate per 100 cigs. = 3.30 g.
[e] Tobacco burned per 100 cigs. = 78.1 g; total condensate per 100 cigs. = 2.33 g.
[f] Tobacco burned per 100 cigs. = 84.1 g; total condensate per 100 cigs. = 2.44 g.

All values in mg per 100 cigarettes.

TABLE 1.47

Retention Times of Various Alkanes on Columns 1 and 2[a]

| | Relative retention time[a] | |
Alkane	Column 1	Column 2
Methane	1.00	
Ethane	1.45	0.039
Ethylene	1.25	0.052
Propane	2.82	0.116
Propene	2.54	0.222
Isobutane	5.41	0.222
n-Butane	8.66	0.362
Acetylene	1.15	0.482
Butene-1	6.73	0.593
Isobutene	6.73	0.637
trans-Butene-2	8.70	0.772
Isopentane		0.772
cis-Butene-2	9.48	0.917
Pentane		1.000
3-Methylbutene-1		1.020
1,3-Butadiene	5.98	1.342
Pentene-1		1.564
Methyl chloride	3.42	1.564
2-Methylbutene-1		1.810
trans-Pentene-2		2.180
Methylpentane		2.180
cis-Pentene-2		2.357
2-Methylbutene-2		2.420
n-Hexane		2.601
Isoprene		3.940

[a] Adapted from Caroff et al. [141].
[b] Column 1: relative to methane; column 2: relative to pentane.

Using these GC columns and conditions, the relative retention times of various alkanes chromatographed on both columns are given in Table 1.47.

Philippe et al. [142] performed a GC analysis on alumina of the liquid air-condensable fraction of cigarette smoke, showing the presence of some 37 compounds, mainly hydrocarbons, in this fraction. Twenty of these hydrocarbons had not been previously reported in cigarette smoke and, in addition to retention times on two different packing materials, infrared and mass spectrometric techniques were used as identification criteria.

In their investigation, two gas chromatographic instruments were used, one being a laboratory-built unit equipped with a thermal conductivity detector and a 9-ft by 1/4-in. copper column filled with Burrell activated alumina and operated at room temperature (about 24°C) without thermostatic control using a helium carrier-gas flow rate of 50 ml/min. Since some fractions eluted from the alumina column contained more than a single component, these were further fractionated at room temperature with a Perkin-Elmer model 154 gas chromatograph (TC detector) using a 15-ft column packed with 18.7% 3,3-oxydipropionitrile coated on Burrell activated alumina; the helium flow rate was maintained at 40 ml/min.

As for the 34 hydrocarbons identified, there were 20 that had not been previously reported as present in cigarette smoke gases: allene, methylacetylene, cyclopentane, iso-pentane, n-pentane, cyclopentene, 3-methylbutene-1, pentene-1, vinylacetylene, 2-methylbutene-1, trans-pentene-2, 2-methylbutene-2, cis-pentene-2, pentadiene-1,4, ethylacetylene, 2-methylpentane, 3-methylpentane, n-hexane, and pentadiene-1,3 (the two geometric isomers).

The retention times of these compounds on both columns using the conditions specified are shown in Table 1.48.

Caroff et al. [143] developed a GC procedure whereby a direct qualitative analysis of the gas phase of cigarette smoke was accomplished by use of an assembly of three classical columns connected to two flame ionization detectors. The basic Aerograph model 202 gas chromatograph (Fig. 1.48) was modified as shown in Figure 1.49. The columns whose characteristics are given below were arranged in two different ways, I and II, according to Figure 1.49.

Column Characteristics[a]

	Column A	Column B	Column C	Column D
Stationary phase	TCEP[b]	Hallcomid	Squalane	TCEP + Hallcomid
Loading (%)	15	14.7	20	17.7 + 2
Support	C-22 brick	C-22 brick	C-22 brick	C-22 brick
Particle size (mm)	0.25-0.21	0.25-0.21	0.25-0.21	0.25-0.21
Column length (m)	2.10	3	3	3
Preconditioning	15 hr 100°C	13 hr 100°C	14 hr 130°C	13 hr 96°C

[a] All columns were prepared from 1/8-in. copper tubing.
[b] TCEP = 1,2,3-tris(cyanoethoxy)propane.

Column position (Fig. 1.49)	Assembly I	Assembly II
1	Column A	Column A
2	Column C	Column D
3	Column B	Column B

Column Assemblies

Note: Common operating conditions: injector temperature, 140°C; col-temperature, 55°C; detector temperature, 123°C; carrier-gas flow rate, 14 ml/min (Assembly I), 15.4 ml/min (Assembly II).

With this type of GC unit, column types, and operating conditions, Caroff et al. were able to confirm results previously obtained for hydrocarbons and to identify 25 oxygenated and nitrogenous compounds. The hydrocarbons identified with the squalane column were methane, acetylene, ethylene, ethane, propane, propene, propadiene, propyne, isobutane, butene-1, isobutene, n-butane, cis-butene-2, cis-butene, 1,3-butadiene, 3-methylbutene-1, methylbutane, pentene-1, 2-methylbutene-1, n-pentane, isoprene, pentene-2 (cis and trans), 2-methylpentene-2, trans-1,3-pentadiene, cis-1,3-pentadiene, 1,2-pentadiene, 2-methylpentane, cyclopentane, 3-methylpentane, and n-hexane. The oxygenated and nitrogen-containing compounds identified were acetaldehyde, furan, propionaldehyde, methyl acetate, acetone, acrolein, isobutanal, 2-methylfuran, methyl ethyl ketone, methacrolein, methanol, n-butanal, 2,5-dimethylfuran, isovaleraldehyde, ethanol, methyl isopropyl ketone, diethyl ketone, diacetyl, acetonitrile, and crotonaldehyde.

In 1970, Gelpi and Oro [144] employed integrated GC-MS to identify a relatively high amount of paraffins and olefins ranging from C_{10} to C_{33} in a concentrate of cigarette smoke. They summarized their findings as follows:

> The paraffins show a predominance of the odd-numbered carbon chains, but the contrary is true for the olefins (C_{13} to C_{22}). The paraffins show a bimodal distribution with maxima centered below C_{12} and at C_{31}. Two series of methyl alkanes were also detected. In the high-molecular-weight region (C_{29} to C_{33}) the anteiso alkanes with even number of carbon atoms predominate over the corresponding isoalkanes, whereas the odd-numbered isoalkanes predominate over their anteiso homologs. The smoke also contained high amounts of C_{19} and C_{20} isoprenoid olefins. The C_{15}, C_{18}, C_{19}, and C_{20} saturated isoprenoids are present in decreasing concentrations.

To analyze the smoke condensate collected in n-pentane, two separate GC-MS procedures were used: for fraction I (C_{10} to C_{23} range), a 195-m by 0.076-cm capillary column coated with Polysev; for Fraction II (C_{20} to C_{33} range), a 1.7-m by 0.3-cm glass column packed with 1% OV-1. In Figure 1.50, the GC separation of the low-molecular-weight hydrocarbons in tobacco smoke with the stainless steel capillary column is illustrated in addition to the column temperature program used with the F&M model 810 gas

TABLE 1.48

Retention Times on Alumina (Column A) and
Alumina/3,3'-Oxypropionitrile (Column B)

	Retention time (min)	
Compound	Column A	Column B
Propadiene	18.0	
Propyene	63.0	11.8
Cyclopentane	77.0	10.8
n-Pentane	90.0	
Cyclopentene	111.0	17.5
3-Methylbutene-1	117.0	5.2
Butene-3-yne-1	143.0	26.0
Pentene-1	145.0	6.8
trans-Pentene-2	153.0	
2-Methylbutene-1	162.0	
cis-Pentene-2	175.0	8.0
2-Methylbutene-2	177.0	8.8
Pentadiene-1,4	190.0	11.5
Butyne-1	195.0	18.5
2-Methylpentane	260.0	5.8
3-Methylpentane	260.0	6.8
Pentadiene-1,3 (two isomers)	291.0	I 21.7 II 25.7
n-Hexane	305.0	7.0
Butadiene-1,3		7.7
2-Methylbutane		3.8

a Adapted from Philippe et al. [142].

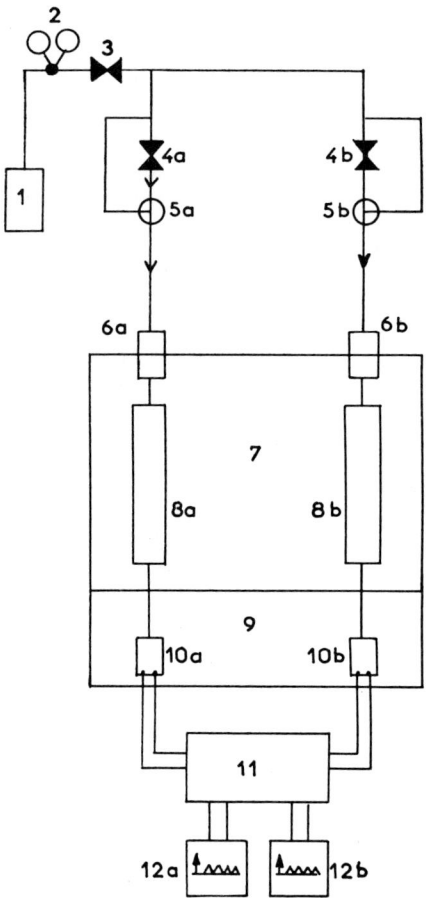

Figure 1.48. Aerograph model 202. 1, Carrier gas tank supply. 2, Tank supply regulator. 3, Carrier gas/carrier flow on-off control. 4a,b, Injector control valves. 5a,b, Differential flow controllers and meters. 6a,b, Injectors. 7, Column oven. 8a,b, Columns. 9, Detector oven. 10a,b, Flame ionization detectors. 11, Integrator. 12a,b, Recorders. From Caroff et al. [143], reproduced from the Journal of Chromatographic Science, by permission of Preston Publications, Inc.

Chemical Composition of Tobacco Smoke

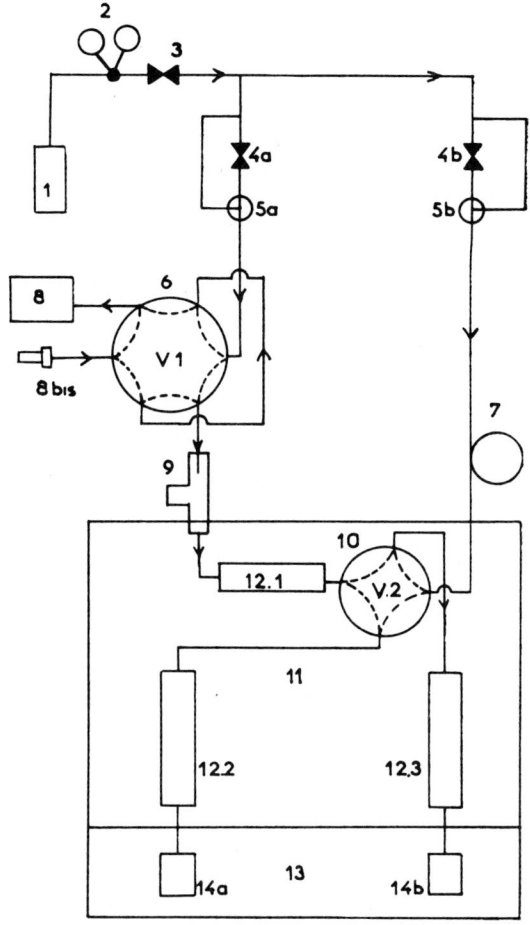

Figure 1.49. Aerograph model 202, modified. 1, Carrier gas tank supply. 2, Tank supply regulator. 3, Carrier gas/carrier flow on-off control. 4a,b, Injector control valves. 5a,b, Differential flow controllers and meters. 6, Six-way gas-sampling valve (V_1). 7, Restrictor. 8, Smoking machine. 8 bis, Cigarette. 9, Modified injector. 10, "Distributing" valve (V_2). 11, Column oven. 12.1,2,3, Columns. 13, Detector oven. 14a,b, Flame ionization detectors. From Caroff et al. [143], reproduced from the Journal of Chromatographic Science, by permission of Preston Publications, Inc.

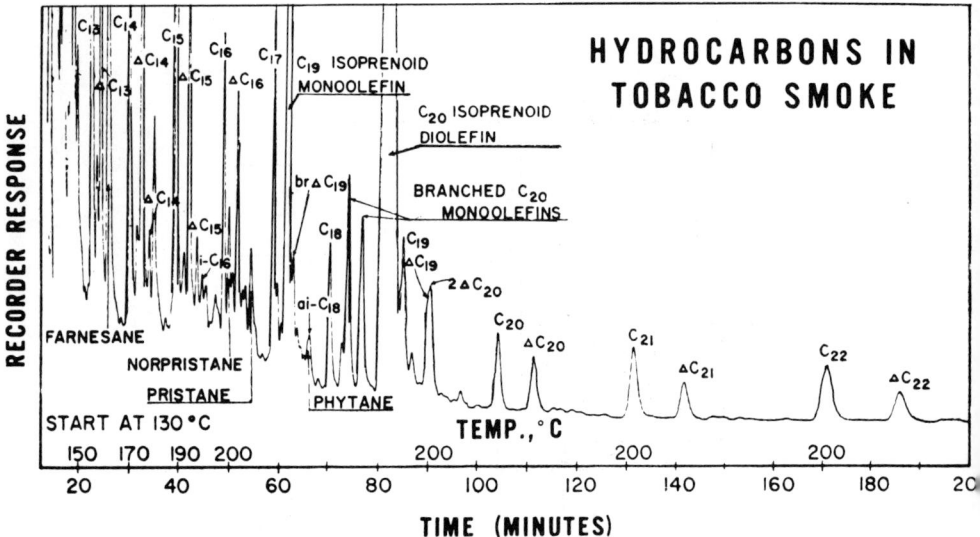

Figure 1.50. Gas chromatographic separation of the low-molecular-weight hydrocarbons in tobacco smoke. Stainless steel capillary column (195 m long by 0.076 cm, i.d.) coated with Polysev [m-bis-m-(phenoxyphenoxy)-phenoxy-benzene]. F&M model 810 gas chromatograph equipped with a flame ionization detector. Range 10; attenuation, 1. Nitrogen pressure 1050 gcm^2. No split. The temperature of injection was held at 130°C for 10 min and then programmed to 200°C at 2°C/min. From Gelpi and Oro [144], reproduced from the Journal of Chromatographic Science, by permission of Preston Publications, Inc.

chromatograph. According to the GC-MS analysis, the isoprenoids, farnesane, norpristane, pristane, and phytane, were also present in this fraction. As noted by Gelpi and Oro, "their respective mass spectra showed molecular ions at m/e 212 ($C_{15}H_{32}$) for farnesane, 254 ($C_{18}H_{38}$) for norpristane, 268 ($C_{19}H_{40}$) for pristane, and 282 ($C_{20}H_{42}$) for phytane. Major fragments were observed in all cases at m/e 113 ($C_8H_{17}^+$) and 183 ($C_{13}H_{27}^+$), both diagnostic ions in the mass spectra of saturated isoprenoid chains. Also other significant fragments were observed at m/e 127 ($C_9H_{19}^+$) in the mass spectrum of farnesane; at m/e 99 ($C_7H_{15}^+$) and 169 ($C_{12}H_{25}^+$) in the spectrum of norpristane; and m/e 127 ($C_9H_{19}^+$) and 197 ($C_{14}H_{29}^+$) in the spectrum of phytane, all supporting the above identifications."

With regard to the analysis of fraction II, the GC-MS results were similar to those already reported for cigarette smoke [42], for the leak wax alkanes of Nicotiana tabacum [44,145] and for tobacco "aroma" [45].

D. Ketones

Exclusive of the low-boiling ketones reported in previous sections of this chapter (low-boiling volatiles and hydrocarbons), Quin et al. [81] continued their investigation of alkaloids of burley tobacco smoke. They showed that previously unidentified fractions 4 and 6 corresponded to methyl-3-pyridyl ketone and ethyl-3-pyridyl ketone, respectively. In addition to comparing their GC behavior on two different columns (column A, 2 m by 6 mm, packed with 20% polyethylene glycol 1025 on Firebrick and operated at 152° C with a helium flow rate of 67 ml/min; column B, 1 m by 6 mm, filled with 20% polyethylene glycol 20M on Firebrick and operated at 153° C with a carrier-gas flow rate of 67 ml/min), they obtained confirmatory identification by comparative paper chromatography in two solvent systems of their 2,4-dinitrophenylhydrazones. Since these findings, Quin et al. have examined other lots of burley tobacco cigarettes from a different source and found neither ketone in the smoke of these cigarettes and concluded that apparently the growth and history of the tobacco prior to smoking is of much importance in determining the concentration, or even the presence, of these ketones in the smoke.

Because of the complexity of cigarette smoke, Bell, Ireland, and Spears [146] found it necessary to employ extensive fractionation to isolate and identify minor components such as aromatic ketones in cigarette smoke condensate. These procedures involved solvent partition, column, paper, and gas chromatography. The gas chromatographic system, which permitted the collection of smoke constituents for subsequent analysis by ultraviolet and infrared spectroscopy and mass spectrometry, consisted of a Perkin-Elmer model 880 gas chromatograph equipped with a dual column system, dual flame ionization detectors, linear temperature programming, and a splitting arrangement to obtain samples of eluted components. For their purpose, a 10-ft by 1/8-in. stainless steel column packed with 8% Apiezon L coated on 80-90 mesh Anakrom ABS gave satisfactory resolution; however, other phases were also examined: OV-1, OV-17, SE-30, and SE-52.

They noted that, in their study of the acetonitrile fraction, fluoren-9-one and a number of alkylated fluoren-9-ones were identified by UV, IR, and MS analyses and by GC retention values as shown in Table 1.49.

E. Phenolic Constituents [88, 90, 147-152]

In 1960, Carruthers and Johnstone [148] reported some of their results obtained in an examination of phenols from the smoke of a variety of British cigarettes. In their investigation, the following GC parameters were employed: a 6-ft by 1-in. column packed with 20% E-301 silicone grease on 80-100 mesh, acid-washed Celite; column temperature, adjusted so that anisole was eluted in approximately 10 min; nitrogen carrier-gas flow rate,

TABLE 1.49

List of the Fluoren-9-ones Isolated and Identified in Cigarette Smoke Condensate Showing Ultraviolet Absorption Maxima and Retention Times[a]

	Ultraviolet absorption maxima (μm) (in cyclohexane)		Retention time (min)		
			I[b]	II[c]	III[d]
Fluorene-9-one	256.5	247.5			6.6
1-Methylfluoren-9-one	259.0	250.0		12.5	
2-Methylfluoren-9-one	259.5	251.0			8.2
3-Methylfluoren-9-one	261.0	251.0	9.5		
4-Methylfluoren-9-one	256.3	248.0	9.9		
Dimethylfluoren-9-one	262.0	253.0	10.4	14.7	
Dimethylfluoren-9-one	262.0	253.0	10.8		9.8
Dimethylfluoren-9-one	264.5	259.5	11.4		
Dimethylfluoren-9-one	260.0	251.0	11.7		
Dimethylfluoren-9-one	262.5	253.0		14.0	9.3
Ethylfluoren-9-one	257.5	251.5	10.7	14.7	
Methylethylfluoren-9-one	261.0	252.0		16.6	

[a] From Bell et al. [146], courtesy of Analytical Chemistry.
[b] Condition I: 10-ft by 1/8-in. stainless steel, 10% OV-17, 170 to 280°C at 6°C/min.
[c] Condition II: 13-ft by 1/8-in. stainless steel, 16% Apiezon L, 200 to 310°C at 6°C/min.
[d] Condition III: 10-ft by 1/8-in. stainless steel, 10% OV-17, 200 to 300°C at 6°C/min.

100 ml/min. When subjecting a fraction (b.p. 150 to 300°C) isolated from the smoke of 5000 cigarettes to GC analysis using the above conditions, the following compounds were found: anisole, o-cresyl methyl ether, m-cresyl methyl ether, p-cresyl methyl ether, veratrole, m-methoxyacetophenone, and p-methoxyacetophenone.

In 1961, Hoffmann and Wynder [149] determined the phenolic components of cigarette smoke by gas chromatography using a 2-m by 6-mm-i.d. stainless steel column packed with 15% di-n-octylsebacate coated on 100-140 mesh Gas Chrom P. With a helium carrier-gas flow rate of 40 to 50 ml/min and a column temperature of 180°C, the relative retention times obtained for some phenols are listed in Table 1.50, the retention time of phenol being about 6 to 8 min depending on the actual carrier-gas flow rate used.

TABLE 1.50

Relative Retention Times for Some Phenols[a]

Phenol	Relative retention time
o-Chlorophenol	0.82
Salicylaldehyde	0.86
Phenol	1.00
o-Cresol	1.24
m-Cresol	1.54
o-Ethyl phenol	1.75
p-Cresol	1.55
2,4-Dimethyl phenol	1.78
2,5-Dimethyl phenol	1.80
2,3-Dimethyl phenol	2.20
3,5-Dimethyl phenol	2.24
m-Ethyl phenol	2.24
p-Ethyl phenol	2.25
3,4-Dimethyl phenol	2.46
2,3,5-Trimethyl phenol	3.32
3,4,5-Trimethyl phenol	4.15

[a] Adapted from Hoffmann and Wynder [149].

Crouse et al. [150] developed analytical methods for phenolics involving condensation of cigarette smoke and isolation of steam-volatile phenols after a sequence of ether extractions from aqueous solutions of controlled alkalinity. The primary GC instrument used for the initial separation of the phenolic extracts to which 2,6-di-tert-butylphenol was added as internal standard was an Aerograph model A350 dual column gas chromatograph equipped with thermal conductivity detectors and two column systems: column A, 4-ft by 1/4-in. copper column packed with 25% DEGS on 60-80 mesh, acid-washed Chromosorb; column B, 6-ft by 1/4-in. copper column packed with 30% Apiezon N coated on the same solid substrate. Whereas the carrier-gas flow rates for column A and column B were adjusted to 46.1 and 63.4 ml/min, respectively, in order that phenol would be eluted from each column in exactly 12.0 min, the injector, column, and detector temperatures used with both columns were the same, these being 225, 160, and 250°C, respectively. However, when the isolated phenols contained less than 5 mg of material, the GC analyses were performed on a Barber-Colman model 10 gas chromatograph equipped with the more sensitive radium ionization detector. Nevertheless, the same set of standards and calibration procedures were used for both instruments.

With the GC conditions cited above, the retention times of pure phenols chromatographed on both columns are listed in Table 1.51. When applied to the determination of phenols in condensed smoke from cigarettes (13 different brands), typical data for the various phenolics such as that given in Tables 1.52 and 1.53 were obtained.

Based on their findings, they concluded that:

Analyses of smoke condensates from 13 major brands of cigarette, filters removed, indicates that the steam-distillable phenols were comprised of 10 to 15% phenol itself, 16 to 32% identified alkyl phenols, primarily cresols, and 55 to 75% other phenols. Smoke obtained with the filters intact showed a reduction in phenols to 30 to 50% for 12 of the cigarettes; the filter in one (cigarette C) permitted but 15% of the identified phenols to pass through.

Spears [151] in 1963 quantitatively determined phenol in cigarette smoke by a method involving solvent partition, steam distillation, and gas chromatography. The GC analysis was performed with a Perkin-Elmer model 154-D Vapor Fractometer equipped with a thermal conductivity detector and a 6-ft by 1/4-in. stainless steel column packed with 8% UCON 50 HB oil coated on 40-60 mesh Firebrick previously covered with 25% Nylon 66. Using 2-hydroxyacetophenone as internal standard and specified GC conditions (helium carrier-gas flow rate, 78 ml/min; column temperature, 168°C), the retention times of identified phenolic compounds are listed in Table 1.54.

Malaterre et al. [152] analyzed the monophenolic fraction of tobacco smoke condensate by a combination of chromatographic techniques. The monophenols were separated from cigarette smoke condensates by thin-layer

TABLE 1.51

Retention Times of Various Phenols on Two Column Systems[a]

Component	Retention time (min)	
	DEGS	Apiezon N
2-Chlorophenol	6.81	9.12
2,6-Di-tert-butylphenol	7.85	134.00
2,6-Dimethylphenol	9.95	23.50
Phenol	12.00	12.00
2-Methylphenol	12.00	17.17
2,4,6-Trimethylphenol	12.00	43.25
2,5-Dimethylphenol	14.17	29.40
2,4-Dimethylphenol	14.65	29.40
3-Methylphenol	15.50	20.35
4-Methylphenol	15.50	20.35
2,3-Dimethylphenol	17.90	33.60
3,5-Dimethylphenol	19.60	34.00
2,4,5-Trimethylphenol	21.75	56.20
2,3,5-Trimethylphenol	21.75	56.20
3,4-Dimethylphenol	23.25	40.00
3,4,5-Trimethylphenol	36.00	78.25
4-Methoxyphenol	46.70	47.70
3-Methoxyphenol	51.50	52.25

[a] Adapted from Crouse et al. [150].

chromatography on Kieselgel G using a 3:1 mixture of cyclohexane and ethylacetate. In turn, the monophenolic fraction was recovered from the support and analyzed by GC on a capillary column. Using this combined analytical approach, the yields for 3,4,5- and 2,3,5-trimethylphenols were 80 ± 2% and 65%, respectively, whereas the concentration determined for phenol, cresol (para and meta), and 2,6-dimethylphenol were 8.0, 3.0, and 0.5 µg/mg of condensate, respectively.

TABLE 1.52
Phenols in Condensed Smoke from Cigarettes, Filters Removed
(μg/cigarette)[a]

Sample	mm	No. of determinations	Phenol	2-Methyl-phenol	3- and 4-Methyl-phenol	2,4-Dimethyl-phenol	2,6-Dimethyl-phenol
A	80	4	82	11.0	50	11.9	12[c]
B	85	4	90	14.5	51	12.8	22
C	85	3	87	17.5	54	9.6	19.7
D	80	2	109	19	69	16.4	29
E	85	2	129	22[d]	82[d]	19.0	31
F	80	1	115	18	68	8.6	32
G	85	2	122	20	74	7.0[c]	32
H	85	2	161[d]	21	74	13.0	46
I	85	1	111	19	68	10.0	31
J	85	1	130	18	74	17.3	29
K	85	1	124	21	77	19.6[d]	30
L	85	2	115	16	56	8.0	32
M	85	1	71[e]	14.1[c]	41[c]	7.1	20

[a]From Crouse et al. [150], reproduced from the Journal of Chromatographic Science, by permission of Preston Publications, Inc.
[b]Gravimetric determinations.
[c]Lowest.
[d]Highest.

Incompletely resolved phenols				Summary				
2-Ethyl- and/or 2-methoxy- phenol	2,3-Dimethyl- and/or 3,5-dimethyl- phenol	3-Ethyl- and/or 4-ethyl- phenol	2,4,5- and 2,3,5- Trimethyl- phenol	Phenols		Other identified phenols		Total phenols[b] (μg)
				μg	% of total	μg	% of total	
9.9	4.4	15	1.6	82	17.2	114.9	24.1	476
9.5	4.2	16	1.7	90	18.6	131.7	27.2	485
12.0	3.6	18	6.1	87	19.3	173.1	38.4[d]	450[c]
8.4	3.6	22	7.2	109	18.4	174.6	31.8	549
14[d]	5.0	28[d]	7.6[d]	129	23.0	198.6[d]	35.5	560
7.9[c]	2.4	22	4.5	115	21.9	163.4	31.1	525
13	Tr	26	Tr	122	23.9[d]	172.0	33.7	511
10	3.3	21	6.2	161[d]	22.9	190.5	27.1	703[d]
8.8	5.2	24	5.5	111	20.1	171.5	31.0	553
13	4.6	24	6.6	130	20.1	196.5	30.3	648
11	6.7[d]	22	3.0	124	19.5	190.3	30.0	635
8.5	2.5	16	3.9	115	20.1	142.9	25.0	571
9.2	0.4[c]	12[c]	Tr[c]	71[c]	15.3[c]	93.8[c]	20.3[c]	462

TABLE 1.53
Phenols in Condensed Smoke from Cigarettes, Filters Intact
(μg/cigarette)[a]

Sample	mm	No. of determinations	Phenol	Simple phenols			
				2-Methyl-phenol	3- and 4-Methyl-phenol	2,4-Dimethyl-phenol	2,6-Dimethyl-phenol
A	80	5	25	3.9	16	3.5	10
B	85	3	29	4.0	17	5.0	12
C	85	3	12.8[c]	1.1[c]	8.1[c]	1.5[c]	2.5[c]
D	80	3	31	4.0	28	6.6	13
E	85	3	43	4.2	30[d]	8.2[d]	16
F	80	2	44	6.0	29	2.1	17
G	85	2	45	7.0	30	2.6	18
H	85	1	64[d]	11.1[d]	29	2.0	19
I	85	1	55	6.1	28	6.8	21[d]
J	85	1	48	7.0	24	7.4	18
K	85	1	53	11.0	26	7.0	19
L	85	2	46	6.5	23	1.8	8
M	85	2	19.5	3.2	22	2.0	8

[a]From Crouse et al. [150], reproduced from the Journal of Chromatographic Science, by permission of Preston Publications, Inc.
[b]Gravimetric determinations.
[c]Lowest.
[d]Highest.

Incompletely resolved phenols				Summary				
2-Ethyl- and/or 2-methoxy- phenol	2,3-Dimethyl- and/or 3,5-dimethyl- phenol	3-Ethyl- and/or 4-ethyl- phenol	2,4,5- and 2,3,5- Trimethyl- phenol	Phenols		Other identified phenols		Total phenols[b] (µg)
				µg	% of total	µg	% of total	
2.4	2.3	8	2	25	8.5	48.1	16.2	296
2.1	2.9	5	2	29	9.8	50.0	16.7	296
0.5	0.9	5	Tr[c]	12.8[c]	8.5[c]	19.6[c]	13.1[c]	150[c]
5.0	4.4	4	7.9	31	8.7	72.9	20.5	355
5.4	5.7[d]	6	4.0	43	11.7	79.5	21.7	367
4.5	Tr[c]	10	5.2	44	14.3	70.8	19.8	358
4.9	1.9	8	4.8	45	15.1	77.2	25.2	297
4.9	Tr[c]	Tr[c]	5.2	64[d]	18.2[d]	71.2	20.2	352
5.6	2.8	18	10.5[d]	55	14.1	98.8	25.3	391[d]
8.1[d]	3.1	14	4.5	48	15.4	96.1	30.8[d]	312
5.4	5.6	20[d]	5.0	53	14.0	99.0[d]	26.2	378
3.7	Tr[c]	5	5.6	46	14.2	53.6	16.5	324
1.0	1.9	6	1.9	19.5	10.0	46.0	23.6	195

TABLE 1.54

Phenolic Smoke Components[a]

Component	Retention time (min)
2-Methoxyphenol	9.5
2,6-Dimethylphenol	13.5
2,4,6-Trimethylphenol	18.2
2-Methylphenol	21.2
Phenol	23.8
2,4-Dimethylphenol	27.2
2,5-Dimethylphenol	27.2
4-Methylphenol	30.0
3-Methylphenol	31.2
3-Ethylphenol	42.0
4-Ethylphenol	42.0
4-Methoxyphenol	90
3-Methoxyphenol	108
Internal standard (2-hydroxyacetophenone)	7.8

[a] Adapted from Spears [151].

Using a gas chromatograph equipped with flame ionization detection, several of the capillary columns found satisfactory for both the identification and resolution of such phenolics as 2-methylphenol, 3-methylphenol, 4-methylphenol, 2,4-dimethylphenol, 2,5-dimethylphenol, 2,6-dimethylphenol, 3,4-dimethylphenol, 2,4,6-trimethylphenol, 2,3,5-trimethylphenol, 2,3,4-trimethylphenol, 3,4,5-trimethylphenol, 2,3,6-trimethylphenol, 2,3,4,6-tetramethylphenol and 2-methoxyphenol were as follows:

	Column A	Column B	Column C
Length (m)	55	35	43
i.d. (mm)	0.4	0.4	0.4
Liquid phase	Didecylphthalate	Carbowax 20M	OV-101
Column temp. (°C)	120	200	105
CG (H_2) inlet pressure (bar)	2.2	0.4	0.4

On the other hand, the identification of eluted components in the smoke condensate was made using an LKB 9000 GC-MS instrument equipped with a 4-m by 2.2-mm-i.d. stainless steel column packed with 5% Carbowax 20M coated on Chromosorb W. To obtain retention times for phenol, p-cresol, 4-methyl-2,6-di-tert-butylphenol, 2,3,4-trimethylphenol, and 3,4,5-trimethylphenol of approximately 3.70, 4.79, 5.98, 8.70, and 8.70 min, respectively, the following settings were used: column temperature, programmed from 140 to 200°C at 2°C/min; ion source temperature, 270°C; separator temperature, 250°C; ionization potential, 70 eV.

In 1976, Brunnemann et al. [90] developed a quantitative analytical method for catechols in tobacco smoke using catechol-^{14}C as internal standard. Main and sidestream smoke particulates were collected in cold traps, the acidic portion of the particulate matter was neutralized, and the catechols were enriched by complexing with boric acid. Finally, the concentrate of the dihydroxybenzenes was analyzed by GC and GC-MS. Figure 1.51 represents a diagram of the analytical procedure developed for the determination of catechols. To obtain chromatograms similar to that shown in Figure 1.52, the following instrumentation was used: For the analysis, a Hewlett-Packard model 5710A gas chromatograph equipped with a flame ionization detector and a recording integrator was used. Mass spectra were determined on a Hewlett-Packard model 5710-598A integrated GC-MS instrument. For the separation of the catechol concentrate aiming at distinct identification of all catechols, a 4-m by 4-mm-i.d. glass column packed with 10% OV-17 on 80-100 mesh Gas Chrom Q (Fig. 1.52) was used. For routine catechol analysis, separations were performed with a 2-m by 4-mm-i.d. glass column packed with 10% UCW-98 on 80-100 mesh Gas Chrom Q. The chromatographic conditions were injector temperature, 200°C; column temperature, 140°C (OV-17) and 120°C (UCW-98); detector temperature,

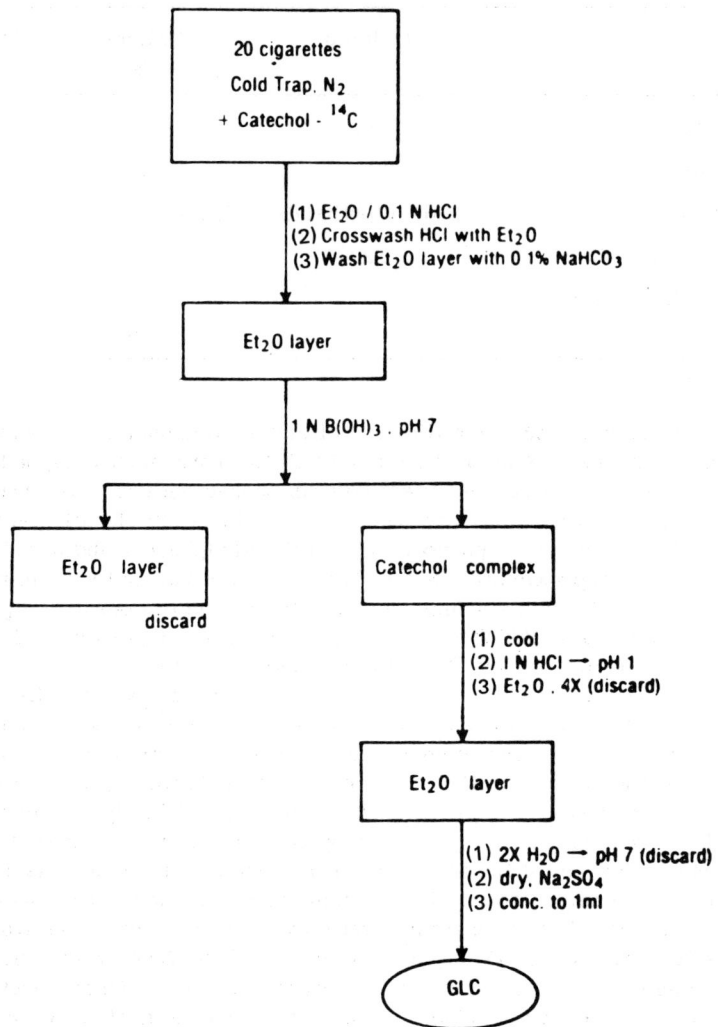

Figure 1.51. Analytical procedure for catechols in tobacco smoke. From Brunnemann et al. [90], courtesy of <u>Analytical Letters</u>.

220°C; helium carrier-gas flow rate, 40 ml/min. Maintaining these conditions, the retention times with the OV-17 column were approximately 6.90, 8.35, 10.00, 10.00, 11.32, 18.20, and 19.50 min for catechol, 3-methylcatechol, hydroquinone, resorcinol, 4-methylcatechol, 4-ethylcatechol, and 4-n-propylcatechol, respectively.

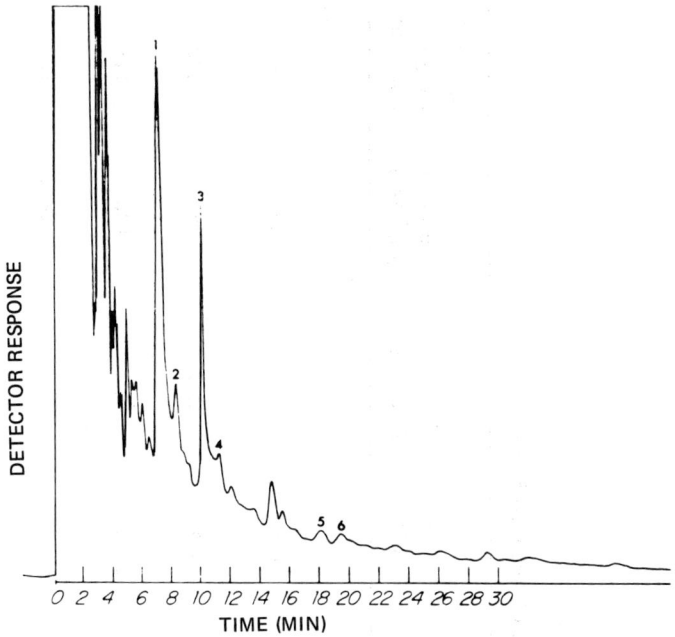

Figure 1.52. Typical gas chromatogram of catechols in tobacco smoke. Peak number 1: catechol; 2: 3-methylcatechol; 3: hydroquinone and resorcinol; 4: 4-methylcatechol; 5: 4-ethylcatechol; 6: 4-n-propylcatechol. From Brunnemann et al. [90], courtesy of Analytical Letters.

Tables 1.55 and 1.56 show a summary of results for catechol, TPM, and nicotine in the smoke of several commercial and experimental cigarettes.

F. Acidic Components [69, 70, 88, 153-156]

In 1958, Quin and Hobbs [153] analyzed the nonvolatile acid fraction of cigarette smoke, after conversion to a mixture of methyl esters with diazomethane, by GC techniques. Of the 16 esters detected, 11 had been identified. Lactic, glycolic, succinic, and malonic acids constituted about 75% of those identified, among which only succinic acid had been previously reported as a smoke constituent.

For the qualitative analysis of the methyl esters, a Perkin-Elmer model 154-B gas chromatograph equipped with a thermal conductivity detector and either of two column systems (A, two 1-m by 6-mm-o.d. glass U-shaped columns in series; B, a 3-m by 1/4-in.-o.d. copper coiled column) packed with the stationary liquid phases shown below, which were coated on 60-100

TABLE 1.55

Catechol in the Mainstream of Tobacco Smoke[a]

No.	Product	Weight (mg)	No. of puffs	TPM (dry) (mg/cig.)	Nicotine (mg/cig.)	Catechol (μg/cig.)	Catechol (μg/g tob. smoked)
	Commercial products						
1.	Cigarette A, 85 mm, NF	1170	9.7	31.7	1.86	272	318
2.	Cigarette B,[b] 70 mm, NF	1091	9.3	28.4	2.11	284	422
3.	Cigarette C, 85 mm, F	1060	10.0	20.3	1.35	136	178
4.	Cigarette C, 50 mm, column smoked with F	1060	8.5	18.7	0.82	118	166
5.	Cigarette C, 50 mm, column smoked w/o F	1060	7.4	25.0	1.59	182	256
6.	Cigarette D, 85 mm, F	970	9.9	19.3	1.36	191	256

7. Cigarette D, 50 mm column smoked with F	970	8.2	15.5	1.20	172	282
8. Cigarette D, 50 mm column smoked w/o F	970	7.1	25.5	1.72	278	456
9. Cigarette E, 70 mm, F	860	6.0	9.1	0.71	89	145
10. Cigarette F,[b] 70 mm, F	1130	9.0	17.3	1.38	174	238
11. Little cigar A, 85 mm, F	956	7.7	20.3	1.60	140	169
12. Little cigar B, 85 mm, F	1078	9.8	31.8	1.82	362	429
13. Cigarette F, 100 mm, F	1148	9.3	20.1	1.39	147	166
14. Nontobacco cigarette 100 mm, F	1598	16.0	29.3	——	276	224

[a] From Brunnemann et al. [90], courtesy of Analytical Letters.
[b] French cigarettes.
F = filter cigarettes; NF = nonfilter cigarettes.

TABLE 1.56
Catechol in the Mainstream of Tobacco Smoke[a]

No.	Product	Weight (mg)	No. of puffs	TPM (dry) (mg/cig.)	Nicotine (mg/cig.)	Catechol (μg/cig.)	Catechol (μg/g tob. smoked)
	Experimental cigarettes						
1.	Reference, IRI, 85 mm, NF	1100	11.0	36.1	2.65	328	409
2.	NCI #17,[b] 85 mm, NF	1158	8.4	19.1	0.93	163	193
3.	NCI, #25,[c] 85 mm, NF	1231	9.8	10.7	0.49	25	28
4.	Cigarette 1, 85 mm, F	1292	9.4	12.8	1.27	45	48
5.	Cigarette 2, 85 mm, F	1257	6.5	9.7	0.34	46	50
6.	Cigarette 3, 85 mm, F	1256	7.3	9.6	0.67	39	43
7.	Marijuana, 85 mm, NF	1115	10.7	20.7		188	231

[a] From Brunnemann et al. [90], courtesy of Analytical Letters.
[b] Standard experimental blend, RTS, slurry process, control.
[c] Standard experimental blend, RTS, slurry process, hexane–ethanol azeotrope extracted.

Celite 545 or 30-60 mesh C-22 Firebrick in a ratio of 1 to 3 (w/w), except for the use of a 1 to 5 ratio with liquid II.

Liquid Stationary Phase Used

Designation	Description
	(a) For 150 and 190° C separations
I	Flexol 8N8 [$(C_7H_{15}COOC_2H_4)_2NCOC_7H_{15}$]
II	Flexol R-2H (polyester)
III	Flexol 4GO (polyethylene glycol dioctanoate)
	(b) For 150° C separations
IV	Flexol A-26 (dioctyl adipate)
V	Flexol TOF (trioctyl phosphate)
VI	Tricresyl phosphate
VII	Didecyl phthalate
VIII	Dinonyl sebacate
	(c) For lactate-glycolate separations
IX	Carbowax 1500 (polyethylene glycol)
X	Carbowax 4000 (polyethylene glycol)

As shown in Tables 1.57, 1.58, and 1.59, the identity of the GC peaks was established by comparing their retention times on different columns with those of standards under the same conditions. On the other hand, some general conditions for the quantitative determination of ten new acids (as methyl esters) found in tobacco smoke by gas chromatography are listed in Table 1.60.

In 1965, Oakley et al. [155] showed that free formic and acetic acids and their salts in cigarette smoke may be quantitatively determined in the presence of formates and acetates after adsorbing the acids on a strong base anion-exchange resin in the fluoride form. The compounds are eluted and methylated in a single step with an anhydrous methanol-hydrochloric acid solution. The methyl esters were subsequently determined with an F&M model 500 gas chromatograph equipped with a thermal conductivity detector and a 4-ft by 1/4-in.-o.d. column packed with 20% Carbowax 20M on 60-80 mesh Chromosorb P. The GC conditions used in this study were injector temperature, 200°C; column temperature, 50°C; detector temperature, 225°C; helium carrier-gas flow rate, 100 ml/min.

TABLE 1.57

Identification of Methyl Esters by Comparative
Gas Chromatography at 150°C[a]

Columns and conditions								
Liquid phase[b]	Dinonyl sebacate		Didecyl phthalate		Flexol 4GO		Flexol A-26	
Solid support	Firebrick		Celite		Firebrick		Firebrick	
Length (m)	2		3		2		2	
Temp. (°C)	154		146		146		138	
He flow (ml/min)	9		11		11		16	
	Retention time (min)							
Ester of	Known	Smoke	Known	Smoke	Known	Smoke	Known	Smoke
Lactic } Glycolic }	7.0	7.1	6.9	6.9	5.0	5.0	6.0	6.0
Oxalic	8.2	8.1	11.4	11.5	7.2	7.2	8.7	8.6
Unknown		11.8		14.8		c		c
Malonic	13.3	13.1	17.1	17.3	11.8	11.6	13.9	13.8
Furoic	20.5	20.4	26.6	26.8	16.8	16.8	20.8	20.9
Levulinic	21.9	22.0	30.1	d	17.9	17.8	23.1	23.2
Succinic	23.5	23.8	30.3	30.7	19.6	19.7	25.6	25.6
Unknown		26.8		34.1		20.8		28.9

[a] From Quin and Hobbs [153], courtesy of <u>Analytical Chemistry</u>.
[b] On solid support 1 to 3 w/w.
[c] Not detected under these conditions.
[d] Not resolved from succinate peak.

TABLE 1.58

Identification of Methyl Lactate and Glycolate[a]

	Columns and conditions			
Liquid phase[b]	Carbowax 4000		Carbowax 1500	
Solid support	Celite		Firebrick	
Length (m)	3		2	
Temp. (°C)	123		122	
He flow (ml/min)	13		20	
	Retention time (min)			
	Known	Smoke	Known	Smoke
Methyl lactate	21.0	21.1	15.2	15.2
Methyl glycolate	29.0	28.9[c]	22.0	21.9[d]

[a] From Quin and Hobbs [153], courtesy of <u>Analytical Chemistry</u>.
[b] On solid support 1 to 3 w/w.
[c] Combined with dimethyl oxalate.
[d] Well separated from dimethyl oxalate peak at 19.5 min.

With the above GC conditions, the retention times for methyl formate and acetate were 55 and 78 sec, respectively, whereas the methanol solvent peak appeared at about 140 sec and was not completely eluted for 15 to 20 min. By adding known quantities of aqueous solutions of formic and acetic acids to distilled water, to filter pads containing smoke particulate matter, and to water traps before smoking, recoveries of 95 to 105% of each acid were obtained. At the 95% confidence level, it was reported that the relative standard deviation of the method for an individual determination of the acids in cigarette smoke was ±18% for formic acid and ±14% for acetic acid.

Wynder and Hoffmann [88] analyzed steam-volatile acids in cigarette smoke condensate by gas chromatography without derivatization. In their method, aliquots of the final acid solution were injected into a Perkin-Elmer model 154D gas chromatograph equipped with a thermal conductivity detector and a 2-m by 6-mm-o.d. stainless steel column packed with 5% Trimer acid on Teflon 6. With the injector and column temperatures maintained at 150 and 100°C, respectively, and the helium carrier gas set at 50 ml/min, the retention times for formic, acetic, and propionic acids were 2.4, 5.2, and 10.9 min, respectively.

TABLE 1.59

Identification of Methyl Esters by Comparative
Gas Chromatography at 190°C[a]

	\multicolumn{6}{c}{Columns and conditions}					
Liquid phase[b]	\multicolumn{2}{c}{Flexol 8N8}	\multicolumn{2}{c}{Flexol 4GO}	\multicolumn{2}{c}{Flexol R-2H}
Liquid to solid, w/w	\multicolumn{2}{c}{1 to 3}	\multicolumn{2}{c}{1 to 3}	\multicolumn{2}{c}{1 to 5}
Temp. (°C)	\multicolumn{2}{c}{189}	\multicolumn{2}{c}{187}	\multicolumn{2}{c}{192}
He flow (ml/min)	\multicolumn{2}{c}{18}	\multicolumn{2}{c}{18}	\multicolumn{2}{c}{8}
Ester of	\multicolumn{6}{c}{Retention time (min)}					
	Known	Smoke	Known	Smoke	Known	Smoke
Unknown		7.0		7.6		7.1
Glutaric	8.1	8.0	8.1	8.2	8.0	8.0
Unknown		9.4		9.4		9.6
Adipic	10.5	10.5	12.4	12.2	12.0	12.0
Malic	13.1	13.1	13.2	13.3	14.1	14.3
Unknown		17.8[c]		16.3[d]		11.7[e]
Phthalic	24.8[c]	25.5[c]	25.0[d]	25.8[d]	16.5[e]	16.7[e]

[a] From Quin and Hobbs [153], courtesy of *Analytical Chemistry*.
[b] On firebrick in glass columns.
[c] At He flow 34 ml/min.
[d] At He flow 38 ml/min.
[e] At He flow 26 ml/min.

In 1974, Guerin et al. [156] determined by GC and GC-MS palmitic, oleic, linoleic, linolenic, and stearic acids (as their silyl derivatives) in cigarette smoke. Equally applicable to analyses of smoke condensate and particulate matter, a 0.4-ml aliquot of the pyridine extract of either smoke condensate or particulate matter (see original paper for details) is transferred to a 3.5-ml capacity septum-capped vial and 0.2 ml of BSTFA-1% TMCS [bis(trimethylsilyl)trifluoroacetamide-trimethylchlorosilane] is added. The vial is heated at 65°C for 30 min, cooled to room temperature, and a 5-μl aliquot is then withdrawn and injected into the gas chromatograph.

TABLE 1.60

Some General Conditions for Quantitative Determination
of Esters by Gas Chromatography[a]

Methyl ester of	Column Liquid[b]	Length (m)	Temp. (°C)	Helium (ml/min)
Glycolic	IX	3	120	20
Lactic	X[c]	3	100	13
Oxalic	VII	3	110	13
Malonic	IV	2	115	13
Furoic				
Levulinic	IV	2	130	13
Succinic				
Glutaric	II[d]	3	170	13
Adipic				
Malic	II[d]	2	185	8
Phthalic	II[d]	2	185	26

[a] From Quin and Hobbs [153], courtesy of <u>Analytical Chemistry</u>.
[b] On Firebrick, 1 to 3 w/w.
[c] On Celite, 1 to 3 w/w.
[d] On Firebrick, 1 to 5 w/w.

GC separations were performed with a Tracor model 220 gas chromatograph equipped with a flame ionization detector and a 6-ft by 0.25-in.-o.d. glass Pyrex column packed with 4% OV-101 on 80-100 mesh Chromosorb G(HP). The other GC conditions employed were injector temperature, 210°C; column temperature, 190°C; detector temperature, 230°C; helium carrier-gas flow rate, 70 ml/min. Operated as indicated, the retention times of the silyl derivatives of palmitic, oleic, linoleic, linolenic, and stearic acids were approximately 11.60, 20.20, 20.20, 20.20, and 23.00 min, respectively. As shown in the chromatogram, the esters of oleic, linoleic, and linolenic acids were not resolved.

On the other hand, mass spectra were obtained with a Varian model 1200 gas chromatograph interfaced with a single-stage mass spectrometer with a 30-cm radius, 90°-sector magnet operated at 4 kV accelerating voltage and 25 eV ionizing voltage. With this integrated GC-MS unit, Guerin et al. used a 60-cm by 4-mm-i.d. glass column packed with 3% OV-101 on 80-100 mesh Chromosorb G(HP). Whereas the column temperature was programmed from 170 to 290°C at 6°C/min, all tubing, separator, and source temperatures were maintained at 250°C. Also capable of obtaining mass scans (m/e 12 to 400) in about 5 sec, the helium flow rate of 30 ml/min was satisfactory. With regard to their mass spectral studies, they noted that:

> The first GC peak gave a mass spectrum agreeing well with that from known TMSi-palmitic acid. The M^+ (m/e 328) and (M^+ - 15)(m/e 313) were prominent. The remainder of this spectrum, as well as those from other TMSi fatty acids, appeared to be common for all, apparently derived from fragments which were the same for all acids.
> The composite peak showed the mass spectra of TMSi-oleic (M^+, 354), linoleic (M^+, 352), and linolenic (M^+, 350) acids, in good agreement with the spectra of the known materials. The last GC peak proved to be the TMSi-stearic acid as expected. The M^+ at m/e 356 and (M^+ - 15) at m/e 341 were prominent.

With regard to the quantitative aspects of the procedure, C-1-^{14}C-palmitic, oleic, linoleic, linolenic, and stearic acids were added to smoked filter pads and to condensate to test the extraction efficiency. Essentially 100% recovery was obtained with overnight soaks. Table 1.61 summarizes the replicate determinations of free fatty acids in the total particulate matter from a 1R1 cigarette, the data being generally comparable to that presented for TPM by Hoffmann et al. [70] for the blended cigarette.

G. Polynuclear Aromatic Hydrocarbons and Related Compounds [157-176]

In 1967, Carugno and Rossi [160] developed a GC procedure for the detection of polynuclear hydrocarbons in cigarette smoke as well as their identification using glass capillary columns with flame ionization or electron capture detectors, the use of capillary columns affording the opportunity of separating a large number of polynuclear hydrocarbons from complex mixtures.

In their investigation, a Carlo Erba model C gas chromatograph equipped with a flame ionization or electron capture detector, a linear programmer, and a 65-m by 0.30-mm-i.d. glass capillary column coated with SE-52 was used. The other chromatographic conditions employed were injector temperature, 320°C; column temperature, either isothermal at 200°C or programmed from 100 to 300°C at 1.8°C/min; nitrogen carrier-gas flow rate, 1 ml/min. The retention times of 38 polynuclear hydrocarbons resolved under isothermal column conditions are listed in Table 1.62 together with

TABLE 1.61

Replicate Determinations of Free Fatty Acids in the
Total Particulate Matter from a 1R1 Cigarette[a]

		Weight % of TPM			
Port no.	Cigs. per port	Palmitic	Oleic, linoleic, linolenic	Stearic	Total
1	4	0.37	0.56	0.13	1.06
2	4	0.35	0.53	0.14	1.02
3	4	0.38	0.55	0.13	1.06
4	4	0.35	0.54	0.12	1.01
5	4	0.37	0.53	0.12	1.02
6	4	0.39	0.57	0.12	1.08
7	4	0.38	0.55	0.14	1.07
Average		0.37	0.55	0.13	1.05
Rel. std. dev. (%)		2.70	1.80	5.40	2.40
CI_{95}		±0.01	±0.01	±0.01	±0.02

[a] From Guerin et al. [156], courtesy of <u>Analytical Chemistry</u>.

the retention indices of these and other compounds separated and identified by column temperature programming. The utility and applicability of the electron capture detector for selective determinations was stressed, considering the fact that many polynuclear hydrocarbons show a considerable affinity for free electrons.

Using an improved polynuclear aromatic hydrocarbon (PAH) separation procedure for cigarette smoke condensate, Snook et al. [170] obtained GC data with a Hewlett-Packard model 5750 gas chromatograph equipped with a 10-ft by 1/8-in. stainless steel column packed with 5% Dexsil 300 on 100-120 mesh Chromosorb W-AW. The following GC conditions were used: injector temperature, 300°C; column temperature, programmed from 100 to 325°C at 2°C/min; flame ionization detector temperature, 350°C; helium carrier-gas flow rate, 48 ml/min. Compounds were identified by GC retention data, with internal spiking techniques, as well as by UV and GC-MS methods.

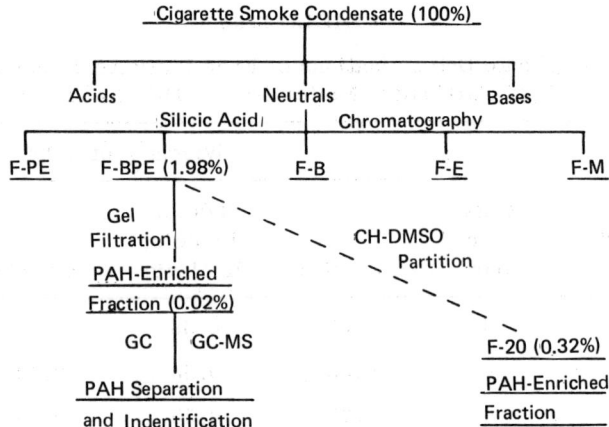

Severson et al. [172] also used a four-step method for the reproducible analysis of PAH components in small quantities in cigarette smoke condensate using the procedure of Snook et al. [170] with slight modifications. PAH were isolated from as little as 1 g of cigarette smoke condensate by solvent partition, column chromatography, and analytical gel filtration as outlined below:

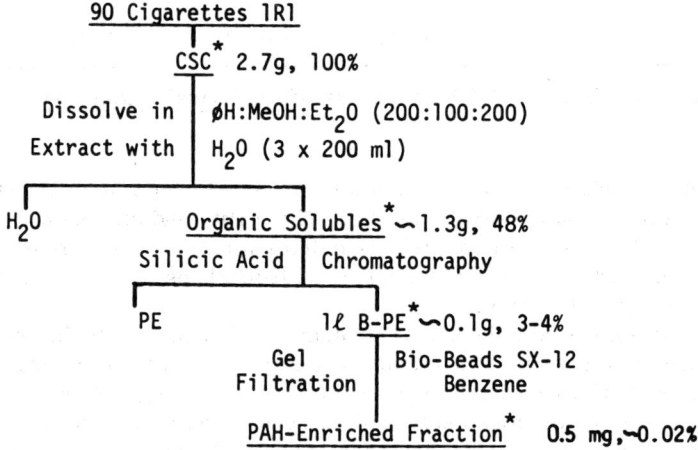

The isolate from the gel filtration step was then analyzed by gas chromatography. In this study, Severson et al. used a Hewlett-Packard model 5830A gas chromatograph equipped with a flame ionization detector and a 15-ft by 1/8-in. stainless steel column packed with 5% Dexsil 300 coated on 100-120 mesh Chromosorb W-AW. Using the specified conditions (injector temperature, 290°C; column temperature, held isothermally at 90°C

TABLE 1.62

Retention Times and Retention Indices of Some Polynuclear Hydrocarbons[a]

Compound	Ret. index[b]	Ret. time (min)[c]
Naphthalene	1150	0.20
Azulene	1292	0.39
Diphenyl	1373	0.47
Acetylnaphthalene	1437	0.67
Acenaphthene	1458	0.75
Fluorene	1555	1.07
2-Methylfluorene	1673	1.07
1-Methylfluorene	1677	1.07
trans-Stilbene	1686	1.59
Phenanthrene	1741	2.04
Anthracene	1750	2.12
2-Methylanthracene	1870	3.10
4,5-Methylenephenanthrene	1875	3.10
1-Methylphenanthrene	1890	3.22
9-Methylanthracene	1920	3.70
Fluoranthene	2020	5.12
Pyrene	2070	6.06
1,2-Benzofluorene	2179	8.23
2,3-Benzofluorene	2195	8.65
3,4-Benzofluorene	2195	8.65
1-Methylpyrene	2215	9.53
3-Methylpyrene	2220	9.77
Benzo(g,h,i)fluoranthene	2326	14.10
3,4-Benzophenanthrene	2332	14.10
1,2-Benzoanthracene	2389	16.90

(continued)

TABLE 1.62 (continued)

Compound	Ret. index[b]	Ret. time (min)[c]
Chrysene	2395	17.50
Triphenylene	2395	17.50
Naphthacene	2425	18.90
7-Methyl-1,2-benzanthracene	2575	29.45
2,3-Benzofluoranthene	2700	43.40
3,4-Benzofluoranthene	2700	43.40
11,12-Benzofluoranthrene	2706	44.10
7,12-Dimethyl-1,2-benzanthracene	2713	44.80
1,2-Benzopyrene	2760	51.30
3,4-Benzopyrene	2773	53.25
Perylene	2800	56.90
3-Methylcholanthrene	2906	78.00
20-Methylcholanthrene	2906	78.00
1,2,7,8-Dibenzanthracene	3078	
1,2,3,4-Dibenzanthracene	3114	
1,2,5,6-Dibenzanthracene	3114	
Benzotetraphene	3136	
Benzo(g,h,i)perylene	3150	
Picene	3150	
Anthanthrene	3186	
1,2,3,4-Dibenzopyrene	3477	
Coronene	3544	
1,2,4,5-Dibenzopyrene	3567	
3,4,9,10-Dibenzopyrene	3600	
3,4,8,9-Dibenzopyrene	3620	

[a] Adapted from Carugno and Rossi [160].
[b] Programmed temperature (100 to 300°C at 1.8°C/min).
[c] Isothermal at 200°C.

for 10 min, then programmed to 325°C at 2°C/min where it was held isothermally again for 45 min; detector temperature, 350°C; helium carrier-gas flow rate at 90°C, 40 ml/min), the gas chromatogram of the major constituents in the GF-C fraction derived from experimental cigarettes is shown in Figure 1.53. Peaks were identified as follows: (1), naphthalene; (2), 2-methylnaphthalene; (3), 1-methylnaphthalene; (4), 1- and 2-methylnaphthalene; (5), 2,6- and 2,7-dimethylnaphthalene as well as 1-vinylnaphthalene; (6), 1,3- and 1,6-dimethylnaphthalene as well as 2-vinylnaphthalene; (7), 2,3-, 1,4-, and 1,5-dimethylnaphthalene; (8), acenaphthylene and 1,7-dimethylnaphthalene; (9), acenaphthene and 1,8-dimethylnaphthalene; (10), dibenzofuran; (11), 1-methylacenaphthylene; (12), fluorene and methylacenaphthylene; (13), 9-methylfluorene and methylacenaphthylene; (14), methylacenaphthylene and methylacenaphthene; (15), benz(f)indene; (16), 2- and 3-methylfluorene and dimethylacenaphthylene; (17), 1- and 4-methylfluorene and dimethylacenaphthylene; (18), phenanthrene; (19), anthracene; (20), 2- and 3-methylphenanthrene and 2-methylanthracene; (21), 1- and 9-methylphenanthrene and 1-methylanthracene; (22-25), dimethylphenanthrenes and dimethylanthracenes; (26) fluoranthene; (27), acephenanthrylene; (28), pyrene; (29), 8-methylfluoranthrene and 1,2-benzofluorene; (30), 1- and 2-methylfluoranthene as well as 2,3- and 3,4-benzofluorene; (31), 2-methylpyrene; (32), 1- and 4-methylpyrene; (33-36), dimethylfluoranthenes, dimethylpyrenes, and methylbenzofluorenes; (37); benzo(g,h,i)fluoranthene, dimethylpyrenes, dimethylfluoranthenes, and methylbenzofluorene; (38), 1,2-benzanthracene, chrysene, and triphenylene; (39), 2- and 3-methylchrysenes, methyl-1,2-benzanthracene, methyltriphenylene, and methyl-(g,h,i)-fluoranthene; (40), 1-, 4-, and 6-methylchrysenes, methyl-1,2-benzanthracenes, and methyltriphenylenes; (41,42), dimethylchrysenes, dimethyl-1,2-benzanthracenes, and dimethyltriphenylenes; (43), benzo-(b,j,k)fluoranthenes; (44), benzo(a)fluoranthene; (45), benzo(a,e)pyrenes; (46), perylene, and methylbenzo(b,j,k)-fluoranthenes; (47), methylbenzo-(a,e)pyrenes; (48), pentatriacontane.

The procedure described above was used to screen levels of PAH from CSC of different origins and to evaluate the effects of filters, tobacco types, tobacco processing methods, and a chemical tobacco additive.

In 1976, Lee et al. [174] carried out with the combination of chromatographic and spectral methods comparative analyses of the polynuclear aromatic hydrocarbon fractions of tobacco and marijuana smoke condensates. The PAH fraction components comprised the largest known group which, as a class, is credited with the major carcinogenic activity of smoke condensates. The constituents of selectively enriched extracts, further purified and fractionated by a combination of column chromatography and high-resolution liquid chromatography, were analyzed with a Hewlett-Packard model 5980A integrated GC-MS (dodecapole) instrument equipped with a 11.0-m by 0.26-mm-i.d. glass capillary column coated with SE-52 liquid stationary phase. A typical chromatogram of the PAH fraction of smoke condensate

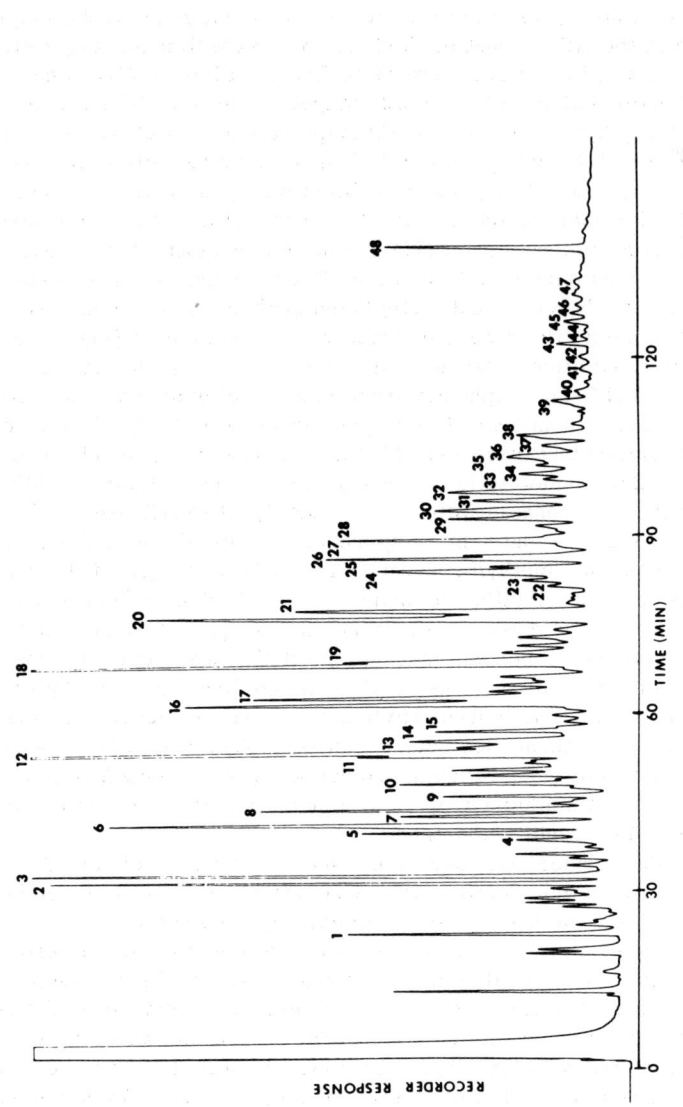

Figure 1.53. Gas chromatogram of PAH constituents in GF-C fraction derived from experimental cigarettes. From Severson et al. [172], courtesy of Analytical Chemistry.

Chemical Composition of Tobacco Smoke

from standard research tobacco is shown in Figure 1.54; the compounds identified by integrated GC-MS, proton NMR, and standard compound retention data are listed in Table 1.63.

Among the aromatic constituents in cigarette smoke identified by preparative gas chromatography (5-ft by 1-in. column packed with 20% Reoplex coated on 30-60 mesh Firebrick maintained at 180°C with a nitrogen carrier-gas flow rate of 200 ml/min) by Cook et al. [158] were naphthalene, 1-methylnaphthalene, 2-methylnaphthalene, 1,6-, 1,8-, 2,6-, and 2,7-dimethylnaphthalene, and 1,3,6-trimethylnaphthalene.

On the other hand, a 15-ft by 1-in. column filled with 20% silicone elastomer E-301 on Firebrick and held isothermally at 120°C with a carrier-gas flow rate of 200 ml/min enabled these analysts to separate and identify o-xylene, m-xylene, ethylbenzene, 2,4-dimethyl-4-vinylcyclohexene, styrene, 3-ethyltoluene, and 4-ethyltoluene.

In 1976, Schmeltz et al. [171] developed a method for the qualitative and quantitative analysis of naphthalenes in cigarette smoke that consists of three solvent distribution steps, column chromatography, and finally integrated gas chromatography-mass spectrometry (see separation scheme below).

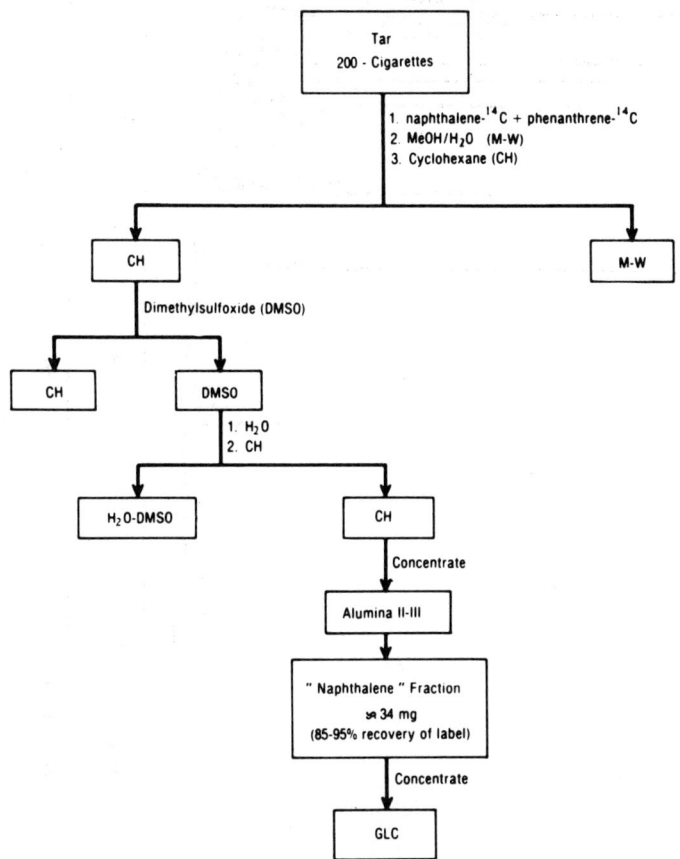

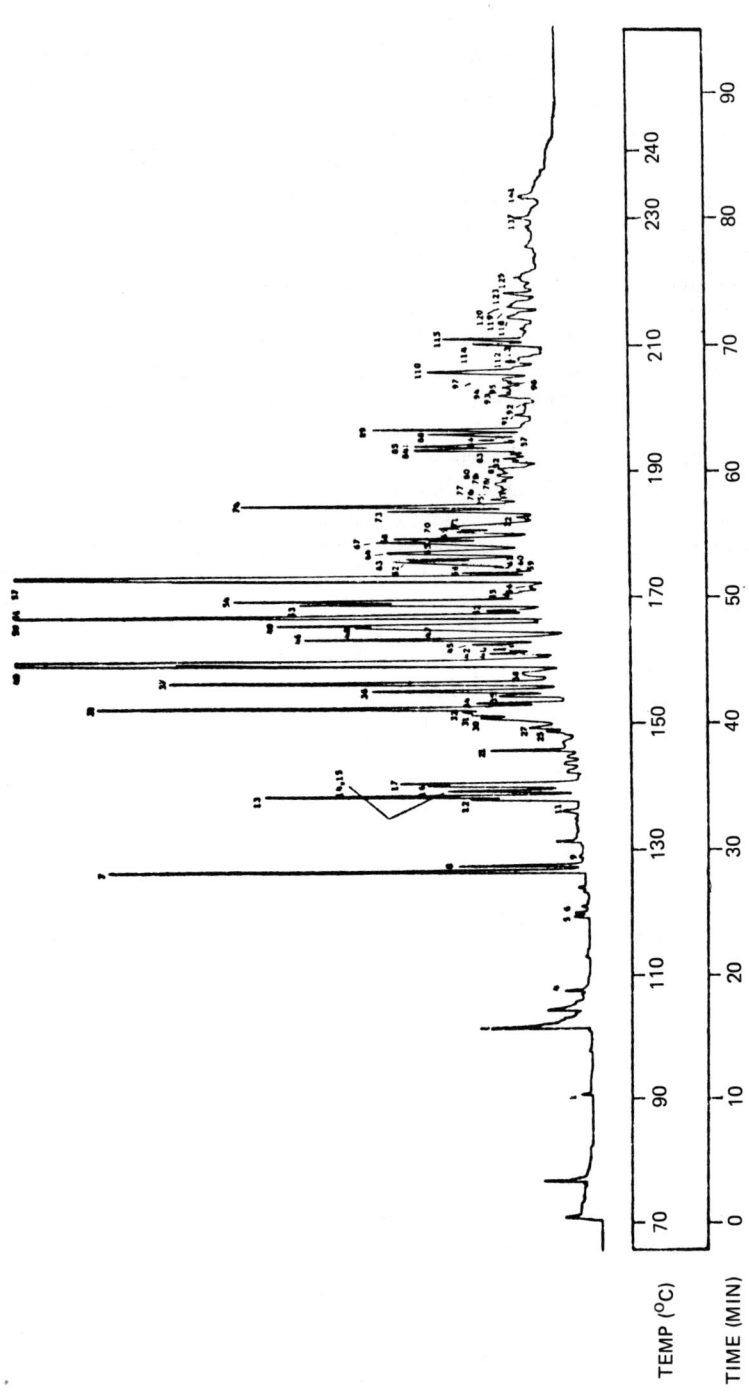

Figure 1.54. Capillary-column gas chromatography of the polynuclear aromatic hydrocarbon fraction of smoke condensate from 100 g of standard research tobacco. From Lee et al. [174], courtesy of Analytical Chemistry.

Using naphthalene-^{14}C and phenanthrene-^{14}C as internal standards, the "naphthalene fraction" from tobacco smoke was analyzed with a Hewlett-Packard 7620A gas chromatograph equipped with a flame ionization detector and a 3-m by 2.16-mm stainless steel column packed with 10% UCW-98 on Chromosorb W. After an initial hold for 32 min at 100°C, the temperature was programmed to 160°C at 2°C/min. The other GC conditions used were injector temperature, 250°C; detector temperature, 250°C; helium carrier-gas flow rate, 32 ml/min. For isothermal runs, conditions were as above, but the column temperature was 130°C.

GC-MS profiles were recorded on a Hewlett-Packard 5980A data system.

Using the above conditions and temperature programming, the retention times of various naphthalenes identified in the "naphthalene fraction" of cigarette smoke (as characterized by GC-MS) are listed in Table 1.64.

Applied to the analysis of the smoke derived from a U.S. blended 85-mm cigarette without filter tip, GC data showed that the smoke contained 2.8 µg of naphthalene, 1.2 µg of 1-methylnaphthalene, 1.0 µg of 2-methylnaphthalene, 33 ng of 1- and 2-ethylnaphthalene, and 220 ng of dimethylnaphthalenes. Furthermore, tri-, tetra-, and pentamethylnaphthalenes were present. The recovery obtained for various naphthalenes varied between 85 and 95%.

Davis [161] and Janini et al. [175] developed GC methods for the determination of benzo(a)pyrene in cigarette smoke.

In the procedure advocated by Davis, sample preparation consisted of the following: (1) smoke condensate collection, (2) addition of the perylene internal standard, (3) partitioning between solvent pairs, (4) silica gel chromatography, (5) acid treatment of fraction residues, and (6) GC analysis of the benzo(a)pyrene (BaP) fraction. The analysis for BaP was carried out with a Beckman GC-5 gas chromatograph equipped with an electron capture detector, a dual-flame ionization detector and a 9-ft by 1/8-in.-o.d. stainless steel column packed with 3% SE-30 on 60-80 mesh Chromosorb W. When operated under the specified conditions (injector temperature, 350°C; column temperature, 245 to 250°C; EC detector temperature, 275 to 280°C; carrier-gas flow rate not specified), the retention time of BaP peak shown in a typical chromatogram was approximately 4.66 min with as little as 1 ng of BaP easily measured with the electron capture detector. Inasmuch as perylene and benzo(e)pyrene have retention times very close to that of BaP, it was shown experimentally that perylene had a low electron affinity; relative to BaP = 100, its value was less than one. Similarly, the value of BeP was about 20. Based on perylene recovered (determined by UV spectrometry), recoveries of BaP ranged from 70 to 90%, depending on whether fractionation method A or B was employed.

Janini et al. [175] performed the gas chromatographic BaP analysis on a nematic crystal as liquid stationary phase [N,N'-bis(p-methoxybenzylidene)-α,α'-bi-p-toluidine]. Having isolated the PAH fraction from cigarette smoke condensate by conventional extraction and column chromatography, Janini et al. determined BaP with a Hewlett-Packard model 7610 gas

TABLE 1.63

PAH Identified in Marijuana and Tobacco Smoke Condensates[a]

Peak no.	Marijuana (µg/100 cigarettes)	Tobacco (µg/100 cigarettes)	MW	Name
1	6.3	0.3	131	Methylindole
2	3.2		145	Ethylindole[b]
3	1.0		168	Dibenzofuran
4	1.4	0.5	166	Methylacenaphthylene
5	0.8	0.3	180	2-Methylfluorene
6	1.4	0.3	180	1-Methylfluorene
7	8.9	8.5	178	Phenanthrene
8	3.3	2.3	178	Anthracene
9	0.4	0.1	196	Ethylmethylbiphenyl[c]
10	6.5		167	Carbazole
11	0.8	0.4	192	3-Methylphenanthrene
12	2.6	2.0	192	2-Methylphenanthrene
13	5.3	5.6	192	2-Methylanthracene
14	3.2	2.4	192	
15	3.2	2.4	190	4H-Cyclopenta[def]phenanthrene

16	2.9	2.7	192	9-Methylphenanthrene
17	4.2	3.2	192	1-Methylphenanthrene
18	3.4		181	Methylcarbazole
19	3.6		181	Methylcarbazole
20	5.1		181	Methylcarbazole
21	3.1	1.6	204	Methyl-4H-cyclopenta[def]phenanthrene
22	3.0		181	Methylcarbazole
23	0.3	0.4	206	Ethylphenanthrene or ethylanthracene[b]
24	0.7	0.6	206	Ethylphenanthrene or ethylanthracene[b]
25	0.6	0.5	206	Ethylphenanthrene or ethylanthracene[b]
26	0.7	0.5	206	Ethylphenanthrene or ethylanthracene[b]
27	1.5	0.8	206	Ethylphenanthrene or ethylanthracene[b]
28	0.7	0.6	206	Ethylphenanthrene or ethylanthracene[b]
29	0.6	0.7	206	Ethylphenanthrene or ethylanthracene[b]
30	3.0	1.6	206	Ethylphenanthrene or ethylanthracene[b]
31	4.3	1.8	206	Ethylphenanthrene or ethylanthracene[b]
32	2.5	1.9	206	Ethylphenanthrene or ethylanthracene[b]
33	8.9	8.3	202	Fluoranthene
34	0.6	1.6	206	Ethylphenanthrene or ethylanthracene[b]
35	2.9	1.2	202	Benzacenaphthylene

(continued)

TABLE 1.63 (continued)

Peak no.	Marijuana (μg/100 cigarettes)	Tobacco (μg/100 cigarettes)	MW	Name
36	4.9	3.4	206	Ethylphenanthrene or ethylanthracene[b]
37	6.6	6.8	202	Pyrene
38	1.9	0.7	218	Ethyl-4H-cyclopenta[def]phenanthrene[b]
39	2.2	0.7	218	Ethyl-4H-cyclopenta[def]phenanthrene[b]
40		X	282	p,p'-TDEE
41	1.3	1.4	218	Ethyl-4H-cyclopenta[def]phenanthrene[b]
42	1.9	0.8	218	Ethyl-4H-cyclopenta[def]phenanthrene[b]
43	0.6	0.5	220	Ethylmethylphenanthrene or ethylmethylanthracene[c]
44	1.4	0.7	220, 218	Ethylmethylphenanthrene or ethylmethylanthracene[c]
				Ethyl-4H-cyclopenta[def]phenanthrene[b]
45	2.4	1.6	218	Ethyl-4H-cyclopenta[def]phenanthrene[b]
46	4.0	4.6	216	Methylfluoranthene
47	1.8	1.8	216	Methylfluoranthene
48	3.8	3.6	216	Methylfluoranthene
49	4.2	4.9	216	Benzo[a]fluorene
50	5.4	5.5	216	2-Methylpyrene and benzo[b]fluorene

Chemical Composition of Tobacco Smoke

51		X	318	o,p'-TDE
52	2.5	1.2	220	Ethylmethylphenanthrene or ethylmethylanthracene[c]
53	4.1	4.4	216	4-Methylpyrene
54	4.8	5.6	216	1-Methylpyrene
55	0.8	0.9	216	Methylfluoranthene
56	0.6	0.3	216	Methylfluoranthene
57		X	318	p,p'-TDE
58	1.1	1.5	230	Ethylfluoranthene or ethylpyrene[b]
59	0.3	0.5	230	Ethylfluoranthene or ethylpyrene[b]
60	0.5	0.9	230	Ethylfluoranthene or ethylpyrene[b]
61	1.1	1.0	230	Ethylfluoranthene or ethylpyrene[b]
62	2.1	2.4	230	Ethylfluoranthene or ethylpyrene[b]
63	2.1	2.4	230	Ethylfluoranthene or ethylpyrene[b]
64	2.5	2.7	230	Ethylfluoranthene or ethylpyrene[b]
65	1.4	1.8	230	Ethylfluoranthene or ethylpyrene[b]
66	1.7	1.6	230, 226	Ethylfluoranthene or ethylpyrene,[b] acefluoranthylene
67	2.4	3.0	230	Ethylfluoranthene or ethylpyrene[b]
68	2.3	2.6	230, 226	Ethylfluoranthene or ethylpyrene,[b] acepyrylene
69	1.2	1.4	230	Ethylfluoranthene or ethylpyrene[b]
70	1.6	1.7	230	Ethylfluoranthene or ethylpyrene[b]

(continued)

TABLE 1.63 (continued)

Peak no.	Marijuana (μg/100 cigarettes)	Tobacco (μg/100 cigarettes)	MW	Name
71	1.4	1.3	230	Ethylfluoranthene or ethylpyrene[b]
72	0.4	0.4	226, 230	Benzo[ghi]fluoranthene, ethylfluoranthene, or ethylpyrene[b]
73	3.3	2.6	228	Benz[a]anthracene
74	5.5	5.1	228	Chrysene
75	0.9	0.8	244	Ethylmethylfluoranthene or ethylmethylpyrene[c]
76	0.7	0.6	244	Ethylmethylfluoranthene or ethylmethylpyrene[c]
77	0.9	0.6	244	Ethylmethylfluoranthene or ethylmethylpyrene[c]
78	1.0	0.7	244	Ethylmethylfluoranthene or ethylmethylpyrene[c]
79	0.8	0.6	244	Ethylmethylfluoranthene or ethylmethylpyrene[c]
80	1.0	0.7	244	Ethylmethylfluoranthene or ethylmethylpyrene[c]
81	0.7	0.7	244	Ethylmethylfluoranthene or ethylmethylpyrene[c]
82	1.0	0.6	242	Methylchrysene or methylbenz[a]anthracene
83	1.0	0.5	242	Methylchrysene or methylbenz[a]anthracene
84	2.7	2.2	242	Methylchrysene or methylbenz[a]anthracene
85	2.1	2.2	242	Methylchrysene or methylbenz[a]anthracene

#				Name	MW
86	1.0	1.1		Methylchrysene or methylbenz[a]anthracene	242
87	0.9	0.7		Methylchrysene or methylbenz[a]anthracene	242
88	2.2	1.9		Methylchrysene or methylbenz[a]anthracene	242
89	2.7	2.9		Methylchrysene or methylbenz[a]anthracene	242
90	X				
91	0.5	0.5		Binaphthyl	254
92	0.5	0.3		Binaphthyl	254
93	0.8	0.7		Ethylchrysene or ethylbenz[a]anthracene[b]	256
94	0.6	0.6		Ethylchrysene or ethylbenz[a]anthracene[b]	256
95	1.0	0.7		Ethylchrysene or ethylbenz[a]anthracene[b]	256
96	0.5	0.6		Ethylchrysene or ethylbenz[a]anthracene[b]	256
97	1.5	0.7		Ethylchrysene or ethylbenz[a]anthracene[b]	256
98	0.7	0.7		Ethylchrysene or ethylbenz[a]anthracene[b]	256
99	0.4	0.3		Ethylchrysene or ethylbenz[a]anthracene[b]	256
100	0.7	0.7		Ethylchrysene or ethylbenz[a]anthracene[b]	256
101	0.6	0.6		Methylbinaphthyl	268
102	0.4	0.4		Methylbinaphthyl	268
103	0.4	0.3		Methylbinaphthyl	268
104	0.6	0.3		Methylbinaphthyl	268
105	0.3	0.3		Methylbinaphthyl	268

(continued)

TABLE 1.63 (continued)

Peak no.	Marijuana (μg/100 cigarettes)	Tobacco (μg/100 cigarettes)	MW	Name
106	0.3	0.6	270	Ethylmethylchrysene or ethylmethylbenz[a]anthracene[c]
107	0.3	0.4	270	Ethylmethylchrysene or ethylmethylbenz[a]anthracene[c]
108	0.4	0.4	282	Ethylbinaphthyl[b]
109	0.3	0.3	282	Ethylbinaphthyl[b]
110	3.0	2.1	252	Benzo[j]fluoranthene
111	1.1	1.2	252	Benzo[k]fluoranthene
112	1.1	0.7	252	Benzofluoranthene
113	0.7	0.5	252	Benzofluoranthene
114	1.8	1.3	252	Benzo[e]pyrene
115	2.9	1.7	252	Benzo[a]pyrene
116	0.9		252	Perylene
117	0.3	0.2	266	Methylbenzopyrene or methylbenzofluoranthene
118	0.8	0.6	266	Methylbenzopyrene or methylbenzofluoranthene
119	0.5	0.5	266	Methylbenzopyrene or methylbenzofluoranthene
120	0.6	0.6	266	Methylbenzopyrene or methylbenzofluoranthene
121	0.6	0.6	266	Methylbenzopyrene or methylbenzofluoranthene

Chemical Composition of Tobacco Smoke

122	1.2	0.6	266	Methylbenzopyrene or methylbenzofluoranthene
123	0.9	0.7	266	Methylbenzopyrene or methylbenzofluoranthene
124		0.6	266	Methylbenzopyrene or methylbenzofluoranthene
125	0.7	0.5	266	Methylbenzopyrene or methylbenzofluoranthene
126	0.5	0.5	266	Methylbenzopyrene or methylbenzofluoranthene
127	0.5	0.3	266	Methylbenzopyrene or methylbenzofluoranthene
128		0.2	266	Methylbenzopyrene or methylbenzofluoranthene
129	0.3	0.4	266, 280	Methylbenzopyrene, ethylbenzopyrene, or ethylbenzofluoranthene[b]
130	0.4	0.5	280	Ethylbenzopyrene or ethylbenzofluoranthene[b]
131	0.3	0.3	280	Ethylbenzopyrene or ethylbenzofluoranthene[b]
132		0.3	280	Ethylbenzopyrene or ethylbenzofluoranthene[b]
133		0.3	280	Ethylbenzopyrene or ethylbenzofluoranthene[b]
134	0.3		276	d
135	0.3		276, 278	d, dibenz[a,i]anthracene
136	0.6		276	d
137	1.0	0.3	276	d
138	0.3		276	d
139	0.3	0.6	278	Dibenz[a,h]anthracene or dibenz[a,c]anthracene
140	0.4	0.2	276	d
141	0.7	0.3	276	Benzo[ghi]perylene

(continued)

TABLE 1.63 (continued)

Peak no.	Marijuana (μg/100 cigarettes)	Tobacco (μg/100 cigarettes)	MW	Name
142	0.4		276	d
143	0.5		276	Anthanthrene
144	0.5		290	e
145	0.2		290	e
146	0.4		290	e
147	0.5		290	e
148	0.4		290	e
149	0.5		290, 302	e, dibenzopyrene
150	0.3		290, 302	e, dibenzopyrene
151	0.4		290	e
152	0.3		304, 306	Diphenylacenaphthylene, quaterphenyl
153	1.2		306	Quaterphenyl

[a] From Lee et al. [174], courtesy of Analytical Chemistry.
[b] Could also be dimethyl-.
[c] Could also be trimethyl- or propyl.
[d] Compounds with molecular weight 276 can be any of the following: indeno[1,2,3-cd]pyrene, indeno[1,2,3-cd]-fluoranthene, aceperylene, phenanthro[10,1,2,3-cdef]fluorene, acenaphth[1,2-a]acenaphthylene, dibenzo[b,mno]-fluoranthene. Further possibilities are the benzo derivatives of acepyrylene and acefluoranthylene.
[e] Compounds with molecular weight 290 are methyl derivatives of those with molecular weight 276.

TABLE 1.64

Retention Times of Compounds in Naphthalene Fraction[a]

Compound	Retention time (min)
1-Methylindene	11.25
Naphthalene	14.42
Ethylbenzofuran(s)	16.15
Ethylindene(s)	21.50
2-Methylnaphthalene	25.20
1-Methylnaphthalene	27.65
Biphenyl	41.40
2-Ethylnaphthalene	44.80
1-Ethylnaphthalene	44.80
2,6-Dimethylnaphthalene	46.80
2,7-Dimethylnaphthalene	46.80
1,3-Dimethylnaphthalene	48.80
1,6-Dimethylnaphthalene	48.80
1,7-Dimethylnaphthalene	48.80
Acenaphthylene	51.00
1,4-Dimethylnaphthalene	51.00
2,3-Dimethylnaphthalene	51.00
1,5-Dimethylnaphthalene	51.00
1,2-Dimethylnaphthalene	52.70
4-Methylbiphenyl	55.10
1,8-Dimethylnaphthalene	55.10
Trimethylnaphthalenes (5)	56.50-63.20

[a] Adapted from Schmeltz et al. [171].

chromatograph equipped with a dual flame ionization detector and a 6-ft by 2-mm-i.d. glass column packed with 2.5% BPhBT (nemantic liquid crystal) on 100-120 mesh Chromosorb W-HP. Using the specified GC conditions (injector temperature, 275° C; column temperature, 270° C; detector temperature, 275° C; helium carrier-gas flow rate, 20 ml/min), the retention times of phenanthrene, anthracene, fluoranthene, pyrene, benzo(m,n,o)-fluoranthene, triphenylene, chrysene, benzo(k)fluoranthene, benzo(e)pyrene, perylene, and BaP were approximately 0.42, 0.51, 1.18, 1.61, 3.56, 3.73, 5.25, 6.95, 13.90, 15.10, 19.17, and 24.75 min, respectively. Capable of detecting as little as 10 ng BaP in a 5-μl injection, the authors noted that the baseline separation of BaP from benzo(e)pyrene in a relatively short time is as yet unmatched by other separation techniques including GC using other liquid crystalline phases.

Hoffman et al. [163] isolated and determined by GC the 9-methylcarbazoles found in cigarette smoke. As noted by Hoffmann et al., "the most satisfactory separation of 9-methylcarbazoles were obtained at 210° C on a 3-mm by 2-m column filled with 5% OV-225 on Gas Chrom P. This column separates also carbazole from 9-methylcarbazoles. The separation of 9-methylcarbazole from 9-ethylcarbazole, however, was best achieved on a 3-mm by 2-m column with 10% OV-1 on Gas Chrom P. The retention times are given in Table 1.65. About 1 μg of 9-methylcarbazole reaches the full scale of a 1-mV recorder with an attenuation of 100. For the gas chromatographic isolation of the N-alkylcarbazoles from a concentrate from cigarette

TABLE 1.65

Retention Times of 9-Alkylcarbazoles[a]

Compound	Retention time (min)	
	Column A[b]	Column B[c]
9-Methylcarbazole	7.9	5.4
1,9-Dimethylcarbazole	13.6	8.6
2,9-Dimethylcarbazole	10.6	7.5
3,9-Dimethylcarbazole	10.6	7.5
4,9-Dimethylcarbazole	11.4	7.6
9-Ethylcarbazole	7.9	5.9

[a] Adapted from Hoffmann et al. [163].
[b] Column A: 2-m by 3-mm stainless steel, 5% OV-225, 210° C isotherm.
[c] Column B: 2-m by 3-mm stainless steel, 10% OV-1, 210° C isotherm.

smoke, a 1:4 glass splitter was installed so that 80% of the effluent was collected in a glass capillary. The effluent was rechromatographed for mass spectral analysis."

Typical EI mass spectra of 9-methylcarbazole, 4,9-dimethylcarbazole, and 9-ethylcarbazole are shown in Figure 1.55.

In 1978, Snook et al. [176] described a new method for the isolation of carbazoles as well as indoles. The four-step procedure involves water extraction, silicic acid chromatography, gel filtration chromatography, and gas chromatography. Silicic acid chromatography effectively separated the indoles and carbazoles from polynuclear aromatic hydrocarbons, compounds that would interfere in the indole/carbazole analysis. Gel filtration chromatography on Bio-Beads SX-12, in benzene, produced a relatively pure indole/carbazole isolate by an adsorption phenomena similar to that reported for PAH. The indole/carbazole isolate was sufficiently refined for identification of its components by GC-UV and GC-MS techniques. In addition to quantitatively determining carbazole by the proposed method, the identification of all three possible isomers of benzocarbazole and their alkyl derivatives was reported for the first time.

The gel filtration fractions were analyzed with a Hewlett-Packard model 5750 gas chromatograph equipped with a flame ionization detector and a 15-ft by 1/8-in. stainless steel column packed with 3% Dexsil 300 on 100-120 mesh Chromosorb W-HP. The other GC conditions used were injector temperature, 290° C; column temperature, held isothermally for 5 min at 90° C, then programmed to 325° C at 2° C/min, and finally constant at 325° C until all compounds of interest had been eluted; detector temperature, 350° C; helium carrier-gas flow rate, 40 ml/min. Compounds were identified by UV spectral analyses of preparative samples and by GC-MS spectral analyses on a Varian 1400 GC instrument interfaced with a DuPont 21-492 mass spectrometer.

Gel fractions 41-55 which contained the indole/carbazole components yielded the GC chromatogram shown in Figure 1.56, the peaks of which are identified below:

Indole/Carbazole Identifications

Peak no.[a]	Compound[b]
1	Indole
2	Skatole
3	3-Ethylindole
4,5	Dimethylindole
6	Propylindole

(continued)

Peak no.[a]	Compound[b]
7,8	Ethylmethylindole
9-11	Trimethylindole
12	Propylmethylindole
13,14	Ethyldimethylindole
15	Tetramethylindole
16	Tetramethylindole/propyldimethylindole
17	Propylethylindole
18	Carbazole
19	1-Methylcarbazole
20	3-Methylcarbazole
21	2-Methylcarbazole
22	4-Methylcarbazole
23	Ethylcarbazole
24	Dimethylcarbazole
25	Dimethylcarbazole/3-phenylindole
26	Dimethylcarbazole
27	Dimethylcarbazole
28,29	Dimethylcarbazole/trimethylcarbazole
30	Trimethylcarbazole
31,32	Trimethylcarbazole/methylphenylindole
33,34	Trimethylcarbazole/tetramethylcarbazole
35	Benzo[a]carbazole
36	Benzo[b]carbazole
37	Benzo[c]carbazole
38,39	Methylbenzocarbazole
40,41	Dimethylbenzocarbazole

[a] Peak numbers refer to Figure 1.56.
[b] Major component.

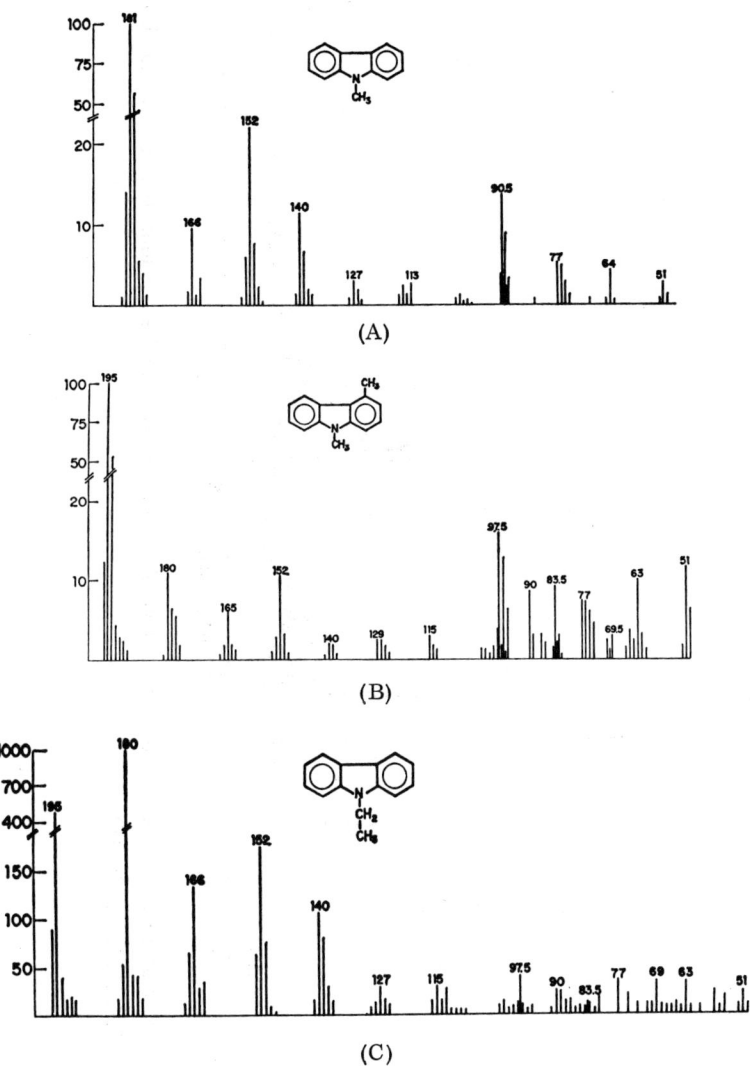

Figure 1.55. Mass spectra of 9-methylcarbazole (A), 4,9-dimethylcarbazole (B), and 9-ethylcarbazole (C). Adapted from Hoffmann et al. [163].

204   Natural, Pyrolytic, and Carcinogenic Products of Tobacco

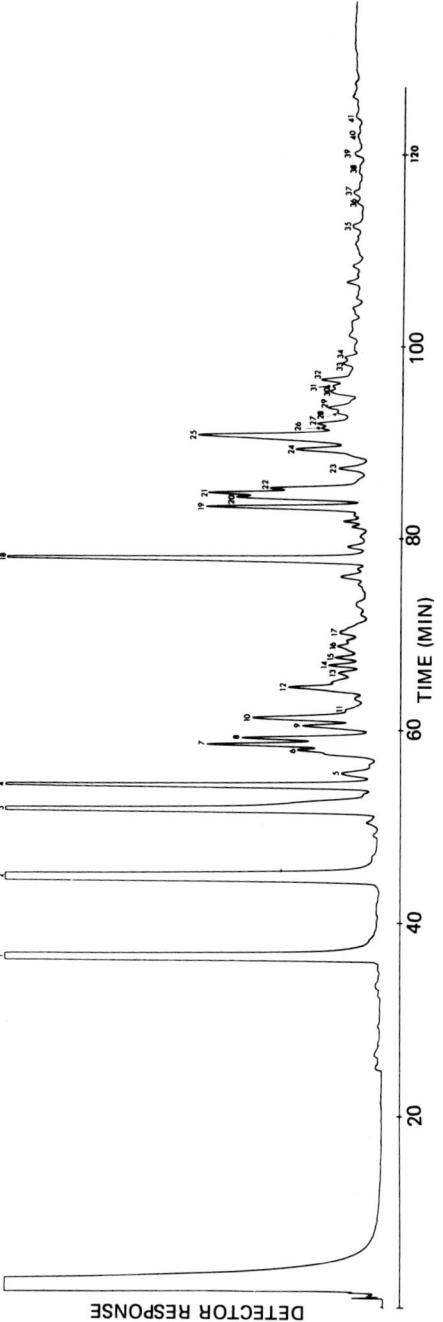

Figure 1.56. Gas chromatogram of the indoles and carbazoles in cigarette smoke condensate. From Snook et al. [176], courtesy of Analytical Chemistry.

In 1970, Hoffmann and Rathkamp [164] reported a portion of a study on the identification of tumorigenic agents in cigarette smoke. For the determination of 1-alkylindoles, the nonvolatile particulate matter of the smoke collected in solvent was distributed between two pairs of solvents and the resulting concentrate chromatographed on alumina and subsequently analyzed by gas chromatography. With their method, it was found that the mainstream smoke of an 85-mm U.S. cigarette without filter tip contained 420 ng of 1-methylindole, 126 ng of 1-ethylindole, 32 ng of 1,2- and 1,7-dimethylindole, 380 ng of 1,3-dimethylindole, and 99 ng of 1,4-, 1,5-, and 1,6-dimethylindole. The isolated components were identified by retention times, ultraviolet spectra, and mass spectra.

For the analysis, a Perkin-Elmer model 800 gas chromatograph equipped with a dual flame ionization detector and a 5-m by 3-mm column packed with 5% XE-60 on Gas Chrom P (no other GC conditions given in text) was used. With this column, gas chromatograms of concentrates of 1-alkylindoles from cigarette smoke were shown and the retention times of 1-alkylindoles as well as non-N-alkylated indoles were given as listed in Table 1.66.

The characteristic MS peaks in mass spectra of 1-methylindole, 1-ethylindole, 1,3-dimethylindole, 1,2-dimethylindole, 1,7-dimethylindole, and 1,4-dimethylindole are shown in Figure 1.57.

Finally, Hoffmann and Rathkamp [168] also described a method in 1972 for the determination of fluorenes in cigarette smoke, the nonvolatile particulate matter being distributed among three solvent pairs, chromatographed on alumina, and the fluorene concentrate analyzed by gas chromatography. For this investigation, mass spectra were determined with a Perkin-Elmer Hitachi RMU-6D instrument at 70 eV, whereas all GC separations and analyses were carried out on a Perkin-Elmer model 800 gas chromatograph equipped with a dual flame ionization detectot and a 6-m by 3-mm column packed with either 5% OV-17 or OV-1 on 60-80 mesh Gas Chrom P. Both systems used a helium carrier-gas flow rate of 40 ml/min, with the most satisfactory separation of the methylfluorenes obtained at 190°C on the OV-17 column and the separation of fluorene from 9-methylfluorene achieved with the OV-1 column at 160°C as shown in Table 1.67.

The EI mass spectra of 1-methylfluorene, 4-methylfluorene, and 2,3-dimethylfluorene are shown in Figure 1.58, and are used for comparative purposes to identify isolated alkylfluorene compounds as well as spectral data given in Table 1.68.

Using this procedure, Hoffmann and Rathkamp reported that the mainstream smoke of an 85-mm U.S. blended nonfilter cigarette contained 417 ng of fluorene, 62.1 ng of 1-methylfluorene, 254 ng of 2- and 3-methylfluorene, 79.4 ng of 4-methylfluorene, traces of 9-methylfluorene, and several dimethylfluorenes.

Pyrolysis experiments suggested that in cigarette smoke the fluorenes are partially formed via the Diels-Alder addition of indenes with volatile dienes. This hypothesis was supported by the correlation between the smoke

TABLE 1.66

Retention Times of 1-Alkylindoles and Other Alkylindoles[a]

Compound	Retention time (min)
1-Methylindole	7.9
1,2-Dimethylindole	14.5
1,3-Dimethylindole	9.9
1,4-Dimethylindole	10.7
1,5-Dimethylindole	10.7
1,6-Dimethylindole	10.8
1,7-Dimethylindole	14.5
1-Ethylindole	9.0
Indole	18.8
2-Methylindole	22.0
3-Methylindole	20.8
4-Methylindole	23.4
5-Methylindole	22.8
6-Methylindole	22.6
7-Methylindole	22.0
3-Ethylindole	28.0
2,3-Dimethylindole	31.2
2,5-Dimethylindole	31.8
2,7-Dimethylindole	31.4

[a] Adapted from Hoffmann and Rathkamp [164].

concentration of fluorenes with that of isoprene. The column conditions used for isoprene analysis were as follows: 6.6-m by 3-mm column of 15% Carbowax 550 on Gas Chrom P, column temperature 65° C isothermal. The retention time for isoprene was 4.5 ± 0.2 min.

In addition to the columns used specifically to separate PAH and related compounds derived from cigarette smoke and particulate matter, other systems have been used by analysts to resolve similar materials found in the atmosphere, coke oven effluents, heavy petroleum fractions, engine oils, airborne particulate matter, and so on. These systems employ various liquid stationary phases in conjunction with either packed or capillary columns. Several systems in each category (packed or capillary) will be discussed.

With regard to packed columns [177-187], in 1965, Gudzinowicz [177, 178] noted that in the work reported by James [188], hydrocarbon types could be further identified by plotting the relative retention volumes obtained for all compounds with one liquid phase versus those derived from another. He noted that all the hydrocarbon species fell on lines of different slopes, which, with some reservation, could be used for identification purposes.

This can be well illustrated by experiments conducted by Gudzinowicz [177] to establish the selectivity or advantages of two high-temperature liquid stationary phases, SE-30 (nonpolar) and QF-1 (polar), for resolving polar-nonpolar organic mixtures and to compare each component's retention times with both phases under identical chromatographic conditions.

In Table 1.69 the compounds investigated are listed with their retention times for SE-30 and QF-1 columns operated isothermally at several temperatures with constant carrier-gas flow rate (100 ml/min), molecular weights, and formulas. Six-foot columns packed with 5% (w/w) of each liquid phase on 80-90 mesh Anakrom AS were conditioned for 72 hr at 300° C to minimize column bleed.

Typical chromatograms of multicomponent mixtures derived with each liquid phase at specified temperatures were reproduced for comparative purposes. From the recorded retention times in Table 1.69 and the chromatograms shown, one obvious conclusion can be drawn: Shorter elution times for most components (with the exception of the polar sulfoxide-sulfone compounds) result with QF-1 packed columns than with SE-30. For some applications, this might be advantageous when resolution can be relegated to a position of secondary importance in contrast to speed of analysis. For example, a 12-ft by 3/16-in.-o.d. glass column packed with 5% (w/w) QF-1 on Anakrom AS, 80-90 mesh, and operated at 308° C with a carrier-gas flow rate of 100 ml/min, elutes 1,3,5-trimethyl-2,4,6-tris(3,5-t-butyl-4-hydroxybenzyl)benzene with a molecular weight of 775 and a $C_{54}H_{78}O_3$ formula in approximately 22 to 23 min, whereas the same compound with SE-30 has a retention time greater than 50 min with excessive peak broadening.

By plotting the SE-30 logarithmic retention times versus those for QF-1, one can determine qualitatively, with some reservations, whether or not some compounds are members of a particular molecular species (Fig. 1.59).

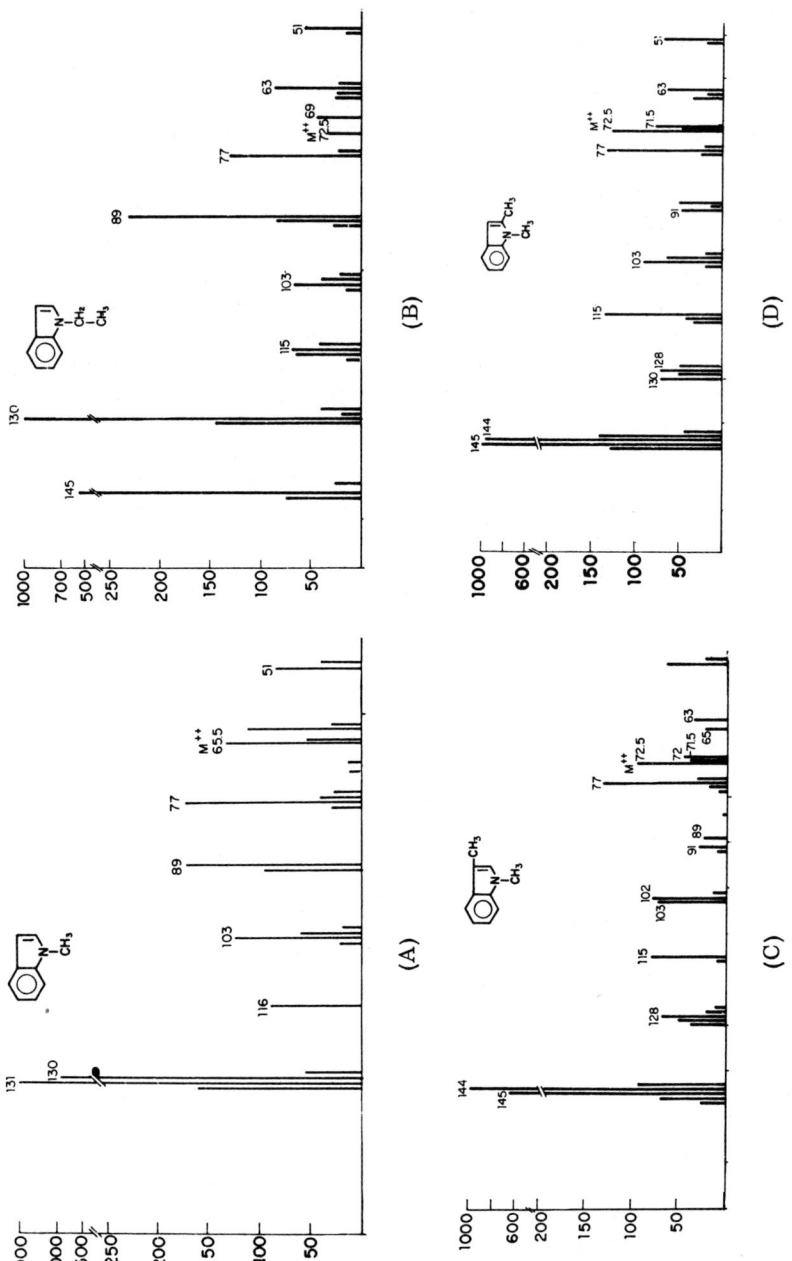

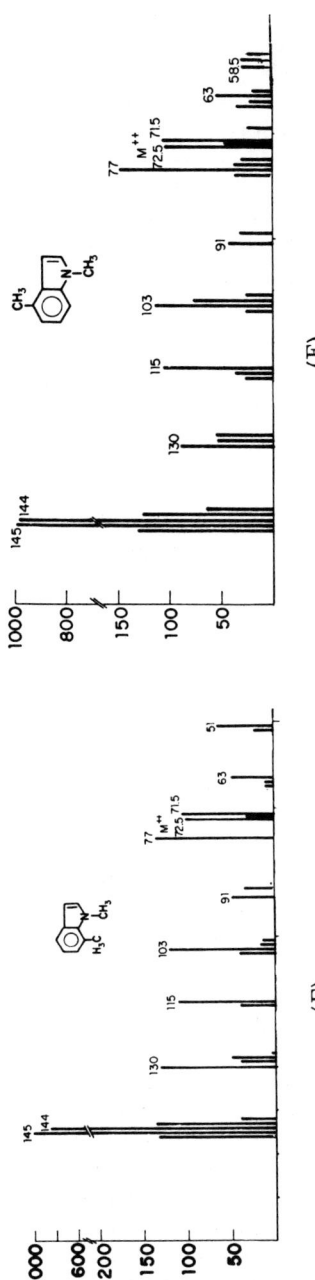

Figure 1.57. Mass spectra of 1-methylindole (A), 1-ethylindole (B), 1,3-dimethylindole (C), 1,2-dimethylindole (D), 1,7-dimethylindole (E), and 1,4-dimethylindole (F). Adapted from Hoffmann and Rathkamp [164].

TABLE 1.67

Retention Times of Fluorenes[a]

Compound	Retention time (min)	
	Column A[b]	Column B[c]
Fluorene	11.3	6.8
1-Methylfluorene	17.2	11.6
2-Methylfluorene	16.4	11.3
3-Methylfluorene	16.2	11.5
4-Methylfluorene	18.2	12.0
9-Methylfluorene	11.5	7.6
1,9-Dimethylfluorene	17.4	12.4
2,3-Dimethylfluorene	27.8	20.6
9,9-Dimethylfluorene	10.8	7.1
2-Ethylfluorene	22.6	14.2

[a] Adapted from Hoffmann and Rathkamp [168], courtesy of Analytical Chemistry.
[b] Column A: 6-m by 3-mm SS column, 5% OV-17, 190°C.
[c] Column B: 4.5-m by 3-mm SS column, 5% OV-1, 160°C.

For example, at 216°C, one would conclude from the plot that compounds 2, 5, 7, 8, and 12 are slightly different than compounds 1, 3, 4, 9, 10, 11, 13, and 14. In fact, a paraffinic compound such as squalane is easily distinguishable. With QF-1, it is eluted at 216°C before tri-o-tolylphosphate, whereas the order is reversed with SE-30. With the phenyl sulfide derivatives, the polarity of QF-1 tends to provide much better separations than SE-30.

By plotting logarithmic retention times for some compounds (numerically identified in Table 1.69) at column temperatures of 100 and 216°C with both phases versus their molecular weights, nearly linear relationships result (Fig. 1.60). With the SE-30 methyl silicone polymer, the condensed ring aromatics and their alkyl derivatives show better correlations than the QF-1 trifluoropropyl methyl silicone phase, QF-1 accentuating the greater nonlinearity observed for compounds with alkyl substituents such as retene (1-methyl-7-isopropyl phenanthrene), methylcholanthrene, and 9,10-dimethyl-1,2-benzanthracene.

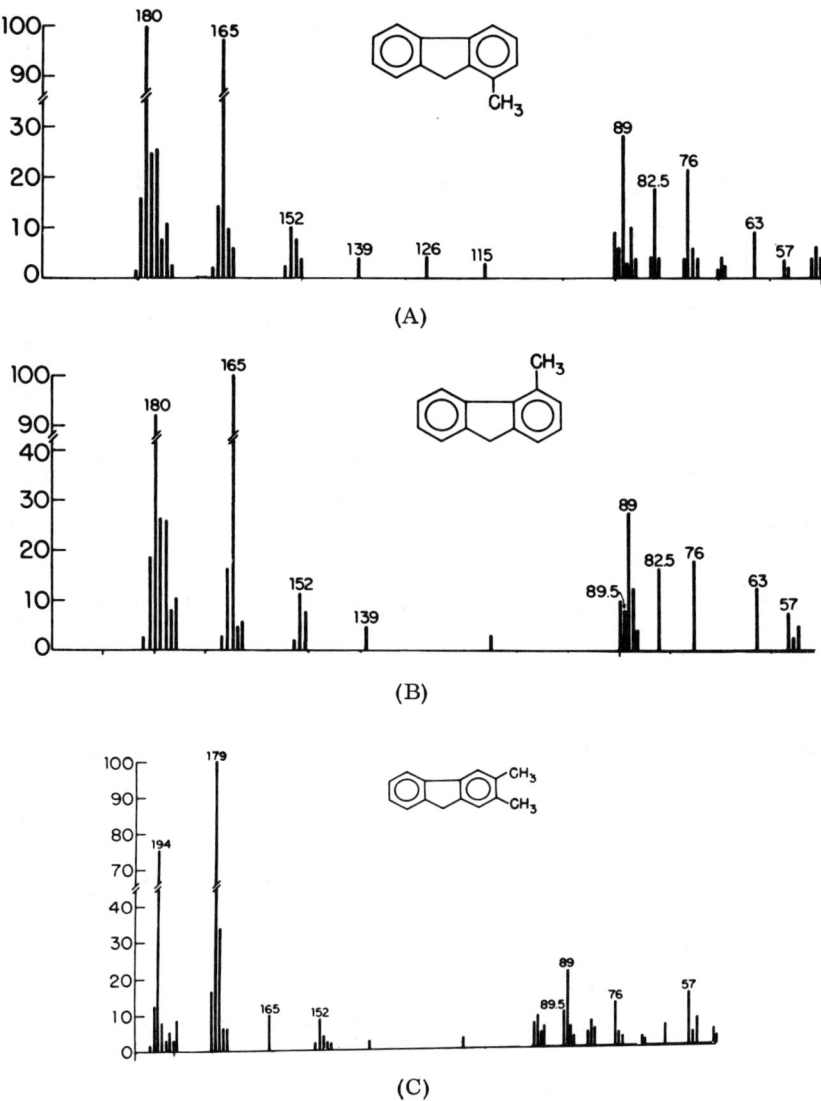

Figure 1.58. Mass spectra of 1-methylfluorene (A), 4-methylfluorene (B), and 2,3-dimethylfluorene (C). Adapted from Hoffmann and Rathkamp [168].

TABLE 1.68

Mass Spectra of Fluorenes
Relative Abundance of Characteristic Ions[a]

m/e	Fluorene	1-Methyl-	2-Methyl-	3-Methyl-	4-Methyl-	9-Methyl-	1,9-Dimethyl-	2,3-Dimethyl-	9,9-Dimethyl-	2-Ethyl-
194							46	76	39	84
180		100	98	91	91	66	14	16	16	17
179		25	28	29	25	16	100	100	100	100
166	100	15	13	16	15	16	0	1	1	14
165	86	98	100	100	100	100	4	9	3	81
164	3	10	4	5	4	6	1	2	1	22
152	0	10	10	10	11	7	8	8	9	10
139	8	4	4	5	5	5	2	2	2	5
115	5	3	3	4	4	3	2	2	2	4
89	3	28	30	33	29	16	13	20	20	21
82.5	23	18	16	18	19	12	3	6	3	6
76	1	21	27	28	25	12	9	12	14	13
63	8	9	10	12	10	6	4	9	6	6
51	3	6	5	7	6	4	3	5	5	3

[a] From Hoffmann and Rathkamp [168], courtesy of *Analytical Chemistry*.

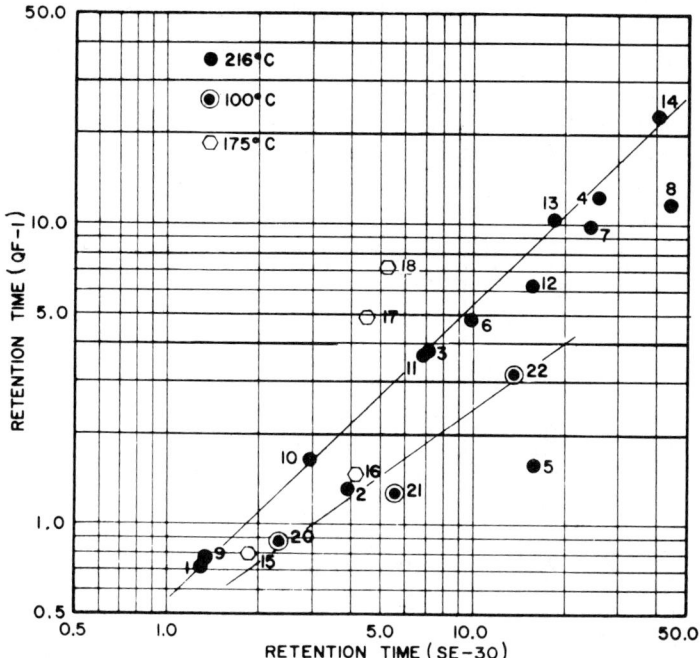

Figure 1.59. Plot of SE-30 retention times versus QF-1 retention times. From Gudzinowicz [177].

DeMaio and Corn [179] performed GC analyses with packed columns of polynuclear aromatic hydrocarbons of air pollution samples. After a single benzene extraction of suspended atmospheric particulate matter collected on a glass-fiber filter paper, the GC analysis was performed with an F&M model 810 gas chromatograph equipped with dual flame ionization detectors and two pairs of columns: set A, composed of two stainless steel columns, 6 ft by 1/8-in. o.d., packed with 2% Apiezon L on 60-80 mesh Diatoport S; set B, two copper columns, 20 ft by 1/8-in. o.d., packed with 2% SE-30 on 60-80 mesh Gas Chrom Z. The two sets of conditions used with each column system are given in Table 1.70. These GC parameters yielded the relative retention times for various PAH compounds listed in Table 1.71.

In 1969, Beeson and Pecsar [180] used a new liquid phase, poly-m-phenoxylene, which permitted PAH compounds to be satisfactorily determined and resolved as illustrated in Figure 1.61, where the GC characteristics are compared to the OV-1 liquid phase. These compounds were analyzed with a Varian Aerograph model 1525-B gas chromatograph equipped with a flame ionization detector and 5-ft by 1/8-in.-o.d. stainless steel columns packed with 3% PMPE or 3% OV-1 on 100-120 mesh Chromosorb

TABLE 1.69

Comparison of Retention Times (min) with SE-30 and QF-1 at Different Column Temperatures[a,b]

Compound	MW	Formula	Ident. no.	Stationary liquid phase					
				QF-1			SE-30		
				216°C	236°C	256°C	216°C	236°C	256°C
Phenanthrene	178.2	$C_{14}H_{12}$	1	0.72	0.55		1.30	0.87	0.70
Retene	234.3	$C_{18}H_{18}$	2	1.30	1.00		3.89	2.18	1.45
Chrysene	228.3	$C_{18}H_{12}$	3	3.77	2.06		7.09	3.78	2.35
Methylcholanthrene	268.3	$C_{21}H_{16}$	4	12.20	5.87		25.72	12.07	6.58
Squalane	422.8	$C_{30}H_{62}$	5	1.57	0.88	0.57	15.65	6.92	3.65
Tri-o-tolyl phosphate	368.4	$C_{21}H_{21}PO_4$	6	4.84	2.40	~1.30	9.96	4.92	2.83
Tri-2,4-xylyl phosphate	410.5	$C_{24}H_{27}PO_4$	7	9.90	4.33	2.27	24.00	10.58	5.53
Bis(ethylhexyl)tetrachlorophthalate	528.4	$C_{24}H_{34}O_4Cl_4$	8	11.90	5.07	2.60	43.90	17.93	8.85
Anthracene	178.2	$C_{14}H_{10}$	9	0.77	0.53		1.33	0.95	0.71
Pyrene	202.2	$C_{16}H_{10}$	10	1.63	1.03	0.73	2.93	1.83	1.23

Compound	MW	Formula						
Benz[a]anthracene	228.3	$C_{18}H_{12}$	11	3.67	2.04	1.27	6.83	3.67	2.32
9,10-Dimethyl-1,2-benzanthracene	256.3	$C_{20}H_{16}$	12	6.24	3.23	1.83	15.52	7.57	4.40
Benzo[a]pyrene	252.3	$C_{20}H_{12}$	13	10.17	5.10	2.84	18.35	8.95	5.20
Dibenz[a,h]anthracene	278.3	$C_{22}H_{14}$	14	22.90	10.50	5.34	40.80	18.38	9.99
					175°C			175°C	
Phenyl sulfide	186.3	$C_{12}H_{10}S$	15		0.80			1.86	
Phenyl disulfide	218.3	$C_{12}H_{10}S_2$	16		1.47			4.09	
Phenyl sulfoxide	202.3	$C_{12}H_{10}SO$	17		4.90			4.53	
Phenyl sulfone	218.3	$C_{12}H_{10}SO_2$	18		7.07			5.29	
					100°C			100°C	
Mesitylene	120.2	C_9H_{12}	19		~0.47			1.13	
Durene	134.2	$C_{10}H_{14}$	20		0.87			2.34	
Pentamethylbenzene	148.2	$C_{11}H_{16}$	21		1.27			5.48	
Hexamethylbenzene	162.3	$C_{12}H_{18}$	22		3.16			13.47	

[a] From Gudzinowicz [177].
[b] Constant carrier-gas flow, 100 ml/min.

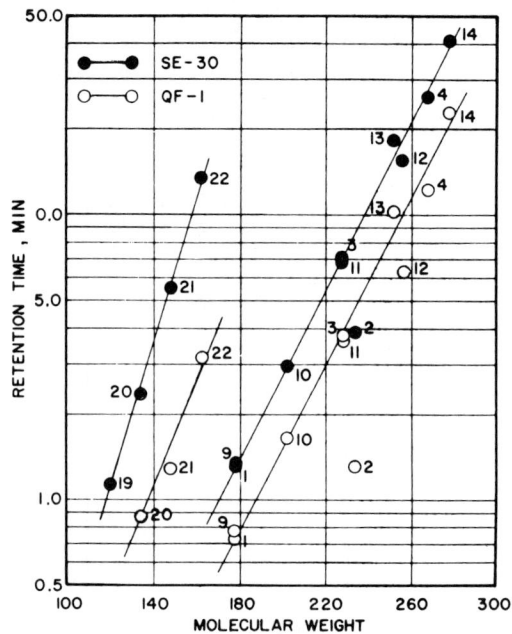

Figure 1.60. Plot of retention time versus molecular weight. From Gudzinowicz [177].

TABLE 1.70

Operating Conditions with Set A/Set B Packed Columns[a]

	Set A		Set B	
	Cond. 1	Cond. 2	Cond. 1	Cond. 2
Detector temp. (°C)	318	320	320	322
Injector temp. (°C)	310	330	305	308
Initial column temp. (°C)	220	240	180	250
Program rate (°C/min)	6	6	4	
Final column temperature (°C)	268	280	250	250
Post injection interval (min)	4	4	10	
Upper limit interval (min)	10	60	20	
Helium carrier-gas flow rate (ml/min)	85	25	55	50

[a] Adapted from DeMaio and Corn [179].

TABLE 1.71

Relative Retention Values for Some PAH Compounds
on Two Column Systems[a]

	Relative retention time			
	Set A		Set B	
Compound	Cond. 1	Cond. 2	Cond. 1	Cond. 2
Pyrene	1.00	1.00	1.00	1.00
Chrysene	1.74	1.71	1.59	1.75
Benz[a]anthracene	2.04	2.19	1.32	1.71
Benzo[k]fluoranthene	2.56	3.28	2.08	3.28
Benzo[e]pyrene	2.81	3.80	2.46	3.79
Benzo[a]pyrene	2.89	3.86	2.51	4.00
Benzo[g,h,i]perylene	5.45	9.87		11.80

[a] Adapted from DeMaio and Corn [179].

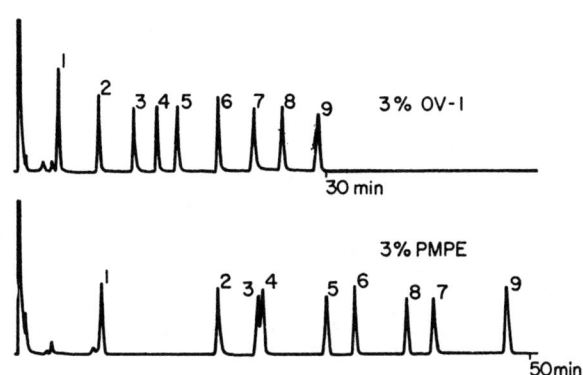

Figure 1.61. Polynuclear aromatic analysis. Five feet of 3% loading, isothermal for 5 min; then OV-1, 150 to 300°C at 4°C/min; PMPE, 180 to 350°C at 4°C/min.

1. Perhydropyrene
2. Pyrene
3. 1-Methylpyrene
4. 1,1'-Binaphthyl
5. Triphenylene
6. 2,2'-Binaphthyl
7. 1,2-Benzpyrene
8. 1,3,5-Triphenylbenzene
9. 1,2,3,4-Dibenzanthracene

From Beeson and Pecsar [180], courtesy of Analytical Chemistry.

W-HP. With the injector and detector temperatures both maintained at 300° C and the helium carrier-gas flow rate set at 33 ± 2 m/min, both columns were programmed as indicated in Figure 1.61. Beeson and Pecsar also noted that a higher initial temperature is used on the PMPE column because of the longer retention times of the samples. Because of the specificity of the phase the 1,2-benzpyrene and 1,3,5-triphenylbenzene elute in the reverse order of their elution from the OV-1 column. When the chromatogram is amplified by a factor of ten, and the minor constituents are observed, the more selective PMPE column resolves 30 components while the OV-1 resolves only 15.

Searl et al. [181] developed a routine method to measure polynuclear aromatic compounds in coke oven effluents. Although other techniques are involved, the method as outlined below is designated as a GC-UV procedure.

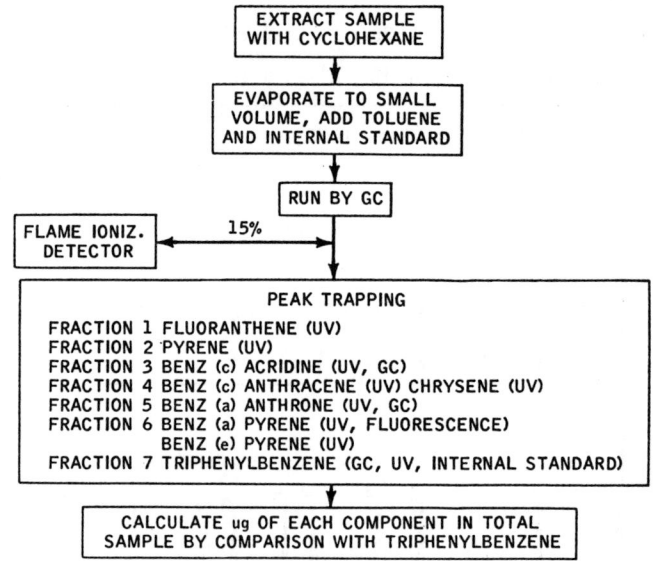

In practice, samples are collected on a filter and the filter is extracted with cyclohexane. An internal standard (1,3,5-triphenylbenzene) is added and a portion of the extract is injected into a gas chromatograph for separation into fractions that are trapped. Ultraviolet spectra of selected fractions provided a quantitative measurement of fluoranthene, pyrene, benz(a)-anthracene, chrysene, benzo(a)pyrene, and benzo(e)pyrene. Benz(c)acridine and benz(a)anthrone were also included in the method but were not observed to be present.

The GC separations were performed with a Perkin-Elmer model 900 gas chromatograph equipped with a flame ionization detector and a 10-ft by

1/8-in.-o.d. column packed with 2% SE-30 on Chromosorb G. Using the specified GC conditions (injector temperature, 315° C; column temperature, programmed from 175 to 275° C at 4° C/min, then held for 15 min at 275° C; detector temperature, 290° C; carrier gas and flow rate not specified), the retention times of pyrene, benz(c)acridine, benz(a)anthracene, benz(a)-anthrone, benzanthracene-7,12-dione, benzo(a)pyrene, and 1,3,5-triphenyl-benzene were approximately 11.72, 17.05, 17.95, 18.55, 20.30, 25.00, and 26.50 min, respectively.

Bhatia [182] determined PAH compounds with a Beckman gas chromatograph, model GC-4, equipped with a dual flame ionization detector and a 20-ft by 1/8-in.-o.d. copper column packed with 0.4% OV-7 coated on 60-80 mesh glass beads. The type of resolution achieved with this column is shown in Figure 1.62.

Lao et al. [184] used both GC and computerized GC-MS methods for the analysis of PAH compounds in environmental samples. Their three-step method consisted of (1) a preliminary separation of PAH by solvent and/or column chromatography, (2) identification by a combination of gas chromatography with quadrupole mass spectrometry and computer, and (3) measurement by computerized gas chromatography using internal standards. Samples of industrial effluents, coke oven emissions, coal tar, and airborne particulates were investigated using the GC and GC-MS conditions given below:

GC Conditions (Packed Column)

Gas chromatograph:	Perkin-Elmer model 990 equipped with flame ionization detector
Injector temp. (° C):	325
Column:	12-ft by 1/8-in.-o.d. stainless steel packed with 6% Dexsil 300, 400, or 410 on 80-100 mesh Chromosorb W-HP
Column temp. (° C):	initially held at 165° C for 2 min, then programmed to 295° C at 4° C/min, then held at 295° C for 50 min
Detector temp. (° C):	300
Helium carrier-gas flow rate (ml/min):	40

GC-MS Conditions (Packed Column)

Instrument:	Finnigan model 9500 gas chromatograph interfaced with a Finnigan model 1015D mass spectrometer; both integrated with a model 6000 data system

Column:	5-ft by 2-mm-i.d. packed with 3% OV-1 on 80-100 mesh Gas Chrom Q
Column temp. (°C):	programmed from 165 to 290°C at 4°C/min
Helium carrier-gas flow rate (ml/min):	20
Filament current (μA):	500
Electron energy (eV):	70
Mass range (amu):	35-350
Scan speed (sec):	1

Using the Dexsil 400 packed column, some of the PAH materials identified in coke oven emissions are listed in Table 1.72. Dexsil 300 and 410 yielded similar RRT data, and all GC patterns were equivalent except that less sample was used for the 400 and 410 columns.

Janini and co-workers [185,186] used nematic liquid crystals for the gas chromatographic separation of PAH compounds. In their 1976 GC evaluation and GC-MS application of these new high-temperature liquid crystal stationary phases for PAH separations [186], N,N'-bis(p-phenylbenzylidene)-α,α'-bi-p-toluidine (BPhBT), with a nematic range of 257 to 403°C, and N,N'-bis(p-hexyloxybenzylidene)-α,α'-bi-p-toluidine (BHxBT), with a smectic range of 127 to 229°C and a nematic range of 229 to 274°C, were synthesized and their use as stationary phases for GC separations of PAH compounds was investigated. GC studies were carried out with a Hewlett-Packard 7610 gas chromatograph equipped with a dual flame ionization detector and 2- to 6-ft by 2-mm-i.d. glass columns packed with 2.5% of the liquid stationary phase coated on 100-120 mesh Chromosorb W-HP. GC-MS analyses were conducted with a Varian 1400 gas chromatograph interfaced with a Finnigan 3300 mass spectrometer equipped with a Finnigan 6000 data system. With the integrated GC-MS unit, the operating conditions used were injector temperature, 250°C; transfer line temperature, 275°C; analyzer temperature, 120°C; helium carrier-gas flow rate, 28 ml/min; emission current, 0.50 μA; electron energy, 70 eV.

Using columns of varying length and different operating conditions, typical retention times obtained for some PAH compounds are listed in Table 1.73. As stressed by Janini et al. [186], a 3-ft by 2-mm-i.d. column packed with 2.5% BPhBT on 100-120 mesh Chromosorb W-HP showed insignificant noise levels when operated at 275°C in the GC-MS system. When the GC-MS instrument was operated in the selective ion-monitoring mode, the detection limit for benzo(a)pyrene was about 4 ng. Isothermal BPhBT column operation at high temperatures (270 to 290°C) permitted the

Chemical Composition of Tobacco Smoke 221

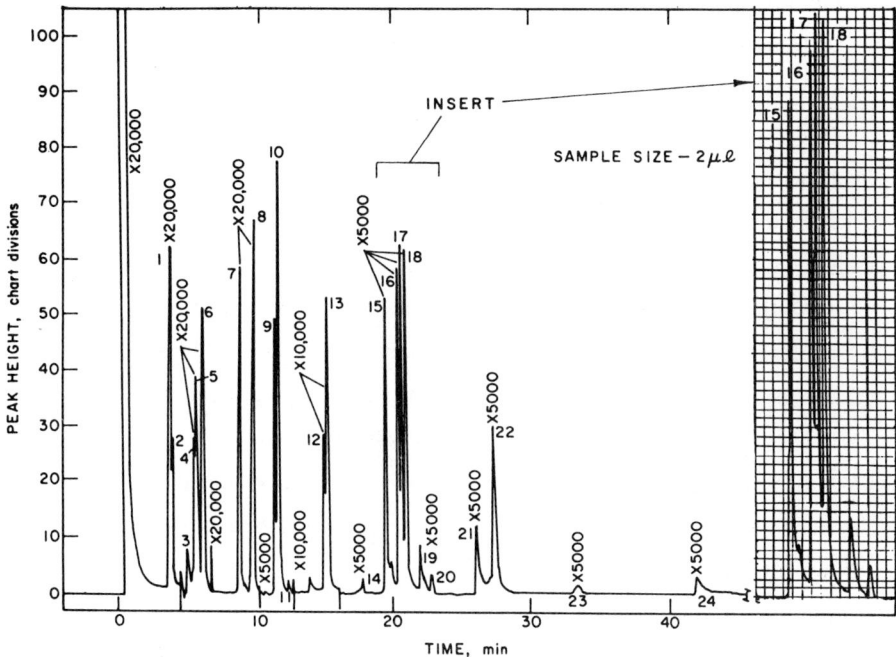

Figure 1.62. Gas chromatogram of a synthetic mixture of polycyclic aromatic hydrocarbons and n-alkanes. Column: 60/80 mesh of glass beads coated with OV-7. Temperature program: 4 min at 170°C; 6°C/min rise to 260°C; at 260°C to end of analysis. (1) phenanthrene, (2) anthracene, (3) carbazole, (4) 2-methylphenanthrene, (5) 3-methylphenanthrene, (6) 1-methylphenanthrene + 1-methylanthracene, (7) fluoranthene, (8) pyrene, (9) chrysofluorene, (10) 2,3-benzfluorene, (11) n-tetracosane, (12) 1,2-benzanthracene, (13) chrysene, (14) n-octacosane, (15) benz(k)fluoranthene, (16) benz(a)pyrene, (17) benz(e)pyrene, (18) perylene, (19) 3-methylcholanthrene, (20) n-dotriacontane, (21) 1,2,5,6-dibenzanthracene, (22) 1,12-benzperylene, (23) n-hexatriacontane, (24) dibenz(a,i)pyrene. Portion of original chromatogram is shown in the insert to indicate more clearly the resolution between peaks 15, 16, 17, and 18. From Bhatia [182], courtesy of <u>Analytical Chemistry</u>.

TABLE 1.72

PAH Compounds Identified in Coke Oven Emissions[a]

Octahydrophenanthrene	Dihydrobenzo(c)fluorene
Octahydroanthracene	Pyrene
Dihydrofluorene	Benzo(a)fluorene
Benzindene	Benzo(b)fluorene
Fluorene	Benzo(c)fluorene
Dihydrophenanthrene	Methylfluoranthene
Dihydroanthracene	Methylpyrene
1-Methylfluorene	Benzo(c)phenanthrene
2-Methylfluorene	Benzo(g,h,i)fluoranthene
4-Methylfluorene	Dihydrobenz(a)anthracene
9-Methylfluorene	Dihydrochrysene
Methylfluorene	Dihydrotriphenylene
Benzoquinoline	Benz(a)anthracene
Acridine	Chrysene
Phenanthrene	Triphenylene
Anthracene	Dihydromethylbenz(a)anthracene
Fluorene carbonitrile	Dihydromethylchrysene
Methylphenanthrene	Dihydromethyltriphenylene
Methylanthracene	Methylbenz(a)anthracene
Ethylphenanthrene	Methyltriphenylene
Ethylanthracene	Methylbenz(a)pyrene
Octahydrofluoranthene	Methylbenz(c)pyrene
Octahydropyrene	Dimethylbenz(a)anthracene
Dihydrofluoranthene	Dimethyltriphenylene
Dihydropyrene	Dimethylchrysene
Fluoranthene	Benzo(j)fluoranthene
Dihydrobenzo(a)fluorene	Benzo(k)fluoranthene
Dihydrobenzo(b)fluorene	Benzo(b)fluoranthene

TABLE 1.72 (continued)

Methylbenzo(k)fluoranthene	Dibenzanthracene
Methylbenzo(b)fluoranthene	o-Phenylenepyrene
Benzo(e)pyrene	Benzo(g,h,i)perylene
Benzo(a)pyrene	Anthanthene
Perylene	Methyldibenzanthracene
Methylbenzo(a)pyrene	Methylbenzo(g,h,i)perylene
Dimethylbenzo(k)fluoranthene	Coronene
Dimethylbenzo(b)fluoranthene	Dibenzopyrene
Dimethylbenzo(a)pyrene	

[a] Adapted from Lao et al. [184].

the analysis of high-molecular-weight PAH (having 22 to 24 carbons), providing, for the first time, baseline separation of some of the 5 to 7 ring PAH geometric isomers.

In addition to the BPhBT and BHxBT liquid crystal stationary phases, Janini et al. [185] have shown that gas chromatography in the nematic region of N,N'-bis(methoxybenzylidene)-α,α'-bi-p-toluidine (BMeBT) yielded baseline separations for geometric isomers of 3 to 5 ring PAH compounds. As noted by the authors, BMeBT is also applicable to 2 to 6 ring PAH compounds, stressing that the unique selectivity of this liquid phase, based on differences in the molecular length-to-breadth ratio of solute geometric isomers, has enabled the complete GC separation of mixtures heretofore not possible. Retention ratios for benzo(a)pyrene/benzo(e)pyrene and chrysene/benz(a)anthracene/triphenylene mixtures at 260°C were 1.6/1.0 and 1.8/1.4/1.0, respectively.

Separations were achieved using a Varian 1440 gas chromatograph equipped with a flame ionization detector and 4- to 6-ft by 0.125-in.-o.d. columns packed with 2.5% BMeBT on 100-120 mesh Chromosorb W-HP.

Using a 4-ft by 0.125-in.-o.d. stainless steel column and specified conditions (injector temperature, 265°C; column temperature, held for 2 min at 185°C, then programmed to 265°C at 4°C/min; detector temperature, 265°C; nitrogen carrier-gas flow rate, 40 ml/min), a synthetic mixture of 16 polycyclic hydrocarbons of wide molecular-weight range were separated. The retention times in minutes were fluorene, 1.16; phenanthrene, 2.31; anthracene, 2.79; fluoranthene, 4.88; pyrene, 5.57; benzo(mno)fluoranthene, 8.82; triphenylene, 9.52; benz(a)anthracene, 10.63; chrysene, 11.57; naphthacene, 14.33; benzo(k)fluoranthene, 18.50; benzo(e)pyrene, 19.41;

TABLE 1.73

GC Conditions and Packed Columns Used for Separating PAH Compounds[a]

PAH compounds	Retention time (min)[b]				
	Column A	Column B	Column C	Column D	Column E
Fluoranthene	1.67				
Pyrene	2.24				
Benzo[m,n,o]fluoranthene	5.76				
Triphenylene	6.78	1.11			
Benzo[a]anthracene	8.20	1.48			
Chrysene	9.98	1.77			
Naphthacene	13.58	2.50			
Benzo[k]fluoranthene		4.22	5.65		
Benzo[e]pyrene		4.78	6.28		
Perylene		5.85	7.99		
Benzo[a]pyrene		6.84	9.98		
3-Methylcholanthene		9.16			

Dibenz[a,c]anthracene	4.44
Benzo[g,h,i]perylene	7.77
Dibenz[a,h]anthracene	9.78
Picene	16.90
Pentacene	24.00
Dibenzo[def,p]chrystene	6.47
4,5,7,8-Dibenzpyrene	9.05
Coronene	10.78
Benzo[rst]pentaphene	20.25
1,2,6,7-Dibenzpyrene	27.60

[a] Adapted from Janini et al. [186].
[b] Column A: 6-ft by 2-mm–i.d. glass; packing, 2.5% BHxBT; injector, 240° C; column, 215° C; detector, 240° C; flow rate, 30 ml/min. Column B: 4-ft by 2-mm–i.d. glass; packing, 2.5% total (w/w) of 1:1 BHxBT/BPhBT mixture; injector, 275° C; column, 265° C; detector, 275° C; flow rate, 30 ml/min. Column C: 3-ft by 2-mm–i.d. glass; packing, 2.5% BPhBT; column, 275° C; flow rate, 28 ml/min; scan rate, 200 amu/sec; m/e range, 40–350. Column D: 2-ft by 2-mm–i.d. glass; packing, 2.5% BPhBT; injector, 280° C; column, 270° C; detector, 280° C; flow rate, 30 ml/min. Column E: 2-ft by 2-mm–i.d. glass; packing, 2.5% BPhBT; injector, 290° C; column, 290° C; detector, 290° C; flow rate, 50 ml/min.

perylene, 21.05; benzo(a)pyrene, 22.60; benzo(ghi)perylene, 27.20; and dibenz(ah)anthracene, 39.00.

In addition to the packed columns discussed above for the separation of polycyclic aromatic compounds, capillary columns in recent years have been utilized where resolution of multicomponent mixtures is essential, having been successfully applied to PAH compounds as well as C_8 to C_{12} aromatics, C_{12} alkylnaphthalenes and methylbiphenyls, and so on [78, 169,189-195].

As early as 1967, Willis [189] established retention time-boiling point correlations during programmed temperature capillary column analysis of C_8H_{10}, C_9H_{12}-indan, C_9H_{10}-indene, $C_{10}H_{14}$-tetralins, and their n-alkyl homologs using an F&M model 810 gas chromatograph modified for on-column injection and equipped with a flame ionization detector and a 200-ft by 0.02-in.-i.d. stainless steel column coated with ditridecylphthalate. Under specified GC conditions (injector temperature, 225°C; column temperature, programmed from 30 to 200°C at 4°C/min; detector temperature, 225°C; carrier-gas flow rate, 6 ml/min), Figure 1.63 shows retention time-boiling point correlations for some C_8 to C_{11} aromatic compounds; compounds 1, 2, 3, 4, 5, 6, 10, 14, 15, 18, 20, 22, 25, 28, 31, 33, 34, 35, 36, 37, and 39 were ethylbenzene, p-xylene, m-xylene, o-xylene, isopropylbenzene, n-propylbenzene, 2-ethyltoluene, 1,2,4-trimethylbenzene, 3-isopropyltoluene,

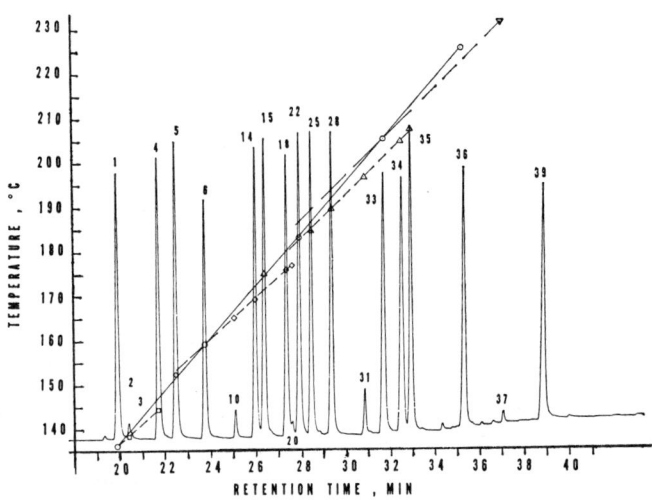

Figure 1.63. Retention time-boiling point correlations for C_8-C_{11} aromatic compounds. ☐ C_8H_{10}, ◊ C_9H_{12}-indan, △ $C_{10}H_{14}$-tetralin, ⊙ n-alkyl homologs. From Willis [189], courtesy of Analytical Chemistry.

1,2,3-trimethylbenzene, indan, n-butylbenzene and 4-n-propyltoluene, 2-n-propyltoluene, 1,2-dimethyl-4-ethylbenzene, 1,2,4,5-tetramethylbenzene, n-pentylbenzene, 1,2,3,4-tetramethylbenzene, tetralin (tetrahydronaphthalene), n-hexylbenzene, pentamethylbenzene, and n-heptylbenzene.

Mostecky et al. [190] in 1970 described the separation of dimethylnaphthalenes, ethylnaphthalenes, and methylbiphenyls using a CHROM 2 gas chromatograph equipped with a flame ionization detection and three capillary columns with different packing selectivities: column A, 100-m by 0.25-mm-i.d. stainless steel coated with Apiezon L (column temperature, 180°C; nitrogen CG flow rate, 0.32 ml/min); column B, 50-m by 0.25-mm-i.d. stainless steel coated with m-bis(m-phenoxyphenoxy)benzene (column temperature, 200°C; nitrogen flow rate, 0.25 ml/min); and column C, 50-m by 0.25-mm-i.d. stainless steel coated with polyethyleneglycol adipate (column temperature, 190°C; nitrogen flow rate, 0.30 ml/min). Table 1.74 lists the retention times of a 20-component mixture containing mainly C_{12} alkylnaphthalenes, biphenyl, acenaphthene, and methylbiphenyls relative to naphthalene.

Gouw et al. [191] showed that a 10-m by 0.010-in.-i.d. capillary column coated with OV-101 was a very versatile column for gas chromatography, noting that it had been used to analyze heavy petroleum fractions with end points above 1000°F and high-boiling waxes up to n-C_{58}, yet it could easily resolve isobutane from n-butane at lower temperatures. In conjunction with temperature programming, it was especially effective for the analysis of wide-boiling-range and high-boiling mixtures. A surprising quality was its exceptional ability to resolve isomers of high-molecular-weight compounds. As reported, the column had been used to separate the diastereomers of 1,3,5-triphenyldecane, to resolve anthracene from phenanthrene, and to separate 1,2-benzopyrene from 3,4-benzopyrene.

For the separation of polycyclic aromatic hydrocarbons, a Perkin-Elmer model 226 gas chromatograph equipped with 10-m, OV-101, stainless steel column and flame ionization detection was used with the following conditions: column temperature, 15 min isothermal at 120°C, then temperature programmed to 220°C at 3.75°C/min; helium inlet pressure, 10 psig; helium outlet flow rate, 1.7 ml/min; sample size, about 5 µl; split ratio, 200:1. With these parameters, it was possible to resolve phenanthrene, anthracene, fluoranthene, 1,2-benzopyrene, 3,4-benzopyrene, and perylene in approximately 50 min.

Using a Perkin-Elmer model 226 gas chromatograph equipped with a hydrogen flame ionization detector and a 300-ft by 0.01-in.-i.d. stainless steel column coated with TCEP (1,2,3-tris-2-cyanoethoxypropane) followed by a 150-ft by 0.01-in.-i.d. stainless steel column coated with silicone DC-550, Stuckey [192] described a GC technique which permitted the analysis of aromatic hydrocarbons in the 375-435°F boiling fraction of crude oil. With the injector temperature set at 120°C and the helium carrier gas adjusted to a head pressure of 52 psig (no other conditions specified), Table 1.75 lists the retention times of the aromatics investigated.

TABLE 1.74

Relative Retention Times[a]

Compound	Column A (180°)	Column B (200°)	Column C (190°C)
Naphthalene	1.00	1.00	1.00
2-Methylnaphthalene	1.60	1.50	1.41
1-Methylnaphthalene	1.77	1.68	1.60
Biphenyl	2.10	2.15	2.11
2-Ethylnaphthalene	2.38	2.10	1.84
1-Ethylnaphthalene	2.43	2.21	1.93
2,6-Dimethylnaphthalene	2.63	2.23	1.98
2,7-Dimethylnaphthalene	2.63	2.23	1.98
1,7-Dimethylnaphthalene	2.76	2.38	2.20
1,3-Dimethylnaphthalene	2.87	2.50	2.27
1,6-Dimethylnaphthalene	2.92	2.51	2.27
2-Methylbiphenyl	2.96	2.51	2.34
2,3-Dimethylnaphthalene	3.16	2.73	2.57
1,4-Dimethylnaphthalene	3.20	2.78	2.57
1,5-Dimethylnaphthalene	3.26	2.84	2.62
1,2-Dimethylnaphthalene	3.32	2.99	2.82
4-Methylbiphenyl	3.32	3.16	2.94
3-Methylbiphenyl	3.42	3.31	3.08
1,8-Dimethylnaphthalene	3.89	3.57	3.42
Acenaphthene	3.95	3.73	3.55
Naphthalene: RT (min)	36.20	29.20	30.00

[a] Adapted from Mostecky et al. [190].

TABLE 1.75
Retention Times of Aromatic Hydrocarbons[a]

Compound	Retention time (min)
n-Tridecane	21.8
1-Methyl-4-butylbenzene	22.3
1-Methyl-4-tert-butylbenzene	22.5
Tert-pentylbenzene	22.9
1,2,3-Trimethylbenzene	24.1
1,4-Dimethyl-2-ethylbenzene	24.0
1,3-Dimethyl-4-ethylbenzene	24.4
1-Methyl-3-tert-butylbenzene	24.4
1,2-Dimethyl-4-ethyl benzene	24.8
1-Methyl-3,5-diethylbenzene	25.3
para-Diisopropylbenzene	25.6
1,3-Dimethyl-2-ethylbenzene	26.5
n-Pentylbenzene	26.5
Naphthalene	26.8
3,4-Dimethyl-isopropylbenzene	27.1
1,2-Dimethyl-3-ethylbenzene	28.0
1,2,4,5-Tetramethylbenzene	28.8
Triethylbenzene[b]	29.4
1,2,3,5-Tetramethylbenzene	29.7
Triethylbenzene[b]	33.3
1,2,3,4-Tetramethylbenzene	35.4
Triethylbenzene[b]	35.5
1,2,3,4-Tetrahydronaphthalene	39.3

[a] Adapted from Stuckey [192].
[b] Standard was of mixed isomers.

In a series of three papers (1974-1976), Lee et al. [169,193,194] described the application of capillary columns coated with SE-52 to the identification of PAH compounds in airborne particulate matter [169,194] and in aromatic fractions derived from engine oils [193].

To obtain profiles of the PAH compounds in engine oils using both flame ionization and nitrogen-selective detection, Lee et al. [193] used a Perkin-Elmer model 900 gas chromatograph equipped with both detectors and a 22-m by 0.26-mm-i.d. glass capillary column coated with SE-52 with the helium carrier gas set at 1 ml/min and the injector and detector temperatures maintained at 250 and 260°C, respectively. With the column temperature programmed from 100 to 260°C at 2°C/min, Figure 1.64 is a capillary column gas chromatogram (using FID) of a 16-component standard PAH mixture, whereas in Figure 1.65 the profiles obtained for three engine oil extracts with both detectors are compared. With the nitrogen-selective detector one clearly notes that the differences among the three engine oils are considerably more dramatic than those observed with the flame ionization detector.

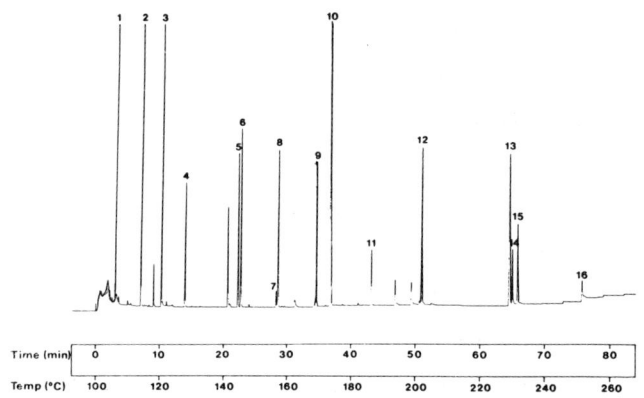

Figure 1.64. Capillary column gas chromatogram (using FID) of PAH standard compounds. Column: 22-m by 0.26-mm-i.d., glass capillary coated with SE-52; injector temperature, 250°C; detector temperature, 260°C. Key: (1) naphthalene, (2) biphenyl, (3) acenaphthene, (4) fluorene, (5) phenanthrene, (6) anthracene, (7) 4,5-methylene phenanthrene, (8) 9-methyl phenanthrene, (9) fluoranthene, (10) pyrene, (11) 1-methyl pyrene, (12) triphenylene and chrysene, (13) benzo[e]pyrene, (14) benzo[a]pyrene, (15) perylene, (16) dibenz[a,c]anthracene. From Lee et al. [193], courtesy of Analytical Chemistry.

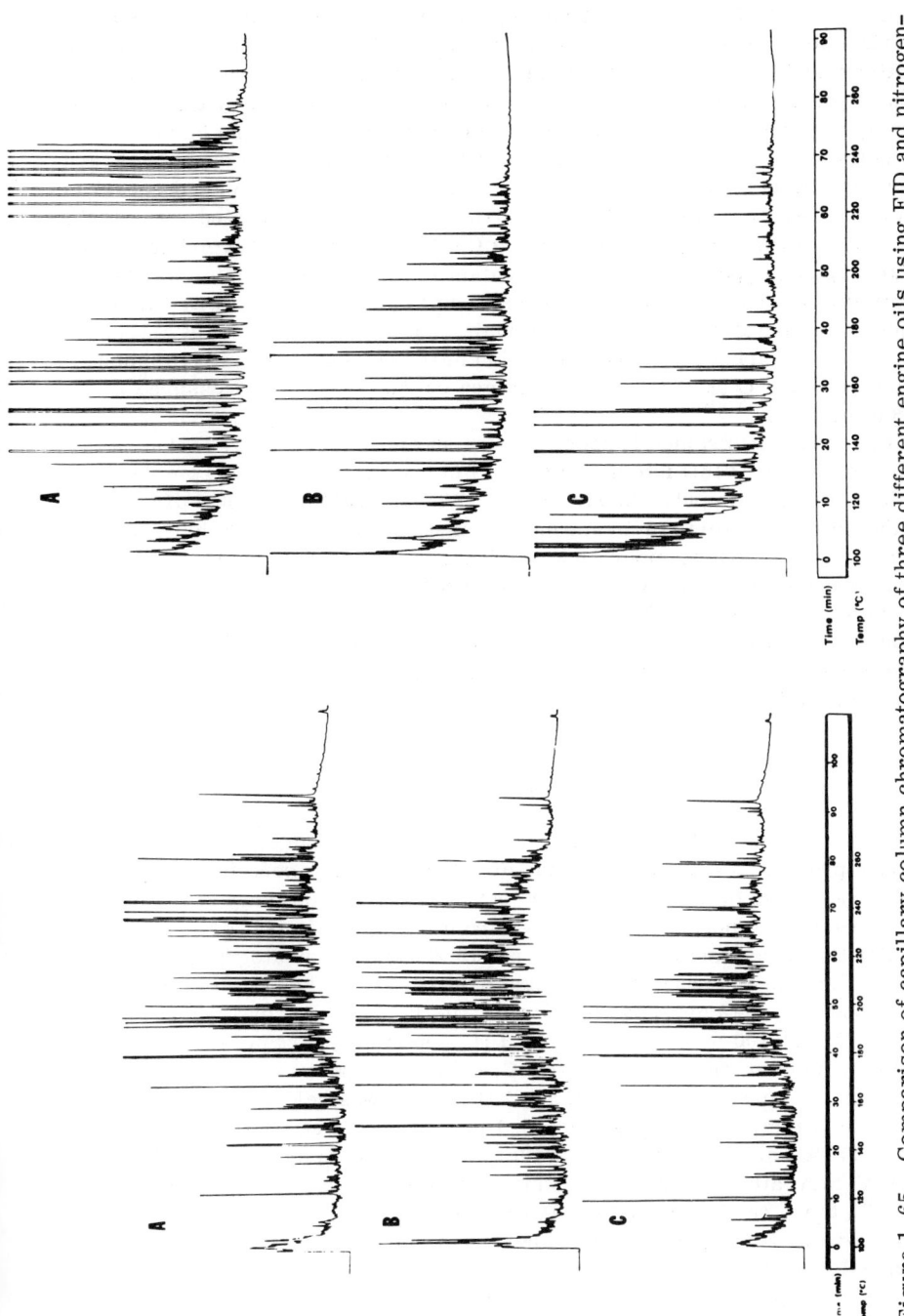

Figure 1.65. Comparison of capillary column chromatography of three different engine oils using FID and nitrogen-selective detectors (GC column same as in Fig. 1.64). Adapted from Lee et al. [193], courtesy of Analytical Chemistry. Left, FID detector; right, nitrogen-selective detector.

In their GC-MS study of polynuclear aromatic hydrocarbons in airborne particulates, the 122 peaks shown in Figure 1.66 are identified in Table 1.76.

Giger and Schaffner [195] also described a procedure for the determination of polycyclic aromatic hydrocarbons in the environment by glass capillary gas chromatography in which the PAH compounds were isolated by a sequence of solvent extraction, gel filtration, and adsorption chromatography. The separation of the PAH concentrates into quantifiable individual constituents was then achieved by GC using 20-m long capillaries coated with SE-52. GC separations were performed with a Carlo Erba model 2101 AC gas chromatograph equipped and operated as follows: 20-m by 0.3-mm-i.d. glass capillary column with barium carbonate interlayer and coated with SE-52; injector temperature, 270°C; flame ionization detector; column at ambient temperature during injection and elution of the solvent, then programmed from 60 to 250°C at 2.5°C/min; hydrogen carrier-gas inlet pressure, 0.8 atm.

For mass spectrometric identification and mass specific detection, a Finnigan model 1015D integrated GC-MS instrument was used, combined with an on-line computer (model 6000). With the column coupled directly to the mass spectrometer by means of a platinum capillary and helium used as carrier gas, the MS operating conditions were electron energy, 70 eV; emission current, 350 µA; preamplifier sensitivity, 10^{-8} A/V.

With the Carlo Erba GC instrument, the retention times of PAH compounds identified in environmental samples (lake sediments, dust, airborne particulates) relative to biphenyl (actual retention time = 9.89 min) are listed in Table 1.77.

H. Sulfur-Containing Compounds [196-200]

In 1971, Guerin [196] showed that the use of GC with flame photometric detection provided a convenient means of detecting sulfur-containing compounds in the gas phase of cigarette smoke, noting that the specificity of the detector is such as to negate interferences from all other compounds in the smoke.

For this investigation, a Tracor model 220 gas chromatograph equipped with a flame photometric detector and 6-ft by 0.25-in.-o.d. glass columns was used. One column was packed with 80-100 mesh Porapak Q and the other with 80-100 mesh Chromosorb 104. Operating parameters were as follows: injector temperature, 160°C; column temperatures, 50°C with Porapak Q whereas the Chromosorb 104 column was held isothermally at 50°C for 5 min, then programmed to 150°C at 5°C/min; detector temperature, 170°C; nitrogen carrier-gas flow rate, 80 ml/min; hydrogen flow rate, 120 ml/min; air flow rate, 35 ml/min; oxygen flow rate, 20 ml/min.

With these conditions and the Porapak Q column, Figure 1.67 illustrates the chromatogram generated by a 5-ml sample of the gas phase of a commercial nonfilter cigarette. Peaks 1 and 4 are of unknown origin, whereas

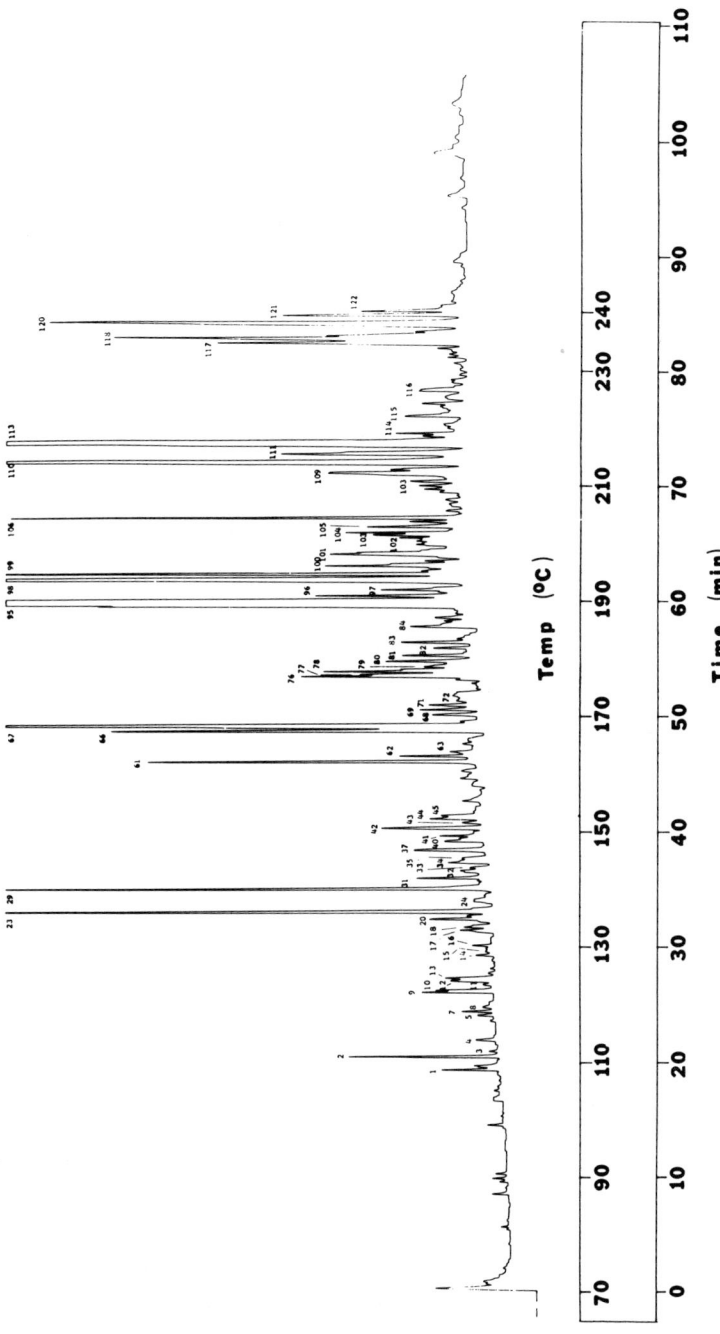

Figure 1.66. Capillary-column gas chromatogram of the total polynuclear aromatic hydrocarbon fraction of air-particulate matter. Conditions: 11-m by 0.26-mm-i.d. glass capillary coated with SE-52 methylphenylsilicone stationary phase; sample introduced through the precolumn technique. From Lee et al. [194], courtesy of Analytical Chemistry.

TABLE 1.76

List of Polynuclear Aromatic Hydrocarbons Identified by
Gas Chromatography/Mass Spectrometry[a]

Peak no.	MW	Compound
1	184	Dibenzothiophene
2	178	Phenanthrene
3	178	Anthracene
4	186	Unknown
5	186	Unknown
6	192,198	Unknown, methyldibenzothiophene
7	198	Methyldibenzothiophene
8	198	Methyldibenzothiophene
9	192	3-Methylphenanthrene
10	192	2-Methylphenanthrene
11	190	4H-cyclopenta[def]phenanthrene
12	192	9-Methylphenanthrene
13	192	1-Methylphenanthrene
14	220	Unknown
15	212	Ethyldibenzothiophene[b]
16	204	Methyl-4H-cyclopenta[def]phenanthrene
17	206	Ethylphenanthrene or ethylanthracene[b]
18	206	Ethylphenanthrene or ethylanthracene[b]
19	206	Ethylphenanthrene or ethylanthracene[b]
20	206	Ethylphenanthrene or ethylanthracene[b]
21	206	Ethylphenanthrene or ethylanthracene[b]
22	206	Ethylphenanthrene or ethylanthracene[b]
23	202	Fluoranthene
24	202	Benzacenaphthylene
25	206	Ethylphenanthrene or ethylanthracene[b]
26	206	Ethylphenanthrene or ethylanthracene[b]

TABLE 1.76 (continued)

Peak no.	MW	Compound
27	208	Benzo[def]dibenzothiophene
28	218	Ethyl-4H-cyclopenta[def]phenanthrene[b]
29	202	Pyrene
30	218	Ethyl-4H-cyclopenta[def]phenanthrene[b]
31	218	Ethyl-4H-cyclopenta[def]phenanthrene[b]
32	218	Ethyl-4H-cyclopenta[def]phenanthrene[b]
33	218	Ethyl-4H-cyclopenta[def]phenanthrene[b]
34, 35	220	Ethylmethylphenanthrene or ethylmethylanthracene[c]
36	216	Methylfluoranthene
37	216	Methylfluoranthene
38	216	Methylfluoranthene
39	216	Methylfluoranthene
40	216	Methylfluoranthene
41	216	Benzo[a]fluorene
42	216	2-Methylpyrene and benzo[b]fluorene
43	220	Ethylmethylphenanthrene or ethylmethylanthracene[c]
44	216	4-Methylpyrene
45	216	1-Methylpyrene
46	230	Unknown
47	232	Ethylmethyl-4H-cyclopenta[def]phenanthrene[c]
48	232	Ethylmethyl-4H-cyclopenta[def]phenanthrene[c]
49	232	Ethylmethyl-4H-cyclopenta[def]phenanthrene[c]
50	232	Ethylmethyl-4H-cyclopenta[def]phenanthrene[c]
51	232	Ethylmethyl-4H-cyclopenta[def]phenanthrene[c]
52	232	Ethylmethyl-4H-cyclopenta[def]phenanthrene[c]
53	232	Ethylmethyl-4H-cyclopenta[def]phenanthrene[c]
54	230	Ethylpyrene or ethylfluoranthene[b]

(continued)

TABLE 1.76 (continued)

Peak no.	MW	Compound
55	230	Ethylpyrene or ethylfluoranthene[b]
56	230	Ethylpyrene or ethylfluoranthene[b]
57	244	Ethylmethylfluoranthene or ethylmethylpyrene[c]
58	230	Ethylpyrene or ethylfluoranthene[b]
59	244	Ethylmethylfluoranthene or ethylmethylpyrene[c]
60	234	Naphthobenzothiophene
61	226	Benzo[ghi]fluoranthene
62	228	Benzo[c]phenanthrene
63	234	Naphthobenzothiophene
64	234	Naphthobenzothiophene
65	244	Ethylmethylpyrene or ethylmethylfluoranthene[c]
66	228	Benz[a]anthracene
67	228	Chrysene
68	258	Unknown
69	242	Methylbenzo[c]phenanthrene
70	240	Methylbenzo[ghi]fluoranthene
71	242	Methylbenzo[c]phenanthrene
72	248	Methylnaphthobenzothiophene
73	258	Unknown
74	248	Methylnaphthobenzothiophene
75	248	Methylnaphthobenzothiophene
76	242	Methylchrysene or methylbenz[a]anthracene
77	242	Methylchrysene or methylbenz[a]anthracene
78	242	Methylchrysene or methylbenz[a]anthracene
79	242	Methylchrysene or methylbenz[a]anthracene
80	242, 240	Methylchrysene or methylbenz[a]anthracene
81	254	Binaphthyl

TABLE 1.76 (continued)

Peak no.	MW	Compound
82	254	Binaphthyl
83	254	Binaphthyl
84	256	Ethylchrysene or ethylbenz[a]anthracene[b]
85	256	Ethylchrysene or ethylbenz[a]anthracene[b]
86	256	Ethylchrysene or ethylbenz[a]anthracene[b]
87	256	Ethylchrysene or ethylbenz[a]anthracene[b]
88	256	Ethylchrysene or ethylbenz[a]anthracene[b]
89	256	Ethylchrysene or ethylbenz[a]anthracene[b]
90	256	Ethylchrysene or ethylbenz[a]anthracene[b]
91	268	Methylbinaphthyl
92	268	Methylbinaphthyl
93	268	Methylbinaphthyl
94	268	Methylbinaphthyl
95	252	Benzo[j]fluoranthene
96	252	Benzo[k]fluoranthene
97	252	Benzofluoranthene
98	252	Benzo[e]pyrene
99	252	Benzo[a]pyrene
100	252	Perylene
101	266	Methylbenzopyrene or methylbenzofluoranthene
102	266	Methylbenzopyrene or methylbenzofluoranthene
103	266	Methylbenzopyrene or methylbenzofluoranthene
104	266	Methylbenzopyrene or methylbenzofluoranthene
105	266	Methylbenzopyrene or methylbenzofluoranthene
106	306	Quaterphenyl
107	264	Unknown
108	276	d

(continued)

TABLE 1.76 (continued)

Peak no.	MW	Compound
109	276	d
110	276	d
111	278	Dibenzanthracene
112	278	Dibenzanthracene
113	276	Benzo[ghi]perylene
114	276	Anthanthrene
115	292	Methyldibenzanthracene
116	304	Diphenylacenaphthalene
117	302	Dibenzopyrene
118	302	Dibenzopyrene
119	300	e
120	300	Coronene
121	302	Dibenzopyrene
122	302	Dibenzopyrene

[a] From Lee et al. [194], courtesy of Analytical Chemistry.
[b] Could also be dimethyl-.
[c] Could also be trimethyl- or propyl-.
[d] Compounds with molecular weight 276 can be any of the following: indeno[1,2,3-cd]pyrene, indeno[1,2,3-cd]fluoranthene, cyclopenta[cd]perylene, phenanthro[10,1,2,3-cdef]fluorene, acenaphth[1,2-a]acenaphthylene, dibenzo[b,mno]fluoranthene, dibenzo[e,mno]fluoranthene, dibenzo[f,mno]fluoranthene. Further possibilities are the benzo derivatives of cyclopenta[cd]pyrene and cyclopenta[cd]fluoranthene.
[e] Possibilities include the cyclopenta derivatives of compounds with molecular weight 276.

peak 2 corresponds to hydrogen sulfide and peak 3 to carbonyl sulfide. Guerin noted that the chromatograms generated in his study suggest that cigarettes deliver approximately equal amounts of H_2S and COS. Equally interesting is that no evidence was found for the presence of SO_2.

The practical utility of the GC method was tested by examining the delivery of sulfur-containing compounds from several filter and nonfilter

TABLE 1.77

Retention Times of PAH Compounds Relative to Biphenyl[a]

Compound	RRT
Biphenyl	1.00
Acenaphthene	1.29
Fluorene	1.73
Phenanthrene	2.56
Anthracene	2.61
Methylphenanthrenes	3.05-3.17
4,5-Methylenephananthrene	3.12
Fluranthene	3.70
Pyrene	3.88
Benzo[a]fluorene	4.27
Benzo[b]fluorene	4.38
Benz[a]anthracene	5.10
Chrysene/triphenylene	5.15
Benzofluoranthenes	6.10
Benzo[e]pyrene	6.27
Benzo[a]pyrene	6.32
Perylene	6.37
Dibenzanthracenes	7.20
Indeno[1,2,3-cd]pyrene	7.26
Benzo[ghi]perylene	7.30
Anthracene	7.40
Coronene	8.37

[a] Adapted from Giger and Schaffner [195].

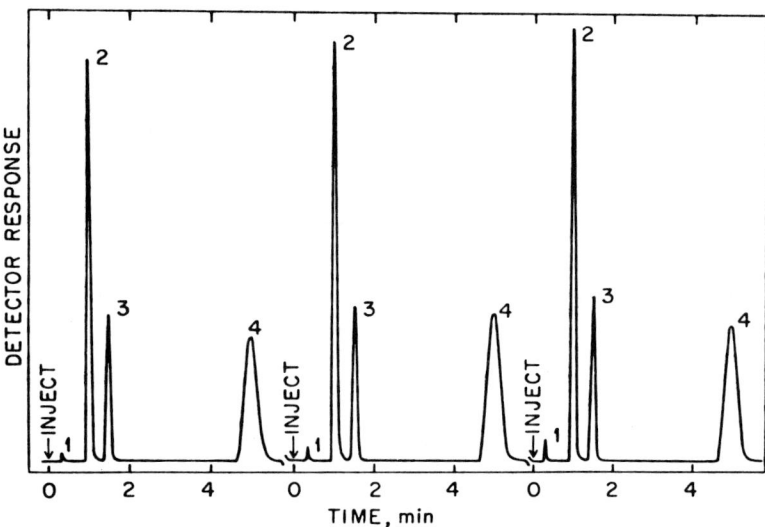

Figure 1.67. Sulfur-specific gas chromatograms from replicate samples of the gas phase of a nonfilter cigarette. From Guerin [196], courtesy of <u>Analytical Letters</u>.

cigarettes. An examination of Figure 1.68 (Porapak Q column) shows that the common filters (presumably cellulose acetate) produce a slight overall reduction in sulfur delivery. The variation between brands may reflect the difference in tobacco blend more than filter efficiency. The influence of charcoal filters is much more pronounced (Fig. 1.69). Interestingly, the design of the charcoal filter appears to be important: Brand G employs charcoal "fibers" while brand H uses charcoal "granules."

Guerin obtained a more complete display of the sulfur-containing compounds delivered by cigarettes by using temperature programming as shown in Figure 1.70, a chromatogram obtained from the Kentucky reference cigarette with the Chromosorb 104 column. The presence of as many as 11 compounds is indicated. An examination of Figure 1.70 shows that the charcoal filter effects the distribution of higher-molecular-weight compounds as well as lower.

Groenen and Van Gemert [197] used a gas chromatograph with flame photometric detection to detect at least 37 sulfur compounds in the gas phase of cigarette smoke. In a comparative study on three types of cigarettes, prepared from flue-cured, air-cured, and sun-cured tobacco, respectively, a great similarity was observed, but there were also some striking differences between air-cured tobacco on the one hand and flue-cured and sun-cured tobacco on the other.

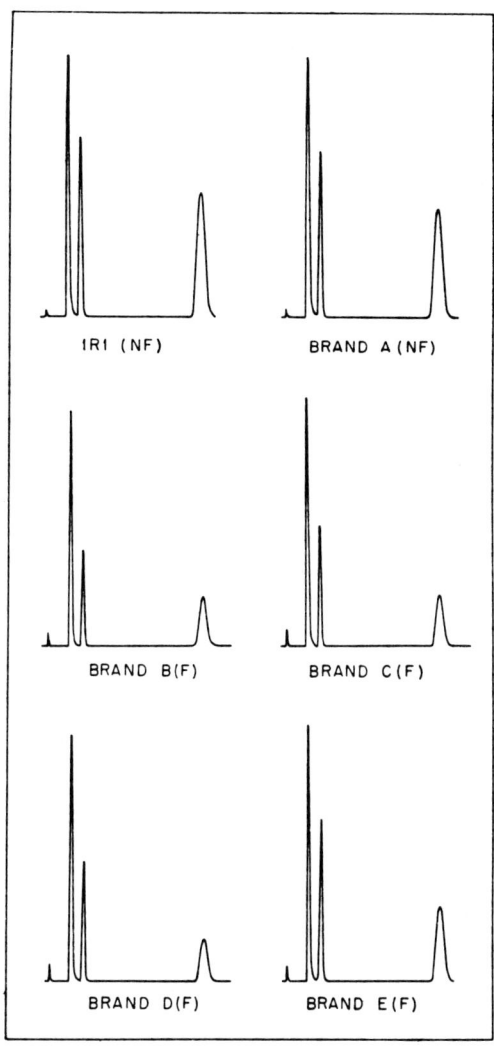

Figure 1.68. Deliveries of gaseous sulfur compounds—gas phases of filter and nonfilter cigarettes. From Guerin [196], courtesy of Analytical Letters.

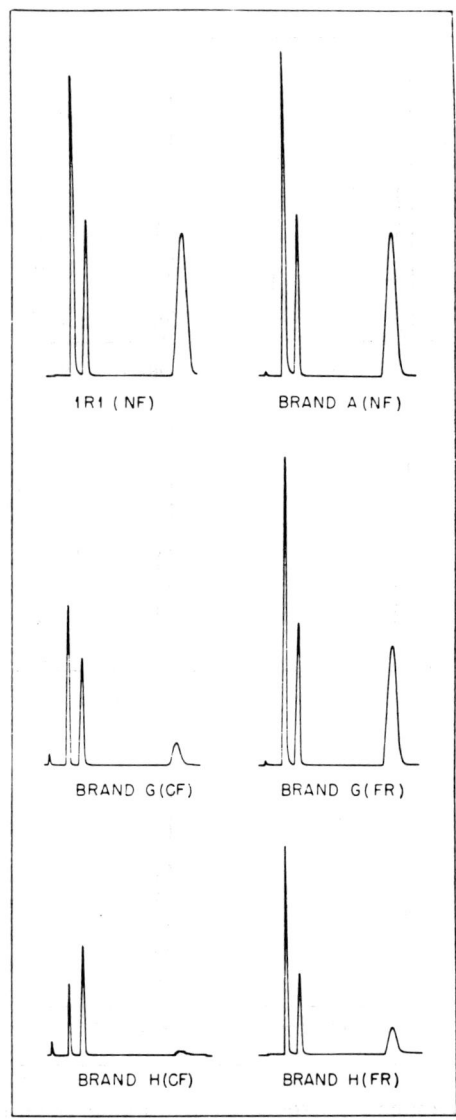

Figure 1.69. Deliveries of gaseous sulfur compounds—gas phases of charcoal filter and nonfilter cigarettes. From Guerin [196], courtesy of Analytical Letters.

Chemical Composition of Tobacco Smoke

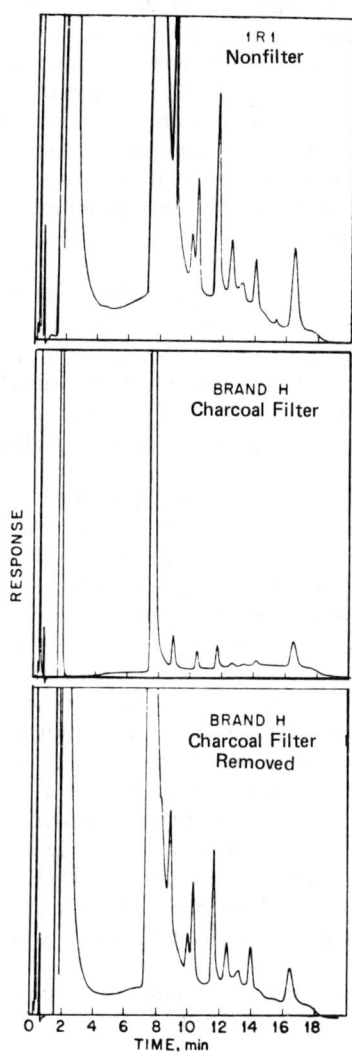

Figure 1.70. Sulfur-specific programmed temperature chromatograms—gas phases of nonfilter and charcoal filter cigarettes. From Guerin [196], courtesy of Analytical Letters.

The gas chromatograph used with a Tracor Microtek model MT 160 equipped with a standard Melpar flame photometric detector and a 5.5-m by 3-mm i.d. glass column packed with 25% 1,2,3-tris-2-cyanoethoxypropane coated on 60-80 mesh Chromosorb W-AW. For the separation and detection of sulfur components in cigarette smoke, the GC conditions employed were column temperature, 75° C; nitrogen carrier-gas flow rate, 80 ml/min; hydrogen flow rate, 144 ml/min; air flow rate, 40 ml/min; oxygen flow rate, 20 ml/min. Some of the compounds postulated as possibly being present in cigarette smoke based on relative retention time data were H_2S, CS_2, methyl mercaptan, ethyl mercaptan, dimethyl sulfide, isopropyl mercaptan, methyl ethyl sulfide, SO_2, n-propyl mercaptan, isobutyl mercaptan, allyl mercaptan, methyl n-propyl sulfide, diethyl sulfide, isoamyl mercaptan, n-butyl mercaptan, methyl n-butyl sulfide, ethyl n-propyl sulfide, methyl allyl sulfide, n-amyl mercaptan, thiophene, di-n-propyl sulfide, diisobutyl sulfide, dimethyl disulfide, n-hexyl mercaptan, n-propyl allyl sulfide, n-propyl n-butyl sulfide, methyl ethyl disulfide, diallyl sulfide, di-n-butyl sulfide, diethyl disulfide, 2,4-dimethyl thiophene, methyl n-propyl disulfide, methyl allyl disulfide, dimethyl trisulfide, and n-propyl allyl disulfide.

In 1974, Horton and Guerin [198] described a GC system which permitted the generation of reliable quantitative results for sulfur components (H_2S, COS, SO_2, and CS_2) in cigarette smokes. These studies were carried with a Tracor model MT 220 gas chromatograph equipped with a flame photometric detector and a 6-ft by 1/8-in.-o.d. fluoroelastomer (FEP) column packed with Tracor Special Silica, which was suitable for the determination of COS, H_2S, CS_2, and SO_2. For these components, the following GC conditions were maintained: column temperature, room temperature or about 24° C; injector temperature, 75° C; detector temperature, 125° C; nitrogen carrier-gas flow rate, 15 ml/min. On the other hand, for the determination of higher-boiling sulfur compounds, a 18-ft by 1/4-in.-o.d. glass column packed with 20% FFAP on 60-80 mesh Chromosorb W-AW was used in conjunction with the other GC settings prescribed: injector temperature, 225° C; column temperature, isothermal for 6 min at 24° C, then programmed at 4° C/min to 225° C; detector temperature, 225° C; nitrogen carrier-gas flow rate, 60 ml/min. For both columns, FPD gases were hydrogen, 180 ml/min; air, 40 ml/min; and oxygen, 20 ml/min.

With the Tracor Special Silica column, the retention times for COS, H_2S, CS_2, and SO_2 were approximately 3.67, 4.08, 6.00, and 9.66 min, respectively.

With the FFAP, approximately 28 peaks were observed, 7 of which were identified and had the following retention times (min): COS, 3.48; dimethyl sulfide, 5.03; H_2S, 9.10; CS_2, 12.40; dimethyl disulfide, 24.80; thiophene, 33.30; diethyl sulfide, 36.00.

In 1976, Bertsch et al. [200] applied the heart-cutting technique in high-resolution GC to sulfur compounds in cigarette smoke. The apparatus used consisted of the following:

A Shimadzu GC 5AP$_5$ gas chromatograph with flame ionization detector and flame photometric detector was used with an inlet modification [201]. Figure 1.71 shows the experimental arrangement of the components. Nickel capillary columns [202], 300 ft by 0.02 in., were coated with Emulphor ON 870 or OV-17 silicone fluid. Inlet and outlet traps were also made of short pieces of coated nickel tubing. The outlet trap was located outside the oven and connected to a vacuum line via an on/off ball valve. Dividers for the column effluent and makeup gas were 1/16-in. stainless steel tees. Two short pieces of stainless steel tubing, 1/2-in. by 0.05 in. served as restrictors between outlet transfer line and detectors. All metal parts were silanized, using standard procedures. Helium served as carrier gas and nitrogen, regulated by fine metering valves, was the makeup gas. The flow rates for the detectors were individually optimized. Both detectors were fitted with low dead-volume tubing leading directly into the jets.

Figures 1.72 and 1.73 show chromatograms of fresh tobacco smoke in the universal and sulfur-sensitive modes. The top chromatograms of Figure 1.72 represent the sulfur profile and FID profile of a standard cigarette puff. The bottom chromatograms represent reconstituted sulfur compounds

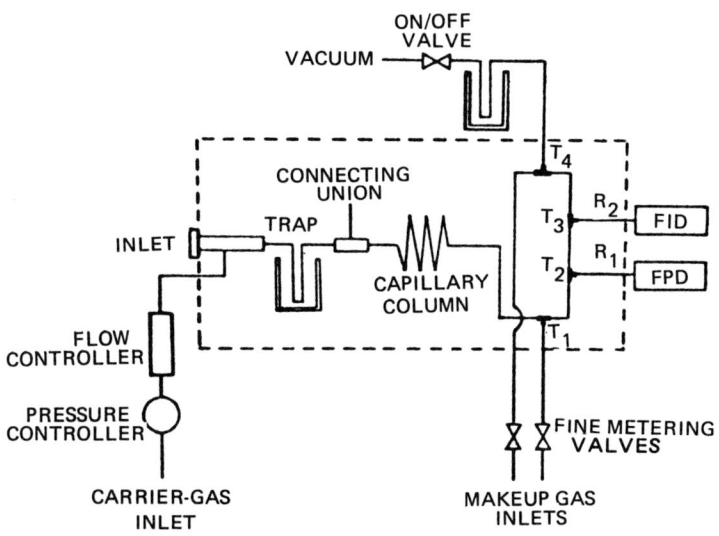

Figure 1.71. Schematic diagram of GC flow system. From Bertsch et al. [200], courtesy of Analytical Chemistry.

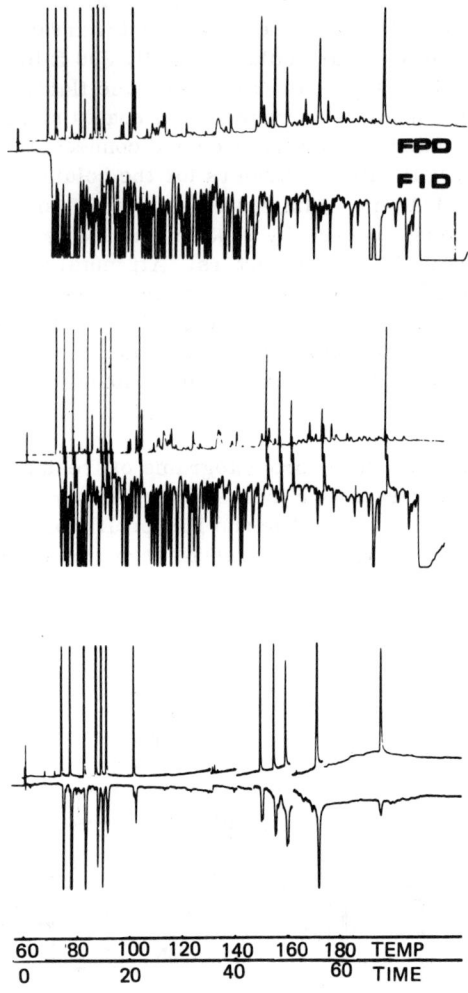

Figure 1.72. Top: Chromatograms of total cigarette smoke condensate; upper recording FPD; lower recording FID. Sampling conditions, see text. Chromatographic conditions: Nickel capillary 300 ft by 0.02 in.; Emulphor ON 870; carrier gas He 12 psi; temp. progr. 80°C (10 min) 80 to 180°C, 2°C/min. Middle: Heart cutting of selected sulfur compounds. Conditions as above. Bottom: Reconstructed chromatograms of selected fractions. Conditions as above. From Bertsch et al. [200], courtesy of Analytical Chemistry.

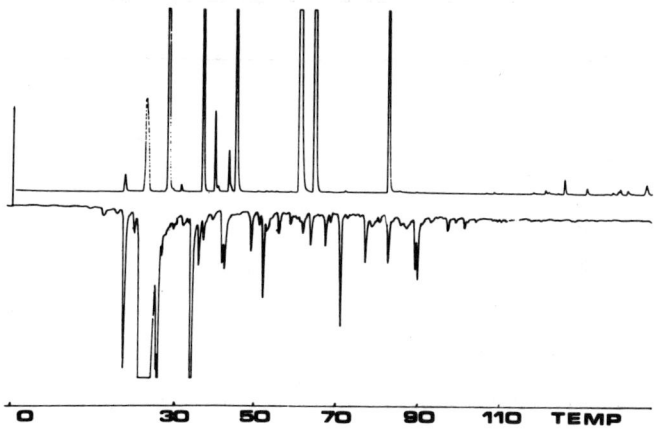

Figure 1.73. Partial chromatogram of isolated fractions on phase of different selectivity. Upper recording FPD; lower recording FID; sampling conditions as above. Chromatographic conditions: nickel capillary column 245 ft by 0.02 in. OV-17; carrier gas He 8 psi; temp. prog. 30°C (20 min) 30 to 180°C, 2°C/min. From Bertsch et al. [200], courtesy of Analytical Chemistry.

which were cut in the previous run. All sulfur compounds selected for cutting were well recovered. The flame ionization detector demonstrates that the majority of interferences, in this case compounds which do not contain sulfur, have been removed at this point. Figure 1.73 represents the first part of the final separation on a phase of different selectivity.

I. Analysis of Specific Miscellaneous Compounds

1. Aryl Methyl Ethers

In 1958, Carruthers, Johnstone, and Plimmer [203] required a GC procedure for the analysis of complex mixtures of phenols; this they successfully accomplished by separating these compounds as their methyl ethers. The analyses were carried out with a 260-cm column packed with 28.5% Apiezon M coated on C-22 Firebrick and operated at 145°C with a nitrogen carrier-gas flow rate of 15 ml/min. With these conditions, the retention volumes of the ethers ($R-O-CH_3$) relative to anisole are given below.

Although the m- and p-cresyl methyl ethers and the 2,5- and 3,5-xylyl methyl ethers could not be separated, the method was nevertheless used to analyze the phenolic constituents of tobacco smoke.

R of R-O-CH₃	RRV
Phenyl	1.00
o-Cresyl	1.56
m-Cresyl	1.67
p-Cresyl	1.67
2,3-Xylyl	2.96
2,4-Xylyl	2.42
2,5-Xylyl	2.50
2,6-Xylyl	2.05
3,4-Xylyl	3.16
3,5-Xylyl	2.50
o-Methoxyphenyl	2.51
m-Methoxyphenyl	3.11

2. Hydroxybenzyl Alcohols and Hydroxyphenyl Ethanols

Using the GC procedure and instrumentation developed by Hecht et al. [50] for the determination of hydroxybenzyl alcohols and hydroxyphenyl ethanols in tobacco leaf (see Section I.D), these investigators also applied this method to the determination of these components in tobacco smoke. The results obtained showed that the mainstream smoke of a typical 85-mm U.S. non-filter cigarette contained 3.3 ± 0.3 µg/cigarette of 2-(p-hydroxyphenyl)-ethanol (PHPE) and 1.0 ± 0.1 µg/cigarette of m-hydroxybenzyl alcohol (MHBA); that filter cigarettes contained from 1.2 to 2.8 µg/cigarette PHPE and 0.28 to 0.66 µg/cigarette MHBA. PHPE, but not MHBA, was detected in unburned tobacco (0.96 to 15.00 µg/g). The major source of PHPE in mainstream smoke was apparently transfer from tobacco, while MHBA was formed, at least partially, from cellulose. PHPE levels in cigarette tobacco and smoke were reduced by extraction of the tobacco with a hexane-ethanol azeotrope.

A typical chromatogram obtained for the analysis of the hydroxyphenyl alcohol fraction of cigarette smoke is shown in Figure 1.74; the results of the smoke analyses of some commercial and experimental cigarettes and cigars are listed in Table 1.78.

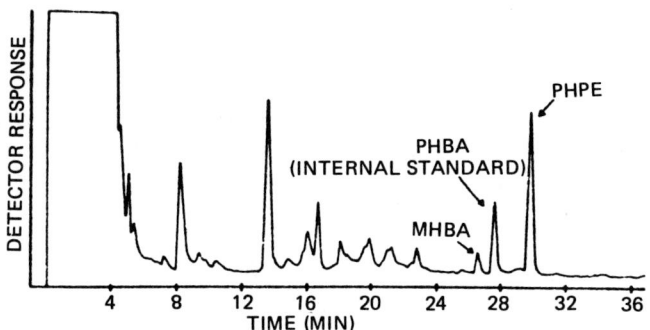

Figure 1.74. Gas chromatogram of the hydroxyphenyl alcohol fraction of cigarette smoke. From Hecht et al. [50], reproduced from the Journal of Analytical Toxicology, by permission of Preston Publications, Inc.

TABLE 1.78

PHPE and MHBA in Mainstream Cigarette and Cigar Smoke[a,b]

Product	PHPE (μg/cig)	MHBA (μg/cig)
Cigarette A (nonfilter)	3.3[c]	1.0[c]
Cigarette B (filter)	2.0	0.28
Cigarette C (filter)	1.2	0.66
Cigarette D (filter)	1.4	0.41
Burley cigarette (nonfilter)	0.9	0.28
Bright cigarette (nonfilter)	2.4	0.78
Standard experimental-blend cigarette (nonfilter)	4.2	0.56
Standard experimental-blend cigarette, extracted (nonfilter)	2.5	1.1
Cellulose cigarette (nonfilter)	ND[d]	0.32
Cigar	5.5[e]	2.5[e]

[a] From Hecht et al. [50], reproduced from the Journal of Analytical Toxicology, by permission of Preston Publications, Inc.
[b] All cigarettes were 85 mm.
[c] Average of four determinations: standard deviations; 0.3 μg/cig (PHPE), 0.1 μg/cig (MHBA); all other values were single determinations.
[d] ND = not detected; detection limit = 0.1 μg/cig.
[e] μg/7.2 g cigar.

3. Terpenes

Ho, Griest, and Guerin [204] reported their findings from a study in which "the blind assay technique was applied to the chromatographic data generated from GC profiles of terpene-enriched fractions from the total particulate matter of eight experimental cigarettes. Correlation of the entire terpene fraction peak area and 16 individual peaks with data from three skin-painting bioassay parameters indicated that the terpene fraction as a whole and at least 12 of its constituents correlate well with biological activity. Six correlating peaks were tentatively identified as d-limonene, damascenone, norphytene, neophytadiene, phytol, and squalene. High correlations were also observed between the TPM concentrations of these constituents and the total terpene fraction. Of the terpenes visualized, d-limonene would act as the best indicator of the tobacco smoke biological activity."

The GC profiling and quantitative analysis of the terpene fraction was carried out with a Perkin-Elmer model 3920 gas chromatograph equipped with a flame ionization detector and a 10-ft by 1/8-in.-o.d. glass column packed with 3% Dexsil 400 on 80-100 mesh Chromosorb G-HP. The GC conditions used for terpene analysis were injector temperature, 330°C; column temperature, held isothermally for 8 min at 100°C, then programmed to 320°C at 2°C/min; detector temperature, 300°C; helium carrier-gas flow rate, 26 ml/min at 100°C. The GC terpene peaks were also identified by integrated GC-MS, the model 3920 gas chromatograph being interfaced via a jet separator to a magnetic deflection-type mass spectrometer.

The high-resolution GC profiles of the terpene-enriched fraction for three of the eight experimental cigarettes are shown in Figure 1.75, in which the top profile corresponds to a cigarette possessing high skin-painting biological activity. From these chromatograms, the retention times of d-limonene, damascenone, norphytene, neophytadiene, phytol, and squalene were approximately 3.40, 24.60, 41.50, 48.30, 64.50, and 93.20 min, respectively.

4. Ammonia

With regard to the determination of ammonia in tobacco smoke, two GC studies have been reported in the literature [75, 205].

Ayers [75] described the instrumentation and procedure used for the GC analysis of ammonia in tobacco (see Sec. I.G of this chapter). This method was also applied to its analysis in cigarette smoke.

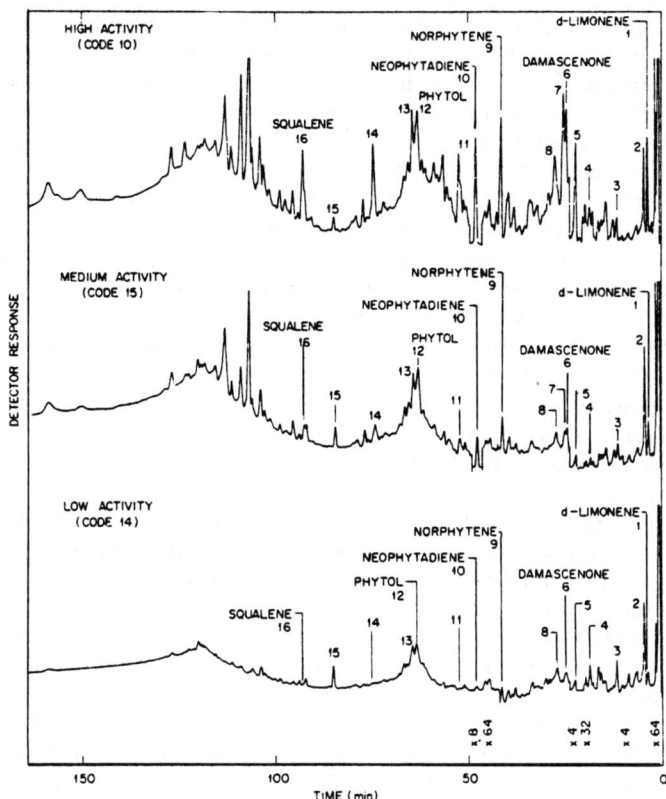

Figure 1.75. High-resolution GLC profiles of the terpene-enriched fraction from three NC1 experimental cigarettes. From Ho et al. [204], courtesy of <u>Analytical Chemistry</u>.

As noted by Ayers, smoke was collected by four different trapping systems: (1) a spiral trap cooled in liquid nitrogen, (2) a U-trap cooled in Dry Ice/acetone, (3) a bubbler trap containing dilute sulfuric acid, and (4) a Cambridge filter. The first three methods gave comparable results, but a small proportion (about 5%) of ammonia passed through the Cambridge filter. Thus, for all subsequent work, bubbler traps were used. On completion of smoking, the contents of the traps (two bubblers in series) were filtered through a sintered glass filter (porosity 2) into a beaker. The traps are washed out with ammonia-free water and the washings also passed through the filter into the beaker. The internal standard (5 ml of a 10% w/v solution of methylamine in 0.05 M sulfuric acid) was added, and the whole solution concentrated to about 1 ml by evaporation. Portions of this concentrated solution were used for GC analysis.

Brunnemann and Hoffmann [205] described in 1975 another method for ammonia determination in cigarette and cigar smoke. After the tobacco products were smoked through acid traps, their contents were also concentrated and then analyzed by GC using a barium oxide precolumn for the removal of water. The GC results were obtained with a Hewlett-Packard model 7620 gas chromatograph equipped with a thermal conductivity detector and a 2-m by 4-mm-i.d. glass column packed with Chromosorb 103. The GC conditions were injector temperature, 200°C; column temperature, 110°C; detector temperature, 140°C with a bridge current of 280 mA; helium carrier-gas flow rate, 45 ml/min. With these settings, the retention time of ammonia was 1 min.

Using this procedure, Brunnemann and Hoffmann found that "the mainstream smoke of the standard cigarette (85 mm without filter tip) contained 131 ± 6.8 µg of ammonia. The mainstream smoke of plain cigarettes, filter cigarettes, little cigars, cigars, and a marijuana cigarette contained between 80 and 300 µg of ammonia; sidestream smoke of cigarettes and little cigars delivered 40 to 73 times higher amounts of ammonia than mainstream smoke; a Manila cigar yielded 670 times more ammonia in the sidestream than in the mainstream smoke. A study of five cigar tobaccos indicated that their mainstream smoke yields highest ammonia values when smoked in the form of cigarettes, medium values when smoked in the form of little cigars, and the lowest values when smoked in the form of cigars.

5. Hydrazine

The GC method of Liu et al. [76] for the determination of hydrazine in tobacco, outlined in Section I.G, was also applied to its analysis in smoke. For smoke analysis, hydrazine was trapped from tobacco smoke by reaction with pentafluorobenzaldehyde. The formation of decafluorobenzaldehyde azine whose mass spectrum is shown below prevented the loss of hydrazine

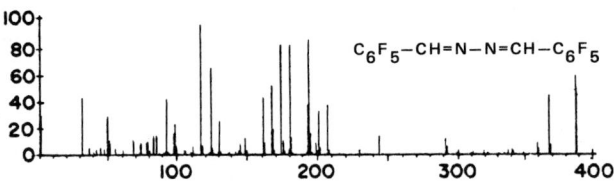

by artificial reaction with other smoke constituents. Decafluorobenzaldehyde azine was enriched by chromatography and subsequently determined by GC with FID and ECD (detection limit for hydrazine, 0.1 ng/cigarette). From the mainstream and sidestream smoke of a popular U.S. nonfilter cigarette were isolated 31.5 and 94.2 ng of hydrazine, respectively.

6. Maleic Hydrazide

Liu and Hoffmann [206] developed a method for the determination of maleic hydrazide (used to control tobacco sucker) using a variety of techniques, including ion-exchange chromatography, reaction with 4-chlorobenzyl chloride, absorption chromatography, and, finally, gas chromatography. To determine the bis(4-chlorobenzyl) derivative of maleic hydrazide, a Hewlett-Packard model 7620A gas chromatograph equipped with flame ionization and ^{63}Ni electron capture detection and a 0.6-m by 3-mm column packed with 3% XE-60 on 100-120 mesh Gas Chrom Q was used. To obtain a retention time of approximately 24 min for the MH derivative as shown in the upper part of Figure 1.76, the following GC conditions were employed: column temperature, 210°C; helium carrier-gas flow rate, 40 ml/min. These conditions were sufficient to provide a detection limit of 1 ng using the ^{63}Ni ECD.

7. 1- and 2-Naphthylamines

Masuda and Hoffmann [207] described a GC method for the analysis of 1-naphthylamine (I) and 2-naphthylamine (II) in cigarette smoke. The basic volatiles of the smoke of 300 cigarettes were reacted with pentafluoropropionic anhydride and the resulting neutral components chromatographed on Florisil. The concentrates of N-pentafluoropropionamides of I and II were analyzed by gas chromatography with an electron capture detector which had a sensitivity limit less than 1 ng.

The GC separations were performed with a Varian Aerograph model 1200 gas chromatograph equipped with an electron capture detector and a 1.7-m by 3-mm glass column filled with 7.5% QF-1 and 5% DC-200 on Gas Chrom Q (80-100 mesh). At a column temperature of 145°C (no other GC conditions given), the retention times of the PFP derivatives of 1- and 2-naphthylamines were 7.1 and 9.5 min, respectively. Their mass spectra, obtained with a Hitachi-Perkin-Elmer RMU-6D mass spectrometer operated at 70 eV, are shown in Figure 1.77.

8. Nitric Oxide

Horton, Stokely, and Guerin [208] developed a method for the determination of nitric oxide (NO) in the gas phase of cigarette smoke using a Coulson

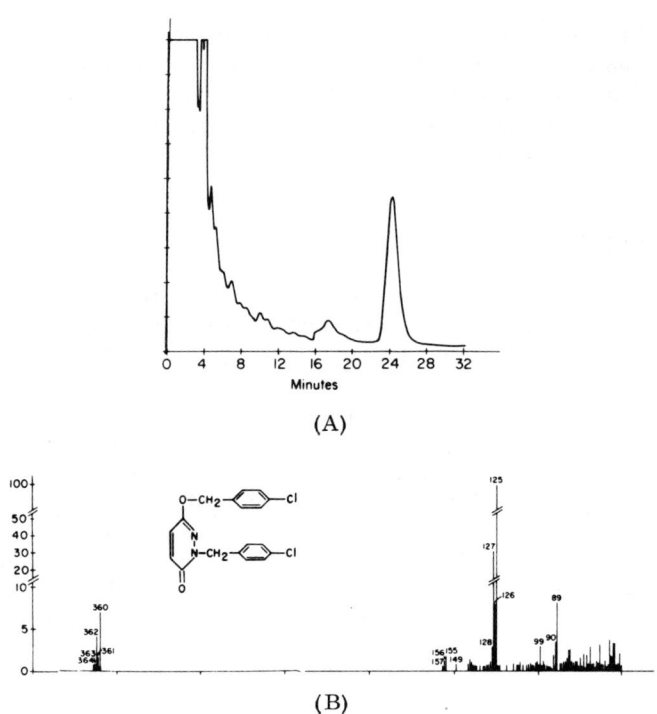

Figure 1.76. Gas chromatogram (A) and mass spectrum (B) of bis(4-chlorobenzyl) derivative of maleic hydrazide. Adapted from Liu and Hoffman [206].

electrolytic conductivity detector. A sample-averaging smoking device was used to sample the gas phase and transfer it to the detector via a gas chromatograph containing a Porapak P column. The NO separated by the GC column was passed to the furnace of the Coulson electrolytic conductivity detector where it was reduced to NH_3 by a hydrogen-activated nickel catalyst.

For this study, a Microtek model 220 gas chromatograph equipped with a Coulson electrolytic conductivity detector and a 4-ft by 1/4-in. Teflon column packed with 80-100 mesh Porapak P was used for the determination of NO. The column was operated at room temperature (24°C) with a helium carrier-gas flow rate of 60 ml/min. The Coulson detector furnace was maintained at 500°C; hydrogen flowed through the combustion tube at 50 ml/min and the helium flow was 10 ml/min, in addition to that from the column. With these conditions, NO was eluted from the column in approximately 50 sec.

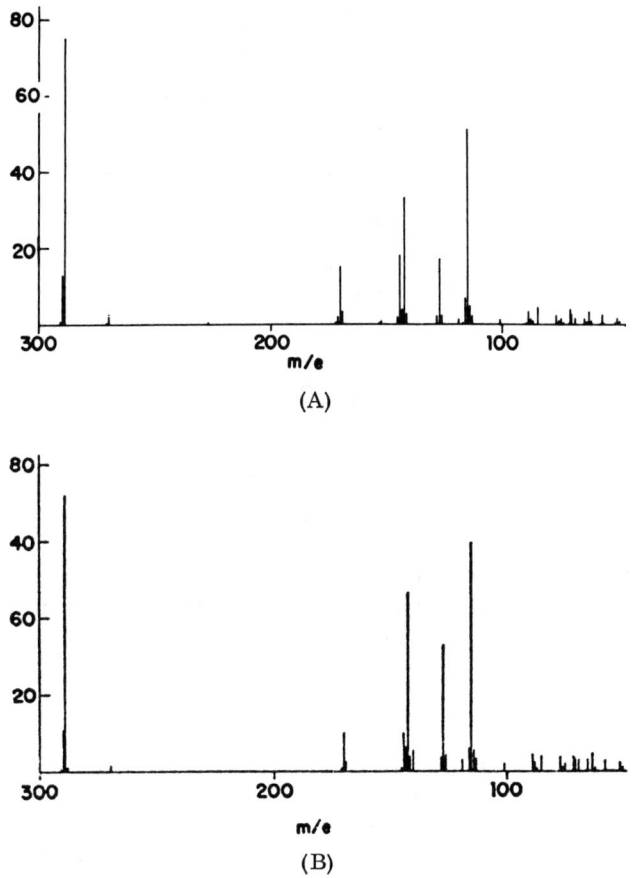

Figure 1.77. Mass spectrum of PFP derivative of 1-naphthylamine (A) and 2-naphthylamine (B). Adapted from Masuda and Hoffmann [207].

A chromatogram of NO in the gas phase of cigarette smoke is shown in Figure 1.78. The small peak on the tail of the NO peak is CO_2. Some of the CO_2 in the sample is converted to CH_4 by the catalyst but the efficiency of the conversion is poor at 500° C, and the alkaline trap is too small to remove all of the remaining CO_2. The CO_2 in no way interferes with the measurement of the NO peak. The CO_2 can be separated by several minutes from NO with a Porapak Q column.

The results of the determination of NO in the average gas-phase sample of five randomly selected, filter-tipped cigarettes are shown in Table 1.79 (column 1). Column 2 shows data obtained by a collaborator who used an

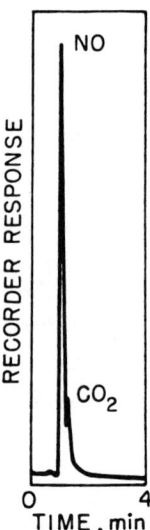

Figure 1.78. Chromatogram of NO in a Kentucky reference cigarette (1R1) using the Coulson electrolytic conductivity detector. From Horton et al. [208], courtesy of Analytical Letters.

TABLE 1.79

Comparative Determinations of Nitric Oxide in the Gas Phase of Cigarette Smoke[a]

Cigarette filter	NO (mg/cigarette)		
	This lab. (GC)	Collaborator (IR)	This lab. (colorimetric)
Control[b]	0.50	0.42	0.28
Cellulose acetate	0.42	0.45	0.27
Cellulose-charcoal-cellulose	0.39	0.40	0.38
Cellulose-oxidant-cellulose	0.26	0.22	0.11
Perforated[c]	0.16	0.17	0.14

[a] From Horton et al. [208], courtesy of Analytical Letters.
[b] Empty filter tip.
[c] Control cigarette with perforated tip.

IR analyzer, while data in column 3 was based on the use of Saltzman reagent. With this system, it was reported that the lower limit for NO detection was 5 ng.

9. Nitrobenzenes

Hoffmann and Rathkamp [209] described a method for the determination of nitrobenzenes in cigarette smoke in which the nonvolatile particulate matter of the smoke collected in solvent was suspended in dilute acid and water steam distilled. The neutral portion of the distillate was chromatographed on alumina, and the concentrate of the nitrobenzenes was analyzed by gas chromatography. For quantitative studies, a Varian Aerograph model 1200 gas chromatograph equipped with an electron capture detector and a 6.1-m by 3-mm column packed with 5% OV-225 on 80-100 mesh Gas Chrom P was used. At a column temperature of 200° C (no other GC conditions specified), the retention times of several nitrobenzenes, these listed in Table 1.80, were obtained. With the electron capture detector, as little as 1 ng of a nitrobenzene could be detected.

In a related MS study (electron energy of 50 eV), the mass spectra of 2-nitrotoluene, 4-nitro-1,3-dimethylbenzene, 4-nitrocumene, 4-nitro-1,2-dimethylbenzene, 4-nitrotoluene, and 2-nitro-1,4-dimethylbenzene are shown in Figure 1.79.

10. Vinyl Chloride

Hoffmann, Patrianakos, Brunnemann, and Gori [210] developed a chemical-analytical method for the quantitative determination of vinyl chloride in tobacco smoke. Vinyl chloride from the mainstream smoke is trapped on charcoal, extracted, and subsequently converted to 1,2-dibromo-1-chloroethane. The latter was enriched by column chromatography and determined by gas chromatography using an electron capture detector. GC analyses were performed with a Hewlett-Packard gas chromatograph (model 7260A) equipped with a ^{63}Ni electron capture detector and a 12-ft by 2-mm-i.d. glass column packed with 10% OV-17 on 80-100 mesh Gas Chrom Q. With the column maintained at 60° C, the argon carrier gas with 5% methane set at 22 ml/min, and the detector temperature held at 200° C, the retention times of dibromopropane, dibromopropene, dibromobutane, 2,3-dibromo-n-butane, dibromobutene and 1,2-dibromo-1-chloroethane were about 10.50, 13.70, 17.10, 21.80, 25.85, and 32.60 min, respectively.

From the mainstream smoke of a popular 85-mm cigarette without filter tip, Hoffmann et al. isolated 12.2 ng of vinyl chloride per cigarette. The VC content in the smoke of some domestic and foreign cigarettes and little cigars ranged from 5 to 27 ng, and that of a marijuana cigarette was 5.4 ng.

TABLE 1.80

Retention Times of Several Nitrobenzenes[a]

Compound	Retention time (min)
Nitrobenzene	5.60
2-Nitro-1,3-dimethylbenzene	5.89
2-Nitrotoluene	6.17
2-Nitroethylbenzene	6.90
3-Nitrotoluene	7.23
2-Nitro-1,3,5-trimethylbenzene	7.68
2-Nitro-1,4-dimethylbenzene	8.00
4-Nitrotoluene	8.00
2-Nitrocumene	8.35
4-Nitro-1,3-dimethylbenzene	8.85
5-Nitro-1,3-dimethylbenzene	9.48
4-Nitroethylbenzene	10.34
4-Nitrocumene	11.63
4-Nitro-1,2-dimethylbenzene	12.60

[a] Adapted from Hoffmann and Rathkamp [209].

11. p,p'-DDT in Tobacco Smoke

Chopra and co-workers conducted several systematic studies on the breakdown of p,p'-DDT in tobacco smokes: (1) the isolation and identification of degradation products from the pyrolysis of p,p'-DDT in a nitrogen atmosphere [211], and (2) investigations into the presence of methyl chloride, dichloromethane, and chloroform in tobacco smokes [212].

In 1971, Chopra and Osborne [211] pyrolyzed p,p'-DDT at 900°C in a nitrogen atmosphere and collected the pyrolysis products in pentane at -80°C, and then isolated them by fractional distillation and chromatography on alumina and Florisil columns. The products isolated were p,p'-DDT, p,p'-DDE, p,p'-TDE, bis-(p-chlorophenyl)chloromethane, bis-(p-chlorophenyl)methane, p,p'-dichlorobiphenyl, α,p-dichlorotoluene, hexachloroethane, chlorobenzene, tetrachloroethylene, trichloroethylene, carbon tetrachloride, chloroform, and dichloromethane. The solid (first eight) pyrolysis products were identified by gas chromatography and IR spectroscopy, and the liquid (the last six) pyrolysis products were identified by gas chromatography and colorimetric tests. Besides these pyrolysis products, p,p'-DDM, and cis- and trans-p,p'-dichlorostilbenes were also detected in the pyrolyzate.

The GC analyses were carried out with a Microtek model MT 220 gas chromatograph equipped with a ^{63}Ni electron capture detector and three column systems: column A, 6-ft by 1/4-in. glass packed with 3% SE-30 on 80-90 mesh Chromoport 30; column B, 6-ft by 1/4-in. glass packed with 5% SE-30 on 80-90 mesh Chromoport 30; and column C, 6-ft by 1/4-in. glass packed with 20% Carbowax 20M on 80-100 mesh Chromoport 30. Depending on the compound being investigated, the column temperatures and carrier-gas flow rates were varied.

Chopra and Sherman [212] investigated the possible presence of methyl chloride, dichloromethane, and chloroform in tobacco smokes. The chlorohydrocarbons were determined with a Microtek 220 gas chromatograph equipped with an electron capture and a microcoulometric detector. The columns used were (1) a 6-ft by 1/4-in. glass packed with 3% OV-17 on 60-70 mesh Chromosorb W, (2) a 6-ft by 1/4-in. glass packed with 20% Carbowax 20M on 80-90 mesh Chromoport 30, and (3) a 6-ft by 1/4-in. glass packed with 3% SE-30 on 80-90 mesh Chromoport 30. Columns 1 and 2 were used for dichloromethane and chloroform, and columns 1 and 3 were used for methyl chloride. The column temperatures for dichloromethane and chloroform were 22°C, and that for methyl chloride, 0°C.

From their data, it was concluded that:

1. There is no chloroform in the smoke of tobacco samples containing no p,p'-DDT; there is a linear relationship between the amount of p,p'-DDT present in tobacco and the amount of chloroform found in the tobacco smoke.

2. The absence of dichloromethane in tobacco smoke condensates suggests that either dichlorocarbene is not formed in tobacco smokes from tobacco treated with p,p'-DDT, or what is more likely, dichlorocarbene if formed reacts with other constituents of tobacco smoke, such as water, in preference to hydrogen.

3. Their value for methyl chloride (about 200 μg of methyl chloride/g of tobacco smoked) in tobacco smokes can be considered to be consistent with literature values.

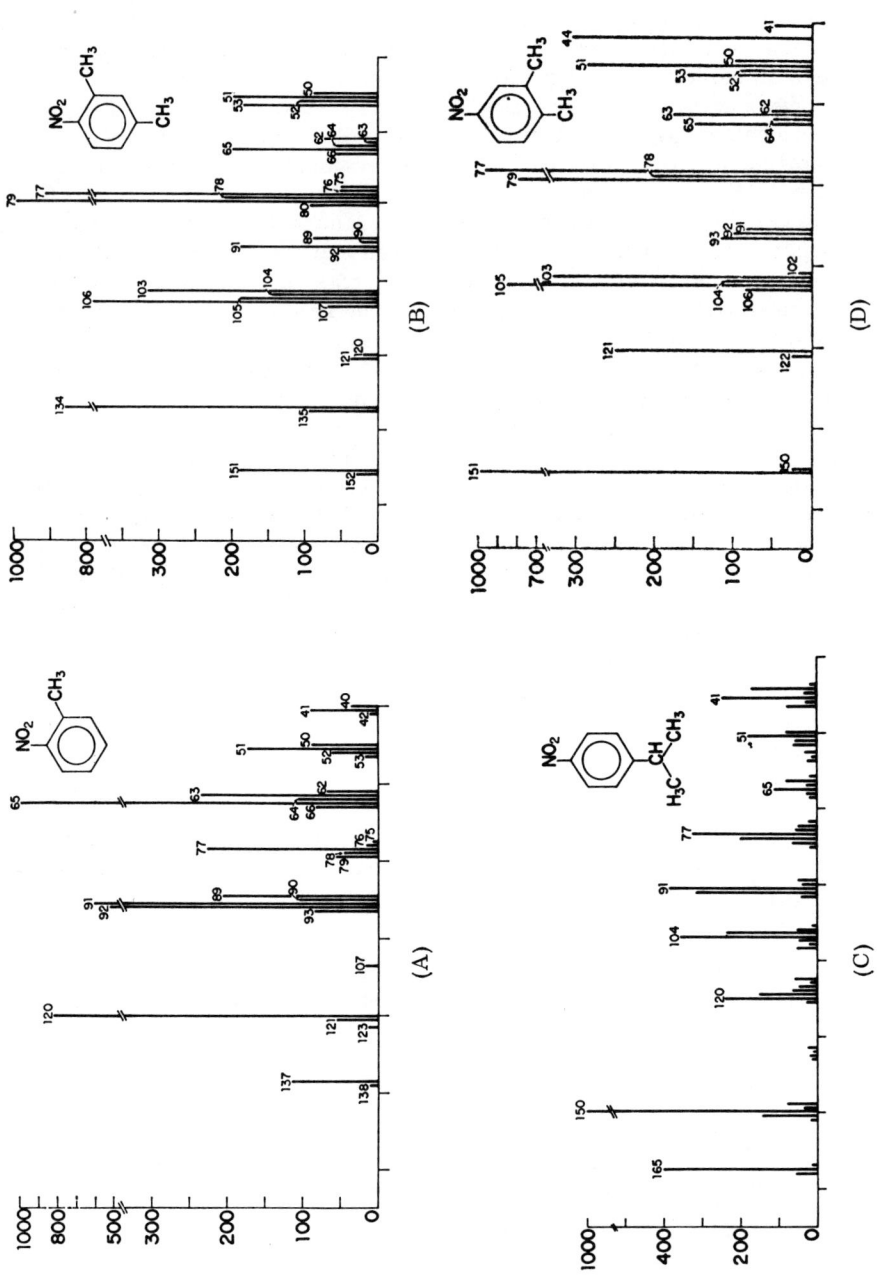

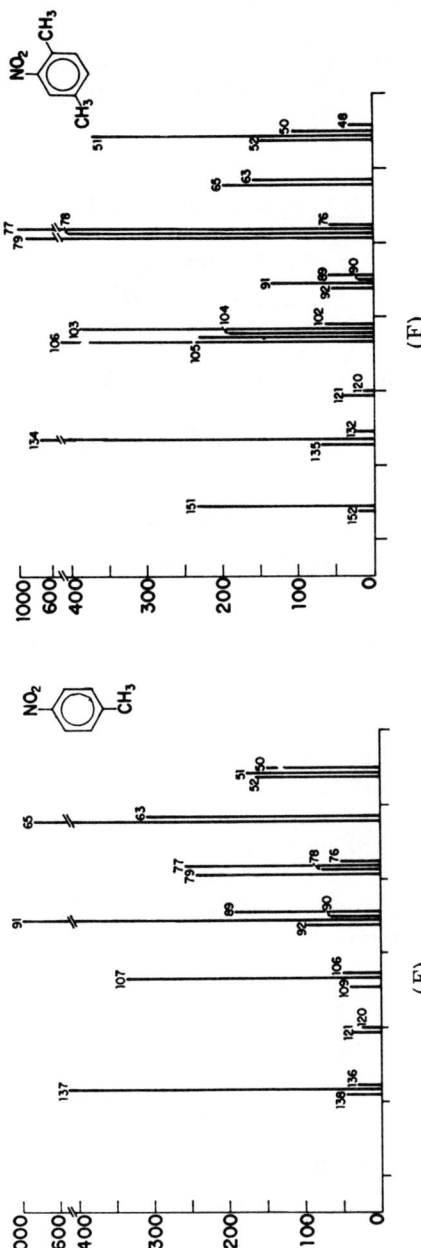

Figure 1.79. Mass spectra of 2-nitrotoluene (A), 4-nitro-1,3-dimethylbenzene (B), 4-nitrocumene (C), 4-nitro-1,2-dimethylbenzene (D), 4-nitrotoluene (E), and 2-nitro-1,4-dimethylbenzene (F). Adapted from Hoffmann and Rathkamp [209].

REFERENCES

1. Jaffe, J. H., in L. S. Goodman and A. Gilman (Eds.), The Pharmacological Basis of Therapeutics, 4th ed., Macmillan, New York, 1970.
2. Stedman, R. L., Chem. Rev., 68, 153 (1968).
3. Quin, L. D., Nature, 182, 865 (1958).
4. Kobashi, Y., Nippon Kagaku Zasshi, 82, 1262 (1961).
5. Kobashi, Y., and Watanabe, M., Nippon Kagaku Zasshi, 82, 1265 (1961).
6. Quin, L. D., and Pappas, N. A., J. Agr. Food Chem., 10, 79 (1962).
7. Cundiff, R. H., and Markunas, P. C., J. Assoc. Offic. Agr. Chem., 43, 519 (1960).
8. Jarboe, C. H., and Rosene, C. J., J. Chem. Soc., 2455 (1961).
9. Craig, J. C., Mary, N. Y., Goldman, N. L., and Wolf, L., J. Amer. Chem. Soc., 86, 3866 (1964).
10. Craig, J. C., Mary, N. Y., and Wolf L., J. Org. Chem., 29, 2868 (1964).
11. Quin, L. D., J. Org. Chem., 24, 911 (1959).
12. Kanazawa, J., and Sato, R., Bunseki Kagaku, 13, 356 (1964).
13. Alworth, W. L., DeSelms, R. C., and Rapoport, H., J. Amer. Chem. Soc., 86, 1608 (1964).
14. Kobashi, Y., and Watanabe, M., Japan Analyst, 13, 1243 (1964).
15. Alworth, W. L., and Rapoport, H., Arch. Biochem. Biophys., 112, 45 (1965).
16. Massingill, J. L., Jr., and Hodgkins, J. E., Anal. Chem., 37, 952 (1965).
17. Yasumatsu, N., Agr. Biol. Chem., 31, 1441 (1967).
18. Jacin, H., Slanski, J. M., and Moshy, R. J., Anal. Chim. Acta, 41, 347 (1968).
19. Anastasov, A., Mokhnachev, I. G., and Sherstyackh, N. A., Bulg. Tyutyun., 13, 24 (1968).
20. Harke, H. P., and Drews, C. J., Z. Anal. Chem., 242, 248 (1968).
21. Weeks, W. W., Davis, D. L., and Bush, L. P., J. Chromatogr., 43, 506 (1969).
22. Yasumatsu, N., and Murayama, T., Hatano Tob. Exp. Sta. Bull., 63, 75 (1969).
23. Cano, J. P., Catalin, J., Badre, R., Dumas, C., Viala, A., and Guillerm, R., Ann. Pharm. Fr., 28, 581 (1970).
24. Jenden, D. J., Roch, M., and Booth, R., J. Chromatogr. Sci., 10, 153 (1972).
25. Bush, L. P., J. Chromatogr., 73, 243 (1972).
26. Stead, A. H., Moffat, A. C., Caddy, B., Fish, F., and Scott, D., J. Chromatogr., 84, 392 (1973).
27. Caddy, B., Fish, F., and Scott, D., Chromatographia, 6, 251 (1973).
28. Caddy, B., Fish, F., and Scott, D., Chromatographia, 6, 335 (1973).

29. Caddy, B., Fish, F., and Scott, D., Chromatographia, 6, 293 (1973).
30. Donike, M., and Stratmann, D., Chromatographia, 7, 182 (1974).
31. Neelakantan, L., and Kostenbauder, H. B., Anal. Chem., 46, 452 (1974).
32. Foltz, R. L., Clarke, P. A., Knowlton, D. A., and Hoyland, J. R., Rapid Identification of Drugs from Mass Spectra, Battelle, Columbus Laboratories, Columbus, Ohio, January 1974.
33. Burns, D. T., and Collin, E. J., J. Chromatogr., 133, 378 (1977).
34. Baker, J. K., Anal. Chem., 49, 906 (1977).
35. Cundiff, R. H., and Markunas, P. C., Anal. Chem., 27, 1650 (1955).
36. Keller, C. J., Bush, L. P., and Grunwald, C., J. Agr. Food Chem., 17, 331 (1969).
37. Finkle, B. S., Foltz, R. L., and Taylor, D. M., J. Chromatogr. Sci., 12, 304 (1974).
38. Moffat, A. C., J. Chromatogr., 113, 69 (1975).
39. Hoffmann, D., Hecht, S. S., Ornaf, R. M., and Wynder, E. L., Science, 186, 265 (1974).
40. Hecht, S. S., Ornaf, R. M., and Hoffmann, D., J. Natl. Cancer Inst., 54, 1237 (1975).
41. Munson, J. W., and Abdine, H., Anal. Lett., 10, 777 (1977).
42. Carruthers, W., and Johnstone, R. A. W., Nature, 184, 1131 (1959).
43. Eglinton, G., Hamilton, R. J., Raphael, R. A., and Gonzalez, A. G., Nature, 193, 739 (1962).
44. Mold, J. D., Stevens, R. K., Means, R. E., and Ruth, J. M., Biochemistry, 2, 605 (1963).
45. Johnston, R. L., and Jones, L. A., Anal. Chem., 40, 1728 (1968).
46. Ellington, J. J., Schlotzhauer, P. F., and Schepartz, A. I., J. Chromatogr. Sci., 15, 295 (1977).
47. Lewis, J. S., Patton, H. W., and Kaye, W. I., Anal. Chem., 28, 1370 (1956).
48. Carruthers, W., and Johnstone, R. A. W., Chem. and Ind., 867 (1960).
49. Severson, R. F., Ellington, J. J., Schlotzhauer, P. F., Arrendale, R. F., and Schepartz, A. I., J. Chromatogr., 139, 269 (1977).
50. Hecht, S. S., Carmella, S., and Hoffmann, D., J. Anal. Toxicol., 2, 56 (1978).
51. Novotny, M., Lee, M. L., Low, C. E., and Maskarinec, M. P., Steroids, 27, 665 (1976).
52. Keller, C. J., Bush, L. P., and Grunwald, C., J. Agr. Food Chem., 17, 331 (1969).
53. Chambaz, E., and Horning, E. C., Anal. Biochem., 30, 7 (1967).
54. Novotny, M., and Zlatkis, A., Chromatogr. Rev., 14, 1 (1971).
55. Bouche, J., and Verzele, M., J. Gas Chromatogr., 6, 501 (1968).
56. Grob, K., and Grob, G., J. Chromatogr. Sci., 7, 584 (1969).
57. Novotny, M., and Farlow, R., J. Chromatogr., 103, 1 (1975).

58. Knights, B. A., *J. Gas Chromatogr.*, 5, 273 (1967).
59. Knights, B. A., and Laurie, W., *Phytochem.*, 6, 407 (1967).
60. Brooks, C. J. W., Henderson, W., and Steel, G., *Biochim. Biophys. Acta*, 296, 431 (1973).
61. Benveniste, P., Hirth, L., and Ourisson, G., *Phytochem.*, 5, 31 (1966).
62. Bergman, J., Lindgren, B. O., and Svahn, C. M., *Acta Chem. Scand.*, 19, 1661 (1965).
63. Schmeltz, I., Miller, R. L., and Stedman, R. L., *J. Gas Chromatogr.*, 1, 27 (1963).
64. Schmeltz, I., Stedman, R. L., and Miller, R. L., *J. Assoc. Offic. Agr. Chem.*, 46, 779 (1963).
65. Burdick, D., Schmeltz, I., Miller, R. L., and Stedman, R. L., *Tobacco Sci.*, 7, 97 (1963).
66. Johnstone, R. A. W., and Plimmer, J. R., *Chem. Rev.*, 59, 885 (1959).
67. Buyske, D. A., Wilder, P., Jr., and Hobbs, M. E., *Anal. Chem.*, 29, 105 (1957).
68. Mold, J. D., Means, R. E., and Ruth, J. M., *Phytochem.*, 5, 59 (1966).
69. Jones, T. C., and Schmeltz, I., *Tob. Sci.*, 12, 10 (1968).
70. Hoffmann, D., and Woziwodzki, H., *Beitr. Tabakforsch.*, 4, 167 (1968).
71. Ellington, J. J., Fisher, P. G., Higman, H. C., and Schepartz, A. I., *J. Chromatogr. Sci.*, 14, 570 (1976).
72. Campbell, I. M., and Naworal, J., *J. Lipid Res.*, 10, 589 (1969).
73. Chortyk, O. T., Severson, R. F., and Higman, H. C., *Beitr. Tabakforsch.*, 8, 204 (1975).
74. Myher, J. J., Marai, L., and Kuksis, A., *Anal. Biochem.*, 62, 188 (1974).
75. Ayers, C. W., *Talanta*, 16, 1085 (1969).
76. Liu, Y. Y., Schmeltz, I., and Hoffmann, D., *Anal. Chem.*, 46, 885 (1974).
77. Gudzinowicz, B. J., Gudzinowicz, M. J., and Martin, H. F., *Fundamentals of Integrated GC-MS, Part 3: The Integrated GC-MS Analytical System*, Vol. 7, Marcel Dekker, New York, 1977.
78. Novotny, M., *Anal. Chem.*, 50, 16A (1978).
79. Arcos, J. C., *Amer. Lab.*, June 1978, pp. 65-73.
80. Quin, L. D., *J. Org. Chem.*, 24, 914 (1959).
81. Quin, L. D., Menefee, B. S., and Pappas, N. A., *J. Org. Chem.*, 26, 267 (1961).
82. Kobashi, Y., Hoshaku, H., and Watanabe, M., *Nippon Kagaku Zasshi*, 84, 71 (1963).
83. Mumpower, R. C., and Kiefer, J. E., *Tobacco Sci.*, 11, 144 (1967).
84. Lyerly, L. A., *Tobacco Sci.*, 11, 49 (1967).

85. Levins, R. J., and Ikeda, R. M., J. Gas Chromatogr., 6, 331 (1968).
86. Grob, K., and Voellmin, J. A., J. Chromatogr. Sci., 8, 218 (1970).
87. Guerin, M. R., Olerich, G., and Horton, A. D., J. Chromatogr. Sci., 12, 385 (1974).
88. Wynder, E. L., and Hoffmann, D., Tobacco and Tobacco Smoke: Studies in Experimental Carcinogenesis, Academic Press, New York, 1967.
89. Newman, R. H., Jones, W. L., and Jenkins, R. W., Jr., Anal. Chem., 41, 543 (1969).
90. Brunnemann, K. D., Lee, H. C., and Hoffmann, D., Anal. Lett., 9, 939 (1976).
91. McNiven, N. L., Raisinghani, K. H., Patashnik, S., and Dorfman, R. I., Nature, 208, 788 (1965).
92. Beckett, A. H., Rowland, M., and Triggs, E. J., Nature, 207, 200 (1965).
93. Beckett, A. H., and Triggs, E. J., Nature, 211, 1415 (1966).
94. McKennis, H., Jr., Srivastava, S. C., and Bowman, E. R., Va. J. Science, 18, 200 (1967).
95. Horning, E. C., Horning, M. G., Carroll, D. I., Stillwell, R. N., and Dzidic, I., Life Sci., 13, 1331 (1973).
96. Harke, H. P., Schuller, D., Frahm, B., and Mauch, A., Res. Commun. Chem. Pathol. Pharmacol., 9, 595 (1974).
97. Dumas, C., Durand, A., Badre, R., Cano, J. P., Viala, A., and Guillerm, R., Eur. J. Toxicol., 8, 142 (1975).
98. Horning, E. C., Carroll, D. I., Dzidic, I., Haegele, K. D., Horning, M. G., and Stillwell, R. N., in R. S. Melville and V. F. Dobson (Eds.), Selected Approaches to Gas Chromatography-Mass Spectrometry in Laboratory Medicine, DHEW Publ. No. (NIH) 75-762, Washington, D.C., 1975.
99. Goldfarb, T., Gritz, E. R., Jarvik, M. E., and Stolerman, I. P., Clin. Pharmacol. Therapeut., 19, 767 (1976).
100. Gritz, E. R., Baer-Weiss, V., and Jarvik, M. E., Clin. Pharmacol. Therapeut., 20, 552 (1976).
101. Veal, J. T., paper presented at 173rd National ACS Meeting, New Orleans, March 21-25, 1977.
102. Wolstenholme, W. A., and Gerber, J. N., Amer. Lab., February 1977, pp. 69-82.
103. Schievelbein, H., and Grundke, K., Z. Anal. Chem., 237, 1 (1968).
104. Isaac, P. F., and Rand, M. J., Eur. J. Pharmacol., 8, 269 (1969).
105. Burrows, I. E., Corp, P. J., Jackson, G. C., and Page, B. F. J., Analyst, 96, 81 (1971).
106. Isaac, P. F., and Rand, M. J., Nature, 236, 308 (1972).
107. Feyerabend, C., Levitt, T., and Russell, M. A. H., J. Pharm. Pharmacol., 27, 434 (1975).

108. Falkman, S. E., Burrows, I. E., Lundgren, R. A., and Page, B. F. J., Analyst, 100, 99 (1975).
109. Pilotti, A., Enzell, C. R., and McKennis, H., Beitr. Takakforsch., 8, 339 (1976).
110. Hengen, N., and Hengen, M., Clin. Chem., 24, 50 (1978).
111. Dow, J., and Hall, K., J. Chromatogr., 153, 521 (1978).
112. Petrakis, N. L., Gruenke, L. D., Beelen, T. C., Castagnoli, N., Jr., and Craig, J. C., Science, 199, 303 (1978).
113. Wartman, W. B., Cogbill, E. C., and Harlow, E. S., Anal. Chem., 31, 1705 (1959).
114. Corcoran, A. C., Halmer, O. M., and Page, I., J. Biol. Chem., 128, 89 (1939).
115. Bowman, E. R., Turnbull, L. B., and McKennis, H., Jr., J. Pharm. Exp. Therapeut., 127, 92 (1959).
116. Beckett, A. H., and Rowland, M., J. Pharm. Pharmacol., 17, 59 (1965).
117. Johnstone, R. A. W., Quan, P. M., and Carruthers, W., Nature, 195, 1267 (1962).
118. Grob, K., Beitr. Tabakforsch., 9, 315 (1962).
119. Grob, K., J. Gas Chromatogr., 3, 52 (1965).
120. Rushneck, D. R., J. Gas Chromatogr., 3, 318 (1965).
121. Newsome, R. J., Norman, V., and Keith, C. H., Tobacco Sci., 9, 24 (1965).
122. Vollmin, J. A., Omura, I., Seibl, J., Grob, K., and Simon, W., Helv. Chim. Acta, 49, 1768 (1966).
123. Grob, K., Beitr. Tabakforsch., 3, 403 (1966).
124. Newman, R. H., Jones, W. L., and Jenkins, R. W., Jr., Anal. Chem., 41, 543 (1969).
125. Grob, K., and Vollmin, J. A., J. Chromatogr. Sci., 8, 218 (1970).
126. Forehand, J. B., and Kuhn, W. F., Anal. Chem., 42, 1839 (1970).
127. Grob, K., Chem. and Ind., 248 (1973).
128. Holzer, G., Oro, J., and Bertsch, W., J. Chromatogr., 126, 771 (1976).
129. Zeldes, S. G., and Horton, A. D., Anal. Chem., 50, 779 (1978).
130. Bertsch, W., Anderson, E., and Holzer, G., J. Chromatogr., 112, 701 (1975).
131. Patton, H. W., and Tuoey, G. P., Anal. Chem., 28, 1685 (1956).
132. Philippe, R. J., and Hobbs, M. E., Anal. Chem., 28, 2002 (1956).
133. Osborne, J. S., Adamek, S., and Hobbs, M. E., Anal. Chem., 28, 211 (1956).
134. Barbezat-Debreuil, S., Compt. Rend., 246, 2907 (1958).
135. Cuzin, J. L., Thor, L. V., and Morel, S., in Actes Second International Science and Tobacco Congress, Brussels, June 1958, p. 507.
136. Carugno, N., and Giovannozzi-Sermanni, G., in Proceedings of the Second International Science and Tobacco Congress, Brussels, June 1958, p. 50.

137. Kozak, A. I., and Swinehart, J. S., J. Org. Chem., 25, 222 (1960).
138. Carugno, N., in E. Sawicki and K. Cassel (Eds.), Analysis of Carcinogenic Air Pollutants, National Cancer Institute, Monograph No. 9, Bethesda, Md., 1962.
139. Norman, V., Newsome, J. R., and Keith, C. H., 17th Tobacco Chemists' Research Conference, Montreal, Canada, September 22-25, 1963.
140. Spears, A. W., Lassiter, C. W., and Bell, J. H., J. Gas Chromatogr., 1, 34 (1963).
141. Caroff, J., Veron, J., Badre, R., and Guillerm, R., J. Gas Chromatogr., 2, 320 (1964).
142. Philippe, R. J., Moore, H., Honeycutt, R. G., and Ruth, J. M., Anal. Chem., 36, 859 (1964).
143. Caroff, J., Veron, J., Badre, R., and Guillerm, R., J. Gas Chromatogr., 3, 196 (1965).
144. Gelpi, E., and Oro, J., J. Chromatogr. Sci., 8, 210 (1970).
145. Eglinton, G., Scott, P. M., Belsky, T., Burlingame, A. L., Richter, W., and Calvin, M., in G. D. Hobson and M. C. Louis (Eds.), Advances in Organic Geochemistry, Pergamon Press, London, 1964, p. 41.
146. Bell, J. H., Ireland, S., and Spears, A. W., Anal. Chem., 41, 310 (1969).
147. Commins, B. T., and Lindsey, A. L., Anal. Chim. Acta, 15, 557 (1956).
148. Carruthers, W., and Johnstone, R. A. W., Nature, 185, 762 (1960).
149. Hoffmann, D., and Wynder, E. L., Beitr. Tabakforsch., 1, 101 (1961).
150. Crouse, R. H., Garner, J. W., and O'Neill, H. J., J. Gas Chromatogr., 1, 18 (1963).
151. Spears, A. W., Anal. Chem., 35, 320 (1963).
152. Malaterre, M., Loheac, J., Sellier, N., and Guichon, G., Chromatographia, 8, 624 (1975).
153. Quin, L. D., and Hobbs, M. E., Anal. Chem., 30, 1400 (1958).
154. Stedman, R. L., Burdick, D., Chamberlain, W. J., and Schmeltz, I., Tobacco Sci., 8, 79 (1964).
155. Oakley, E. T., Weissbecker, L., and Resnik, F. E., Anal. Chem., 37, 380 (1965).
156. Guerin, M. R., Olerich, G., and Rainey, W. T., Anal. Chem., 46, 761 (1974).
157. Carugno, N., and Waltz, P., Proceedings of the 3rd World Tobacco Scientific Congress, Salisbury, 1963.
158. Cook, J. W., Johnstone, R. A. W., and Quan, P. M., Israel J. Chem., 1, 356 (1963).
159. Ayres, C. I., and Thornton, R. E., Beitr. Tabakforsch., 3, 285 (1965).
160. Carugno, N., and Rossi, S., J. Gas Chromatogr., 5, 103 (1967).

161. Davis, H. J., Anal. Chem., 40, 1583 (1968).
162. Hoffmann, D., Rathkamp, G., and Woziwodzki, H., Beitr. Tabakforsch., 4, 253 (1968).
163. Hoffmann, D., Rathkamp, G., and Nesnow, S., Anal. Chem., 41, 1256 (1969).
164. Hoffmann, D., and Rathkamp, G., Anal. Chem., 42, 366 (1970).
165. Bell, J. H., Ireland, M. S., Schultz, F. J., and Spears, A. W., paper presented at 24th Tobacco Chemists' Research Conference, 1970.
166. Rathkamp, G., and Hoffmann, D., paper presented at 24th Tobacco Chemists' Research Conference, 1970.
167. Chakraborty, B. B., Kilburn, K. D., and Thornton, R. E., Chem. and Ind., 672 (1971).
168. Hoffmann, D., and Rathkamp, G., Anal. Chem., 44, 899 (1972).
169. Novotny, M., Lee, M. L., and Bartle, K. D., J. Chromatogr. Sci., 12, 606 (1974).
170. Snook, M. E., Chamberlain, W. J., Severson, R. F., and Chortyk, O. T., Anal. Chem., 47, 1155 (1975).
171. Schmeltz, I., Tosk, J., and Hoffmann, D., Anal. Chem., 48, 645 (1976).
172. Severson, R. F., Snook, M. E., Arrendale, R. F., and Chortyk, O. T., Anal. Chem., 48, 1866 (1976).
173. Severson, R. F., Snook, M. E., Chortyk, O. T., and Arrendale, R. F., Beitr. Tabakforsch., 8, 273 (1976).
174. Lee, M. L., Novotny, M., and Bartle, K. D., Anal. Chem., 48, 405 (1976).
175. Janini, G. M., Shaikl, B., and Zielinski, W. L., Jr., J. Chromatogr., 132, 136 (1977).
176. Snook, M. E., Arrendale, R. F., Higman, H. C., and Chortyk, O. T., Anal. Chem., 50, 88 (1978).
177. Gudzinowicz, B. J., The Analysis of Pesticides, Herbicides and Related Compounds Using the Electron Affinity Detector, Jarrell-Ash Co., Waltham, Mass., June 1965.
178. Gudzinowicz, B. J., Gudzinowicz, M. J., and Martin, H. F., Fundamentals of Integrated GC-MS, Part 1: Gas Chromatography, Vol. 7, Marcel Dekker, New York, 1976.
179. DeMaio, L., and Corn, M., Anal. Chem., 38, 131 (1966).
180. Beeson, J. H., and Pecsar, R. E., Anal. Chem., 41, 1678 (1969).
181. Searl, T. D., Cassidy, F. J., King, W. H., and Brown, J. B., Anal. Chem., 42, 954 (1970).
182. Bhatia, K., Anal. Chem., 43, 609 (1971).
183. Burchfield, W. H., Wheeler, R. J., and Bernos, J. B., Anal. Chem., 43, 1976 (1971).
184. Lao, R. C., Thomas, R. S., and Monkman, J. L., J. Chromatogr., 112, 681 (1975).

185. Janini, G. M., Johnston, K., and Zielinski, W. L., Jr., Anal. Chem., 47, 670 (1975).
186. Janini, G. M., Muschik, G. M., Schroer, J. A., and Zielinski, W. L., Jr., Anal. Chem., 48, 1879 (1976).
187. Dong, M., Locke, D. C., and Ferrand, E., Anal. Chem., 48, 368 (1976).
188. James, A. T., Research (London), 8, 8 (1955).
189. Willis, D. E., Anal. Chem., 39, 1324 (1967).
190. Mostecky, J., Popl, M., and Kriz, J., Anal. Chem., 42, 1132 (1970).
191. Gouw, T. H., Whittemore, I. M., and Jentoft, R. E., Anal. Chem., 42, 1394 (1970).
192. Stuckey, C. L., J. Chromatogr. Sci., 9, 575 (1971).
193. Lee, M. L., Bartle, K. D., and Novotny, M. V., Anal. Chem., 47, 540 (1975).
194. Lee, M. L., Novotny, M., and Bartle, K. D., Anal. Chem., 48, 1566 (1976).
195. Giger, W., and Schaffner, C., Anal. Chem., 50, 243 (1978).
196. Guerin, M. R., Anal. Lett., 4, 751 (1971).
197. Groenen, P. J., and Van Gemert, L. J., J. Chromatogr., 57, 239 (1971).
198. Horton, A. D., and Guerin, M. R., J. Chromatogr., 90, 63 (1974).
199. Guerin, M. R., Olerich, M. R., and Horton, A. D., J. Chromatogr. Sci., 12, 385 (1974).
200. Bertsch, W., Hsu, F., and Zlatkis, A., Anal. Chem., 48, 928 (1976).
201. Zlatkis, A., Lichenstein, H. A., and Tishbee, A., Chromatographia, 6, 67 (1973).
202. Bertsch, F., Shunbo, F., Chang, R. C., and Zlatkis, A., Chromatographia, 7, 128 (1974).
203. Carruthers, W., Johnstone, R. A. W., and Plimmer, J. R., Chem. and Ind., 331 (1958).
204. Ho, C. H., Griest, W. H., and Guerin, M. R., Anal. Chem., 48, 2223 (1976).
205. Brunnemann, K. D., and Hoffmann, D., J. Chromatogr. Sci., 13, 159 (1975).
206. Liu, Y. Y., and Hoffmann, D., Anal. Chem., 45, 2270 (1973).
207. Masuda, Y., and Hoffmann, D., Anal. Chem., 41, 650 (1969).
208. Horton, A. D., Stokely, J. R., and Guerin, M. R., Anal. Letters, 7, 177 (1974).
209. Hoffmann, D., and Rathkamp, G., Anal. Chem., 42, 1643 (1970).
210. Hoffmann, D., Patrianakos, C., Brunnemann, K. D., and Gori, G. B., Anal. Chem., 48, 47 (1976).
211. Chopra, N. M., and Osborne, N. B., Anal. Chem., 43, 849 (1971).
212. Chopra, N. M., and Sherman, L. R., Anal. Chem., 44, 1036 (1972).

Chapter 2

NATURAL, PYROLYTIC, AND METABOLIC
PRODUCTS OF MARIJUANA

In the past decade, several excellent review articles have appeared in the literature relative to the chemistry and pharmacology of marijuana [1-4]. The various <u>Cannabis sativa</u> preparations are the most widely used illicit drugs throughout the world, with the consumption of marijuana, hashish, dagga, charas, bhang, and others, mostly for hedonistic purposes, ascribed or attributed to 200 to 300 million people (a conservative estimate based on 1970 information) [1].

With regard to its pharmacology, Casarett [4] notes that "of the more than 30 cannabinoids present in cannabis resin, there is little doubt that the major psychoactive ingredient is delta-9-tetrahydrocannabinol (Δ^9-THC). In addition to the euphoria, referred to as the "high," marijuana has a significant effect on short-term memory in that it seems to affect the ability to concentrate rather than interfering with the retrieval of information already present in the memory. The more common pharmacological effects of marijuana include the increase in heart rate, conjunctival vascular congestion, dryness of the mouth and throat, and excessive hunger, especially for sweets. Occasionally dizziness, nausea and vomiting, and diarrhea are experienced. Recent research also indicates that Δ^9-THC decreases intraocular pressure about 25%, suggesting possible therapeutic application in the treatment of glaucoma."

With regard to nomenclature, two numbering systems for this chemical species are used in publications: One utilizes the formal chemical rules for numbering of pyran-type compounds, whereas the second is based on the monoterpene numbering system, where one readily notes the relationship of the cannabinoids to their biogenic terpene precursors without the necessity of changing the numbering of specific carbon atoms [2]. The formal and monoterpenoid numbering systems are indicated below for tetrahydrocannibinol, where Δ^9-THC is equivalent to Δ^1-THC. Furthermore, based on both numbering systems, Δ^8-tetrahydrocannabinol may be designated as Δ^8-THC, Δ^6-THC, or $\Delta^{1(6)}$-THC.

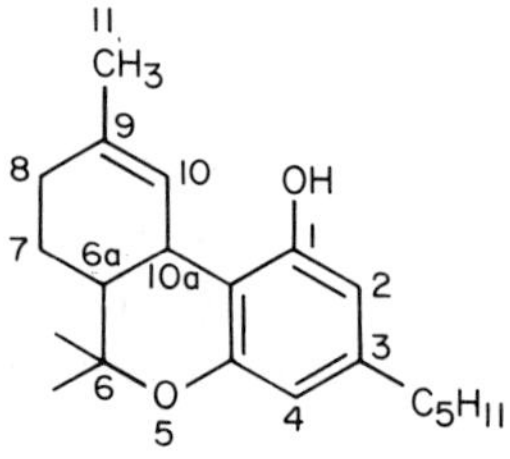

Formal Chemical
Numbering System
Δ^9 - THC

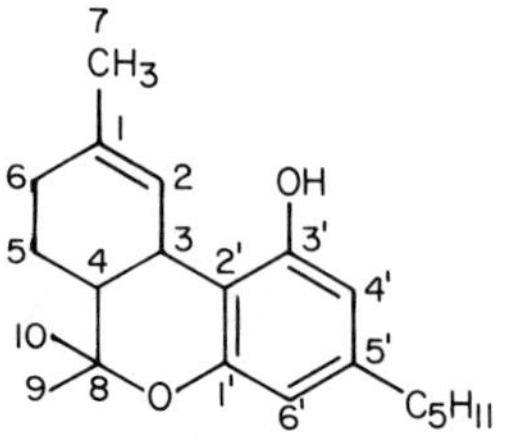

Monoterpenoid
Numbering System
Δ^1 - THC

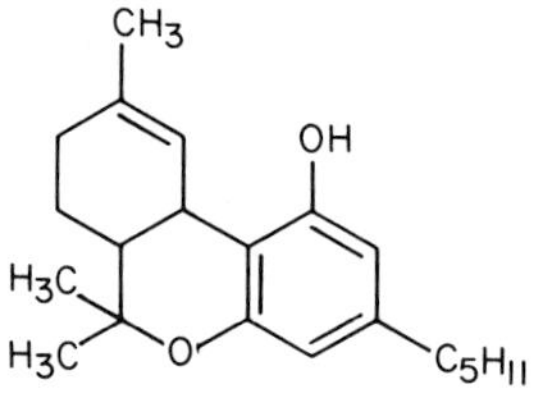

Δ^9 - Tetrahydrocannabinol

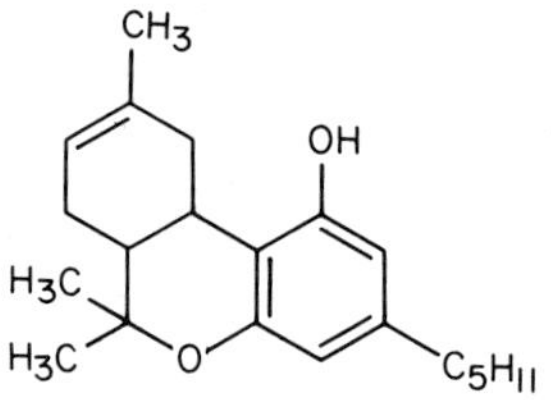

Δ^8 - Tetrahydrocannabinol

In addition to the above, others have used a numbering system as noted below for the n-pentyl side chain in cannabinol (CBN), so that 1", 7-dihydroxycannabinol and 4", 7-dihydroxycannabinol could be identified by the following structures:

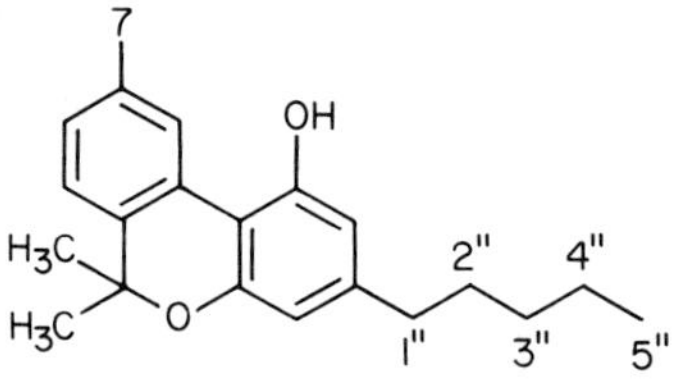

CBN

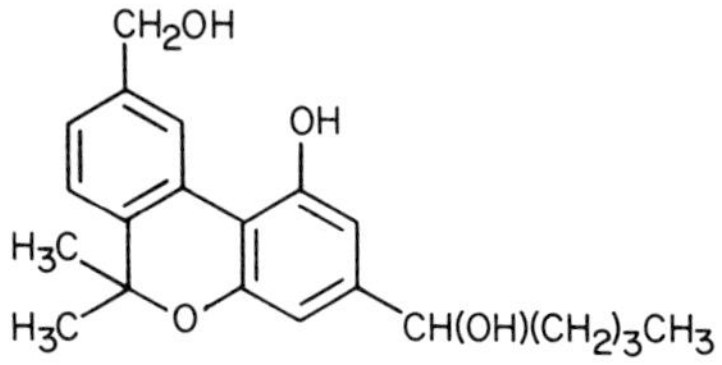

1″,7-di-OH-CBN

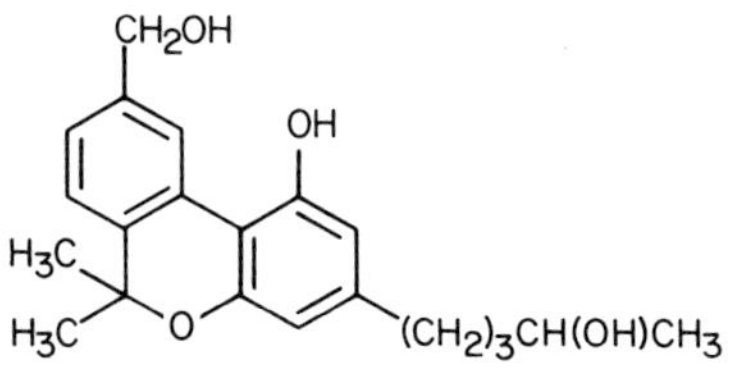

4″,7-di-OH-CBN

With regard to the structures of many naturally occurring cannabinoids, Mechoulam [1] and Mechoulam and Gaoni [5] suggested a biogenetic scheme such as that shown in Figure 2.1, based on a condensation of a terpene derivative with olivetol.

A few of the many other cannabinoids and their metabolic products to be discussed in the text are illustrated in Table 2.1.

The metabolism of Δ^9-THC has been investigated in the rabbit, rat, and man, with the major metabolite being 11-hydroxy-Δ^9-THC. In rabbit liver microsomal preparations, other metabolic products have been characterized, as shown in Figure 2.2.

In addition to the GC and/or GC-MS identification and origin determinations of cannabis appearing in the literature, the procedures used to characterize the chemical composition of the marijuana plant, the various constituents, and their metabolic products in biological fluids, as well as their pyrolysis products in smoke, will be discussed in this chapter.

I. CANNABINOID PATTERNS AND THEIR USE IN DETERMINING CHEMICAL RACE AND ORIGIN

In recent years, marijuana samples of different origins have been investigated by GC and/or GC-MS with the hope of correlating sample composition with origin [6-34, 58].

As noted by Novotny et al. [33]*:

Chromatographic methods have been implicated in at least three forensic applications concerning abuse of marijuana: (1) determination of whether an unknown sample of plant material contains marijuana;

---

*Reproduced with permission from Analytical Chemistry.

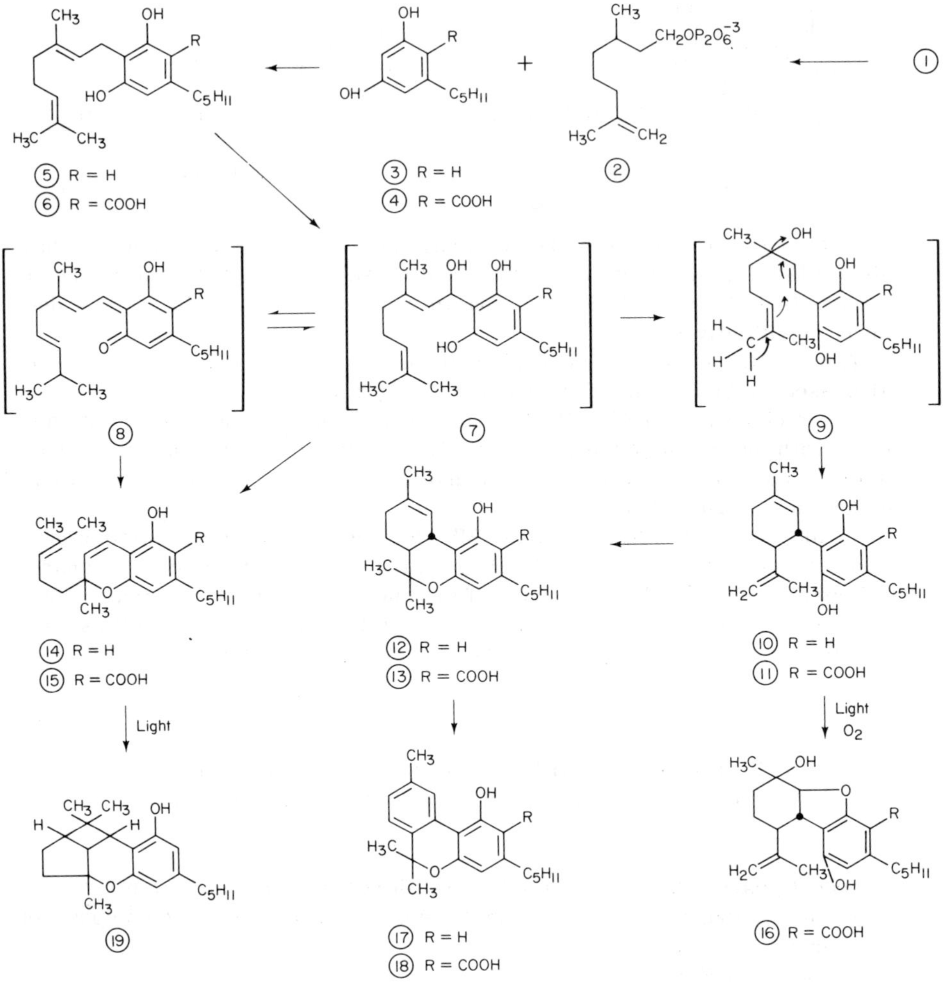

Figure 2.1. Biogenesis of cannabinoids (proposed). 1 = Melvalonate, 2 = geranyl pyrophosphate, 3 = olivetol, 4 = olivetolic acid, 5 = cannabigerol, 6 = cannabigerolic acid, 7 = hydroxycannabigerol, 8 = symmetric intermediate, 9 = allylic rearrangement, 10 = cannabidiol, 11 = cannabidiolic acid, 12 = Δ^9-or Δ^1-trans-tetrahydrocannabinol, 13 = Δ^9 or Δ^1-trans-tetrahydrocannabinol acid, 14 = cannabichromene, 15 = cannabichromenic acid, 16 = cannabielsoic acid, 17 = cannabinol, 18 = cannabinolic acid, 19 = cannabicyclol. Adapted from Mechoulam [1] and Neumeyer and Shagoury [2].

TABLE 2.1

Structures of Some Cannabinoids and Their Metabolic Products

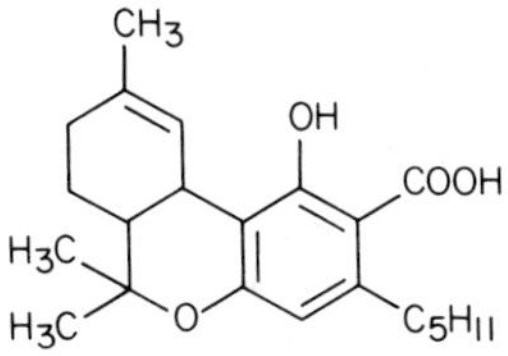

Δ^9-Tetrahydrocannabinolic Acid A

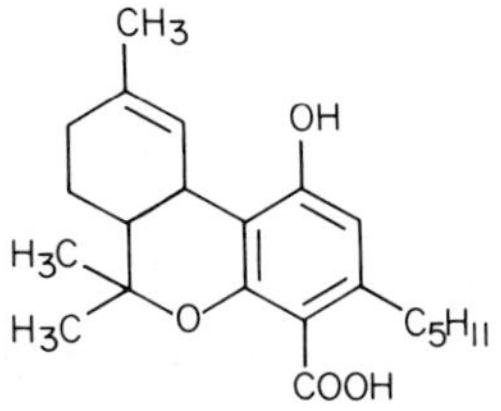

Δ^9-Tetrahydrocannabinolic Acid B

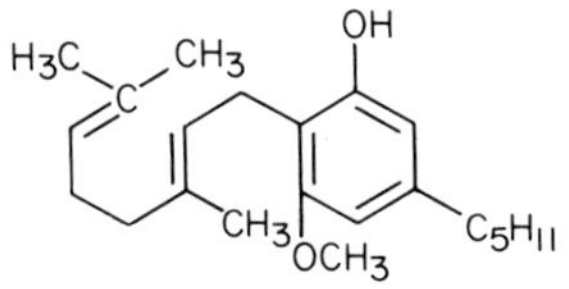

Cannabigerol Monomethyl Ether

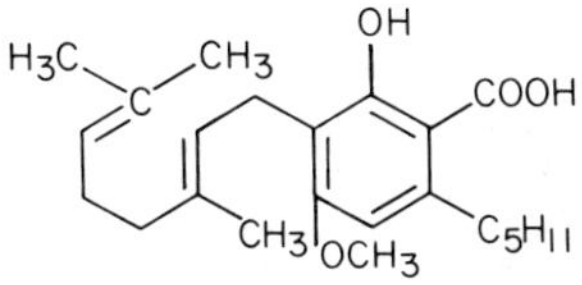

Cannabigerolic Acid Monomethyl Ether

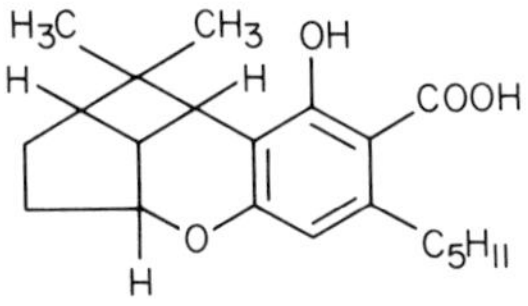

Cannabicyclolic Acid

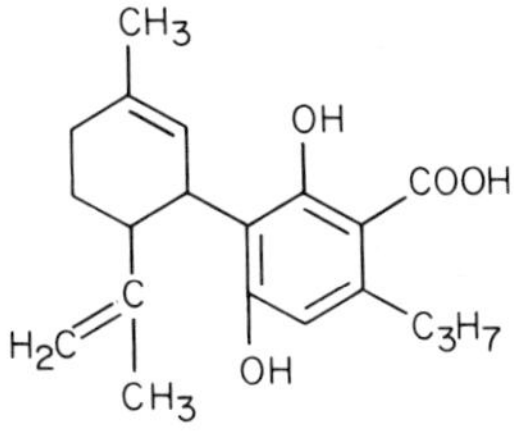

Cannabidivarinic Acid

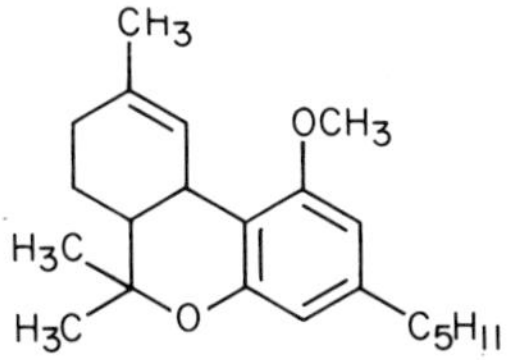

Δ^9-THC Methyl Ether

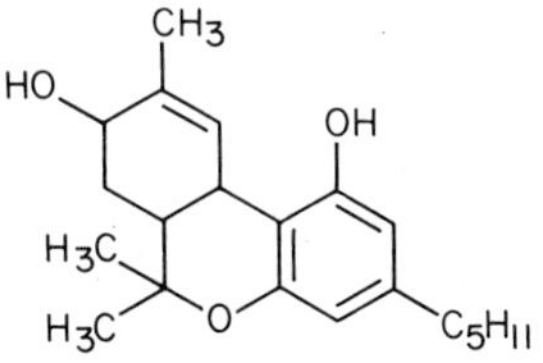

8-OH-Δ^9-THC (or 6-OH-Δ^1-THC)

(continued)

TABLE 2.1 (continued)

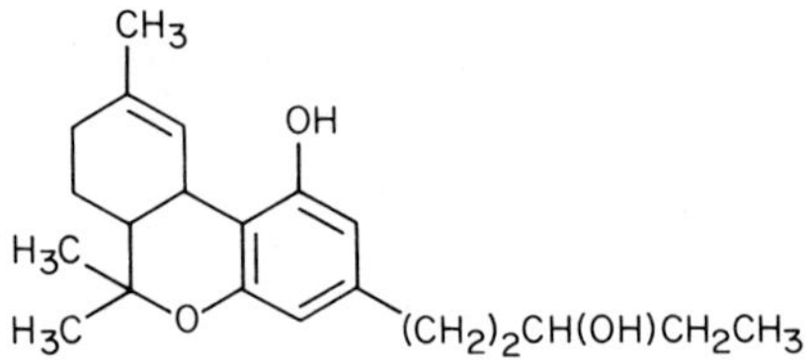

3"-OH-Δ¹-THC

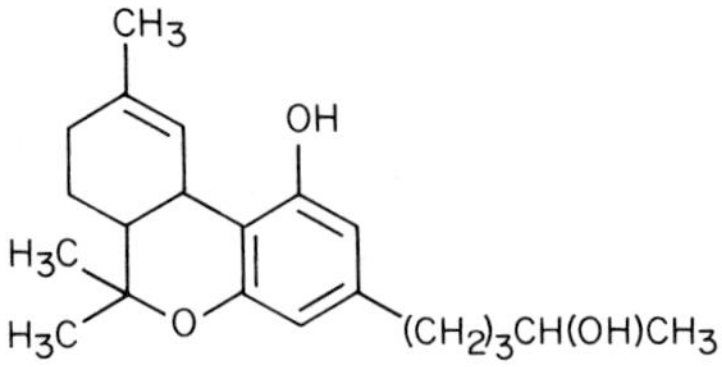

4"-OH-Δ¹-THC

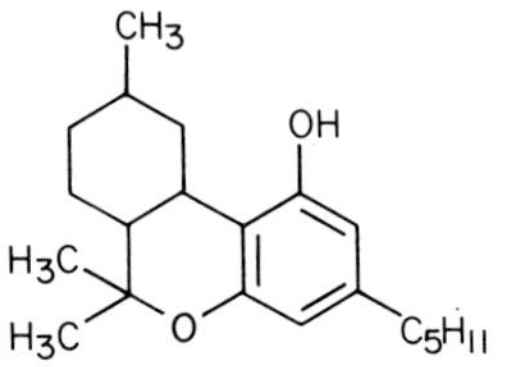

Hexahydrocannabinol

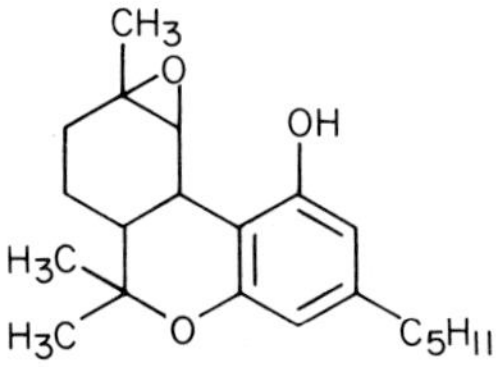

1,2- or 9,10-Epoxyhexahydrocannabinol

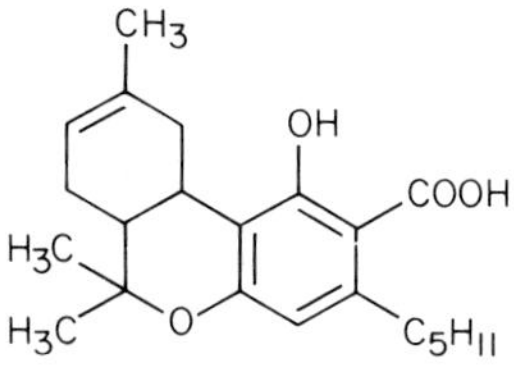

Δ⁸-Tetrahydrocannabinol Acid A

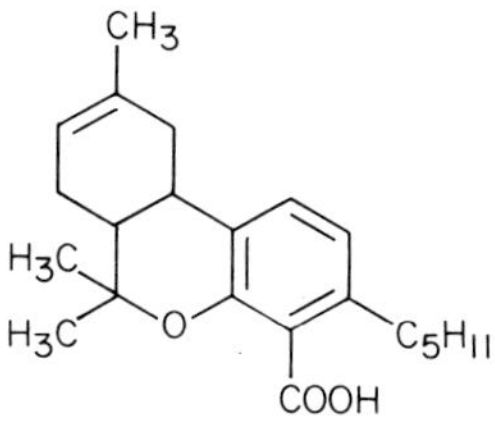

Δ⁸-Tetrahydrocannabinol Acid B

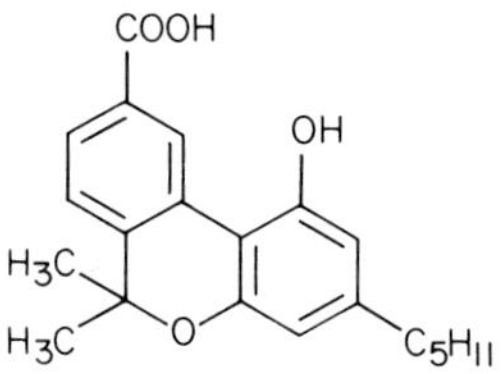

Cannabinol-7-or-11-oic Acid

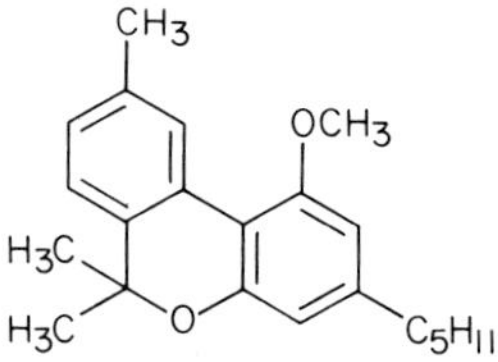

Cannabinol Methyl Ether

Cannabinoid Patterns

TABLE 2.1 (continued)

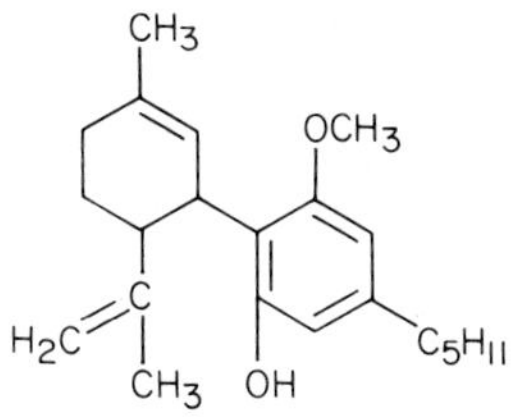

Cannabidiol Methyl Ether

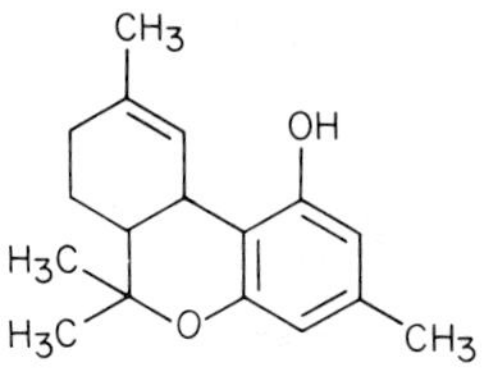

Tetrahydrocannabiorcin
(Δ^9-THC-C_1)

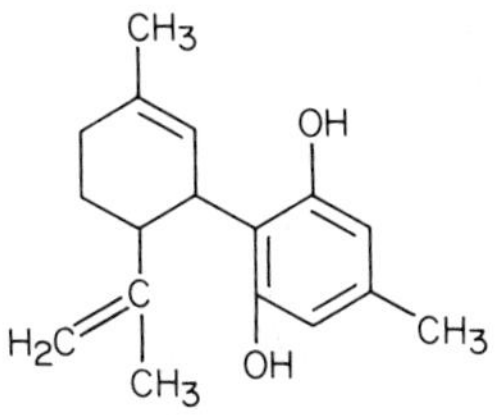

Cannabidiorcin
(Cannabidiol-C_1)

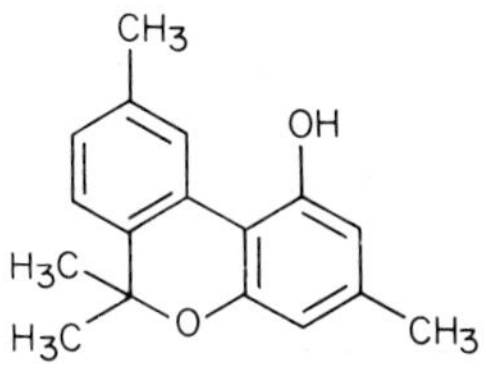

Cannabiorcin
(Cannabinol-C_1)

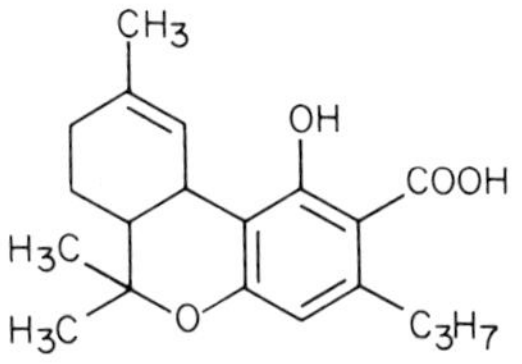

Cannabinolic Acid-C_3
(Tetrahydrocannabivarinic Acid)

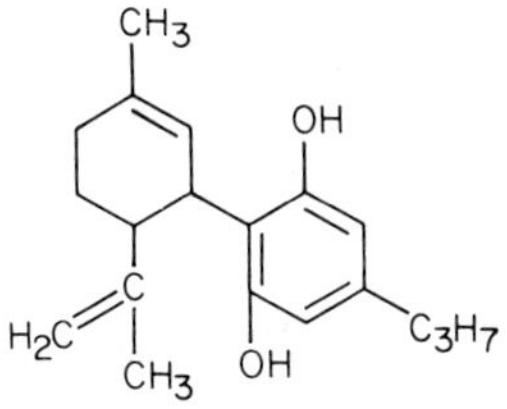

Cannabidivarin
(Cannabidiol-C_3)

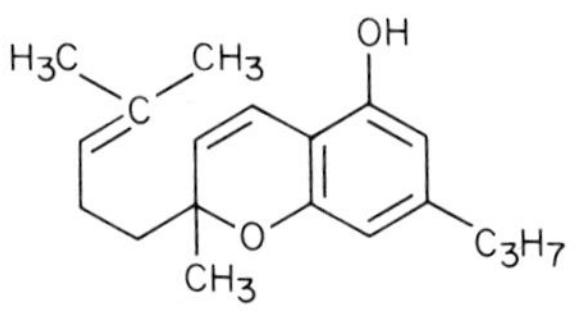

Cannabichromene-C_3

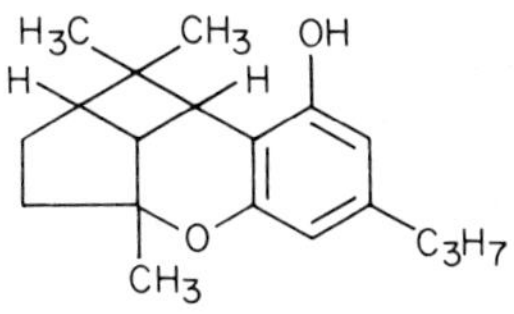

Cannabicyclol-C_3

(continued)

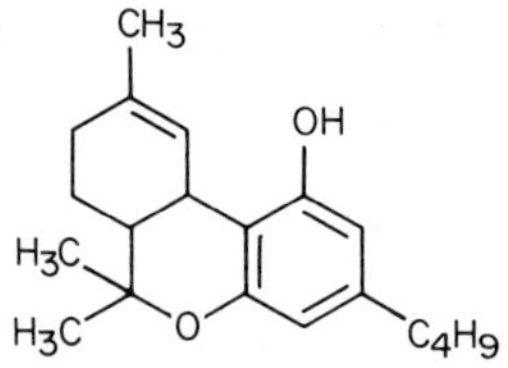

Δ^9-Tetrahydrocannabinol-C$_4$
(Δ^9-THC-C$_4$)

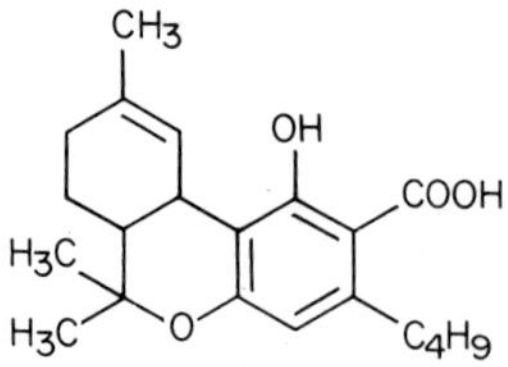

Δ^9-Tetrahydrcannabinolic Acid A-C$_4$
(Δ^9-THC Acid A-C$_4$)

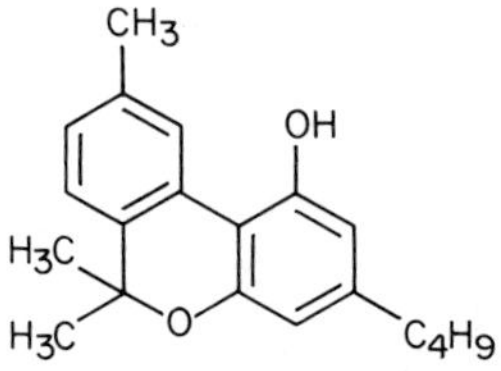

Cannabinol-C$_4$

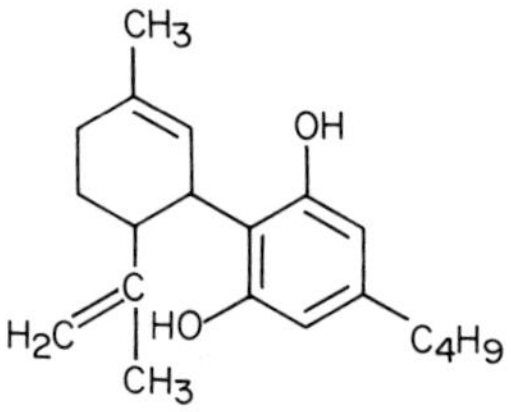

Cannabidiol-C$_4$

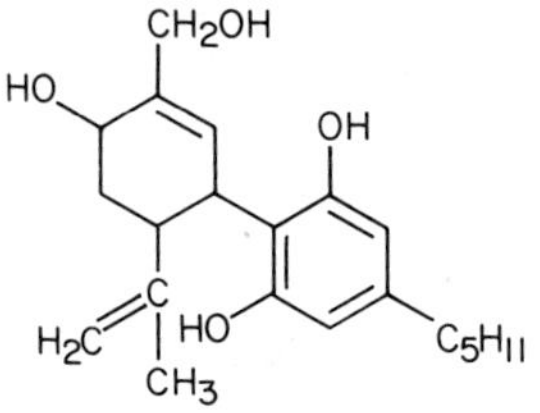

6,7-di-OH-CBD

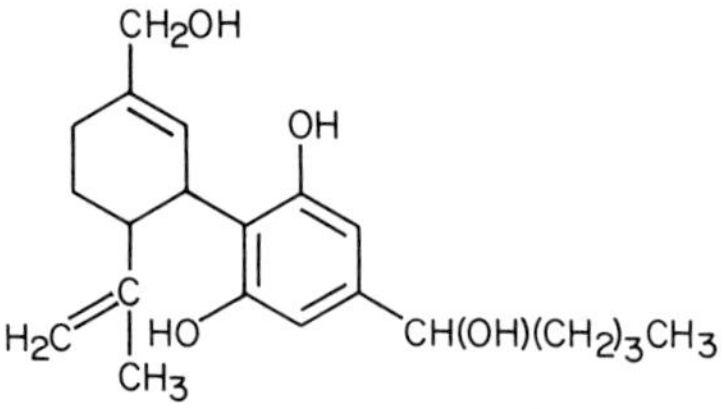

1",7-di-OH-CBD

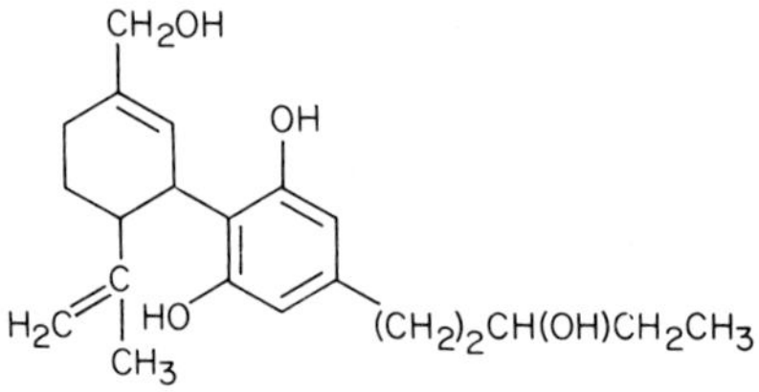

3",7-di-OH-CBD

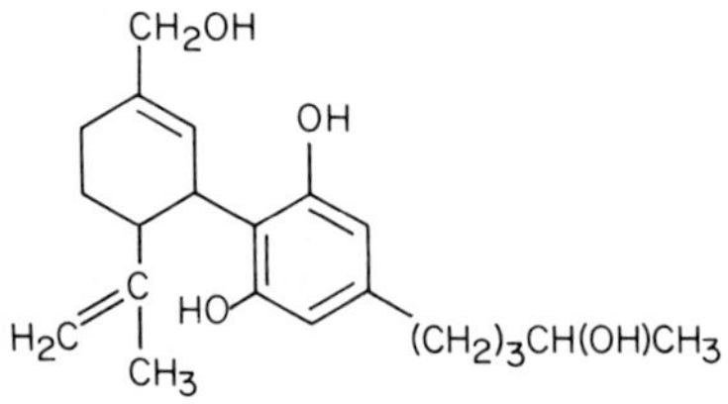

4",7-di-OH-CBD

TABLE 2.1 (continued)

5",7-di-OH-CBD	2",6-di-OH-CBD
3",6-di-OH-CBD	4",6-di-OH-CBD
3"-OH-6-OXO-CBD	4"-OH-6-OXO-CBD

(2) determination of whether cannabis samples confiscated at different locations originate from a common lot; and (3) tracing of illicit marijuana samples to their geographic origin.

Generally, the first case is the most straightforward one and does not require sophisticated separation methods. Even when marijuana is blended with noncannabis plants used as adulterants, combination of a simple histological technique with thin-layer chromatography appears to be sufficient [35] for positive identification. Tracing of illicit marijuana samples to their origin has been a considerably more complicated task; the literature concerning this problem is indeed abundant, with many attempts to correlate sample composition with its origin.

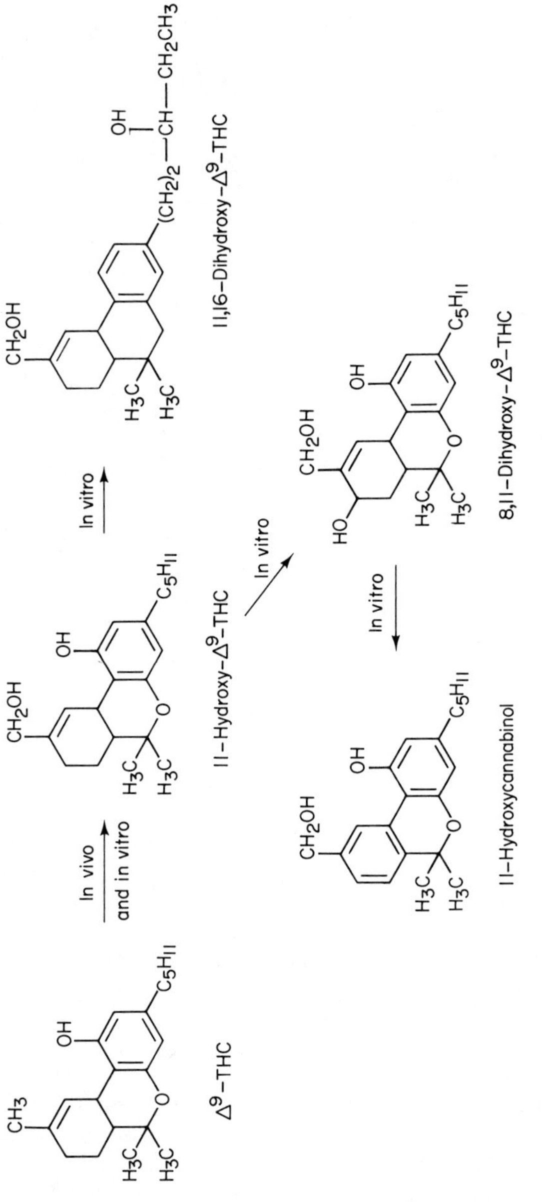

Figure 2.2. Biotransformation of Δ^9-THC.

It has been well established that the various types of cultivated and wild marijuana or hemp differ considerably in their respective cannabinoid content. Cannabis plants from different parts of the world may vary from those producing predominantly Δ^9-tetrahydrocannabinol. Consequently, several workers [9,14,29,36,37] used the measurement of the relative concentrations of cannabidiol, Δ^9-tetrahydrocannabinol, and cannabinol (so-called main cannabinoids) with the objective in mind of determining from which country each sample originates. In the course of such studies, several problems have become apparent that seriously limit this analytical approach: (1) It has been determined that, at least for the first several generations, the content of major cannabinoids produced by the plant is dependent upon the inherited properties of the seed, and that the genotype appears to be far more important than the influence of immediate geographical location and climate. Consequently, it has been suggested [14] that if cannabis seeds are shipped from one country to another for illegal cultivation, there is little valid basis for attempts to correlate the cannabinoid content with the place of origin. (2) Phillips et al. [13] have further observed a cyclic variation of cannabidiol and Δ^9-THC during the growing season of an Indian variety. The variation of cannabidiol content ranged from 0 to almost 70 mg/g of dry plant material within the period of 1 month. The variation in Δ^9-THC was not as great, but ranged from 0 to 15 mg/g during a different period. Another trend observed was that Δ^9-THC content was usually low on the same day that cannabidiol was high, and vice versa. It is obvious that, with such seasonal variations, the evaluation task becomes even more complicated. (3) Although the effects of plant processing and storage have not been investigated in great detail, they cannot be overlooked. Chemical changes of some cannabinoids with time have been observed [29,38,39].

While there seems to be a general agreement on the inherited composition of the major cannabinoids, stronger environmental effects on minor cannabinoids or noncannabinoid constituents of the plant have not been ruled out. In fact, Turner et al. [28] pointed out that the propyl homologues of major cannabinoids might be more indicative of some geographical differences, and Stromberg [20] and de Zeeuw et al. [25] suggest that other constituents should also be considered for analysis.

With the above overview of the complexity of the problem, let us examine some of the GC and/or GC-MS methods used and results obtained from such investigations and why such correlations were proposed.

In 1961, Martin et al. [6], on the basis of previous experience where essential oils were shown to be useful in plant characterization and origin determination [40,41], examined 13 samples of male and female fresh Cannabis sativa L. by GC. The oils obtained via steam distillation of the various plants at different stages of maturity were resolved with a 2.5-m column packed with 20% Apiezon M coated on Chromosorb W and operated at 220°C with a helium carrier-gas flow rate of 75 ml/min. Of the 18 peaks observed in both distillates of male and female plants, relative retention times were used to identify four peaks, these being myrcene, limonene, and α- and β-caryophyllene. Based on these initial findings, other GC analyses of oils from dried cannabis, hashish, and charas showed distinct differences in the GC profiles. They concluded by stressing that "the 14 peaks which are constantly present in the gas chromatograms of the oil serve to characterize, beyond a reasonable doubt, fresh, green Cannabis sativa L. found in Quebec and Ontario. Dried cannabis, charas, and hashish may also be distinguished and characterized on this basis."

As an extension of previous studies, Lerner [8] analyzed the major marijuana constituents, cannabidiolic acid, cannabidiol, tetrahydrocannabinol, and cannabinol, by treating the petroleum extract with diazomethane. The methyl esters formed, together with the unchanged components, were subjected to gas chromatography. Lerner noted that renewed interest in marijuana resulted from three separate but related developments: (1) the isolation of the major constituent, cannabidiolic acid [42-45], an antibacterial substance which has been postulated as the biological precursor of both cannabidiol and tetrahydrocannabinol, (2) use by criminals of an extract of marijuana plant on tobacco to reduce the ease with which illegal cigarettes are recognized by law enforcement officers [46], and (3) the possibility that detailed analyses could be correlated with its origin [7,47].

With diazomethane as the mild methylating agent and an argon ionization gas chromatograph operated at 180°C with a cyanoethyl silicone gum impregnated on 120-mesh silanized Chromosorb W to give a final concentration of 0.5%, the retention times for the major components at an argon carrier-gas flow rate of 80 ml/min are listed in Table 2.2. The relative amounts of these components in certain samples of marijuana, red oil, and hashish are also indicated in the table. Whereas the cannabidiolic acid was converted to its corresponding methyl ester, cannabinol, tetrahydrocannabinol, and cannabidiol were not affected by the diazomethane reaction. By the use of calibration data, quantitative results were obtained by integrating the areas under the peak.

In 1963, a more extensive study was performed and reported by Davis et al. [9] on the identification and origin determinations of cannabis by gas chromatography and paper chromatography. Of the four cannabinoids contained in Cannabis sativa L., tetrahydrocannabinol (THC), the physiologically active component [48,49], is assumed to be a mixture of isomers whose structures differ both in the position of the alicyclic double bond and in optical and steric properties [50-52].

In this investigation, a gas chromatograph equipped with a ^{90}Sr β-ray ionization detector and a katharometer in series was used; the latter detector was used with helium carrier gas for fraction collection. A 7-ft by

TABLE 2.2

Major Components in Marijuana from Various Sources[a,b]

Component[b]	Hashish	Red oil[c]	Marijuana		
			America	Africa	Thailand
Unknown A, 2.7	S	S	S	S	S
Unknown B, 7.1	S	S	L	M	S
Cannabidiol, 10.6	L	L		S	S
Unknown C, 12.2			L	S	S
Tetrahydrocannabinol, 14.6	L	L	L	L	L
Cannabidiolic acid, methyl ester, 17.4			S	S	L
Cannabinol, 24.8	M	M	M	S	M
Unknown D, 28.3	S	S	S	S	S

[a] From Lerner [8], courtesy of <u>Science</u>.
[b] The number after the component is the retention time (in minutes from solvent emergence) on the chromatograph. The retention time of the methyl ester of cannabidiolic acid diacetate was 35.7. Relative contents: S, small; M, medium; L, large.
[c] Red oil is a marijuana concentrate.

1/4-in.-o.d. stainless steel column was packed with 3% SE-30 on 60-80 mesh Chromosorb W, and the operating conditions employed were flash heater, 360° C; column, 212° C; detector, 240° C; argon flow rate, 125 ml/min; inlet pressure, 27 psi; cell voltage, 1275 V.

The gas chromatographic analysis of the benzene extract of the leaves and flowering tops showed the presence of cannabidiol (CBD), cannabinol (CBN), and tetrahydrocannabinol (THC), and that cannabidiol acid (CBDA) was decarboxylated to cannabidiol in the gas chromatograph even at flash heater and column temperatures of 260 and 180°C, respectively. Gas chromatograms of plant extracts and reference standards are shown in Figure 2.3; the retention times relative to cannabidiol for these compounds are tetrahydrocannabinol, 1.36; cannabinol, 1.68; Pyrahexyl (a commercially available THC homolog), 1.89. Quantitative analyses of dry cannabis leaf by gas chromatography are shown in Table 2.3, and an examination of the gas chromatograms in Figure 2.3 clearly indicates the differences in the relative amounts of CBD, THC, and CBN between plants of different

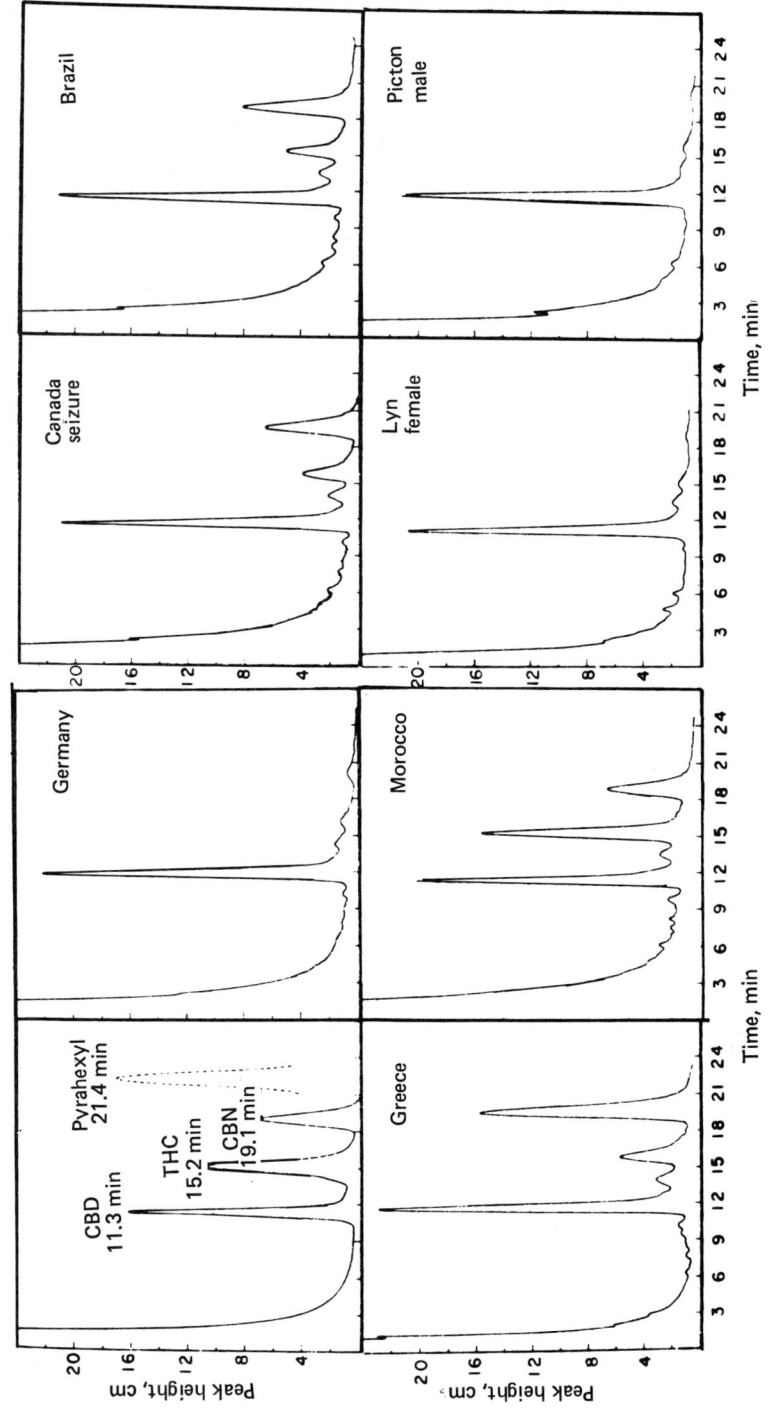

Figure 2.3. Gas chromatograms of standard substances (CBD, THC, CBN, and Pyrahexyl) and extracts of cannabis of different geographical origin. From Davis et al. [9], courtesy of Analytical Chemistry.

TABLE 2.3

Phenolic Content of Dry Cannabis Leaf by Gas Chromatography[a]

Origin[b]	CBD (%)[c]	THC (%)	CBN (%)	Volume injected (μl)[d]
Greece	2.81	0.68	3.30	1.6
Brazil	0.63	0.10	0.44	6.0
Morocco	0.48	0.31	0.24	8.0
Germany	4.70	0.10	0.36	1.0
Switzerland	1.60	0.09	0.11	2.8
Mexico? (Seizure, Canada)	0.90	0.19	0.27	4.8
Canada (Lyn, Ont.)	0.46	0.03	0.04	5.0
Canada (Picton, Ont. ♂)	0.65	0.002	0.003	6.6

[a] From Davis et al. [9], courtesy of <u>Analytical Chemistry</u>.
[b] Female plants except as noted.
[c] Original CBD + CBD from decarboxylation of CBDA.
[d] To give approximately full-scale peak height for CBD peak.

geographic origins. The investigators reported a further correlation, shown in Figure 2.4, if the area under the CBD, THC, and CBN peaks is taken as 100% and the area of the CBD peak, expressed as a percent of this total area, is plotted against the sum of areas of the THC and CBN peaks expressed as a percent of the total. They showed that the position of the points on the straight line produced is related to the geographical origin of the sample and, furthermore, that the climatic factors, sunshine, temperature, and rainfall, yield an apparent rough correlation with the amount of THC.

In 1968, Aramaki et al. [11] reported results derived from a detection method of the principal constituents of marijuana by thin-layer and gas chromatography. With regard to their TLC procedure, a new solvent system, benzene:n-hexane:diethylamine (25:10:1), was found to show a good separation of cannabidiol, tetrahydrocannabinol, and cannabinol in hemp resin by TLC, in which the R_f values of the three constituents were 0.45, 0.35, and 0.25, respectively. Furthermore, the same TLC eluent system was successfully applied to silica gel column chromatography for the isolation of the three components of hemp resin.

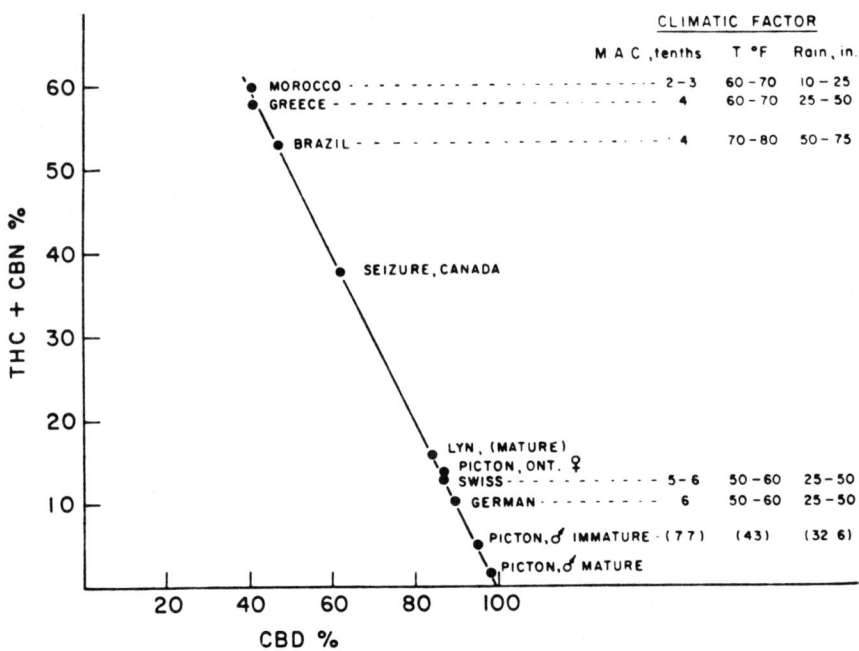

Figure 2.4. Relative percentage of CBD plotted against relative percentage of sum of THC + CBN for cannabis of different geographical origins. Analysis by gas chromatography: M.A.C. = mean annual cloudiness; T = mean temperature; Rain = annual rainfall, inches. From Davis et al. [9], courtesy of Analytical Chemistry.

On the other hand, using cocaine hydrochloride as an internal standard, these three cannabinoids were separated by GC with a Shimadzu GC-1B gas chromatograph equipped with a flame ionization detector and a 2.25-m by 4-mm U-shaped stainless steel column packed with 1.5% SE-30 on 60-80 mesh Chromosorb W. With the given GC conditions (injector temperature, 290° C; column temperature, 220° C; detector temperature, 230° C; nitrogen carrier-gas flow rate, 35 ml/min), the retention times for cannabidiol, THC, and cannabinol were 4.92, 6.58, and 8.17 min, respectively.

Using calibration curves based on peak area measurements and cocaine hydrochloride as internal standard (retention time = 2.82 min), the three cannabinoids contained in six kinds of hemp were quantitatively analyzed as indicated in Table 2.4.

With regard to their GC data, Aramaki et al. noted:

As can be seen in [Table 2.4], total amount of three constituents are much more in Indian hemps than in hemps yielded in other countries.

TABLE 2.4

Percent Cannabinoids in Dried and Powdered Materials[a]

	Constituent concentration (%)			
Source	Cannabidiol	THC	Cannabinol	Total
Japanese				
1. Hemp I	1.05	1.17	0.06	2.28
2. Hemp II	0.41	1.68	0.15	2.24
3. Hemp III	0.20	0.63	Trace	0.83
American Hemp	0.14	0.81	0.54	1.49
Indian				
1. Ganja	0.57	1.42	1.49	3.48
2. Charas	None	0.37	4.45	4.82

[a] Adapted from Aramaki et al. [11].

Content in Japanese hemp III cultivated in a pot is exceptionally low level, probably owing to insufficient nourishment. The percentage of each constituent in different hemps vary considerably. Japanese hemps contain cannabidiol comparatively more than those in Indian and American hemps. Indian hemps contain mostly cannabinol, whereas Japanese hemps do not contain appreciable amounts of it, but a much larger quantity of THC, the physiologically active constituent in hemp resin. Against a common opinion that Japanese hemps have very low psychosomatic effect, the present study has strongly suggested that Japanese hemps have contained the active constituent considerably.

It has been widely believed that cannabidiol was isomerized to THC which was then oxidized to cannabinol biogenetically or spontaneously [43,55]. It must be, therefore, noticed that the total amount of these constituents and ratio of each constituent in hemp resin, might be changeable according to when the hemps were harvested, how long they passed since harvest, and how to store them.

In 1972, Vree et al. [17,22] identified by GC-MS and GC in hashish THC, cannabidiol, and cannabinol analogs with a methyl [17,22] and a propyl [17] side chain. The structures of the parent compounds are shown in Table 2.1, whereas their C_1 and C_3 analogs are illustrated in Table 2.5.

TABLE 2.5

Structures of C_1 and C_3 Analogs of THC, Cannabidiol, and Cannabinol

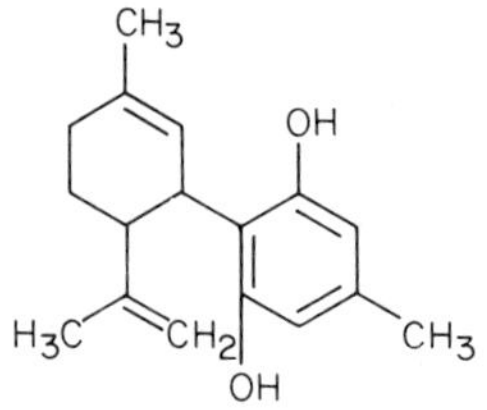 Cannabidiorcin (CBD-C_1)	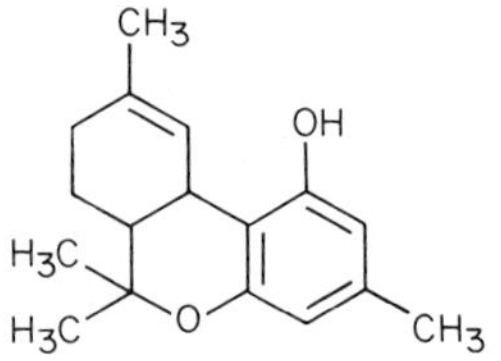 Δ^9-Tetrahydrocannabiorcin (Δ^9-THC-C_1)	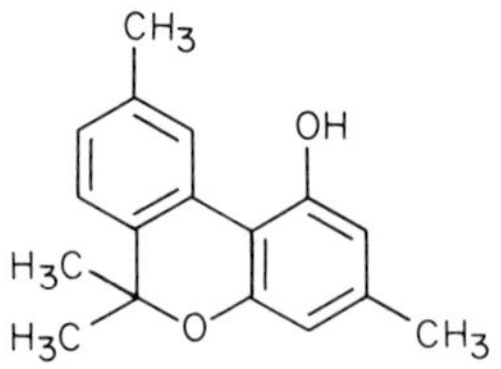 Cannabiorcin (CBN-C_1)
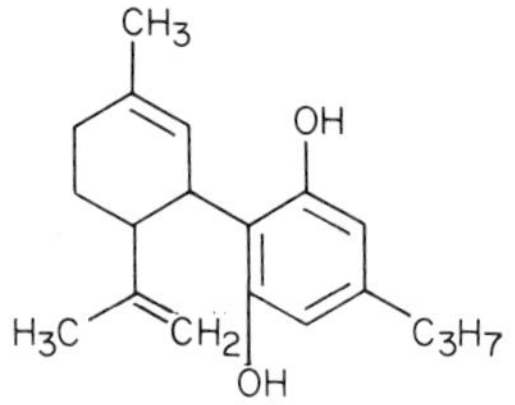 Cannabidivarin (CBD-C_3)	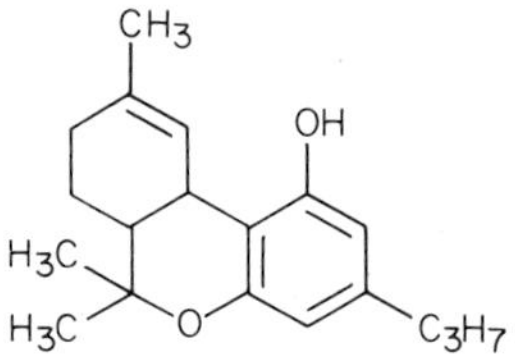 Δ^9-Tetrahydrocannabivarin (Δ^9-THC-C_3)	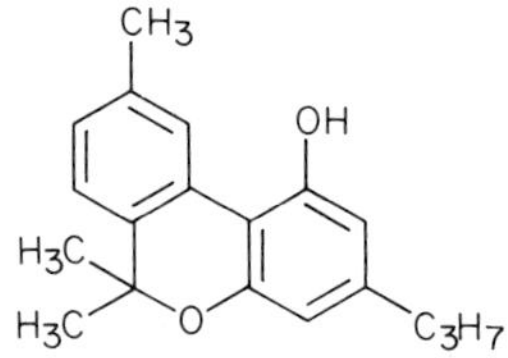 Cannabivarin (CBN-C_3)

Sample preparation consisted of the following:

Lebanese and Nepalese hashish and Brazil marijuana were powdered and extracted with n-hexane by placing the mixture for 10 min in a homogenizer. After filtration of the extracts, most of the solvent was evaporated in order to provide suitable concentrations for GC-MS.

The GC investigations were performed with a Hewlett-Packard 402 gas chromatograph equipped with a flame ionization detector and a 1.8-m by 4-mm-i.d. glass column packed with 3% OV-17 coated on 60-80 mesh Gas Chrom Q. For the separation of the cannabinoids, the GC conditions employed were injector temperature, 250° C; column temperature, 200° C; detector temperature, 250° C; nitrogen carrier-gas flow rate, 20 ml/min; hydrogen flow rate, 30 ml/min; air flow rate, 150 ml/min.

For GC-MS investigations, Vree et al. used a LKB 9000 integrated GC-MS instrument equipped with a 1.8-m by 4-mm-i.d. glass column packed with 3% OV-17 on 60-80 mesh Gas Chrom Q. Other operating features were column temperature, 200° C; separator temperature, 200° C; ion source temperature, 250°C; helium carrier-gas flow rate, 20 ml/min; mass spectra scans, repetitive and taken at 10, 12, 14, 16, 18, and 20 eV during the elution of the GC peak; trap current, 60 μA; acceleration voltage, 3.5 kV; analyzer tube pressure, 2×10^{-6} torr; total ion current monitoring at 20 eV. With this GC-MS unit, the gas chromatograms showed the same elution pattern as the gas chromatograph. The mass spectra obtained were normalized and the relative abundance of a particular mass fragment was plotted against electron voltage. Typical graphs were obtained as shown in Figures 2.5, 2.6, and 2.7.

On the other hand, the cannabinols in the mass spectrometer show a highly characteristic behavior as reported by Korte et al. [53], who found that CBD yielded different mass spectra at different ion source temperatures. Figure 2.8 is the mass spectrum of Δ^9-THC at constant ion source temperature (250°C) but taken at different ionization energy levels. As noted, both mass spectra are different.

In Figure 2.9 (typical energy/mass intensity graphs of Δ^9-THC and Δ^8-THC), it is clearly evident that there is a marked difference in the rate of formation of the mass fragments 231 and 299.

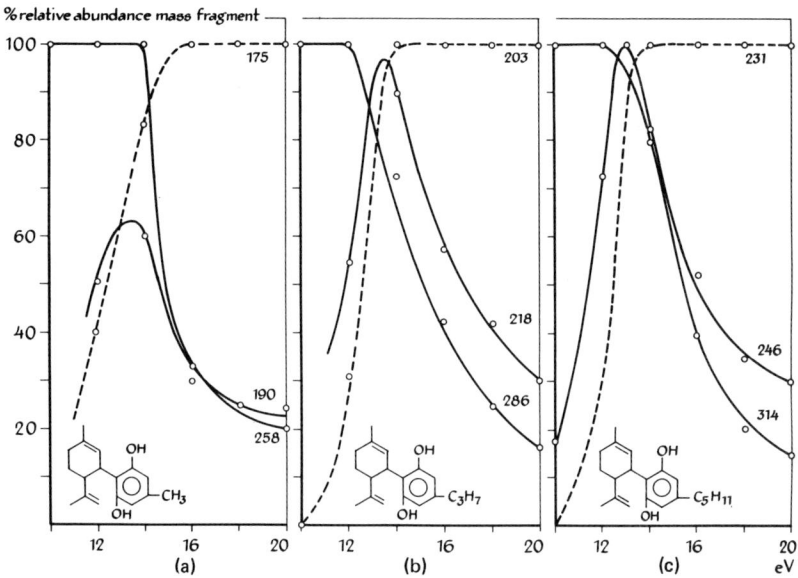

Figure 2.5. Electron voltage-mass fragment intensity graphs of (a) cannabidiorcin, (b) cannabidivarin, and (c) cannabidiol. From Vree et al. [17], courtesy of LKB.

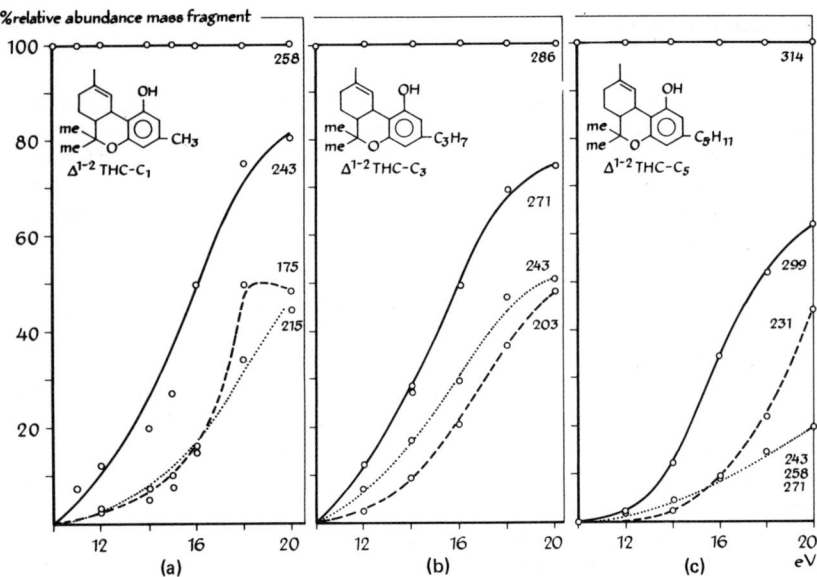

Figure 2.6. Electron voltage-mass fragment intensity graphs of (a) Δ^9-tetrahydrocannabiorcin, (b) Δ^9-tetrahydrocannabivarin, and (c) Δ^9-tetrahydrocannabinol. From Vree et al. [17], courtesy of LKB.

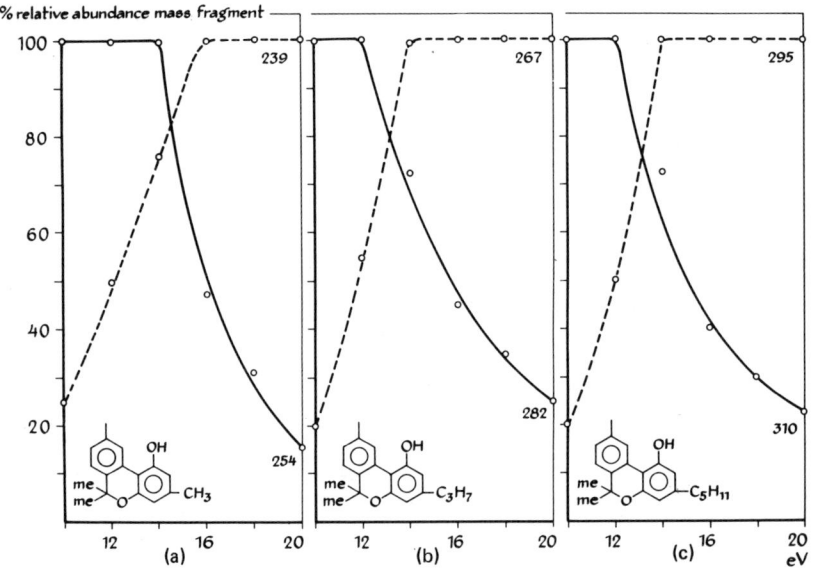

Figure 2.7. Electron voltage-mass fragment intensity graphs of (a) cannabiorcin, (b) cannabivarin, and (c) cannabinol. From Vree et al. [17], courtesy of LKB.

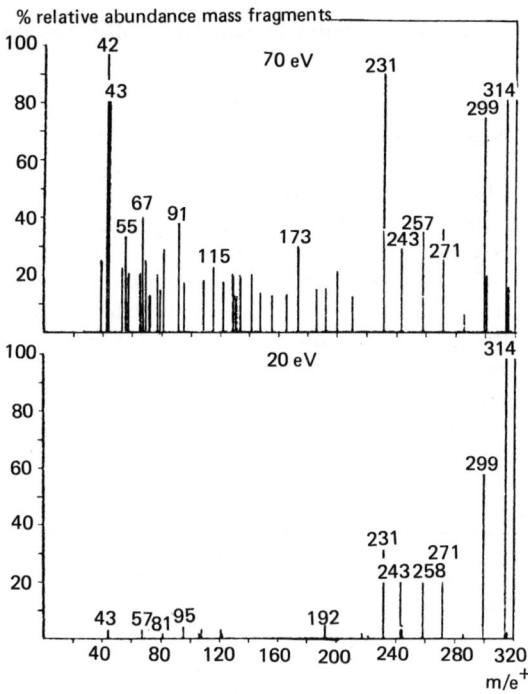

Figure 2.8. Mass spectrum of Δ^9-tetrahydrocannabinol taken at 70 eV (250°C) and 20 eV (250°C). From Vree et al. [17], courtesy of LKB.

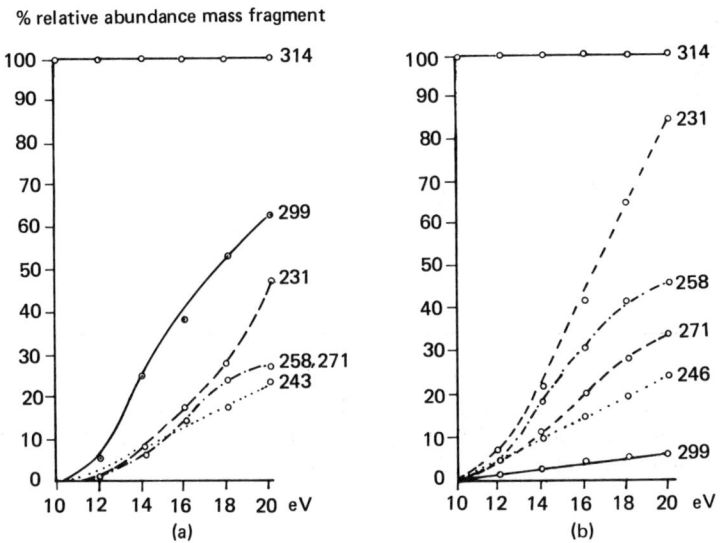

Figure 2.9. Electron voltage-mass fragment intensity graphs of (a) Δ^9-THC and (b) Δ^8-THC. From Vree et al. [17], courtesy of LKB.

With regard to their findings for the C_1 and C_3 analogs of THC, CBD, and CBN, Vree et al. summarized their data as follows*:

1. Methyl Analogs of THC, CBD, and CBN

To prove the existence of the methyl homologs of CBD, THC, and CBN, there are two necessary requirements: (1) The concentration of the C_1 compounds must be high enough to enable construction of mass-intensity graphs, and (2) The compounds must be sufficiently separated from other well-known, but dominating structures.

A gas chromatogram of an extract of Brazilian marijuana shows a distinct peak when eluted from the column before the known structure of C_3 and C_5 cannabinols emerge [peak 1, Figure 2.10].

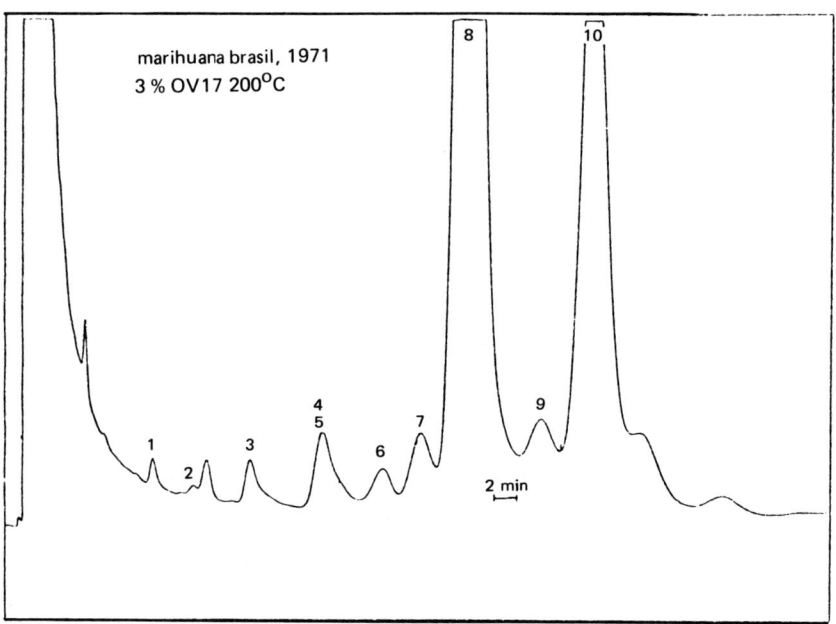

Figure 2.10. Gas chromatogram of an extract of Brazilian marijuana: 1 = Δ^9-THC-C_1, 2 = CBN-C_1, 3 = Δ^9-THC-C_3, 4 = CBN-C_3, 5 = cannabichromene, 6 = cannabigerol methyl ether, 7 = Δ^8-THC-C_5, 8 = Δ^9-THC-C_5, 9 = cannabigerol, and 10 = CBN-C_5. From Vree et al. [17], courtesy of LKB.

---

*Reproduced by permission of LKB.

The molecular weight was found to be 258 and the most significant mass fragments were 243, 215, 187, and 175.

The energy-mass intensity graph [Fig. 2.6] shows the same pattern as those of the known compounds $THC-C_3$ and $THC-C_5$. As reasoned out below for the $THC-C_3$, we now must conclude that peak 1 in Figure 2.10 is the $THC-C_1$.

Compound 2 in the gas chromatogram of Brazilian marijuana shows the same fragmentation pattern as $CBN-C_3$ and $CBN-C_5$, but the corresponding fragments are 56 and 28 less in mass [Fig. 2.7a].

It can be concluded from this observation that when the $THC-C_5$ and $CBN-C_5$ are present in relatively high amounts, the corresponding C_1 and C_3 compounds are also present. The concentrations of the C_1 compounds, in general, are lower than those of the C_3 homologs, which in turn are much lower than those of the C_5 homologs. Probably the biosynthesis of the side chain follows the acetate fragment synthesis, with C_1 as the starting point and C_5 as the end. In the gas chromatogram of an extract of Lebanese hashish [Fig. 2.11], peak 1 is

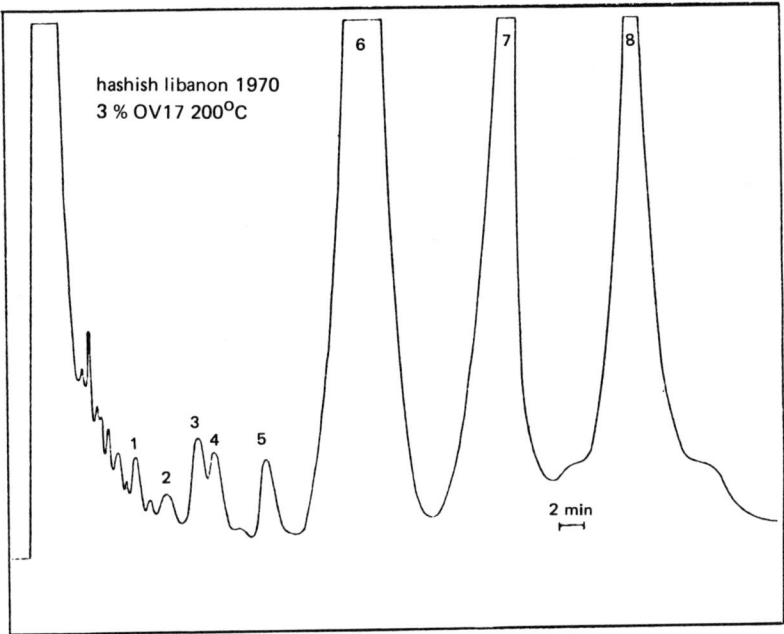

Figure 2.11. Gas chromatogram of an extract of Lebanese hashish: 1 = $CBD-C_1$, 2 = Δ^9-$THC-C_1$, 3 = $CBD-C_3$, 4 = $CBN-C_1$ and compound with MW = 314, 5 = Δ^9-$THC-C_3$, 6 = $CBD-C_5$ and $CBN-C_3$, 7 = Δ^9-$THC-C_5$, and 8 = $CBN-C_5$. From Vree et al. [17], courtesy of LKB.

shown which gives an energy-mass intensity graph as shown in [Fig. 2.5a]. Again, the fragmentation pattern is similar to those of the corresponding C_3 and C_5 derivatives of CBD and it must be concluded that this compound is $CBD-C_1$.

In hashish, the major constituents CBD, THC, and CBN, which bear a pentyl side chain, are accompanied by their homologs with a propyl and methyl side chain.

2. Propyl Analogs of THC, CBD, and CBN

As noted by Vree et al., when a gas chromatogram is taken at low temperature, around 20 compounds can be distinguished and identified which have shorter retention times than CBD [see Fig. 2.12]. In this Nepalese sample, the concentrations of these three compounds are relatively high, compared with the same compounds found in Moroccan, Turkish, and Lebanese hashish. The energy-mass intensity graphs of these three compounds [Figs. 2.5b, 2.6b, and 2.7b] show great similarity to the graphs of CBD, THC, and CBN. The molecular weights of the three compounds are 286, 286, and 282, respectively. This means a constant factor of minus 28 to the corresponding

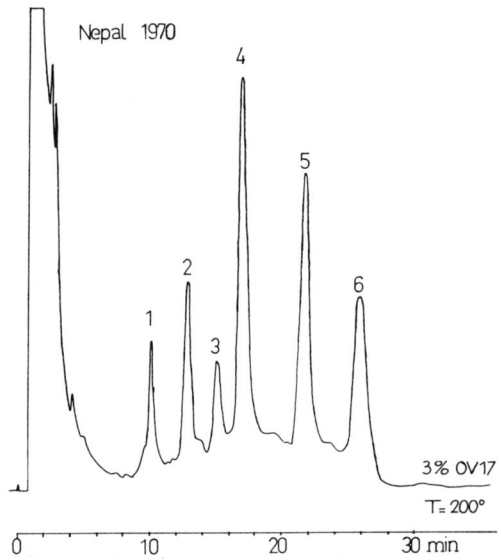

Figure 2.12. Gas chromatogram of an extract of Nepalese hashish: 1 = cannabidivarin, 2 = Δ^9-tetrahydrocannabivarin, 3 = cannabivarin, 4 = cannabidiol, 5 = Δ^9-tetrahydrocannabinol, and 6 = cannabinol. From Vree et al. [17], courtesy of LKB.

molecular weights 314, 314, and 310. From the graphs in [Figs. 2.5b, 2.6b, and 2.7b], it can be seen that every mass fragment mentioned may be converted into a corresponding mass fragment of the known structures CBD, THC, and CBN by adding 28 to the mass. The mass fragment 231 of CBD-C_5 and Δ^9-THC-C_5 reached 50% of the relative abundance at 13 eV and 20 eV, respectively. CBN-C_5 does not have a 231 fragment.

Fetterman et al. [15] reported their findings on chemical definition of phenotype and variations in THC content versus age, sex, and plant part; the seeds of nine strains of Cannabis sativa L. examined were obtained from Iowa, Minnesota, Mexico, Turkey, Italy, France, and Sweden but grown for research in Mississippi. The cannabinoid content was determined by gas chromatography, and the material was divided into two chemical phenotypes according to cannabinoid content. These phenotype categories were used to differentiate between drug-type and fiber-type Cannabis sativa. In addition, the (-)-Δ^9-trans-THC content was determined for both male and female plants, various parts, and a Turkish variety during various stages in its growth.

Using an extraction procedure which was basically that proposed by Lerner [54], analyses were performed with Beckman GC-5 and GC-45 gas chromatographs equipped with flame ionization detectors and 10-ft by 0.125-in. stainless steel columns packed with 2% OV-17 coated on 100-120 mesh Gas Chrom Q. Using the given operating parameters (injector temperature, 230° C; column temperature, 210° C; nitrogen carrier-gas flow rate, 30 ml/min), the retention times of cannabidiol, (-)-Δ^8-trans-THC, (-)-Δ^9-trans-THC, cannabinol, and 4-androstene-3,17-dione (internal standard) were approximately 12.95, 16.95, 20.00, 24.60, and 40.00 min, respectively. Using peak area measurements, quantitative data obtained from these various studies are listed in Table 2.6.

As noted by Sperling [23], Ohlsson et al. [14] used GC, mass spectrometry, as well as thin-layer chromatography to analyze cannabinoids with a view toward determining a correlation between chemical constituents and geographic origin of a cannabis sample. In their study, gas chromatography was carried out with columns packed with 5% SE-30 on Gas Chrom Q. The conclusions drawn from analyses of petroleum ether extracts of fresh plant material as shown in Table 2.7 were as follows:

> The samples Morocco/Sweden and Unknown Sweden show that it is possible to grow in Sweden marijuana rich in the psychotomimetically active Δ^9-THC, provided one has the proper seed material.
> Thus, seeds of a "Δ^9-tetrahydrocannabinol producing Cannabis strain" can produce Δ^9-THC in cooler (Sweden) as well as warmer (Morocco) climates. The plausible implication of this fact is that the type of cannabinoid produced by the plant is dependent upon the

TABLE 2.6

Analysis and Phenotype Classification of Various Samples of Marijuana[a]

Sample	From	Percent			Phenotype ratio	Phenotype
		Cannabidiol	$(-)-\Delta^9$-trans-Tetrahydro-cannabinol	Cannabinol		
Mexican, 1968:						
Female	UM[b]	0.075	1.0	0.54	21	I
Male	UM	0.32	1.2	0.59	5.6	I
Mexican, 1969:						
Female	UM	0.12	1.4	0.073	12	I
Male	UM	0.40	1.5	0.070	3.9	I
Immature	UM	0.063	0.60	0.002	9.5	I
Mexican female plant parts:						
Bracts	UM	0.15	3.7	0.18	26	I
Small leaves	UM	0.085	1.4	0.051	17	I
Seeds	UM	t	0.01	0.01	≫1.0	I
Mexican male plant parts:						
Flowers	UM	0.88	1.6	0.078	1.9	I
Leaves	UM	0.079	1.0	0.047	13	I
Stems	UM	0.055	0.89	0.076	18	I
Turkish, 1968:						
Female[c]	UM	~$6 \times \Delta^9$-THC	0.059	0.023	N.Q.	II
Male	UM	0.24	0.0070	t	0.029	II
Turkish, 1969:						
Female	UM	1.7	0.18	0.062	0.14	II

		~15 × Δ^9-THC	5 × Δ^9-THC			
Turkish female plant parts:						
Bracts[c]	UM	0.37	0.038		N.Q.	=
Leaves[c]	UM	0.32	0.088		N.Q.	=
Stems:						
1-mm. diameter	UM	0.19	0.02	t	0.11	=
2–4 mm.	UM	0.03	0.007	t	0.23	=
10–15 mm.	UM	0.003	t	t	<1.0	=
Roots	UM	0.015	0.0020	0.00074	0.18	=
Seeds	UM	0.0087	0.00057	t	0.066	=
Turkish male at various weeks after planting:						
8 weeks	UM	0.11	0.02	0.02	0.36	=
11 weeks	UM	0.21	0.03	0.04	0.33	=
15 weeks	UM	0.28	0.02	0.02	0.14	=
18 weeks	UM	0.53	0.04	0.01	0.094	=
Turkish female at various weeks after planting:						
8 weeks	UM	0.15	0.02	0.02	0.27	=
11 weeks	UM	0.21	0.03	0.04	0.33	=
15 weeks	UM	0.28	0.02	0.03	0.18	=
18 weeks	UM	0.87	0.04	0.04	0.092	=
19 weeks	UM	1.00	0.05	0.02	0.07	=
Minnesota, 1968:						
1	Minn.	0.77	0.073	0.028	0.13	=
2	Minn.	1.2	0.074	0.016	0.075	=
Minnesota, 1969	UM	0.71	0.054	0.0095	0.089	=
Minnesota female plant parts:						
Bracts	UM	1.3	0.054	0.0033	0.044	=
Leaves	UM	1.0	0.043	t	0.043	=
Iowa, 1968	Iowa	0.95	0.061	0.026	0.092	=

TABLE 2.6 (continued)

Sample	From	Cannabidiol	(-)-Δ^9-trans-Tetrahydro-cannabinol (Percent)	Cannabinol	Phenotype ratio	Phenotype
Des Moines, 1968	Des Moines, Iowa	1.2	0.071	0.010	0.068	II
Illinois male, 1968[d]	Dr. Susiana and Dr. Dunbar, Stanford Univ.	0.26	1.1	0.085	4.6	I
Seized marijuana[d]	Dr. L. Way, Univ. of California	0.88	0.084	t	0.095	II
Carmagnola, 1968	UM	1.2	0.32	0.085	0.34	II
Fibranova, 1969	UM	1.55	0.11	0.040	0.097	II
Unknown[d]	Dr. A. Yuwiler, Vet. Hosp., Los Angeles, Calif.	0.71	0.077	t	0.11	II
Confiscated cigarette[d]:						
a	Dr. H. Isbel, Univ. of Kentucky	1.08	0.15	0.049	0.18	II
b		0.48	0.51	0.10	1.3	I
Monophyllous bracts	UM	6.1	0.5	t	0.082	II
Charas tincture[d]	Dr. C. C. Pfeiffer, Princeton	3.8	1.4	4.0	1.4	I

Sample	Source					Type
Unknown—laboratory grown[d]	UM, grown by Dr. Walter	2.1	0.19	t	0.091	II
Hashish cake and powder[d]	Athens, Greece, Dr. M. Fink, New York Univ.	9.8	2.1	3.5	0.57	II
Red oil or marijuana extract distillate	Research Triangle Institute	0.88	10	3.5	15	I
NIMH 1 (confiscated)[d]	NIMH	0.095	0.58	0.37	10	I
NIMH 12[d]	NIMH	1.0	0.052	0.020	0.072	II
USP fluid extract[d] of *Cannabis sativa*, manufactured by H. K. Mulford Co., 40 years old	Dr. K. Redman, South Dakota State Univ.	2.7	0.43	5.2	2.1	I
Thailand QCD-65472, 1969	Dr. R. Forney, School of Medicine Indianapolis, Ind.	0.16	2.2	t	14	I
Thailand QCD-65169, 1969		0.11	1.3	t	12	I
Carmagnola (Italy), 1969	UM	1.4	0.37	0.077	0.32	II
Fibranova (Italy), 1969	UM	1.6	0.11	0.04	0.094	II
Unknown[d]	UM	0.19	0.025	0.018	0.23	II
Turkish extract, 1969	UM	28	1.4	t	0.050	II

[a] From Fetterman et al. [15], courtesy of *Journal of Pharmaceutical Sciences*.
[b] UM = University of Mississippi.
[c] Cannabidiol peak offscale, roughly estimated. ~ = approximately. N.Q. = not quantitative.
[d] History unknown. t = trace.

TABLE 2.7

Content of Cannabinoids of Fresh Male and Female Cannabis[a]

Sample and plant part	Cannabinoids (% of dry weight)			
	Male		Female	
	CBD	Δ^9-THC	CBD	Δ^9-THC
Beirut 26.4.1969				
Flowers	1.6	0.03	2.6	0.06
Upper leaves	0.6	0.01	1.8	0.00
Large leaves	0.2	0.01	0.2	0.03
Stem	0.08	0.00	0.2	0.00
Bekaa 19.6.1969				
Flowers	1.3	0.1	2.4	0.5
Upper leaves	0.6	0.2	1.1	0.1
Large leaves	0.4	0.1	0.3	0.1
Stem	0.04	0.01	0.1	0.0
Bekaa 26.6.1969				
Flowers	1.6	0.2	2.8	0.04
Upper leaves	0.9	0.1	1.4	0.04
Large leaves	0.2	0.1	0.4	0.02
Stem	0.2	0.06	0.2	0.01
Hizzine 3.9.1969				
Flowers (fruits)	0.7	0.6	0.3	0.4
Upper leaves	0.2	1.2	1.0	0.7
Large leaves	0.1	0.4	0.2	0.2
Stem	0.02	0.3	0.1	0.1
Caucasus/Sweden 1.10.1969				
Flowers (fruits)	0.6	0.1	0.5	0.1
Upper leaves	0.3	0.1	0.5	0.1
Large leaves	0.4	0.01	0.1	0.01
Stem	0.05	0.04	0.02	0.01
Turkey/Sweden 1.10.1969				
Flowers (fruits)	0.7	0.01	0.8	0.02
Upper leaves	0.7	0.02	0.5	0.01
Large leaves	0.5	0.01	0.7	0.02
Stem	0.04	0.00	0.03	0.00
Bratislava/Sweden 1.10.1969				
Flowers (fruits)	0.7	0.01	0.7	0.02
Upper leaves	0.2	0.00	0.5	0.02

TABLE 2.7 (continued)

Sample and plant part	Cannabinoids (% of dry weight)			
	Male		Female	
	CBD	Δ^9-THC	CBD	Δ^9-THC
Large leaves	0.2	0.00	0.4	0.01
Stem	0.00	0.00	0.00	0.00
Morocco/Sweden 8.1969 Flowering tops				
grown outdoors			0.04	0.4
grown indoors			0.04	0.1
Unknown/Sweden 8.1969 Flowering tops			0.1	0.1

[a] From Sperling [23], reproduced from Journal of Chromatographic Science, by permission of Preston Publications, Inc.

inherited properties of the seed and that the influence of the climate is limited.

Knowing that "good marijuana quality" Cannabis seeds have been shipped from one country to another, and will continue to do so, for illegal cultivation, and considering the point just made above, it is evident that there is no valid basis for attempts to correlate the cannabinoid content with country of origin for a cannabis sample.

It is also evident that in nature there is a variety of "chemotypes" of C. sativa from one extreme, producing almost exclusively cannabidiol, over intermediate forms, to forms producing predominantly Δ^9-THC.

De Zeeuw et al. [18] in 1972 reported and confirmed that the natural occurrence of the neutral cannabinoids with a pentyl side chain—for example, cannabidiol, tetrahydrocannabinol, and cannabinol—was generally accompanied by homologs with a propyl side chain, of which at least one has psychotropic activity (Δ^9-THC-C$_3$). Samples from different sources of hashish and marijuana were used. Following a solvent extraction of resin or herb, the identity of the cannabinoids was confirmed by GC and MS. For this particular investigation, de Zeeuw et al. carried out their GC analyses with a Becker 409 gas chromatograph equipped with a flame ionization detector and a 2-m by 2-mm stainless steel columns packed with 5% SE-30 on 80-100 mesh, DMCS-treated, acid-washed Chromosorb W. Maintaining specified

GC conditions (injector temperature, 275°C; column temperature, 230°C; detector temperature, 275°C; nitrogen carrier-gas flow rate, 20 ml/min), TMSi derivatives of CBD-C_3, Δ^9-THC-C_3, CBN-C_3, CDB-C_5, Δ^9-THC-C_5, and CBN-C_5 had retention times of approximately 9.25, 12.12, 14.35, 16.25, 21.40, and 25.50 min, respectively.

As noted by de Zeeuw et al., the occurrence of the propyl cannabinoids in nature seems to depend on the origin of the samples. For example, samples from Asian countries such as India, Nepal, and Pakistan usually contained abundant amounts of propyl constituents. In several samples the C_3 cannabinoids were more predominant than their accompanying pentyl analogs, but there was no correlation between the amount of C_3 analog and that of the accompanying pentyl homolog. Middle Eastern and Mediterranean samples also contained C_3 cannabinoids, but in much lower concentration than Asian samples. In all samples investigated, propyl cannabinoids were detected.

In 1972, Stromberg [19-21] examined by gas chromatography and mass spectrometry the minor components of cannabis resin with retention times shorter than cannbidiol [19], reported his findings on comparative GC analyses of hashish where, apart from the geographical origin, a question of great forensic interest is whether cannabis samples seized in different places can be assigned to a common lot [20], and described the data generated in a study where cannabinoids in micro quantities of cannabis were introduced into the GC unit via solid injection [21].

In his investigation of both the light and heavy fractions of hashish extract [19], GC analyses were performed with a Perkin-Elmer F-11 gas chromatograph equipped with a flame ionization detector and two column systems: (1) For light fraction: 4.5-m by 2-mm-i.d. glass column packed with 3% OV-101 on 60-80 mesh Gas Chrom Q; (2) For heavy fraction: 2.7-m by 2-mm-i.d. glass column packed with 5% Dexsil 300 on 60-80 mesh Gas Chrom Q.

Using a nitrogen carrier-gas flow rate of 30 ml/min, each column was temperature programmed in the following manner:

1. For light fraction: start at 70°C, 0 to 22 min at 3°C/min, 22 to 29 min at 136°C, 39 to 50 min at 1°C/min, 56 to 74 min at 4°C/min, and finally 74 to 90 min at 225°C.

2. For heavy fraction: 0 to 24 min at 210°C, 24 to 50 min at 5°C/min, 56 to 64 min at 370°C.

To confirm component structures, mass spectra were obtained with an integrated LKB 9000 GC-MS instrument. In the light fraction, some 60 peaks were easily observed.

In the comparative investigation [20], the GC analyses were performed with the same Perkin-Elmer F-11 gas chromatograph equipped with a 1.9-m by 2-mm glass column packed with 5% JXR coated on 80-100 mesh,

DMCS-treated, acid-washed Chromosorb W. The other GC conditions employed were injector temperature, 225°C; column temperature, programmed at different heating rates from 100 to 300°C; nitrogen carrier-gas flow rate, 30 ml/min.

The GC data obtained indicated that in several cases the chemical composition within the same lot of hashish showed very small variations. Furthermore, if two samples, A and B, are compared in respect to the amounts of the 40 components and the corresponding peak heights of the i-th component are designated as h_i^A and h_i^B, respectively, the correlation could be expressed by

$$d_{AB} = \left[\sum_{i=1}^{40} (h_i^A - h_i^B)^2 \right]^{1/2} \tag{2.1}$$

On the other hand, using the solid injection system, the separations were carried out with a Becker model 402 gas chromatograph equipped with a flame ionization detector and a 2-m by 2-mm-i.d. glass column packed with 3% OV-17 coated on 60-80 mesh, acid-washed Chromosorb W. The other GC conditions were injector temperature, 300°C; column temperature, 235°C; detector temperature, 300°C; nitrogen carrier-gas flow rate, 30 ml/min. With these GC settings, cannabidiol, Δ^9-THC, and cannabinol were eluted in approximately 12.20, 17.30, and 22.60 min, respectively. Comparative GC studies were performed on cannabis grown in Turkey, Holland, Norway (of Swiss origin) and Norway (of South African origin). In all samples, the peaks of the three main cannabinoids were present but in varying amounts and different ratios.

Fetterman and co-workers developed a GC procedure for the determination of cannabinolic acids in Cannabis sativa L. [16] of Mexican origin and the propyl homolog constituents of cannabinoids from an Indian variant of Cannabis sativa L. [24].

For the determination of the cannabinolic acids, the residues from the chloroform extracts of 1 g of Cannabis sativa L. samples were reacted with bis-(trimethylsilyl)-trifloroacetamide reagent containing 1% trimethylchlorosilane. Following a 5-min reaction period at 60°C, the TMSi derivatives Δ^9-THC and Δ^9-THCA as well as the internal standard (not specified) were separated with Beckman GC-5 and GC-45 gas chromatographs equipped with flame ionization detectors and 10-ft by 0.125-in. stainless steel columns packed with 2% OV-17 coated on 100-120 mesh Gas Chrom Q. With a column temperature of 210°C and a nitrogen carrier-gas flow rate of 30 ml/min, the retention times of Δ^9-THC-TMSi, Δ^9-THCA-TMSi, and internal standard were about 12.37, 34.70, and 56.70 min, respectively.

With regard to the propyl homologs of cannabinoids from an Indian variant, Fetterman and Turner [24] performed GC separations with the Beckman GC instruments equipped with 1.8-m by 2-mm-i.d. glass columns packed with 2% OV-17 on Chrom WHP; the other operating conditions were injector temperature, 230°C; column temperature, 210°C; detector temperature, 250°C; nitrogen carrier-gas flow rate, 30 or 10 ml/min for close calculation of relative retention times. With these conditions, the retention times of cannabidivarin, (-)-Δ^9-trans-tetrahydrocannabivarin, cannabidiol, (-)-Δ^8-trans-THC, (-)-Δ^9-trans-THC, cannabigerol, and cannabinol relative to 4-androstene-3,17-dione were 0.18, 0.26, 0.34, 0.44, 0.49, 0.56, and 0.63, respectively. For further identification, GC-MS was used: a Varian series 1400 gas chromatograph interfaced with a DuPont 21-492 high-resolution mass spectrometer.

Figure 2.13 is a gas chromatogram of a 1971 variant of Indian Cannabis sativa L., whereas Figure 2.14 shows an overlay of chromatograms of a Mexican and a Turkish variant of Cannabis sativa L. where it is noted that the C_3 homologs are not present.

In 1973, de Zeeuw et al. [25] showed by GC that the leaves of young marijuana plants (Cannabis sativa L.) contained appreciable amounts of two

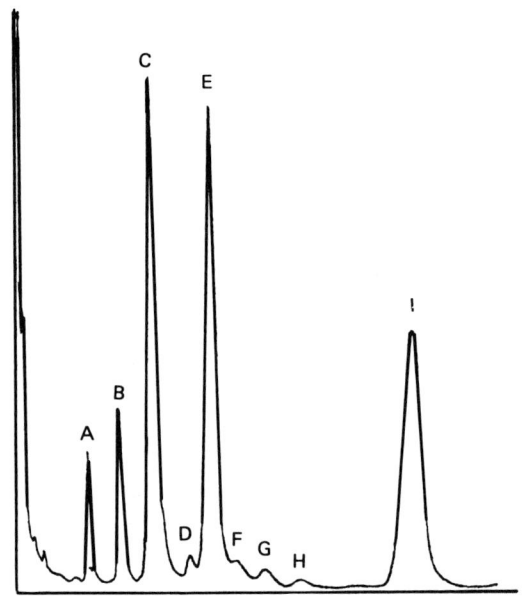

Figure 2.13. Chromatogram of 1971 variant of Indian C. sativa L. Key: A, cannabidivarin; B, (-)-Δ^9-trans-tetrahydrocannabivarin; C, cannabidiol; D, (-)-Δ^8-trans-tetrahydrocannabinol; E, (-)-Δ^9-trans-tetrahydrocannabinol; F, cannabigerol; G, cannabinol; H, unknown peak; and I, 4-androstene-3,17-dione (internal standard). From Fetterman and Turner [24], courtesy of Journal of Pharmaceutical Sciences.

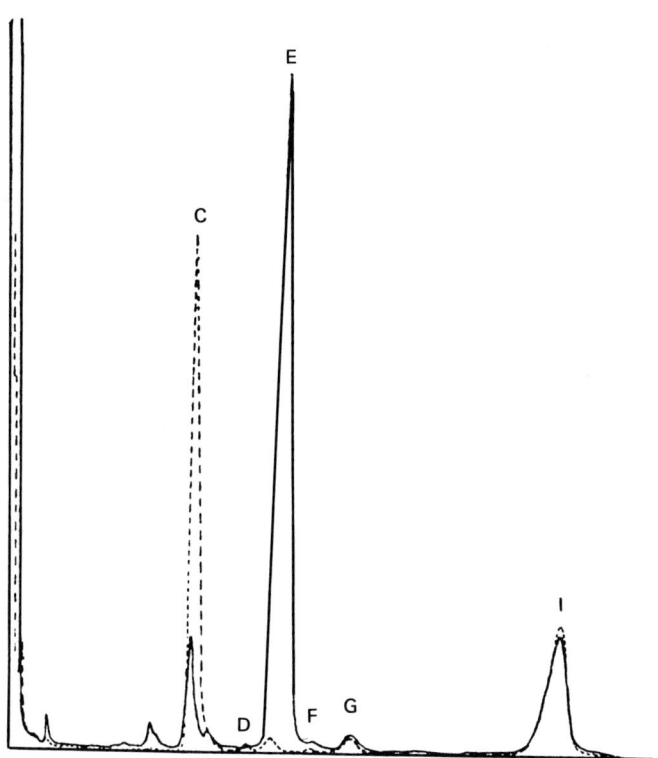

Figure 2.14. Overlay of chromatograms of a Mexican and of a Turkish variant of C. sativa L. Key: ---, Turkish variant; and ——, Mexican variant. Peaks are lettered as in Fig. 2.13. From Fetterman and Turner [24], courtesy of Journal of Pharmaceutical Sciences.

long-chain alkanes, n-heptacosane and n-nonacosane. These alkanes, together with other straight-chain alkanes ranging from C_{19} to C_{32}, could also be detected as minor components in a variety of other marijuana and hashish samples. Depending on the polarity of the GC liquid stationary phase used, the alkanes may have retention times similar to those of the major cannabinoids and thus interfere with qualitative and quantitative determinations of the latter. They further emphasize that the findings indicate that GC alone cannot be used as an accurate and reliable technique for cannabinoid analysis, unless the alkanes are previously removed. Of greater impact, however, is the fact that the alkane composition may be of additional advantage in determining the origin of seized cannabis samples.

Following a chloroform extraction and a specified sample work-up, de Zeeuw et al. carried out their GC analysis with a Becker 409 gas chromatograph equipped with flame ionization detectors and 2-m by 4-mm-i.d. stainless steel columns packed with four different liquid stationary phases coated on 80-100 mesh, DMCS-treated Chromosorb W and operated as follows:

Column	Liquid phase	Temperature (°C)			Nitrogen flow rate (ml/min)
		Injector	Column	Detector	
A	5% OV-17	300	250	300	60
B	5% OV-25	300	250	300	60
C	3% QF-1	275	200	275	50
D	5% SE-30	300	240	300	50

With columns A, B, and C and the conditions given above, the retention times of major alkanes and cannabinoids in cannabis are listed in Table 2.8.

In 1973, Turner and co-workers [26-28] showed cannabidiol to be absent in an African variant of Cannabis sativa L. grown in Mississippi whose absence questions or challenges the validity of published biosyntheses of the cannabinoids [26], developed a GC procedure in which cannabidiol and cannabichromene were discretely separated as their TMSi derivatives [27], and studied the propyl homologs in samples of known geographical origin [28].

With regard to the determination of cannabidiol in an African variant, the extraction procedure was essentially that of Lerner [54] and Fetterman [15]. Analyses were performed with Beckman GC-45 and GC-72-5 gas chromatographs equipped with hydrogen flame ionization detectors and 2.43-m by 2-mm-i.d. glass columns packed with 2% OV-17 coated on 100-120 mesh Chromosorb W-HP. With the specified GC conditions (injector temperature, 240°C; column temperature, 210°C; detector temperature, 260°C; nitrogen carrier-gas flow rate, 10 to 30 ml/min, depending on separation requirements), the retention times of olivetol, cannabidivarin, tetrahydrocannabivarin, cannabicyclol, cannabichromene, cannabivarin, cannabidiol, cannabigerol monomethyl ether, $\Delta^{9,11}$-tetrahydrocannabinol (exocyclic), Δ^8-THC, Δ^9-THC, cannabigerol, and cannabinol relative to 4-androstene-3,17-dione were 0.04, 0.18, 0.26, 0.26, 0.34, 0.34, 0.34, 0.38, 0.41, 0.44, 0.49, 0.57, and 0.63, respectively.

As for the discrete separation of cannabidiol and cannabichromene [27], the GC instruments, columns, packings, and conditions were the same as given above with one exception: the nitrogen carrier-gas flow rate was maintained between 10 and 16 ml/min, depending on the instrument used.

The silylation of the plant extract prior to GC analysis was performed as follows:

TABLE 2.8

Retention Times of Major Alkanes and
Cannabinoids on Three Columns[a]

Compound	Retention time (min)		
	Column		
	A	B	C
Cannabidivarol	12.20	14.20	10.85
Δ^9-Tetrahydrocannibivarol	16.30	19.40	13.90
Cannabivarol	20.35	24.75	19.80
Cannabichromene	18.70	21.75	17.70
Cannabidiol	20.35	22.85	19.05
Δ^9-THC	27.50	31.50	24.00
Cannabigerol	31.00	34.80	26.30
Cannabinol	34.40	40.25	34.40
n-Alkane C-22	4.68		5.33
C-23	6.10	4.58	7.25
C-24	8.15	6.18	9.50
C-25	10.78	8.00	12.57
C-26	14.25	10.30	16.75
C-27	18.10	13.05	22.30
C-28	23.40	16.70	29.30
C-29	31.40	22.20	39.00
C-30	41.30	28.40	51.20
C-31	54.60	27.10	67.50

[a] Adapted from de Zeeuw et al. [25].

A 1-g sample was extracted with 40 ml of spectrograde chloroform. The resulting solution was refrigerated at 6° C and shaken at 10-min intervals for 1 hr. The plant material then was removed by filtration, and the mother liquor was concentrated in vacuo at ambient temperature to a greenish paste void of solvent.

Anhydrous pyridine, 0.5 ml, was added, followed by continuous vibration from an ultrasonic vibrator until all resin was in solution. At this point, 0.5 ml of N,O-bis(trimethylsilyl)-trifluoroacetamide with 1% trimethylchlorosilane was added. The resulting reaction mixture was heated, using a heating mantle, for about 10 min at 80°C. Then 0.1 µl of the reaction mixture was injected into the gas chromatograph.

The retention times of cannabidiol bis(trimethylsilyl) ether, cannabichromene trimethylsilyl ether, cannabidiol mono(trimethylsilyl) ether, (-)-Δ^9-trans-THC trimethylsilyl ether, cannabidiolic acid trimethylsilyl ester-bis(ether), cannabichromene, cannabidiol, and (-)-Δ^9-tetrahydrocannabinolic acid trimethylsilyl ester-ether relative to 4-androstene-3,17-dione (internal standard) were 0.11, 0.17, 0.18, 0.22, 0.28, 0.34, 0.34, and 0.64, respectively.

With the GC method described, Turner and Hadley note that TMSi derivatives of cannabinoids can be used to differentiate if a plant sample of Cannabis sativa L. contains cannabidiol, cannabichromene, or a mixture of the two. However, caution must be observed to ensure complete silylation, and/or the absence of the internal standard will lead to erroneous results.

As for the analysis of the propyl homologs in samples of Cannabis sativa L. of known georgraphical origin, Turner et al. [28] used the GC as well as GC-MS procedures previously developed to identify the C_3 homologs. Summarizing their data, they noted:*

Our findings using only those variants from exact geographical locations confirm and extend the previously thought abundance of propyl cannabinoids in freshly grown Cannabis sativa L. [see Table 2.9]. The percentages of cannabidivarin (I) and (-)-Δ^9-trans-tetrahydrocannabivarin (II) given in [Table 2.9] are normalized reports. Each cannabinoid is reported as its percentage in regard to total cannabinoid content. These data were obtained by a GC-computer analysis based on relative retention times of routine cannabis analysis [26].

Just as the C_5H_{11} (olivetyl group) side-chain neutral cannabinoids exist predominantly as their carboxylic acid derivatives in fresh C. sativa L. [16], so do the C_3H_7 (divarinyl group) side chains. GC-MS analysis of an Indian (1N-B) variant, silylated according to a reported procedure [27], contained a significant m/e at 546, corresponding to the trisilylated [TMSi ester-bis-(ether)] derivative of cannabidivarin. Additionally, a significant m/e at 474 is indicative of disilylated (TMSi ester-ether) (-)-Δ^9-trans-tetrahydrocannabivarin. These data

---

*Reproduced by permission of the Journal of Pharmaceutical Sciences.

TABLE 2.9

Variants Analyzed for Propyl Homologs[a]

Geographical origin	Seed code	Sex[b]	Cannabinoids[c]	
			I	II
Afghanistan	AF-A	F	8.33	5.34
		M	t	+
	AF-B	F	+	7.72
		M	+	+
	AF-C	Y	t	48.23
Brazil	BZ-A	Y	−	t
		F	−	t
Chile	CH-A	M	−	t
Canary Islands	CI-A	Y	t	t
Czechoslovakia	CZ-A	Y	t	t
		F	t	t
		M	t	t
Ethiopia	ET-A	M	t	t
France	FR-A	Y	+	t
		F	t	t
		M	t	t
	FR-B	Y	t	t
	FR-C	F	−	−
Ghana	GH-A	X	−	t
	GH-B	Y	−	t
		X	−	t
	GH-C	Y	−	t
	GH-E	F	−	t
		X	−	+
India	IN-A	Y	−	−
		F	−	t
		M	−	t
	IN-B	Y	+	+
		F	+	10.63
		M	+	+
	IN-D	Y	−	+
	IN-E	F	−	+
		M	−	t
	IN-F	Y	−	+
	IN-I	Y	+	t
Iran	IR-A	F	−	+
		M	−	t
Jamaica	JA-A	Y	−	+

(continued)

TABLE 2.9 (continued)

Geographical origin	Seed code	Sex[b]	Cannabinoids[c]	
			I	II
Japan	JP-A	Y	−	+
	JP-B	Y	−	+
Korea	KO-A	Y	−	+
		F	−	t
		M	−	t
Kenya	KE-A	M	−	t
Lebanon	LE-A	F	+	+
		M	−	−
Mexico	ME-A	Y	t	t
		F	t	t
		M	t	t
Mauritius	MA-A	F	−	+
Morocco	MO-A	F	−	−
	MO-B	F	−	−
	MO-C	F	−	−
Manchuria	MN-A	F	−	−
Nepal	NE-C	Y	−	+
		F	t	+
		M	t	+
Nigeria	NI-D	Y	t	8.80
Pakistan	PK-A	F	+	+
		M	t	t
Peru	PU-A	Y	−	t
		M	−	t
Poland	PO-A	Y	−	−
		F	−	−
		M	−	−
	PO-B	Y	−	−
		F	−	−
		M	−	−
Senegal	SE-A	Y	−	t
South Africa	SA-A	Y	t	+
		F	t	53.69
Sierra Leone	SL-A	F	−	t
Sudan	SU-A	F	+	+
Thailand	TI-B	Y	−	t
		F	−	t
		M	−	t
	TI-C	Y	t	t

TABLE 2.9 (continued)

Geographical origin	Seed code	Sex[b]	Cannabinoids[c]	
			I	II
Thailand (continued)		F	t	t
		M	t	t
		Y	t	t
		F	t	t
		M	t	t
	TI-D	Y	t	t
		F	t	t
		M	–	t
Turkey	TU-A	Y	–	–
		F	–	–
		M	–	–
	TU-C	Y	–	–
		F	–	–
		M	–	–
USA	US-A	Y	–	t
Viet Nam	VN-B	Y	–	t
		F	–	t
		M	–	t

[a] From Turner et al. [28], courtesy of the Journal of Pharmaceutical Sciences.
[b] Y = young plants prior to sexual differentiation, F = female, M = male, and X = mixture of male and female.
[c] – = possible (very small peaks), t = trace amounts (approximately 1%), and + = 1-5%.

agree with the mass spectral data obtained for silylated acid derivatives of cannabidiol and (-)-Δ^9-trans-tetrahydrocannabinol, having m/e values of 574 and 502, respectively. Compound cannabivarin was not clearly observed in fresh samples of cannabis but were observed when samples containing (-)-Δ^9-trans-tetrahydrocannabivarin were heated. Thus, no data are presently available on the acid derivative of cannabivarin.

Masoud and Doorenbos [29] reported a procedure for the assay of Δ^9-tetrahydrocannabinol (I), its two corresponding acids Δ^9-tetrahydrocannabinoic acid A (II) and Δ^9-tetrahydrocannabinoic acid B (III), cannabidiol (IV), and its corresponding acid cannabidiolic acid (V) (see Fig. 2.1 and Table

2.1 for structures) qualitatively and quantitatively by chemical fractionation prior to GC and TLC analyses.

The cannabinoids cited above were determined with either the Beckman GC-5 or GC-45 gas chromatograph equipped with flame ionization detectors and a 10-ft by 0.125-in. stainless steel column packed with 2% OV-17 coated on 100-120 mesh Gas Chrom Q. With the specified GC conditions (injector temperature, 250°C; column temperature, 210°C; detector temperature, 280°C; nitrogen carrier-gas flow rate, 40 ml/min; 4-androstene-3,17-dione added to each extract as internal standard), calibrations were made with reference samples of I and IV. The areas under the peaks were directly correlated with the concentration of the compounds. The retention time of I was used for the analysis of I and its two respective acids. Similarly, the retention time of IV was used in the analysis of IV and V.

As noted by Masoud and Doorenbos, various samples of foreign and domestic, wild and cultivated marijuana were analyzed; their data demonstrated the following:

1. Cannabinoids occur in nature as acids or nonacids with comparable abundance.

2. Tetrahydrocannabinoic acids undergo decarboxylation upon storage or exposure to heat.

3. Plants vary significantly in their cannabinoid composition due to heredity.

4. Change in environment does not change the cannabinoid pattern in plants.

In 1973, Small and Beckstead [58] analyzed for their cannabidiol (CBD), cannabinol (CBN), Δ^9-THC, Δ^8-THC, and cannabigerol monomethyl ether (CBGM) by gas chromatography 350 diverse seed acquisitions of cannabis which were grown outdoors under uniform conditions in Ottawa, Canada. These marijuana constituents were determined quantitatively with a gas chromatograph equipped with flame ionization detectors and 6-ft by 3.5-mm-i.d. glass columns packed with 3% OV-7 coated on 80-100 mesh Chromosorb W-HP. Using the specified operating conditions (injection temperature, 275°C; column temperature, 210°C; detector temperature, 290°C; nitrogen carrier-gas flow rate, 60 ml/min; hydrogen flow rate, 40 ml/min; air flow rate, 300 ml/min), the retention times of cannabichromene, CBD, CBGM, Δ^8-THC, Δ^9-THC, and CBN were approximately 27.90, 32.15, 36.90, 38.80, 44.25, and 55.80 min, respectively. Quantitative analyses of each cannabinoid (neutral plus acidic fraction) were conducted using standardized marijuana, calibrated against dibenzyl phthalate as an external standard.

Sample preparation for GC analysis consisted of shaking 1 g of plant material with 10 ml of n-hexane in a 20- by 125-mm culture tube for 15 min; this was then followed by centrifugation. A 1.5-μl aliquot of the n-hexane

Cannabinoid Patterns

extract solution was then withdrawn and injected into the gas chromatograph.

Based on their investigation, Small and Beckstead noted that several patterns of association recurred frequently. Plants originating from countries north of latitude 30° N almost always had notably higher contents of cannabinoids in the females than in the males. Considerable amounts of CBD were present. Less frequently, moderate or high amounts of THC were also present in the females. In plants originating from countries south of latitude 30° N, high amounts of THC and low amounts of CBD were frequently present in both sexes. Plants probably conforming to this latter type frequently failed to reach the flowering stage in the relatively short growing season of Ottawa. CBN was rarely present in freshly harvested plants, and then in only trace amounts. Δ^8-THC was usually present in trace amounts. Trace amounts of a compound having the same retention time as CBGM were consistently present in plants originating from northeastern Asia.

In a prolific and vigorous continuing research effort, Holley et al. [31] reported their GC findings of the presence of cannabidiol (I) and cannabichromene (II) in freshly grown cannabis from known geographical origin. The procedure described by Turner et al. [56] in which silyl derivatives were used in routine analysis provided a method for quantitating I and II as well as (-)-Δ^9-trans-tetrahydrocannabinol (III).

With regard to GC separations and quantitative data obtained, Holley et al. noted:

> [Figure 2.15] shows the routine chromatogram used in determining the concentration of each of the cannabinoids. Here I and II are combined and are under one peak. The same Moroccan variant is shown in [Fig. 2.16] after silylation, one method for the separation of I and II. However, with this method, four peaks must be taken into consideration: I monosilylated, I disilylated, cannabidiolic acid, and II. Silylated cannabichromenic acid has yet to be identified. However, an excellent separation of I and II in the Moroccan variant is easily achieved by the 6% methyl silicone column described previously [56].

Presented as percentages by dry weight, Table 2.10 consists of variants grown in Mississippi; these data were obtained by a GC-computer analysis based on relative retention times of routine cannabis analyses and synthetic standards.

In 1976, Novotny et al. [33] developed a highly specific procedure for "fingerprinting" marijuana samples. The method consists of extraction of a marijuana sample, a single partition step, and the use of a precolumn concentration technique prior to gas chromatography. The high resolving power of glass capillary columns is essential for developing complex chromatographic profiles that are unique for a given sample. In their report,

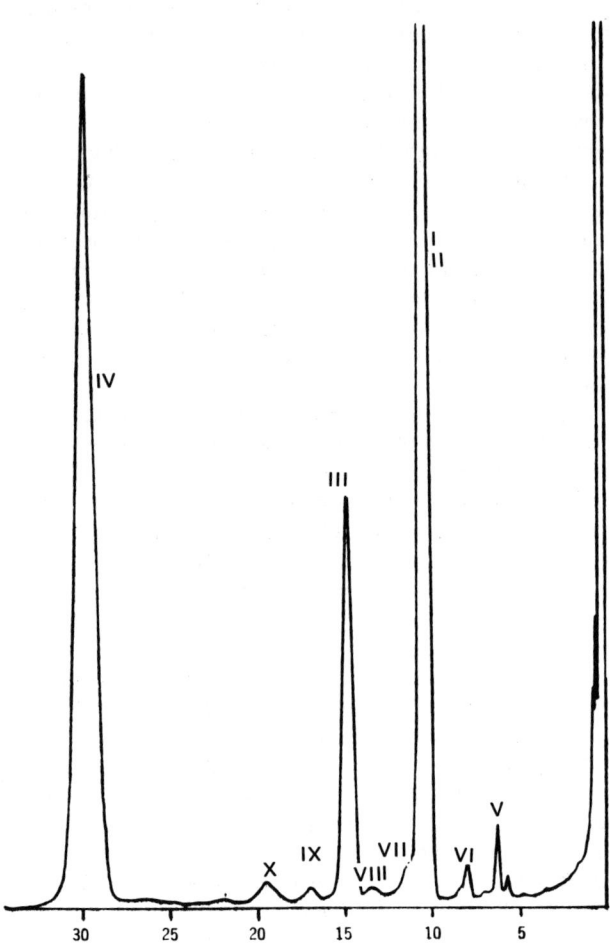

Figure 2.15. Chromatogram of the Moroccan variant, MO-A male, showing: I, cannabidiol; II, cannabichromene; III, (-)-Δ^9-trans-tetrahydrocannabinol; X, cannabinol; and IV, androst-4-ene-3,17-dione, the internal standard. Also shown are: V, cannabidivarin; VI, combined cannabicyclol and (-)-Δ^9-trans-tetrahydrocannabivarin; VII, cannabigerol monomethyl ether; VIII, (-)-Δ^8-trans-tetrahydrocannabinol; and IX, cannabigerol. From Holley et al. [31], courtesy of the Journal of Pharmaceutical Sciences.

Cannabinoid Patterns

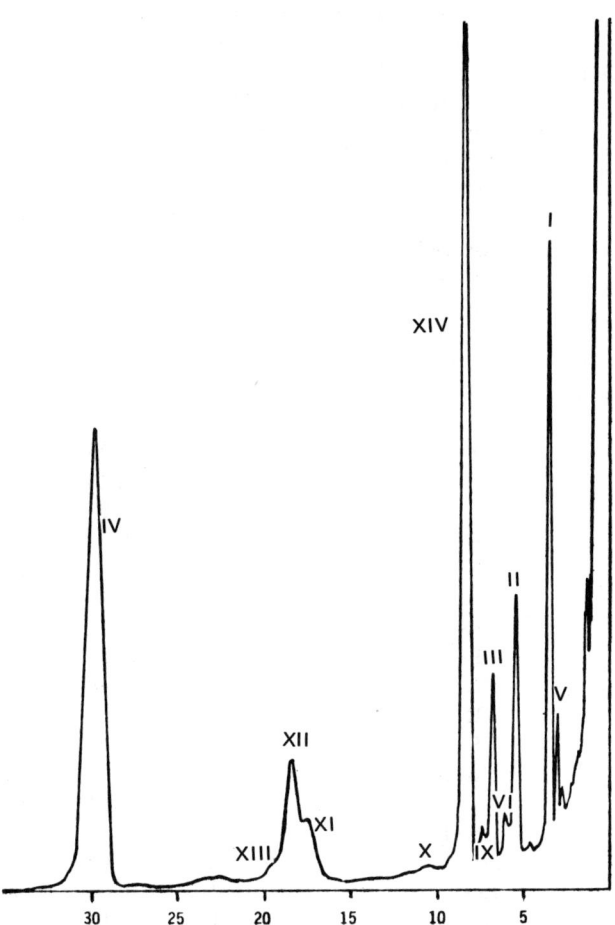

Figure 2.16. Chromatogram of the silylated plant material MO-A, Moroccan male variant, showing the first method for the separation of cannabidiol (I) and cannabichromene (II). Also shown are: III, (-)-Δ^9-trans-tetrahydrocannabinol; IV, androst-4-ene-3,17-dione, the internal standard; V, cannabidivarin monosilylated; VI, cannabicyclol; IX, cannabigerol disilylated; XIV, cannabidiolic acid; X, cannabinol; XI, unknown; XII, (-)-Δ^9-trans-tetrahydrocannabinolic acid A; and XIII, (-)-Δ^9-trans-tetrahydrocannabinolic acid B. From Holley et al. [31], courtesy of the Journal of Pharmaceutical Sciences.

TABLE 2.10

Variants Grown in Mississippi[a]

Geographical origin	Seed code	Sex[b]	I[c]	II	III
Afghanistan	AF-B	M	1.94	0.50	2.68
	AF-B	F	1.26	0.01	0.59
	AF-B(1)/C-71	F	0.19	0.83	2.11
	AF-B(1)/C-71	M	4.58	0.27	2.63
	AF-A(1)/C-71	M	0.11	0.77	4.41
	AF-C	Y	0.03	0.11	2.34
Brazil	BR-A	F	t	0.03	2.16
Canary Islands	CI-A	M	1.48	0.03	0.84
Czechoslovakia	CZ-A	F	1.22	0.04	1.11
	CZ-A	M	1.08	0.08	0.54
	CZ-B	F	0.29	0.02	1.03
	CZ-B	M	1.28	0.65	0.08
Ethiopia	ET-A	M	3.05	0.15	1.29
Fibriman	FI-A	F	0.94	0.00	0.05
	FI-A	M	1.12	0.18	0.06
France	FR-B	F	0.95	0.13	1.49
	FR-C	F	0.11	0.38	3.20
	FR-F	F	1.36	0.00	0.07
	FR-F	M	0.48	0.24	0.38
	FR-G	F	0.90	0.00	0.06
	FR-G	M	1.07	0.00	0.06
	FR-H	M	1.35	0.84	0.08
Ghana	GH-A	Y	0.01	0.08	1.68
	GH-B	Y	t	0.13	2.10
	GH-C	Y	0.02	0.09	2.60
India	IN-A	F	0.03	0.16	1.78
	IN-A	M	0.02	0.10	4.30
	IN-A	Y	0.01	0.15	0.72
	IN-A(1)/C-70	F	0.02	0.10	2.21
	IN-A(1)/C-70	M	0.30	0.20	1.70
	IN-A(2)/C-71	F	0.10	1.72	0.86
	IN-B(2)/C-71	F	0.02	0.26	2.72
	IN-B(2)/C-71	M	2.24	0.00	0.11
	IN-B(3)/C-72	F	t	0.64	1.95
	IN-B(3)/C-72	M	0.00	0.47	1.39
	IN-D	Y	0.01	0.40	1.68
	IN-E	F	0.02	0.52	3.31
	IN-F	M	0.03	1.36	0.72

TABLE 2.10 (continued)

Geographical origin	Seed code	Sex[b]	I[c]	II	III
India	IN-F	Y	0.08	0.83	3.37
(continued)	IN-I	Y	3.40	0.26	0.99
Iowa	IO-A	F	1.70	0.02	0.10
	IO-A	M	1.66	0.04	0.09
	IO-A	Y	t	0.20	1.68
Iran	IR-A	Y	0.91	0.00	0.06
	IR-A(1)/C-71	F	0.73	0.00	0.72
	IR-A(1)/C-71	M	t	0.07	1.40
	IR-A(2)/C-72	F	0.07	0.01	0.33
	IR-A(2)/C-72	M	1.63	0.03	0.18
Jamaica	JA-A	Y	t	0.22	1.84
	JA-B	Y	0.01	0.41	2.20
	JA-C	Y	0.81	0.01	0.04
Kenya	KE-A	M	0.01	0.07	1.84
Korea	KO-A	Y	0.03	0.13	1.17
	KO-A(1)/C-70	Y	0.36	0.17	1.39
	KO-A(2)/C-71	M	0.02	0.91	8.83
	KO-A(2)/C-71	Y	t	0.09	0.34
	KO-B(1)/C-70	M	0.01	0.23	0.94
	KO-B(1)/C-70	Y	t	0.11	0.62
Lebanon	LE-A(1)/C-71	F	1.68	0.05	1.07
Manchuria	MN-A	F	1.93	0.23	1.99
	MN-A	M	0.00	0.56	1.48
Morocco	MO-A(1)/C-70	F	1.61	0.00	0.08
	MO-A(1)/C-70	M	0.95	0.38	0.39
	MO-B	F	1.84	0.02	0.54
	MO-C	F	0.70	0.11	0.71
	MO-C	M	0.87	0.08	0.16
Nepal	NE-C	Y	0.02	0.28	2.75
Nigeria	NI-D	Y	0.02	0.09	1.86
Pakistan	PK-A	F	t	0.05	1.32
	PK-A	M	0.40	0.19	1.51
	PK-A(1)/C-71	F	1.27	0.23	0.71
	PK-A(1)/C-71	M	1.20	0.04	1.37
Peru	PU-A	M	0.49	0.07	0.04
	PU-A	Y	t	0.08	2.06
Poland	PO-A	F	1.09	0.01	0.06
	PO-A	M	0.69	0.28	0.04

(continued)

TABLE 2.10 (continued)

Geographical origin	Seed code	Sex[b]	I[c]	II	III
Russia	RU-A(1)/C-70	F	1.79	0.03	0.10
	RU-A(1)/C-70	M	0.04	0.21	1.05
Senegal	SE-A	Y	0.06	0.20	3.55
Sierra Leone	SL-A	F	0.02	0.03	1.23
South Africa	SA-A	X	0.00	0.11	1.60
	SA-A(1)/C-70	X	0.00	0.27	1.18
	SA-A(1)/C-71	F	0.03	0.61	3.02
	SA-A(2)/C-71	F	0.55	0.01	2.00
	SA-A(2)/C-71	M	0.05	0.68	2.89
	SA-A(2)/C-71	Y	0.00	0.06	0.84
	SA-D	F	t	0.15	1.84
	SA-D	Y	0.02	0.40	6.09
	SA-E	F	0.06	0.01	0.63
	SA-F	F	0.01	0.01	0.33
Sudan	SU-A	F	0.03	0.08	2.10
Thailand	TI-B(1)/C-Mon-70	F	0.02	0.42	2.91
	TI-B(1)/C-Mon-70	M	2.27	0.31	0.93
	TI-C(1)/C-Mon-70	F	2.40	0.06	1.36
	TI-C(1)/C-Mon-70	M	2.20	0.14	1.91
	TI-D(1)/C-Mon-70	F	1.25	0.11	1.56
	TI-D(1)/C-Mon-70	M	0.01	0.05	1.68
	TI-F	Y	0.87	t	0.33
	TI-G	Y	0.06	t	1.33
Turkey	TU-A	F	1.28	0.03	0.05
	TU-A	M	1.62	0.08	0.07
	TU-A(1)/C-68	F	1.91	0.04	2.79
	TU-A(2)/C-69	F	0.78	0.02	0.42
	TU-A(2)/C-69	M	2.79	0.16	1.59
	TU-A(2)/C-71	F	2.22	0.03	0.10
	TU-A(2)/C-71	M	1.87	0.24	0.84
	TU-A(3)/C-H-70	F	1.32	0.01	0.92
	TU-A(3)/C-H-70	M	2.02	0.14	0.09
	TU-C	Y	1.17	0.03	0.07
Viet Nam	VN-A(1)/C-71	M	0.54	0.23	3.23
	VN-A(1)/C-71	Y	t	0.04	0.99
	VN-B	Y	0.02	0.14	4.02

[a] From Holley et al. [31], courtesy of the Journal of Pharmaceutical Sciences.

[b] F = female, M = male, Y = young plant, and X = mixture of male and female.

[c] Data are reported as percent by dry weight in all tables. t = trace amounts (less than 0.01% by dry weight).

characteristic profiles of nonpolar marijuana constituents are shown for selected samples from different geographical origins; this profile method demonstrates considerably higher specificity over the conventional measurement of the relative concentrations of major cannabinoids. Using integrated GC-MS, 38 profile constituents have been identified.

Following a solvent extraction of the marijuana samples, the GC profiles were performed in the following manner:

> Four-microliter aliquots of the methylene chloride solution were transferred by means of a 10-μl syringe to a concentration precolumn containing 2 mg of a highly deactivated support [57]. The solvent was flushed out of the precolumn with helium gas at room temperature for 5 min. The precolumn was transferred to the injection port [57] of a Varian 1400 gas chromatograph which was modified to accommodate the precolumn. Concentration of the sample into the first part of the analytical column (kept at room temperature) was accomplished by thermal stripping of organics from the precolumn held at 250°C. Sample trapping time of 20 min was found sufficient for reproducible and quantitative sample transfer into the capillary column. The oven temperature was subsequently increased to 70°C and programmed to 240°C at 2°C/min for recording the GC profiles.
>
> The column used in this study was an 11-m by 0.26-mm-i.d. glass capillary column coated with SE-52 methylphenylsilicone phase.

On the other hand, GC-MS investigations were carried out with the same column system, which was connected directly to the ion source of a Hewlett-Packard model 5980A integrated GC-dodecapole mass spectrometer. Again, the GC was modified for the use of the precolumn and all EI mass spectra were obtained with an electron energy of 70 eV.

Using this analytical approach, Figure 2.17 demonstrates that differences can be observed with samples from different plants; analytical data and structure identification of 38 peaks observed in these chromatograms are listed in Table 2.11.

In 1977, Rowan and Fairbairn [34] examined by GC and GC-MS the cannabinoid contents of seedlings from 12 strains of cannabis of known chemical race. Fourteen days after emergence of the shoot, those of the tetrahydrocannabinol (THC) type could be distinguished from those of the cannabidiol (CBD) type. The true leaves of the THC type contained a relatively high content of THC and also contained cannabichromene (CBC), sometimes as the major cannabinoid, whereas the CBD type had much lower amounts of THC and no CBC, CBD being the major component. A strain from China corresponded to the THC type but showed some unusual features. Seeds purchased at seven outlets in Britain were also examined. Only two batches germinated, but these proved to be of a THC type which resembled the Chinese strain.

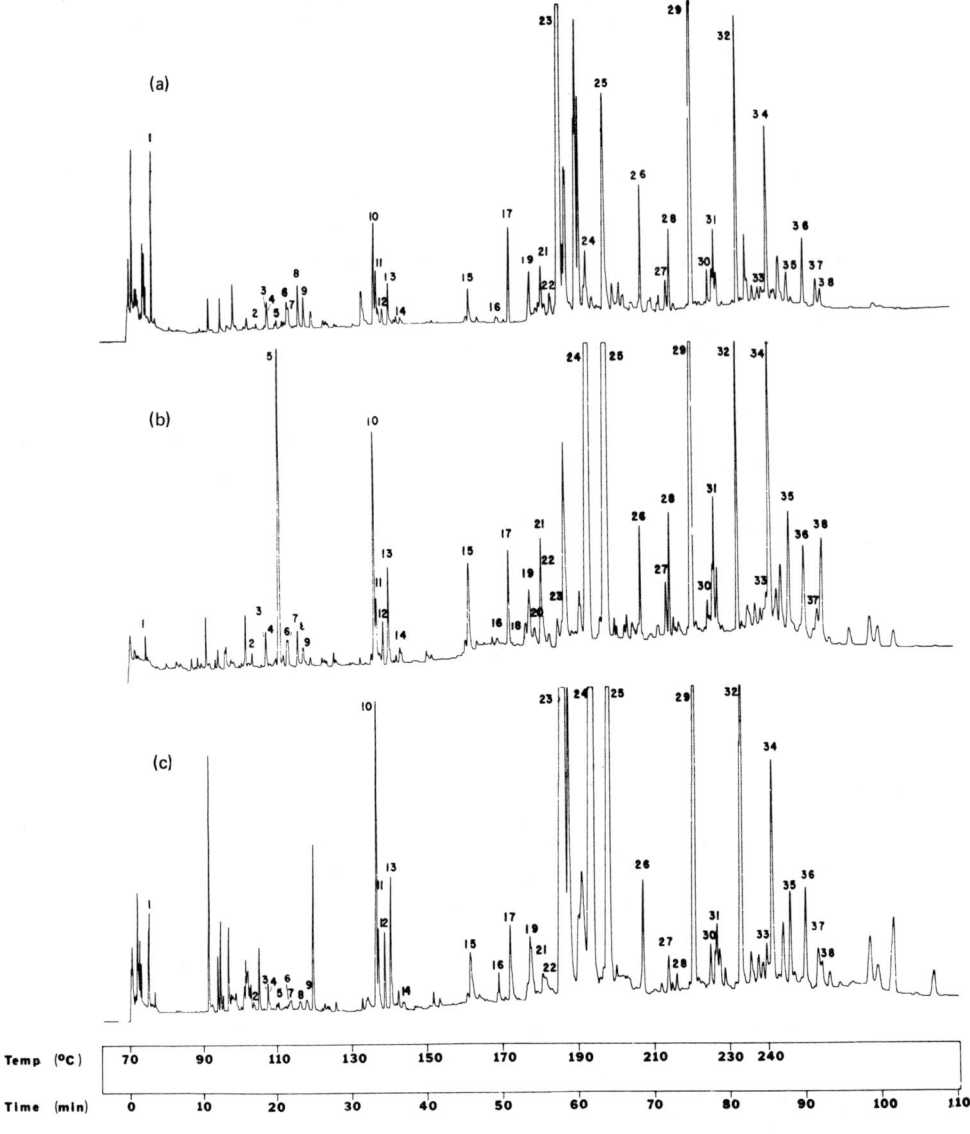

Figure 2.17. Capillary gas chromatograms of the cyclohexane fraction of three different marijuana samples, (a) Turkish marijuana, (b) Mexican marijuana, and (c) Indiana-grown Mexican marijuana. Column: 11 m × 0.26-mm-i.d., glass capillary column coated with SE-52 methylphenylsilicone stationary phase. For peak identifications, see Table 2.11. From Novotny et al. [33], courtesy of Analytical Chemistry.

TABLE 2.11

Analytical Data and Structure Identification[a]

Peak no.	MW	Formula	Significant m/e (rel. int.)	Compound
1	96	C_6H_8O	96(100), 67(54), 95(45), 81(25), 53(22)	2,4-Dimethylfuran
2	204	$C_{15}H_{24}$	69(100), 41(80), 135(42), 149(21), 163(20)	6-Methyl-2-(1-methyl-3-cyclohex-1-enyl)hepta-1,5-diene
3	202	$C_{15}H_{22}$	43(100), 134(70), 91(64), 69(63), 202(14)	3,8,10(15)-Cadinatriene
4	220	$C_{15}H_{24}O$	43(100), 161(70), 187(44), 202(21), 220(4)	12-Hydroxy-2,8-eudesmadiene
5	222	$C_{12}H_{14}O_4$	149(100), 177(22), 105(8), 93(4), 222(1)	Diethyl phthalate[b]
6	220	$C_{15}H_{24}O$	43(100), 91(55), 93(53), 136(38), 220(6)	2-Hydroxy-3,7-cadinadiene
7	220	$C_{15}H_{24}O$	136(100), 41(96), 79(75), 91(68), 220(18)	2-Hydroxy-3,7(11)-cadinadiene
8	202	$C_{15}H_{22}$	43(100), 91(74), 93(73), 131(37), 202(31)	2-(3-Methyl-2-butenyl)-p-mentha-2,6-diene
9	202	$C_{15}H_{22}$	91(100), 131(52), 159(23), 187(26), 202(8)	Eudesma-2,6,8-triene

(continued)

TABLE 2.11 (continued)

Peak no.	MW	Formula	Significant m/e (rel. int.)	Compound
10	250	$C_{15}H_{22}O_3$	43(100), 58(87), 95(39), 109(27), 249(8)	5(9-Ketodecyl)-2-furfuraldehyde
11	250	$C_{17}H_{30}O$	43(100), 58(87), 95(31), 137(7), 250(2)	1-(2,6-Dimethylheptyl)-p-menthene-8(9)
12	278	$C_{20}H_{38}$	81(100), 95(91), 123(62), 278(21), 137(19)	1-[4(β,γ,γ-trimethyl-γ-valerolactonyl)]-p-menthene-8(9)
13	278	$C_{20}H_{38}$	81(100), 95(85), 68(76), 123(68), 278(26)	1-(2,7-Dimethyl-octyl)-p-menthene-8(9)
14	262	$C_{18}H_{30}O$	43(100), 84(80), 97(59), 125(42), 262(14)	Farnesyl acetone
15	256	$C_{16}H_{32}O_2$	73(100), 129(67), 213(42), 256(39), 227(19)	Ethyl-3,10-dimethyl-undecanoate
16	286	$C_{19}H_{26}O_2$	203(100), 218(19), 286(10), 243(5), 271(3)	Cannabidivarin
17	314	$C_{21}H_{30}O_2$	231(100), 314(25), 299(11), 271(10), 246(6)	$\Delta^{4(8)}$-iso-Tetrahydrocannabinol
18	328	$C_{21}H_{28}O_3$	246(100), 257(80), 300(64), 285(60), 328(36)	2-[5-(2-keto-n-pentyl)resorcinolyl]-p-mentha-1,8(9)-diene
19	314	$C_{21}H_{30}O_2$	231(100), 314(17), 299(6), 271(5), 246(3)	Cannabicyclol

#	MW	Formula	Fragments m/z (%)	Name
20	312	$C_{17}H_{16}O_4$	149(100), 91(51), 206(29), 257(12), 312(6)	Benzyl butyl phthalate[b]
21	314	$C_{21}H_{30}O_2$	231(100), 314(8), 299(6), 271(2), 245(2)	Cannabichromene
22	282	$C_{19}H_{22}O_2$	267(100), 238(14), 282(11), 223(5), 209(3)	Cannabivarin
23	314	$C_{21}H_{30}O_2$	231(100), 246(12), 314(6), 299(3), 271(3)	Cannabidiol
24	314	$C_{21}H_{30}O_2$	299(100), 314(79), 231(66), 271(46), 258(22)	Δ^9-Tetrahydrocannabinol
25	310	$C_{21}H_{26}O_2$	295(100), 238(12), 310(10), 239(4), 251(4)	Cannabinol
26	380	$C_{27}H_{56}$	57(100), 71(88), 43(60), 351(5), 380(3)	Heptacosane
27	394	$C_{28}H_{58}$	57(100), 71(85), 43(55), 365(4), 394(3)	Octacosane
28	402	$C_{27}H_{46}O_2$	295(100), 402(87), 312(76), 231(48), 387(21)	1-[1-Methylcyclohex-3-enyl]-18-methyl-nonadeca-1,5,9-trien-4,5-diol
29	408	$C_{29}H_{60}$	57(100), 71(78), 85(64), 365(2), 408(2)	Nonacosane
30	408	$C_{29}H_{60}$	57(100), 295(19), 218(18), 393(17), 408(7)	9-Methyloctacosane
31	422	$C_{30}H_{62}$	57(100), 71(70), 85(62), 295(9), 422(1)	Triacontane

(continued)

TABLE 2.11 (continued)

Peak no.	MW	Formula	Significant m/e (rel. int.)	Compound
32	436	$C_{31}H_{64}$	57(100), 71(80), 85(63), 351(3), 436(2)	Hentriacontaine
33	424	$C_{30}H_{48}O$	218(100), 203(63), 295(23), 424(15), 409(11)	6-Keto-Δ^{12}- or $\Delta^{13(18)}$-oleanene or -ursene
34	426	$C_{30}H_{50}O$	218(100), 203(54), 426(6), 295(5), 365(4)	α-Amyrin
35	466	$C_{32}H_{50}O_2$	218(100), 203(30), 466(17), 295(14), 390(9)	6-Hydroxy-23-aceto-Δ^{12}- or $\Delta^{13(18)}$-oleanene or -ursene
36	468	$C_{32}H_{52}O_2$	218(100), 203(43), 295(11), 468(8), 408(6)	α-, β-, or γ-Amyrin acetate
37	468	$C_{31}H_{48}O_3$	218(100), 203(35), 396(27), 295(20), 468(8)	6-Keto-10-nor-23-aceto-Δ^{12}- or $\Delta^{13(18)}$-oleanene or -ursene
38	426	$C_{30}H_{50}O$	408(100), 393(91), 302(69), 218(62), 426(18)	β- or γ-Amyrin

[a] From Novotny et al. [33], courtesy of Analytical Chemistry.
[b] Most likely a contaminant from a plastic container.

II. CHEMICAL COMPOSITION OF THE MARIJUANA PLANT

In addition to the various GC and/or GC-MS investigations performed in an attempt to correlate cannabinoid patterns with the race and origin of different marijuana samples, others have been concerned with the identification and quantitation of the chemical composition of the marijuana plant [1,2,6, 28,38,39,54,56,59-162].

In 1961, Farmilo and Davis [59] proposed both paper and gas chromatographic methods for the analysis of cannabis which might prove helpful and assist in determining its origin. Following a methanol extraction of powdered, dried cannabis, the GC studies were performed with a Research Specialties gas chromatograph equipped with a beta-ray ionization detector using the following conditions: A 3% SE-30 coating on Chromosorb W solid phase was activated by heating at 225° C for 12 hr in an argon gas stream in the GC oven, and then at 325° C for 24 hr without the gas flow. In their study, a 1.67- or 2.5-ft column was used (internal dimensions not specified), at 174 to 190°C, for the assay of the samples. With an argon carrier-gas flow rate of about 100 ml/min, they reported:

> The gas chromatographic retention values (RT values) for cannabidiol, tetrahydrocannabinol, and cannabinol are 9.3, 13.5, and 18.3 min at 174° C (2.5-ft) and 109 ml/min. On standing, the cannabidiol standard developed a new material which gave a peak at 8.5 min. The material was also found in an extract from Canadian hemp. Pyrahexyl has an RT value of 23.5 min for the main band at 180° C (2.5-ft) and 100 ml/min. The relative retention time values (RRT values) for tetrahydrocannabinol in terms of cannabidiol are 1.83, 1.58, and 1.42 for Pyrahexyl, natural tetrahydrocannabinol, and synthetic tetrahydrocannabinol, respectively. When using light petroleum or methanol extracts of cannabis, it is recommended that RRT values relative to cannabinol be used for identification purposes, i.e., 0.55 and 0.76 for cannabidiol and tetrahydrocannabinol, respectively.

In that same year, Kingston and Kirk [60], faced with the chemical identification of Cannabis sativa (marijuana) and the separation of the components found in the resin produced by the plant, showed that suitable separations of these high-boiling resinous materials could be performed by GC with a nonpolar silicone rubber liquid phase. The column employed with an Aerograph G-10 gas chromatograph with a standard katharometer was 5-ft by 1/4-in.-o.d. copper filled with 2% SE-30 on 80-100 mesh Chromosorb W. With an injection of 20 µl of a petroleum ether extract of the leaves and flowering tops of the Cannabis plant concentrated to a 10 to 20% solution, six fractions were initially detected at a column temperature of 250° C and a helium carrier-gas flow rate of 40 ml/min. However, by trapping these fractions and rechromatographing at 225° C, twelve fractions were observed as shown in

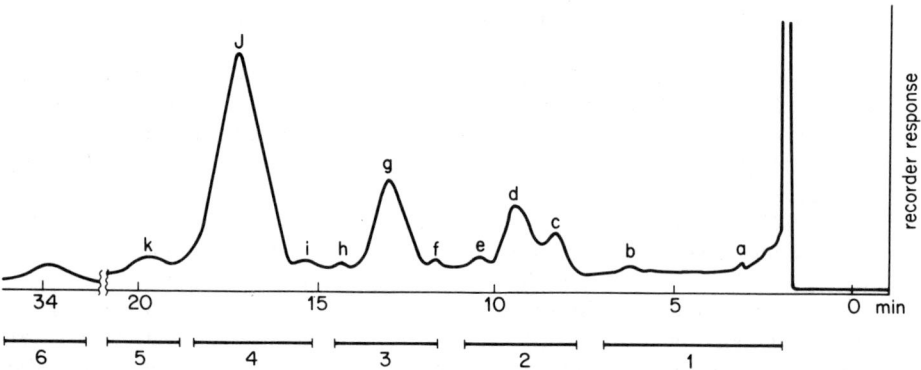

Figure 2.18. Fractions detected in <u>Cannabis sativa</u> resins. From Kingston and Kirk [60], courtesy of <u>Analytical Chemistry</u>.

Figure 2.18, with the initially detected six fractions represented by the numbered horizontal lines.

With ultraviolet spectroscopy, fraction j was identified as cannabinol by forming its acetylated derivative with acetic anhydride and comparing its ultraviolet spectrum with that reported for cannabinol acetate by Adams et al. [163]. Furthermore, fractions d and e were tentatively identified as tetrahydrocannabinol isomers when the main product of a pulegone-orcinol condensation which was synthesized by the procedure of Adams et al. [164] had a retention time of 6.85 min.

In the analysis of the essential oil of marijuana, Nigam et al. [62] investigated by GC the composition of genuine Indian <u>Cannabis sativa L.</u> Using a 20% Reoplex 400 on acid-washed Chromosorb column at either 110 or 170°C and a helium carrier-gas flow rate of 75 ml/min, they were able to separate and identify the compounds whose retention times are listed in Table 2.12. They also noted that the essential oil obtained by hydrodistillation of freshly harvested Indian <u>Cannabis sativa L.</u> was found to contain the following constituents that had not been previously reported: α-pinene, camphene, β-pinene, α-terpinene, β-phellandrene, γ-terpinene, linalool, trans-linalool oxide, sabinene hydrate, α-bergamotene, terpinene-4-ol, β-farnesene, α-terpineol, β-humulene, α-selinene, curcumene, an α,β-unsaturated ketone, and caryophyllene oxide.

In 1966, Taylor et al. [64] synthesized one of the physiologically active constituents of marijuana; they developed a one-step synthesis that produces dl-Δ^6-3,4-trans-tetrahydrocannabinol and two additional isomers, dl-Δ^1-cis-3,4-tetrahydrocannabinol and dl-Δ^6-3,4-cis-tetrahydrocannabinol. They reacted citral with olivetol at 5 to 10°C in benzene in the presence of 10% boron trifluoride etherate and, after a preliminary separation of the reaction mixture on Florisil, separated the resin eluted from the Florisil column

TABLE 2.12

Essential Oil Composition of Cannabis sativa L.[a]

Constituent	Retention time (min)
α-Pinene	1.23
Camphene	1.56
β-Pinene	1.56
Myrcene	1.56
α-Terpinene[b]	2.04
Limonene	2.04
β-Phellandrene[b]	2.18
γ-Terpinene[b]	2.48
p-Cymene	2.72
Alcohol A	3.13
Linalool oxide	3.74
Linalool	4.28
Sabinene hydrate	5.02
α-Bergamotene	5.50
Terpinene-4-ol	6.31
Caryphyllene	6.80
β-Farnesene	6.80
α-Terpineol	8.30
β-Humulene	8.85
α-Selinene	10.07
Curcumene	11.70
α,β-Unsaturated ketone	11.70
Alcohol B	24.80

[a] Adapted from Nigam et al. [62].
[b] Sabinene hydrate degradation artifact under experimental conditions employed.

with hexane-ethyl ether by preparative gas chromatography into dl-Δ^6-cis-tetrahydrocannabinol and dl-Δ^6-trans-tetrahydrocannabinol.

[Reaction scheme: Citral + Olivetol → (Acidic conditions) → d,l-Δ^1-3,4-trans-tetrahydrocannabinol → (Acid or heat) → dl-Δ^6-3,4-trans-tetrahydrocannabinol]

Taylor et al. further noted that the separation of the Δ^1-trans and Δ^6-trans isomers by preparative GC is hindered by thermal isomerization of the Δ^1-trans isomer to the Δ^6-trans isomer and postulated that, in view of this rapid thermal isomerization, the physiological effects from smoking marijuana believed to be caused by the Δ^1-isomer might actually be due to the Δ^6-isomer.

In 1967, Heaysman et al. [67] examined GC methods for the identification of <u>Cannabis sativa L.</u> (Indian hemp). A column of 1% Carbowax 20M on silanized Chromosorb G was found to yield better separation of the cannabinols than columns previously used. Hence, the method could be used to distinguish cannabis in the presence of other vegetable materials that are known to give positive reactions to the usual colorimetric tests. Where doubt exists, this could be resolved by chromatographing the TMSi derivatives. As noted, the method was also applicable to the determination of the resinous extract of cannabis in mixtures of cannabis and tobacco.

These studies were performed with a Pye gas chromatograph equipped with a β-ionization detector operated at 1250 V and a 4-ft by 0.15-in. glass column packed with 1% Carbowax 20M on DMCS-treated, 70-80 mesh Chromosorb G. Using an argon carrier-gas flow rate of 70 ml/min, dibenzyl phthalate and anthracene as internal standards, and column temperatures of 230 and 180°C for the parent compounds and their TMSi derivatives, respectively, the retention times for these materials are listed in Table 2.13.

TABLE 2.13

Retention Times of Cannabinols and Their TMSi Derivatives[a]

Compound	Retention time (min)	
	230° C (as is)	180° C (TMSi)
Cannabidiol	19.63	2.91
Tetrahydrocannabinol	22.00	11.00
Cannabinol	42.25	20.00
Dibenzyl phthalate	59.50	
Anthracene		6.76

[a] Adapted from Heaysman et al. [67].

Plants reported to give positive reactions with the usual cannabis colorimetric reagents were also examined under the conditions for the analysis of the cannabinols (underivatized). The results obtained are shown in Table 2.14, in which retention times relative to dibenzyl phthalate are given.

Caddy et al. [68] also applied GC to the detection of cannabis preparations as a complement to thin-layer chromatography. In their undertaking, a Perkin-Elmer F-11 gas chromatograph equipped with flame ionization detection and spiral 6-ft by 0.3-mm-i.d. columns packed with either 4% XE-60 on acid-washed, DMCS-treated, 100-120 mesh Chromosorb W or 1% XE-60 on acid-washed, DMCS-treated, 100-120 mesh Chromosorb G was used. In Table 2.15, the retention times of unmodified CBD, THC, and CBN as well as their TMSi and trifluoroacetyl derivatives are given using the various GC operating conditions specified. As indicated, the resolution is improved by use of either TMSi ethers or trifluoroacetyl esters.

For the chromatographic identification of cannabis, Betts and Holloway [69] also prepared TMSi derivatives of CBD, THC, and CBN with n-eicosane added as internal standard. They prepared cannabis extracts for GC analysis in the following manner.

After triturating 0.1 g of resin or 0.5 g of tops (freed from seeds) with sand and light petroleum in successive 10 and 2 × 5 ml portions, the combined, filtered light petroleum extracts were concentrated to approximately 1 ml in volume, evaporated to dryness, and the residue dissolved in 0.8 ml of anhydrous pyridine in a 1-ml volumetric flask at room temperature. Following the addition of 0.1 ml of hexamethyldisilazane and 1 drop of trimethylchlorosilane, the contents was made up to volume with anhydrous pyridine (containing n-eicosane as internal standard) and shaken for 1 min. After about 18 hr at room temperature, 1-μl aliquots were injected into the gas chromatograph.

TABLE 2.14

Relative Retention Times of Cannabis Constituents
Compared with Those of Other Vegetables[a]

Constituent	Rel. ret. time
Hyoscyamus niger L.	0.029
Rosmarinus officinalis L.	0.041
Thymus vulgaris L.	0.050
Lavandula officinalis Chaix	0.050
Atropa Belladonna L.	0.050
Salvia officinalis L.	0.059
Rosmarinus officinalis L.	0.059
Datura Stramonium L.	0.059
Thymus vulgaris L.	0.070
Hyoscyamus niger L.	0.070
Lavandula officinalis Chaix	0.079
Rosmarinus officinalis L.	0.092
Hyoscyamus niger L.	0.092
Salvia officinalis L.	0.103
Datura Stramonium L.	0.103
Thymus vulgaris L.	0.122
Hyoscyamus niger L.	0.122
Lavandula officinalis Chaix	0.131
Salvia officinalis L.	0.162
Rosmarinus officinalis L.	0.162
Atropa Belladonna L.	0.181
Papaver somniferum L.	0.210
Datura	0.228
Stramonium L.	0.295
CBD	0.318
Atropa Belladonna L.	0.356

TABLE 2.14 (continued)

Constituent	Rel. ret. time
THC	0.369
Datura Stramonium L.	0.378
Ficus elasticus Roxb.	0.428
Papaver somniferum L.	0.628
CBN	0.697
Rosmarinus officinalis L.	0.730
Ficus elasticus Roxb.	0.753
Rosmarinus officinalis L.	0.840
Dibenzyl phthalate	1.000
Papaver somniferum L.	1.020

[a] Adapted from Heaysman et al. [67].

TABLE 2.15

Retention Times of Cannabis Resin Constituents[a]

Constituent	Retention time (min)		
	Unmodified[b]	TMSi[c]	TFA[d]
CBD	5.07	4.00	4.58
THC	6.74	8.85	11.43
CBN	11.25	15.70	16.60

[a] Adapted from Caddy et al. [68].
[b] Column temperature, 200°C; injector temperature, 250°C; nitrogen carrier-gas flow rate, 65 ml/min at Pi 45 psig; hydrogen, 18 psig; air, 40 psig; column packing, 4% XE-60.
[c] Column temperature, 165°C; injector temperature, 200°C; column packing, 1% XE-60; other conditions same as b.
[d] Column temperature, 150°C; injector temperature, 200°C; column packing, 4% XE-60; other conditions same as b.

GC separations and analyses were carried out with a Pye 104 gas chromatograph equipped with a flame ionization detector and 5-ft by 4-mm-i.d. glass columns packed with 2% SE-30 coated on silanized 80-100 mesh Chromosorb W. With the column temperature maintained at 170°C and the nitrogen carrier-gas flow rate set at 40 ml/min (no other GC conditions specified), the retention times for n-eicosane (I.S.) and the TMSi ethers of CBD, THC, and CBN were 3.90, 10.58, 14.82, and 19.60 min, respectively.

In 1967, Yamauchi et al. [70] isolated Δ^9-tetrahydrocannabinolic acid A (see Table 2.1), a genuine substance of THC, from Mexican hemp cultivated in Japan and obtained its physical constants as well as its chromatographic behavior using a Shimadzu GC-1B gas chromatograph equipped with a 2.25-m by 4-mm-i.d. column packed with 1.5% SE-52 (packing not specified). With the specified GC conditions (injector temperature, 280°C; column temperature, 225°C; nitrogen carrier-gas flow rate, 22.5 ml/min), the retention times of the TMSi derivatives of CBD, Δ^9-THC, CBN, and Δ^9-THCA were approximately 5.33, 7.10, 9.34, and 17.32 min, respectively. On the other hand, the retention times of unmodified constituents (as is) relative to CBD (RRT = 1.000) for Δ^9-THCA, Δ^9-THC, Δ^8-THC, and CBN were 1.23, 1.23, 1.15, and 1.51, respectively.

Yamauchi et al. also noted that quantitative separation of the phenol carboxylic acids and phenols according to the method of Schultz and Haffner [165] (including basic extraction of the acids), followed by gas chromatography, indicated that after the storage of a dried sample of Δ^9-THCA for 4 months at room temperature, 93% of marijuana components were found as phenol carboxylic acid, in which the ratio of CBDA, Δ^9-THCA, and CBNA was 6:89:5, whereas 70% of the components were converted into phenols during the storage of dried Δ^9-THCA at 35°C for 2 months.

Shoyama et al. [74] in 1968 reported the isolation and the structural elucidation of cannabichromenic acid (CBCA) (see Fig. 2.1), a genuine substance of cannabichromene (CBC) in living plants, which could be separated by gas chromatography. Separations were performed with a Shimadzu GC-1B instrument equipped with flame ionization detection and a 2.25-m by 4-mm-i.d. column packed with 1.5% SE-52; the other reported operating conditions were column temperature, 229°C; injector temperature, 290°C; nitrogen carrier-gas flow rate, 20 ml/min. With these GC settings, the retention times obtained for unmodified and silylated derivatives are given below:

Compound	Unmodified	TMSi derivative
CBCA	3.75	7.70
CBD	3.54	2.03
CBC	3.75	2.62

Compound	Unmodified	TMSi derivative
Δ^9-THC	4.67	3.00
CBG (cannabigerol)	5.80	3.39
CBN	5.80	3.96
CBDA		4.69
THCA		7.60

A similar investigation was carried out by Yamauchi et al. [75] for the isolation of cannabigerol monomethyl ether (see Table 2.1) from the domestic hemp "Minamioshihara No. 1." After the methylation of CBG with diazomethane in methanol at 0°C, followed by TLC separation, CBG monomethyl ether was obtained together with CBG dimethyl ether in a ratio of about 1:1. GC separations were run in the same manner as previously described by Shoyama et al. [74].

Cannabigerol Dimethyl Ether

In 1970, Backer et al. [79] described a simple method for the infrared identification of cannabinoids resolved by gas chromatography, specifically, cannabidiol, tetrahydrocannabinol, and cannabinol in marijuana. The method involves extraction of the cannabinols with petroleum ether, filtration, evaporation, and analysis of the residue by GC. Further identification is made by infrared spectrophotometry following their collection from the GC effluent directly onto KBr powder. The GC separations were conducted with a Varian Aerograph model 1860 gas chromatograph equipped with a flame ionization detector, a 10-ft by 1/4-in. stainless steel column packed with 3% SE-30 on acid-washed, DMCS-treated Chromosorb W, and a 10:1 effluent splitter to permit effluent trapping on KBr. Operated at a column temperature of 204°C, an injection temperature of 250°C, and a nitrogen carrier-gas flow rate of 100 ml/min, the retention times obtained for CBD, Δ^9-THC, and CBN were 27.0, 40.5, and 46.5 min, respectively. Typical IR spectra of these three constituents via the KBr disk technique are shown in Figure 2.19.

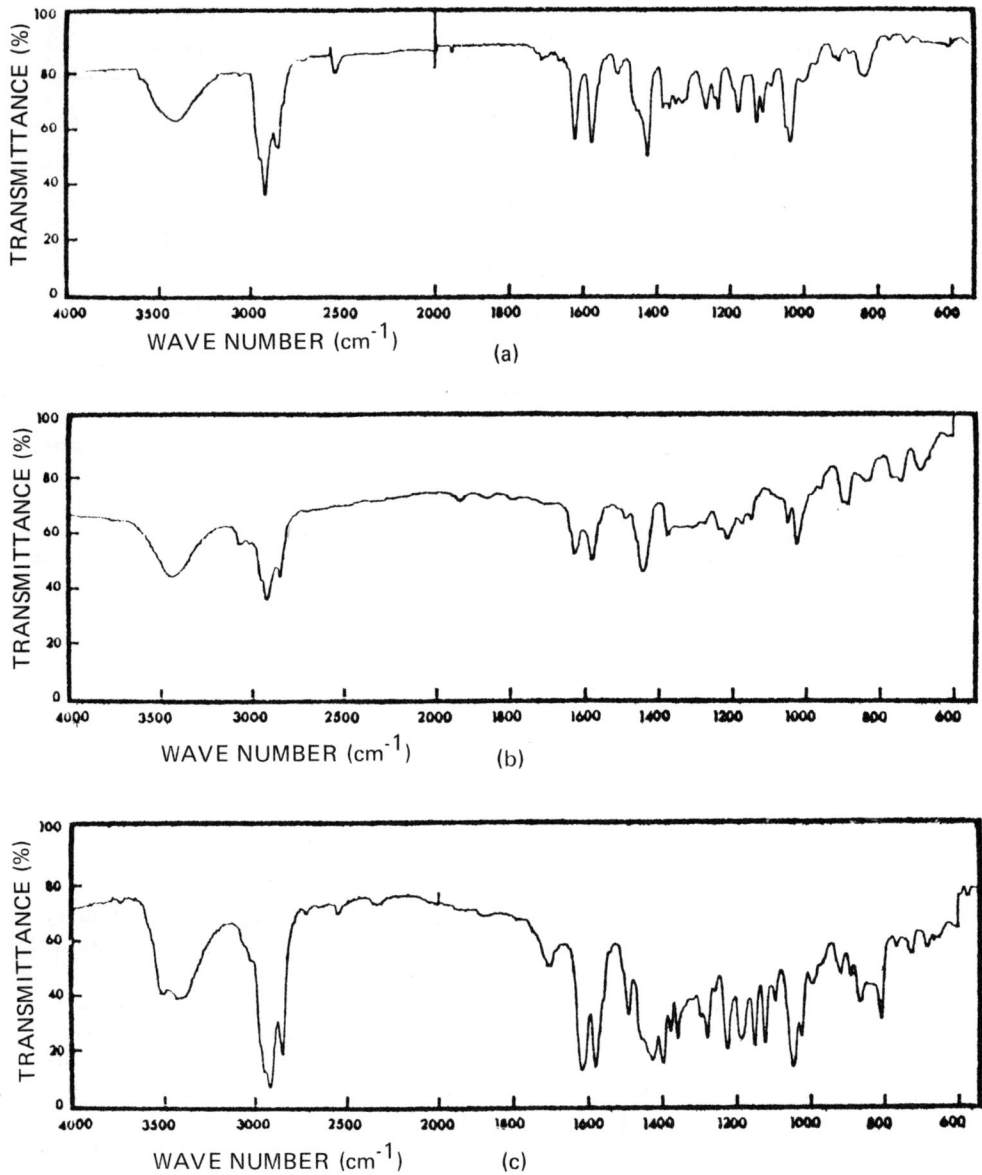

Figure 2.19. Infrared spectrum of (a) Δ^9-tetrahydrocannabinol, (b) cannabidiol, and (c) cannabinol. From Backer et al. [79], reproduced from the Journal of Forensic Sciences, by permission of the American Society for Testing and Materials.

In 1970, Mechoulam [1], discussing marijuana in particular and recent advances in cannabinoid chemistry which have opened the area to more sophisticated biological research, listed (see Table 2.16) the retention times of some neutral cannabinoids on an OV-17 column (Gas Chrom Q as solid stationary support) operated at 235° C with a nitrogen carrier-gas flow rate of 30 ml/min. It was further noted that all cannabinoid acids undergo decarboxylation at the high temperatures necessary (200 to 250° C). For a routine analysis this may be an advantage, for this reaction parallels the smoking process; GC analysis will thus give directly all the THC available to a smoker in a certain sample and, when an exact determination of the content is required, decarboxylation can be prevented by esterification.

In 1970, Patterson and Stevens [80] described a procedure for analysis of cannabis that is advantageous in combining two independent techniques for the detection of CBD, THC, and CBN and which offers positive identification in a reasonably short time (20 min). It consists of extraction of the suspected cannabis or cannabis resin sample with a stock solution of dibenzyl phthalate in light petroleum, the extract then being analyzed without further purification by GC and also by paper chromatography. In their studies they used a Pye 104 gas chromatograph equipped with flame ionization detection and a 5-ft by 4-mm-i.d. glass column packed with AW, siliconized, 80-100 mesh Diatomite C coated with 1% cyclohexanedimethanol succinate (CDMS). Other GC settings employed were nitrogen carrier-gas flow rate, 50 ml/min; column temperature, 220° C. With these conditions (no others given), the retention times of CBD, THC, and CBN relative to dibenzyl phthalate (RRT = 1.00) were 0.26, 0.39, and 0.64, respectively.

TABLE 2.16

Retention Times of Some Neutral Natural Cannabinoids[a]

Compound	Retention time (min)
Cannabicyclol	4.55
Cannabidiol	5.67
Δ^8-THC	7.17
Δ^9-THC	7.87
Cannabinol	10.20
Cannabichromene	5.58
Cannabigerol	9.33

[a] Adapted from Mechoulam [1].

Song et al. [81] developed an extraction procedure and GC method for the quantitative determination of THC in marijuana. In their GC method, 0.2 ml of ethanolic extract of marijuana was transferred to a small vial, and then the ethanol was evaporated with a stream of nitrogen. Following the addition of 0.5 ml of bis-(trimethylsilyl)trifluoroacetamide as silylating reagent and a 1-hr reaction period at room temperature, the excess BSTFA was volatilized off with the aid of a nitrogen stream and the resulting dry residue was dissolved in 2.0 ml of 2,2,4-trimethylpentane. Appropriate aliquots were withdrawn and injected into a Varian Aerograph model 201-B gas chromatograph equipped with a hydrogen flame ionization detector and a 5-ft by 1/8-in. stainless steel column packed with 5% SE-30 on 100-120 mesh Chromosorb W or 3% OV-17 on 100-120 mesh Chromosorb W-HP. Other operating parameters employed were injector temperature, 270 to 280°C; column temperature, 205 to 210°C; detector temperature, 280 to 285°C; nitrogen carrier-gas flow rate, 30 ml/min. Using this method, the retention times of the TMSi derivatives of CBD, THC, and CBN were approximately 13.25, 17.80, and 23.75 min, respectively.

Using a standard curve of peak area versus micrograms of THC, quantitative data on the stability of THC in ethanol was obtained. Over a 14-day testing period, the average integrator response for a solution of Δ^9-THC in ethanol at -15°C was 1245 ± 58, whereas the average amount of THC in an extract of hashish determined 9 times over a 28-day testing period was 21.4 ± 0.2 μg THC/ml of extract. Based on this undertaking, their results indicated that THC in the standard and the extract was stable for the respective time periods tested.

Shoyama et al. [86] isolated with a simple column chromatographic technique from fresh cannabis of the domestic strain the original cannabinoids including Δ^9-tetrahydrocannabinolic acid A (THCA) (Table 2.1), cannabidiolic acid (CBDA) (Fig. 2.1), cannabichromenic acid (CBCA) (Fig. 2.1), and a new component, cannabigerolic acid monomethyl ether (CBGAM) (Table 2.1). Cannabinolic acid (CBNA) (Fig. 2.1), observed in the stored cannabis, was synthesized from THCA by means of ultraviolet light irradiation, and artificial conditions to yield CBNA from THCA, were investigated.

Using the GC procedure described previously [70,74], the retention times of unmodified components and some of their TMSi derivatives are listed in Table 2.17.

In 1970, Davis et al. [88] presented extraction methods suitable for use on a large scale, and the use of falling film molecular distillation was shown to be a practical means of preparing sizable amounts of marijuana concentrates. Also described were analytical GC procedures suitable for qualitative and quantitative estimation of the major cannabinoids and their carboxylic acid derivatives.

In these investigations, the conventional GC analyses were performed with a Varian model 1200 gas chromatograph equipped with a flame ionization detector and a 5-ft by 1/8-in. stainless steel column containing 1.9% OV-17

TABLE 2.17

Retention Times of Various Cannabinoids[a]

Compound	Retention time (min)	
	Unmodified	TMSi derivative
Cannabidiol	3.30	2.00
Cannabidioloc acid	3.30	4.76
Δ^9-THC	4.06	
Δ^9-THC acid A	4.06	7.44
Δ^8-THC	3.80	
Δ^8-THC acid A	3.80	7.12
Cannabigerol	4.99	
Cannabigerolic acid	4.99	
Cannabigerol monomethyl ether	3.56	
Cannabigerolic acid monomethyl ether	3.56	9.12
Cannabichromene	3.30	
Cannabichromenic acid	3.30	7.68
Cannabinol	4.99	
Cannabinolic acid	4.99	9.42
Cannabicyclol	2.69	
Cannabicyclolic acid	2.69	

[a] Adapted from Shoyama et al. [86].

on Gas Chrom Q (mesh size not given). In addition to using androst-4-ene-3,17-dione as internal standard, the other operating parameters employed were injector temperature, 250°C; column temperature, 200 to 230°C; detector temperature, 250°C; nitrogen carrier-gas flow rate, 25 ml/min; hydrogen flow rate, 25 ml/min; air flow rate, 250 ml/min.

For marijuana plant materials containing cannabinolic acids, silylation with bis(trimethylsilyl)trifluoroacetamide followed by GC analysis was used to differentiate the acids from their respective cannabinoids. The TMSi-ether derivatives employed the same conditions as their parent nonacidic compounds with one exception—the column was operated 10°C cooler.

Calibration curves were prepared using various standard solutions of either cannabidiol, Δ^9-THC, or cannabinol with androst-4-ene-3,17-dione as internal standard, all based on peak area measurements. Using the GC conditions given above for nonsilanized compounds, the retention times obtained for cannabidiol, Δ^9-THC, cannabinol, and the internal standard were approximately 8.29, 11.11, 13.40, and 20.70 min, respectively. On the other hand, the TMSi derivatives of CBD, Δ^9-THC, CBN, and Δ^9-THC acid were eluted in 6.12, 9.07, 10.58, and 20.85 min, respectively.

Identification of the various cannabinoids was performed with a LKB 9000 integrated GC-MS instrument equipped with a 9-ft by 1/4-in. glass column packed with 3.8% OV-17 coated on Chromosorb AWS. With this instrument, the ionizing voltage was kept at 20 eV whereas the column was programmed from 180 to 230°C.

Turk et al. [38] were concerned with the identification, isolation, and preservation of Δ^9-tetrahydrocannabinol. (-)-trans-Δ^9-Tetrahydrocannabinol (Δ^9-THC) was isolated from marijuana plant extract by absorptive column and gas chromatography. The absorptive column chromatography method consisted of chromatographing marijuana extract on a column packed with a mixture of silica gel GC-grade (100-120 mesh), silver nitrate, and calcium sulfate in a 3:1:0.5 ratio with benzene as the eluting solvent.

The analytical assay of cannabinoids was carried out with a Hewlett-Packard model 402 gas chromatograph equipped with a flame ionization detector and a 4-ft by 1/4-in. U-shaped, silanized glass column packed with three different liquid stationary phases coated on 100-120 mesh, AW-DMCS, Chromosorb G: The first 1.5 ft were packed with 10% QF-1, the second 1.5 ft with 1% OV-1:1% OV-17, and the last 10 in. (nearest the detector) with 2% OV-17. The other operating parameters maintained for these assays were injector port, 285°C; column temperature, 240°C; detector temperature, 285°C; helium carrier-gas flow rate, 60 ml/min; hydrogen flow rate, 35 ml/min; and oxygen flow rate, 260 ml/min. Using these settings, the retention times of cannabicyclol, CBD, Δ^8-THC, Δ^9-THC, and CBN were were about 4.02, 4.90, 5.59, 6.46, and 8.61 min, respectively.

This analytical GC procedure was used to study the decomposition of Δ^9-THC stored under various conditions. Based on GC data, they noted that "the findings were that the major decomposition product was cannabinol. Δ^9-THC stored in acetone at room conditions decomposed at a more rapid rate than did the material stored under nitrogen or exposed to air at room conditions. Decomposition was significantly inhibited by storage under nitrogen, air, or in acetone at 0°C in the dark."

On the other hand, preparative GC was performed with a Varian Aerograph Autoprep model 712 instrument equipped with an effluent splitter (10 to 1), collection system, flame ionization detector, and a single 3-ft by 3/8-in.-o.d. silanized glass column packed with 1.5 ft of 2% QF-1 and 1.5 ft of 2% OV-17 on 30-60 mesh, AW-DMCS, Chromosorb W. The GC operating conditions employed which yielded retention times for CBD, Δ^8-THC,

Δ^9-THC, and CBN of 7.65, 11.20, 15.55, and 20.00 min, respectively, were injector temperature, 285° C; column temperature, 230° C; detector temperature, 285° C; exit tip temperature, 250° C; helium carrier-gas flow rate, 182 ml/min; hydrogen flow rate, 33 ml/min; air flow rate, 220 ml/min.

In 1971, Knight [93] studied a variety of often-used drugs by computerized GC-MS and solid-probe mass spectrometry, some of which required the formation of more suitable derivatives prior to GC analysis. All of the GC-MS data were acquired using a 5-ft by 2-mm-i.d. glass column packed with 3% OV-1 on 100-120 mesh Gas Chrom Q. Using a column temperature, ionization potential, and emission current of 220° C, 70 eV, and 600 µA, respectively. Figure 2.20 is a reconstructed chromatogram of Δ^8-tetrahydrocannabinol whereas Figure 2.21 is its mass spectrum using the conditions specified above.

In 1971, Gaoni and Mechoulam [94] described the isolation and elucidation of the structures of Δ^9-THC, cannabigerol, cannabichromene, and cannabicyclol as well as a facile conversion of CBD into Δ^9-THC that takes place upon treatment with boron trifluoride etherate. For the quantitative determination of some natural neutral cannabinoids in hashish, these

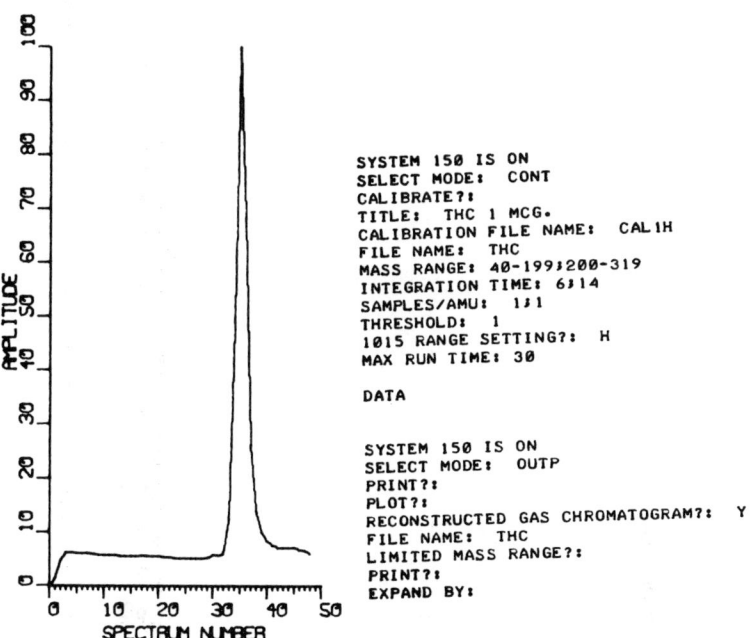

Figure 2.20. Reconstructed chromatogram by computer of Δ^8-THC. From Knight [93], courtesy of Finnigan Corporation.

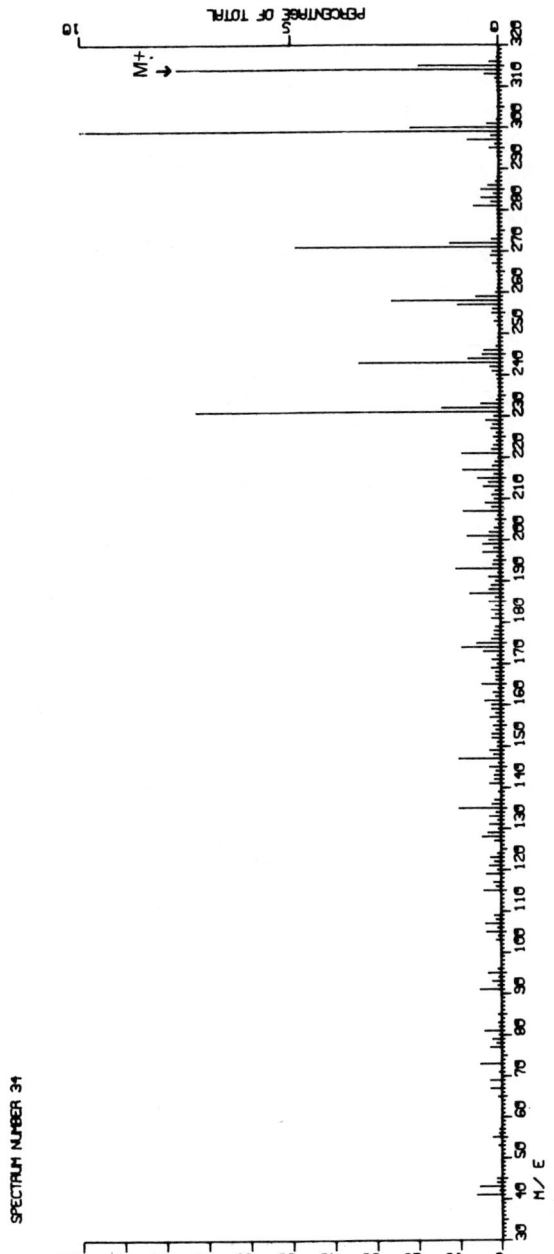

Figure 2.21. Mass spectrum of Δ^8-THC. From Knight [93], courtesy of Finnigan Corporation.

TABLE 2.18

Yields and Retention Times of Some Neutral Cannabinoids[a]

Cannabinoid	Retention time (min)	Yield (%)
Cannabicyclol	4.55	0.11
Cannabidiol	5.67	3.74
Δ^8-THC	7.17	Not detected
Δ^9-THC	7.87	3.30
Cannabinol	10.20	1.30
Cannabichromene	5.58	0.19
Cannabigerol	9.33	0.30

[a] Adapted from Gaoni and Mechoulam [94].

investigators used a Packard model 803 gas chromatograph equipped with a flame ionization detector and 6-ft by 1/8-in. glass columns packed with 2% OV-17 on Gas Chrom Q (mesh size not specified). With the nitrogen carrier-gas flow rate and column temperature maintained at 30 ml/min and 235° C, respectively, the retention times and yields of cannabinoids in hashish are given in Table 2.18.

Vree et al. [95] developed an integrated GC-MS method to identify in hashish the cannabis constituents cannabidiol, tetrahydrocannabinol, and cannabinol, which are also accompanied by their propyl analogs (see Table 2.5), cannabidivarin, tetrahydrocannabivarin, and cannabivarin.

For strictly GC investigations, a Hewlett-Packard model 402 gas chromatograph equipped with a flame ionization detector and a 1.8-m by 3-mm column packed with 3% OV-17 on 60-80 mesh Gas Chrom Q was employed; the other conditions specified for the separation of the cannabinoids were injector temperature, 220° C; column temperature, 190° C; detector temperature, 220° C; nitrogen carrier-gas flow rate, 40 ml/min; hydrogen flow rate, 40 ml/min; air flow rate, 200 ml/min.

The GC-MS studies of Vree et al. were carried out on a LKB 9000 integrated GC-MS instrument equipped with a 1.8-m by 3-mm-i.d. glass column packed with 3% OV-17 on 60-80 mesh Gas Chrom Q. Other operating conditions maintained were column temperature, 180° C; separator temperature, 220° C; ion source temperature, 250° C; helium carrier-gas flow rate, 30 ml/min; mass spectra scans, repetitive and taken at 10, 12, 14, 16, 18, and 20 eV during the elution of the GC peak; trap current, 60 μA; acceleration voltage, 3.5 kV. With this integrated GC-MS unit, the gas chromatograms yielded the same elution pattern as the H-P instrument. The mass spectra were normalized and the relative abundance of a particular mass fragment was plotted against electron voltage. Using the above operating parameters, the retention times of cannabidivarin, Δ^9-tetrahydrocannabivarin, cannabivarin, cannabidiol, and Δ^9-THC relative to cannabinol (RRT = 1.000) were 0.378, 0.489, 0.567, 0.645, and 0.835, respectively.

Merkus [96] in 1971 also reported cannabivarin and tetrahydrocannabivarin as two new constituents of hashish. The retention times of the predominant constituents of Nepal hashish samples relative to cannabinol were obtained on a SE-52 column (length and i.d. not listed, but its packing was 5% SE-52 on Gas Q) with the GC instrument conditions given as column temperature, programmed from 225 to 280° C at 6° C/min; nitrogen carrier-gas flow rate, 40 ml/min. The RRT values for cannabidivarin, tetrahydrocannabivarin, cannabivarin, cannabidiol, and Δ^9-THC were 0.480, 0.612, 0.691, 0.765, and 0.900, respectively. Based on the analysis of three different samples of Nepal hashish, the average percentages found for CBN, THC, CBD, cannabivarin, tetrahydrocannabivarin, and cannabidivarin were 25.7 ± 3.0, 6.8 ± 3.1, 39.4 ± 2.3, 13.3 ± 1.7, 3.7 ± 1.4, and 11.1 ± 1.3%.

In 1971, Stromberg [99] investigated the minor components of cannabis resin and their separation by gas chromatography, thermal stability, and protolytic properties. As noted, "the present method was developed both for forensic purposes and for MS analysis, which is now in the process of development. In this study, the method was applied to two materials, of the hashish and marijuana type, respectively. Some of the protolytic properties were studied for about 40 minor components of the hashish material, while for the marijuana material, the behavior of some 50 minor components under smoking conditions was investigated."

These separations were performed with a Perkin-Elmer F-11 gas chromatograph equipped and operated as indicated below:

Column	1.9-m by 2-mm-i.d. coiled glass packed with 3% OV-17 on 80-100 mesh Gas Chrom Q
Injector temperature (°C)	275
Column temperature (°C)	Programmed, 0 to 2 min isothermal at 100° C; 2 to 18 min at 8° C/min; 18 to 30 min isothermally at 228° C; 30 to 39 min at 8° C/min; 39 to 45 min isothermally at 300° C
Detector temperature (°C)	Not specified
Detector	Flame ionization
Nitrogen carrier-gas flow rate (ml/min)	30

Hydrogen inlet pressure (atm)   1.3

Air inlet pressure (atm)   2.0

With this proposed method of GC operation, the retention times of the major cannabinoids, CBD, THC, and CBN, were approximately 24.47, 28.04, and 30.91 min, respectively.

Schou et al. [104] developed a highly sensitive GC procedure for the determination of tetrahydrocannabinol and cannabinol using a Pye model 104 gas chromatograph equipped with a 10-mCi ^{63}Ni electron capture detector and a glass column (dimensions not given) packed with 2% XE-60 on AW-DMCS-treated Chrom W. Following a derivatization step in which the cannabis resin extract was reacted with trichloroacetic anhydride (1% in benzene), the excess reagent was removed by the addition of 0.25 N sodium hydroxide. After the removal of trichloroacetate with the water phase after 15 min, a few microliters of the benzene layer was withdrawn and injected into the GC unit, whose operating conditions were injector temperature, 250° C; column temperature, 235° C; detector temperature, 250° C; carrier-gas flow rate (not identified), 30 ml/min; quench-gas flow rate, 10 ml/min. Using peak height versus nanograms of component injected calibration curves, the standard curves for Δ^9-THC and CBN were linear at concentration ranges of 0 to 15 and 1 to 5 ng, respectively. The chloroacetyl derivatives of Δ^8-THC, Δ^9-THC, and CBN were eluted in 0.500, 0.539, and 0.788 min, respectively, using the GC conditions given above.

In 1972, Vree et al. [107] developed a GC-MS procedure capable of identifying 25 natural and 45 synthetic cannabinoids. Based on this extensive investigation, Vree et al. summarized their findings as follows:

> It appeared that the retention times of groups of cannabinoids showed a characteristic pattern. An increase in the side chain increases the retention time by a fixed amount of 42% per carbon atom. When the position of the side chain is shifted from the ortho to the para position of the aromatic ring, the retention time is increased by a factor of 1.3. Reduction of the polarity by methylation and silylation reduces the retention time by a factor of 0.53. Branching of the side chain results in an increase in the retention time by 12%. Saturation of the double bonds leads to a decrease in the retention time by a factor of 0.80.

Using marijuana and hashish samples obtained from Columbia, Congo, Laos, Indonesia, and Brazil and Nepal, Afghanistan, Turkey, Lebanon, Morocco, South Africa, and the Netherlands, respectively, much of the entire investigation was performed with GC instruments equipped with flame ionization detectors and 1.8-m by 3-mm-i.d. columns packed with 3% OV-17, 3.8% UCW-98, 3% OV-1, 3% XE-60, or 3% Apiezon L. GC-MS studies, on the other hand, were carried out with a LKB 9000 instrument. The various

TABLE 2.19

Relative Retention Times of Natural Cannabis Constituents[a]

Compound	RRT	Structure[b]
Cannabigerol-C_3	0.143	T 2.1
CBD-C_1	0.201	T 2.1
Δ^9-THC-C_1	0.230	T 2.1
Cannabichromene-C_3	0.270	T 2.1
CBN-C_1	0.280	T 2.1
Cannabicyclol-C_5	0.300	F 2.1
Cannabigerol-C_3-O-methyl	0.348	T 2.20
CBD-C_3	0.425	T 2.1
CBD-C_5-O-methyl	0.445	T 2.1
Δ^9-THC-C_3	0.500	T 2.5
CBN-C_3	0.600	T 2.5
Cannabichromene-C_5	0.600	F 2.1
Cannabinodiol-C_3	0.710	T 2.20
Cannabigerol-C_5-O-methyl	0.770	T 2.20
CBD-C_5	0.810	F 2.1
ortho-CBD-C_5	0.810	T 2.20
Δ^8-THC-C_5	0.845	(see p. 272)
Δ^9-THC-C_5	1.000	F 2.1
Cannabigerol-C_5	1.170	F 2.1
CBN-C_5	1.260	F 2.1
Cannabinodiol-C_5	1.500	T 2.20

[a] Adapted from Vree et al. [107].
[b] T = Table, F = Figure.

Chemical Composition of the Marijuana Plant 345

GC and/or GC-MS operating parameters used were given as follows: column temperature, 180 to 200° C; separator temperature, 240° C; injector temperature, 230° C; detector temperature, 250° C; ion source temperature, 290° C; GC nitrogen carrier-gas flow rate, 30 ml/min; hydrogen flow rate, 30 ml/min; air flow rate, 150 ml/min; GC-MS helium carrier-gas flow rate, 30 ml/min; ionization potential, 20 to 40 eV for GC and 10 to 20 eV for MS; acceleration potential, 3.5 kV; trap current, 60 μA.

Using these GC and/or GC-MS conditions with the 3% OV-17 column, the retention times of 20 natural cannabinoids relative to Δ^9-THC-C_5 are listed in Table 2.19 as well as their structures, which can be located in either the table or figure designated. In Table 2.20 are given the structures of several compounds not previously referred to in this chapter.

From the data given in Table 2.19, the influence of the length of the side chain on the relative retention values is very evident, whereas the influence of the position of the side chain in the aromatic ring system and the influence of reducing the polarity of the aromatic hydroxyl group by methylation (trimethylanilinium hydroxide) and silylation (trimethylsilylimidazole) on the relative retention times can be deduced from the data listed in Tables 2.21 and 2.22, respectively.

Vree et al. [112] also identified cannabicyclol with a pentyl or propyl side chain by means of integrated GC-MS using a LKB 9000 instrument equipped with columns packed with 3% OV-17 on 60-80 mesh Gas Chrom Q.

TABLE 2.20

Several C_3 and C_5 Cannabinoid Derivatives

Ortho-CBD-C_5

Cannabinodiol-C_3

Cannabinodiol-C_5

Cannabigerol-C_3-O-methyl

Cannabigerol-C_5-O-methyl

TABLE 2.21

Influence of the Position of the Side Chain in the Aromatic
Ring System on the Relative Retention Times[a]

Compound	RRT
ortho-Δ^9-THC-C_5	0.760
Δ^9-THC-C_5	1.000
ortho-Δ^8-THC-C_5	0.670
Δ^8-THC-C_5	0.845
ortho-Δ^9-THC-C_5-O-methyl	0.408
Δ^9-THC-C_5-O-methyl	0.545
ortho-Δ^8-THC-C_5-O-methyl	0.336
Δ^8-THC-C_5-O-methyl	0.470

[a] Adapted from Vree et al. [107].

When operated under specified conditions (column temperature, 180°C; separator temperature, 240°C; ion source temperature, 290°C; accelerating voltage, 3.5 kV; trap current, 60 μA; electron energy, 10, 12, 14, 16, 18, and 20 eV during the elution of a peak; helium carrier-gas flow rate, 20 ml/min), the retention times of cannabicyclol-C_3, cannabicyclol-C_5, Δ^9-THC-C_3, CBN-C_3, and CBN-C_5 relative to Δ^9-THC-C_5 were reported to be 0.143, 0.300, 0.500, 0.600, and 1.260, respectively.

At 20 eV, the relative abundance of specific m/e fragment ions for cannabicyclol-C_3 and -C_5 in their mass spectra were as indicated below:

	Relative abundance (%)											
m/e Fragment ion:	174	203	218	230	231	243	246	258	271	286	299	314
Cannabicyclol-C_3	100	2	2						6	20		
Cannabicyclol-C_5	5				100	5	3	7	8		7.5	33

They noted that "cannabicyclol-C_5 can also be distinguished from cannabichromene-C_5, e.g., by the presence of the fragments m/e 299, 271, 258, and 243 in equal relative abundances (6 to 8%). The latter fragments are not present in cannabichromene-C_5. The same observation holds for

TABLE 2.22

Effect of Methylation and Silylation on
Relative Retention Times[a]

Compound	RRT
Δ^8-THC-C_5-O-methyl	0.470
Δ^8-THC-C_5	0.845
Δ^9-THC-C_5-O-methyl	0.545
Δ^9-THC-C_5	1.000
CBD-C_5-O,O-dimethyl	0.243
CBD-C_5-O-methyl	0.445
CBD-C_5	0.810
ortho-Δ^8-THC-C_5-O-methyl	0.336
ortho-Δ^8-THC-C_5	0.670
ortho-Δ^9-THC-C_5-O-methyl	0.408
ortho-Δ^9-THC-C_5	0.760
CBN-C_5-O-methyl	0.740
CBN-C_5	1.260
Cannabigerol-C_5-O-methyl	0.770
Cannabigerol-C_5	1.170
Δ^9-THC-O-TMSi	0.450
Δ^9-THC-C_5	1.000
CBN-C_3-O-TMSi	0.282
CBN-C_3	0.600
CBN-C_5-O-TMSi	0.625
CBN-C_5	1.260

[a] Adapted from Vree et al. [107].

the presence of the mass fragments m/e 271, 243, 230, and 218 in cannabicyclol-C_3, these being absent in cannabichromene-C_3."

In 1972, Hoffman and Yang [113] demonstrated on a variety of silicone columns quantitative gas chromatography of small amounts of Δ^9-THC; retention data in the form of methylene unit values were reported for each column investigated.

For this investigation, Hoffman and Yang performed their GC analyses with a Barber-Colman model 5000 gas chromatograph equipped with a flame ionization detector and 6-ft by 4-mm U-shaped glass columns packed with 5% OV-1, OV-17, Dexsil 300 GC, OV-210, or OV-225 coated on 80-100 mesh, acid-washed, Chromosorb W. When measured with n-alkanes at 240°C, the minimum column efficiency was 2000 theoretical plates. In conjunction with a nitrogen carrier-gas flow rate of 25 ml/min, the temperatures of the various columns used to obtain retention data were OV-1, 260 to 270°C; OV-17, 260 to 270°C; Dexsil 300 GC, 260 to 270°C; OV-210, 210°C; OV-225, 220 to 250°C.

With regard to standard and sample preparation for GC analysis, the alkanes n-$C_{22}H_{46}$, n-$C_{24}H_{50}$, n-$C_{25}H_{52}$, n-$C_{28}H_{58}$, n-$C_{30}H_{62}$, and n-$C_{32}H_{66}$ and Δ^9-THC were dissolved in isooctane in varying ratios so as to cover a range of weight ratios of Δ^9-THC to alkane from 0.4 to 5.0. The concentration of Δ^9-THC varied from 0.2 to 0.4 mg/ml. With columns that gave coincidence or serious overlap of peaks of the above alkanes and Δ^9-THC, the interfering n-alkane was not placed in the solution.

The trimethylsilyl ether of Δ^9-THC was prepared by combining the following proportions by weight of bis-(trimethylsilyl)-trifluoroacetamide to pyridine to trifluoroacetic acid to Δ^9-THC: 120 to 60 to 2 to 1 in isooctane solution. The Δ^9-THC concentrations were approximately 0.2 to 0.7 mg/ml.

The trifluoroacetyl ester of Δ^9-THC was prepared by combining the following proportions by weight of trifluoroacetic anhydride to pyridine to Δ^9-THC; 15 to 3 to 1. After mixing, bis-(trimethylsilyl)-trifluoroacetamide was added in an amount of 1.5 times the weight of pyridine. This addition removed the precipitate of pyridinium trifluoroacetate. The Δ^9-THC concentrations were approximately 0.2 to 0.7 mg/ml.

The experimental procedure used for this investigation was as follows:

Solutions of n-alkanes and Δ^9-THC or its derivatives were injected into the gas chromatograph in amount of 0.2 to 1.0 μl. Chromatographic peak heights were measured, and from the weights of n-alkanes and Δ^9-THC used, the following equation was tested.

$$\frac{\text{peak height } \Delta^9\text{-THC}}{\text{peak height n-}C_nH_{2n+2}} = k \frac{\text{wt } \Delta^9\text{-THC}}{\text{wt n-}C_nH_{2n+2}} \qquad (2.2)$$

A least squares program was used for the IBM 7040 digital computer. Slope, intercept, and correlation coefficient were obtained. Intercepts near zero and correlation coefficients near one were considered sufficient evidence that Eq. (2.2) was obeyed.

An examination of intercept and correlation coefficient data showed no superiority of one alkane over another as an internal standard (except for n-$C_{22}H_{46}$, which was frequently eluted on the solvent band). Nor were there better results in using the TMSi ether or TFA ester over Δ^9-THC itself. Therefore, comparisons were made according to column. Table 2.23 gives the results of fitting the GC data to Eq. (2.2). Two or three alkanes were used together with Δ^9-THC or its derivatives, with n-$C_{24}H_{50}$, n-$C_{28}H_{58}$, and n-$C_{30}H_{62}$ being the most commonly used.

The results show that all columns gave a straight line with an intercept near zero. OV-1 gave the best intercept results and might be considered superior. However, not shown in the table is the temperature of the column. Where low temperatures might be required for a given sample, OV-210 would be the column of choice. The retention times of both the standards and the Δ^9-THC or its derivatives were distinctly lower. With their equipment and chromatographic conditions, the smallest amount injected into the column was 50 ng. However, they noted that one could quantitate even less, about half this amount.

Table 2.24 shows the methylene unit values on the columns used for Δ^9-THC, its TMSi ether, and its TFA ester. These were obtained from locating the Δ^9-THC or its derivative's retention time on a curve of carbon number of the n-alkane versus its retention time.

When 3% columns were used, lower correlation coefficients and very large intercept values relative to those shown in Table 2.23 were obtained.

TABLE 2.23

Proportionality Between Peak Height and Weight Ratios[a]

Column	No. of lines[b]	Av. correlation coefficient ± std. dev.	Av. intercept ± std. dev.
OV-1	7	0.999 ± 0.0001	-0.021 ± 0.042
OV-17	9	0.997 ± 0.001	-0.007 ± 0.111
OV-210	8	0.999 ± 0.0002	0.031 ± 0.100
OV-225	9	0.998 ± 0.001	-0.041 ± 0.098
Dexsil 300 GC	7	0.997 ± 0.001	0.039 ± 0.092

[a] From Hoffman and Yang [113], courtesy of Analytical Letters.
[b] A line was obtained from five points, each of which represented ratios of one alkane and one Δ^9-THC derivative or Δ^9-THC itself.

TABLE 2.24

Methylene Unit Values[a]

Column	Compound chromatographed		
	Δ^9-THC	Δ^9-THC ether	Δ^9-THC ester
OV-1	25.2	22.0	24.0
OV-17	28.6	26.4	26.1
OV-210	28.2	25.5	26.1
OV-225	36.1[b]	25.6	27.5
Dexsil 300 GC	26.1	24.3	22.5

[a] From Hoffman and Yang [113], courtesy of <u>Analytical Letters</u>.
[b] Estimated.

Therefore, it is suggested that columns of silicone no lower than about 5% by weight should be used in determining Δ^9-THC in samples where its concentration is low.

In 1972, Ek et al. [115] determined by gas chromatography the cannabinoid (CBD, Δ^9-THC, and CBN) content in 50 hashish samples seized by Swedish police or customs authorities during the years 1970-1972. GC analyses were carried out with a Perkin-Elmer F-11 gas chromatograph equipped with a flame ionization detector and a 2.7-m by 3-mm-i.d. coiled glass column packed with 5% JXR methyl silicone liquid stationary phase coated on 100-120 mesh Gas Chrom Q. The other conditions employed were injector temperature, 230°C; column temperature, 230°C; nitrogen carrier-gas inlet pressure, 1.7 atm; hydrogen pressure setting, 1.4 atm; air pressure setting, 1.8 atm.

For this investigation, sample preparation consisted of extracting 1 g of the powdered hashish material with methylene chloride in a semimicro Soxhlet extractor until the extract appeared colorless (2 to 4 hr). The extract was then evaporated to a small volume and transferred to a 2-ml volumetric flask containing 10 mg of triphenylcarbinol (internal standard). One microliter of this solution was injected into the GC unit for the determination of the three cannabinoids.

The average content of these cannabinoids in the 50 samples analyzed was CBD, 3.4% (SD = 3.1); Δ^9-THC, 3.2% (SD = 2.1); CBN, 0.7% (SD = 0.6).

Ek et al. noted that "since cannabinoid acids readily decarboxylate in the injection system of the gas chromatograph, the figures actually represent the sums of the content of the respective acids and their corresponding free phenols."

De Zeeuw et al. [116] investigated the occurrence and some analytical properties of Δ^9-tetrahydrocannabinolic acid A (see Table 2.1), including its chromatographic behavior in the presence of other cannabinoids. The acid is inactive but is converted upon smoking into active Δ^9-THC. The acid is present in abundant amounts in various cannabis samples, marijuana in particular, and these will be more active upon smoking than when administered by injection or orally.

The unmodified and TMSi derivatives of the various cannabinoids extracted from cannabis products were evaluated with a Becker model 409 gas chromatograph equipped with a flame ionization detector and a 2-m by 2-mm stainless steel column packed with 5% SE-30 on acid-washed, DMCS-treated, 80-100 mesh Chromosorb G. TMSi derivatives of the cannabinoids were prepared by evaporating small extract samples to dryness under a stream of nitrogen, then redissolving these samples in 0.5 ml of Tri-Sil silylating reagent. After shaking and standing for 5 min, the solution was ready for injection into the chromatograph, which was operated as follows: injector temperature, 275°C; column temperature, 230°C; detector temperature, 275°C; nitrogen carrier-gas flow rate, 20 ml/min. Because Δ^9-THC acid, like other cannabinoid acids, is instantaneously decarboxylated on the column at the temperatures used, silylation was mandatory if these compounds were to be quantitatively determined. These TMSi derivatives separate well, and their retention times together with those of other major cannabis constituents are listed in Table 2.25.

However, when exact quantitation of the acid is required, De Zeeuw et al. noted that "GC still presents a difficulty. A stable and sufficiently pure

TABLE 2.25

Retention Times of Major Cannabis Constituents[a]

	Retention time (min)	
Compound	Unmodified	Silylated
CBD	16.08	12.00
Δ^9-THC	20.60	15.70
CBN	24.40	20.50
CBD acid	(16.08)	25.95
Δ^9-THC acid A	(20.60)	35.85
CBN acid	(24.40)	42.60

[a] Adapted from De Zeeuw et al. [116].

sample of Δ^9-THC acid A is unavailable as a reference standard so that the surface under the peak cannot be related to the exact amount of acid present. Therefore, in our opinion, the best way of quantitating both Δ^9-THC acid A and Δ^9-THC is by determining their sum as Δ^9-THC by normal GC, followed by determining Δ^9-THC alone by GC after silylation. From the difference between these determinations the amount of Δ^9-THC acid A can be derived."

In 1973, Adams and Jones [120] identified in cannabis and its smoke by gas chromatography and mass spectrometry a series of long-chain hydrocarbons. As noted, the level of hydrocarbons was determined to be about half that found in tobacco and its smoke, although the effect of smoking on the paraffins in the cannabis material was comparable to analogous studies of tobacco and its smoke.

The plant extract which contained the hydrocarbons was prepared for GC analysis in the following manner:

The cannabis samples were extracted with 95% ethanol, concentrated and diluted with water, and then the mixture was exhaustively extracted with hexane. After concentration, the hexane solution was made to volume and aliquots were used for analysis. These were chemically separated into basic, acidic, and neutral fractions and the neutral solution was concentrated to a thick syrup. This was chromatographed on silicic acid and initial waxy fractions were eluted with hexane. Treatment with urea in hot methanol formed an adduct which was washed with hexane and decomposed with water and extraction with hexane gave a clean mixture of long-chain hydrocarbons. Separation of the straight-chain paraffins from those with branched-chains was affected with Linde 5A molecular sieves.

Adams and Jones obtained retention time data for these straight-chain, iso-, and anteiso- paraffins using a Beckman GC-4 gas chromatograph equipped with a flame ionization detector and a 10-ft by 1/8-in. column packed with 2% OV-17 on Gas Chrom Q. For plant hydrocarbons, the column temperature was programmed from 200 to 280°C at 5°C/min, whereas the injector and detector temperatures were maintained at 350°C and the helium carrier-gas flow rate at 26 ml/min. For GC-MS analysis, the LKB 9000 instrument was equipped with a 9-ft by 1/4-in. glass column packed with 1% OV-17 on Chromosorb W, which was programmed from 150 to 240°C at 5°C/min. With the integrated GC-MS unit, the ionizing voltage was set at 20 eV, with the separator temperature at 260°C and the ion source temperature at 270°C.

With the GC conditions given above, the retention times of the various paraffin constituents of plant waxes are listed in Table 2.26. As noted, the plant waxes are mixtures of long-chain hydrocarbons, with the odd-carbon paraffins predominating over the even-numbered and branched paraffins.

TABLE 2.26

Retention Times of Hydrocarbons of Cannabis Plant Extract[a]

Carbon number	Assignment	Ret. time (min)
22	N	5.38
23	N	6.37
24	N	7.45
25	Iso	8.08
25	N	8.58
26	Anteiso	9.41
26	N	9.75
27	Iso	10.44
27	N	10.96
28	Anteiso	11.81
28	N	12.51
29	Iso	12.87
29	N	13.42
30	Anteiso	14.18
30	N	14.48
31	N	15.63

[a] Adapted from Adams and Jones [120].

The straight-chain paraffins comprise 88.2% of the paraffin wax, which is characterized by the predominance of nonacosane (55.3%). The other 11.8% of the paraffins are those with methyl branching at the 2 (iso) and 3 (anteiso) positions.

In 1973, Fairbairn and Liebmann [121] devised a convenient method for the complete extraction of the cannabinoids from fresh plant material, herbal cannabis, cannabis resin, and reefers. Based on their extensive study, they noted that chloroform was a more suitable solvent than light petroleum or ethanol, and that simple shaking of powdered material with the solvent was effective. Fresh material should be air dried and powdered before extraction. The main cannabinoids in the extract were determined by gas chromatography using androst-4-ene-3,17-dione as internal standard. In the GC method used, the coefficient of variation for repeated determinations of THC on a single extract was 1.4%; for all operations, including sampling and extraction, it was 2.7%. Duplicate analyses of 24 samples of herbal cannabis and of 20 reefers, all of varying potency, showed that the errors fell within the expected limits for THC, CBD, and CBN.

For the determination of THC, CBD, and CBN, Fairbairn and Liebmann used a Pye 104 gas chromatograph equipped with a flame ionization detector and a 5-ft by 4-mm-i.d. column packed with 2% OV-17 on Chromosorb W (AW-DMCS, 80-100 mesh). To obtain retention times for CBD, Δ^9-THC, and CBN relative to the internal standard (RRT = 1.00) of 0.37, 0.51, and 0.64, respectively, the following GC operating conditions were maintained: nitrogen carrier-gas flow rate, 45 ml/min; column temperature, 235 to 240°C. Quantitative data were obtained using calibration curves based on peak area measurements and, per unit weight, the ratio of the area of the peaks for CBD:THC:CBN was 104:93:84.

Bercht et al. [123] also used integrated GC-MS to identify one of the minor constituents of a hashish sample as cannabinol methyl ether (CBNM) (structure shown in Table 2.1). The GC-MS analyses were performed with a LKB 9000 instrument equipped with 1.5-m by 4-mm-i.d. glass columns packed with 3% OV-17 on 60-80 mesh Gas Chrom Q. The other GC-MS conditions for analysis were column temperature, 200°C; separator temperature, 220°C; ion source temperature, 250°C; ionization potential for repetitive mass spectra during the elution of a component from the GC column, 10, 12, 14, 16, 18, and 20 eV; total ion current for representation of the reconstructed chromatogram, 20 eV; trap current, 60 μA; acceleration potential, 3.5 kV; helium carrier-gas flow rate, 20 ml/min.

Using the above conditions, the retention times of $CBD-C_3$, $CBD-C_5$, CBE (cannabielsoin), $CBG-C_5$ (cannabigerol), and $CBN-C_5$ were approximately 9.00, 16.70, 22.35, 26.00, and 29.65 min, respectively, whereas cannabinol monomethyl ether and cannabidiol monomethyl ether were eluted in about 20.70 and 12.35 min, respectively. The respective retention times of CBNM and CBDM relative to $CBD-C_5$ (used as internal standard) were 1.24 and 0.74. Bercht et al. further noted that "it seems likely that the methyl ether of Δ^9-THC is also present as a natural product. Unfortunately, the relative retention time of this compound, as measured on the synthetic methyl ether, is 1.14 (RT = 19.05 min), which is in the region that in most chromatograms is generally obscured by large amounts of CBD."

In 1973, Vree et al. [124] studied the gas chromatographic behavior of cis- and trans-tetrahydrocannabinol and isotetrahydrocannabinol, noting that the GC behavior of the THC isomers was very characteristic and that there are fixed ratios of the retention times for the cis-trans (1.85), ortho/para isomers (1.20) and iso-THC/Δ^1-THC (2.67 and 4.80).

Some of the structures of the many synthetic cannabinoids and iso-THC compounds examined are shown below:

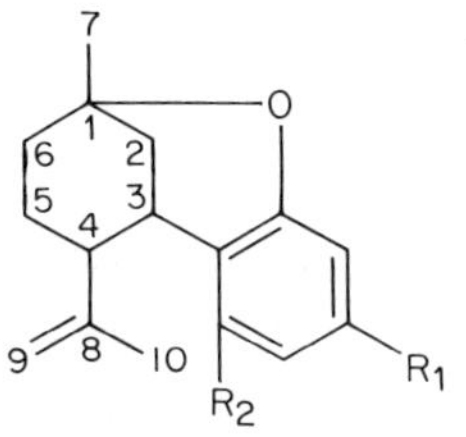

Synthetic iso-THC Compounds

Compound	Double bond	Position of H atoms at C_3 and C_4	R_1	R_2
cis-ortho-8,9-iso-THC-C_5	8,9	cis	OH	C_5H_{11}
cis-para-8,9-iso-THC-C_5	8,9	cis	C_5H_{11}	OH
trans-ortho-8,9-iso-THC-C_5	8,9	trans	OH	C_5H_{11}
trans-para-8,9-iso-THC-C_5	8,9	trans	C_5H_{11}	OH

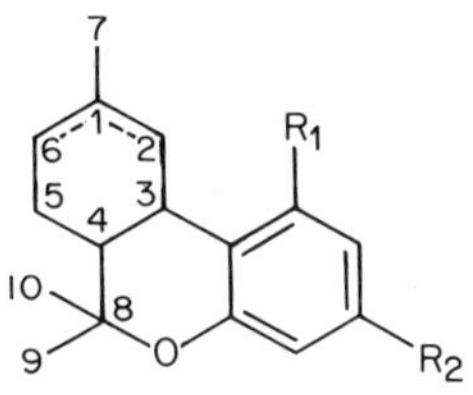

Synthetic Cannabinoids

Compound	Double bond	Position of H atoms at C_3 and C_4	R_1	R_2
cis-ortho-1,6-THC-C_5	1,6	cis	OH	C_5H_{11}
cis-para-1,6-THC-C_5	1,6	cis	C_5H_{11}	OH
cis-ortho-1,6-THC-C_5-α-methyl	1,6	cis	C(CH$_3$)CCCC	OH
cis-para-1,6-THC-C_5-α-methyl	1,6	cis	OH	C(CH$_3$)CCCC
trans-ortho-1,6-THC-C_5	1,6	trans	C_5H_{11}	OH
trans-para-1,6-THC-C_5	1,6	trans	OH	C_5H_{11}
trans-ortho-1,6-THC-C_5-α-methyl	1,6	trans	C(CH$_3$)CCCC	OH
trans-para-1,6-THC-C_5-α-methyl	1,6	trans	OH	C(CH$_3$)CCCC
cis-ortho-1,2-THC-C_5	1,2	cis	C_5H_{11}	OH
cis-para-1,2-THC-C_5	1,2	cis	OH	C_5H_{11}
trans-ortho-1,2-THC-C_5	1,2	trans	C_5H_{11}	OH
trans-para-1,2-THC-C_5	1,2	trans	OH	C_5H_{11}
cis-ortho-HHC		cis	C_5H_{11}	OH
cis-para-HHC		cis	OH	C_5H_{11}
trans-para-HHC		trans	OH	C_5H_{11}

Note: cis-ortho-1,6-THC-C_5 is the same as cis-ortho-Δ^6-THC-C_5.
cis-ortho-1,2-THC-C_5 is the same as cis-ortho-Δ^1-THC-C_5.

For this extensive analytical undertaking, Vree et al. used two instruments as noted below:

1. Gas Chromatography

Gas chromatograph	Hewlett-Packard model 402
Detector	Flame ionization
Column	1.8-m by 3-mm-i.d. glass packed with 3% OV-17 on 60-80 mesh Gas Chrom Q
Injector temperature (°C)	230
Column temperature (°C)	200
Detector temperature (°C)	250
Nitrogen carrier-gas flow rate (ml/min)	30

Chemical Composition of the Marijuana Plant

Hydrogen flow rate (ml/min)	30
Air flow rate (ml/min)	150

2. Gas Chromatography-Mass Spectrometry

GC-MS instrument	LKB 9000
Column	Same as indicated above
Injector temperature (°C)	230
Column temperature (°C)	200
Separator temperature (°C)	230
Ion source temperature (°C)	290
Helium carrier-gas flow rate (ml/min)	20
Ionization potential (eV)	For GC, 20; for GC-MS, 10 to 20
Acceleration potential (kV)	3.5
Trap current (μA)	60

With the GC and GC-MS conditions cited above, the retention times of synthetic and natural cannabinoids relative to trans-para-1,2-THC-C_5 (RRT = 1.00) are listed in Table 2.27.

Bailey et al. [125] also prepared a series of Δ^8-THC in which the side chain varied continuously in methylene units from H to C_7H_{15}, and in which R was made more complex by branching as illustrated below:

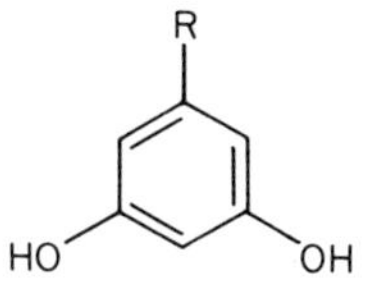

Series A 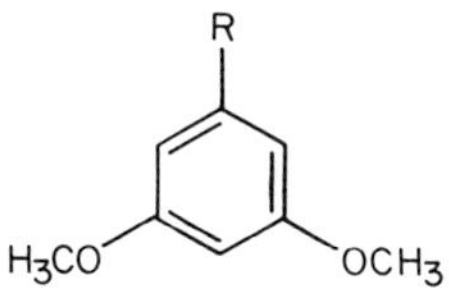   Series B 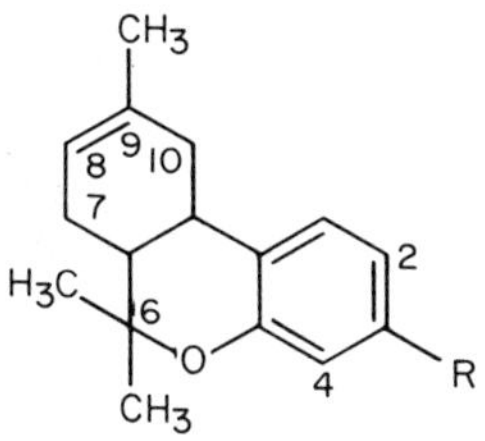   Series C

Structure	R (substituent)
A	H
B	CH_3
C	C_2H_5

Structure	R (substituent)
D	n-C_3H_7
E	n-C_4H_9
F	n-C_5H_{11}
G	n-C_6H_{13}
H	n-C_7H_{15}
I	HC·OH·C_4H_9
J	COC_4H_9
K	CH_2·cyclohexyl
L	CH_2·phenyl
M	HC·CH_3·phenyl

In addition to studying the Δ^8-THC series (series C), the influence of the side chain on the retention time of corresponding 1-alkyl-3,5-dihydroxy (series A) and 1-alkyl-3,5-dimethoxy benzene (series B) compounds were investigated at several column temperatures.

The retention times of these three series of compounds were obtained with a gas chromatograph equipped with a flame ionization detector and a 6-ft by 3/16-in.-i.d. glass column packed with 5% OV-7 on 80-100 mesh Chromosorb W. The retention times of compounds A through H relative to compound F (n-C_5H_{11} side chain) given in Table 2.28 were determined using the following operating conditions: injector temperature, 275°C; column temperature, see Table 2.28; detector temperature, 275°C; nitrogen carrier-gas flow rate, 30 ml/min.

From a plot of log retention time versus the number of carbon atoms in the side chain of the homologous series, they noted that "it is at once apparent that the curves are neither linear nor parallel to one another. The first three members of a series (H to C_2H_5) seem to obey a relationship different from that of the last four (C_4H_9 to C_6H_{15}). There appear to be two almost straight lines for a series at constant temperature, which intercept one another at or about the position for the propyl-substituted compound (an almost equally good fit to either line). The effect is emphasized by lower column temperatures."

To circumvent the discrepancies noted above, they proposed a new method for the identification of derivatives of Δ^8-THC, including those with functionalized side chains, in crude synthesis mixtures. In this system the retention times of the 1-substituted 3,5-dihydroxybenzenes were plotted against the corresponding THC retention times at a convenient temperature.

TABLE 2.27

Retention Times of Synthetic and Natural Cannabinoids
Relative to trans-para-1,2-THC-C_5[a]

Compound	S or N[b]	RRT
Cannabicyclol-C_3	N	0.14
trans-iso-ortho-8 9-THC-C_5	S	0.17
CBD-C_1	N	0.20
trans-iso-para-8,9-THC-C_5	S	0.20
trans-iso-ortho-8,9-THC-C_5-α-methyl	S	0.22
trans-para-1,2-THC-C_1	N	0.23
Cannabichromene-C_3	N	0.27
trans-iso-para-8,9-THC-C_5-α-methyl	S	0.27
CBN-C_1	N	0.28
cis-ortho-HHC	S	0.29
Cannabicyclol-C_5	N	0.30
Cannabigerol-C_3-O-methyl	N	0.34
cis-ortho-1,6-THC-C_5-α-methyl	S	0.34
cis-ortho-THC-C_5	S	0.41
cis-para-HHC	S	0.41
CBD-C_3	N	0.42
cis-ortho-1,2-THC-C_5	S	0.46
cis-para-1,6-THC-C_5	S	0.47
trans-para-1,2-THC-C_3	N	0.50
cis-para-1,6-THC-C_5-α-methyl	S	0.53
cis-para-1,2-THC-C_5	S	0.53
CBN-C_3	N	0.60
Cannabichromene-C_5	N	0.60
trans-ortho-1,6-THC-C_5	S	0.67
trans-ortho-1,6-THC-C_5-α-methyl	S	0.69
Cannabinodiol-C_3	N	0.71

(continued)

TABLE 2.27 (continued)

Compound	S or N[b]	RRT
Cannabigerol-C_5-O-methyl	N	0.77
trans-para-HHC	S	0.77
trans-ortho-1,2-THC-C_5	S	0.78
CBD-C_5	N	0.81
trans-para-1,6-THC-C_3	N	0.84
trans-para-1,6-THC-C_5	S	0.84
trans-para-1,6-THC-C_5-α-methyl	S	0.95
trans-para-1,2-THC-C_5	N	1.00
CBN-C_5	N	1.26

[a] Adapted from Vree et al. [124].
[b] S = synthetic, N = natural.

As noted, this effectively compared the influence of the basic C_{16} cannabinoid skeleton with that of the benzene ring on the retention times of the various compounds. The results shown in two figures indicate that smooth, almost straight-line plots result even when the side chain is branched or functionalized.

In 1973, Hood et al. [126] investigated the composition of the headspace volatiles of marijuana by gas chromatography. To obtain aroma profiles of Cannabis sativa L. which contain compounds in the headspace at concentrations as low as 0.1 ng/ml, the marijuana sample was prepared for vapor analysis by placing 1 g in a microvial fitted with an on-off valve and septum. After equilibration for 1 hr at 65°C, a 5-ml aliquot of the headspace air was withdrawn with a gas-tight syringe and immediately injected into either a Perkin-Elmer 900 or Hewlett-Packard 7610A chromatograph equipped with flame ionization detectors and 6-ft by 2-mm-i.d. glass columns packed with 3% OV-101 on 100-120 mesh Gas Chrom Q or 20% Reoplex 400 on 80-100 mesh, acid-washed Chromosorb W. With helium as carrier gas (50 ml/min), the columns were programmed from 35 to 130°C (OV-101) or 60 to 130°C (Reoplex 400) at 6°C/min to obtain headspace chromatograms. As noted by Hood et al., a third column packed with Chromosorb 101 (80-100 mesh) was operated isothermally at 90°C to resolve low-boiling components.

Based on their data, Hood et al. reported that:

Typical headspace chromatograms show three separate fractions based on ascending order of component boiling points: Fraction 1

TABLE 2.28

Relative Retention Times of 1-Alkyl-3,5-dihydroxybezenes (Series A), 1-Alkyl-3,5-dimethoxybenzenes (Series B), and Δ^8-THC Homologs (Series C)[a]

	Rel. ret. times							
	Series A			Series B		Series C		
Structure	175° C	200° C	220° C	150° C	180° C	200° C	220° C	250° C
A	0.18	0.28	0.34	0.17	0.25	0.20	0.25	0.31
B	0.22	0.34	0.41	0.23	0.32	0.27	0.33	0.38
C	0.32	0.43	0.50	0.32	0.40	0.36	0.42	0.48
D	0.44	0.55	0.59	0.44	0.53	0.48	0.54	0.59
E	0.67	0.75	0.76	0.66	0.70	0.69	0.74	0.77
F	1.00	1.00	1.00	1.00	1.00	1.00	1.00	1.00
G	1.54	1.38	1.37	1.55	1.45		1.48	1.33
H	2.32	1.96	1.80	2.40	2.12		1.87	1.72
F (RT in min)	12.7	4.7	3.2	9.4	4.0	64.0	29.5	12.0

[a] Adapted from Bailey et al. [125].

consists of oxygenated compounds (MW < 100); fraction 2 consists of monoterpene hydrocarbons and oxygenated compounds (MW > 100); fraction 3 consists of sesquiterpene hydrocarbons (MW > 200). Separated constituents were identified by comparison of their relative retention times (RRT) obtained on the adsorption columns of different polarity with the RRT of authentic standards, and by their mass spectra obtained in separate experiments using a Finnigan 3000 GC-MS system with computer data acquisition. Examination of fraction 1 using the Chromosorb 101 column confirmed the presence of acetone (approximately 75% of this fraction) and smaller amounts of methanol, acetaldehyde, ethanol, methyl acetate, and iso-butyraldehyde.

On the other hand, using the OV-101 and Reoplex 400 columns, the RRT of components found in fractions 2 and 3 are listed in Table 2.29, which also includes 2-methyl-2-heptene-6-one, a compound that had not been previously reported as a constituent of marijuana.

Turner et al. [39] in 1973 studied the stability of cannabinoids in stored plant material. A summary of their data indicated that:

The $(-)$-Δ^9-trans-THC content of Cannabis sativa L. stored at -18, +4, and +22 ± 1°C decomposed at a rate of 3.83, 5.38, and 6.92%, respectively, per year, whereas the material stored at 37 and 50°C showed considerable decomposition. C. sativa L. stored in the absence of direct light at -18, +4, and +22 ± 1°C was more stable than cannabis stored under nitrogen. These data indicate that for normal research use, storage under nitrogen at 0°C is not mandatory. Cannabinol is not the only decomposition product of $(-)$-Δ^9-trans-THC. Tentative evidence supports the possible formation of hexahydrocannabinol as a decomposition product in stored C. sativa L.

Hexahydrocannabinol

Using 4-androstene-3,17-dione as internal standard, analyses were performed with gas chromatographs equipped with flame ionization detectors and 8-ft by 2-mm-i.d. columns (glass and stainless steel) packed with 2% OV-17 on 100-120 mesh Chromosorb W-HP or Gas Chrom Q. The GC conditions employed were nitrogen carrier-gas flow rate, 10 to 30 ml/min, depending on the instrument used; injector temperature, 240°C; column temperature, 210°C; detector temperature, 260°C.

For the silylation of plant extracts, 0.5 ml of bis-(trimethylsilyl)trifluoroacetamide with 1% trimethylchlorosilane was added to the extract residue and the resulting reaction mixture was then heated, using a heating mantle, for 10 to 12 min at 80°C, and 0.1 µl was injected into the GC unit.

Using the above GC conditions, the retention times of unmodified $(-)$-Δ^9-trans-tetrahydrocannabivarin, cannabidiol, cannabigerol monomethyl ether, $(-)$-Δ^8-trans-THC, $(-)$-Δ^9-THC, cannabigerol, and cannabinol relative to 4-androstene-3,17-dione were 0.256, 0.330, 0.376, 0.435, 0.481, 0.541, and 0.639, respectively, whereas the retention times relative to the internal standard of $(-)$-Δ^9-trans-THC TMSi ether, cannabinol TMSi ether, $(-)$-Δ^9-trans-tetrahydrocannabinolic acid TMSi ether, and cannabinolic acid TMSi ether were approximately 0.215, 0.282, 0.607, and 0.659, respectively. On the other hand, hexahydrocannabinol had a RRT of 0.370; this being very close to that of cannabigerol monomethyl ether. Silylation of hexahydrocannabinol yielded RRTs of 0.17 and 0.22, respectively. Two peaks were expected, since there existed the possibility of a mixture of C-9 equatorial and axial methyl isomers. The relative retention time of 0.17 was assigned to the C-9 methyl

TABLE 2.29

Relative Retention Times of Components of Marijuana Headspace[a]

Component	RRT			
	OV-101 32° C	OV-101 90° C	Reoplex 400 80° C	Reoplex 400 120° C
α-Pinene	0.46		0.37	
Camphene	0.51		0.49	
β-Pinene	0.63		0.60	
2-Methyl-2-heptene-6-one	0.62			0.32
Myrcene	0.77		0.78	
Δ^3-Carene	0.86		0.69	
α-Terpinene	0.90		0.89	
Limonene	1.00[b]		1.00[b]	
β-Phellandrene	1.00		1.07	
cis-Ocimene	1.14		1.19	
trans-Ocimene	1.24		1.32	
γ-Terpinene	1.28		1.29	
Terpinolene	1.63		1.59	
Linalool	1.84			0.73
β-Caryophyllene		1.00[b]		1.00[b]
trans-α-Bergamotene		1.12		0.87
β-Farnesene		1.26		1.23
Humulene		1.19		1.39

[a] From Hood et al. [126], courtesy of Nature.
[b] Reference material.

axial isomer; the equatorial methyl isomer was assigned the RRT of 0.22.

C-9 Equatorial Hexahydrocannabinol

C-9 Axial Hexahydrocannabinol

In Figure 2.22 the percent cannabinoids found in C. sativa L. material stored at -18°C, ambient temperature (22 ± 1°C), and 50°C over a 104-week testing period are given using the GC method described above.

In 1973, De Zeeuw et al. [128] reported the occurrence in hashish and marijuana of the propyl homolog of cannabichromene (CBC-C_3) (structure given in Table 2.1). Combined GC-MS was performed on a LKB 9000 instrument as previously described [95]. As noted by these investigators, "the occurrence of CBC-C_5 and CBC-C_3 in nature seems to depend on the origin of the sample. In most samples we have investigated so far, the two cannabinoids were minor components in comparison to Δ^9-THC-C_5, CBD-C_5, CBN-C_5 and their propyl homologs. The present Asian sample, however, had rather high quantities of CBC-C_5 and CBC-C_3. In addition, the sample contained small amounts of their respective acids."

Skinner et al. [129] discussed the use of chemical ionization mass spectrometry for the determination of hashish constituents. The cannabinoids were extracted from the hashish with hexane and chromatographed on a 5-ft by 2-mm-i.d. glass column packed with 3% OV-1 on 100-120 mesh Gas Chrom Q at 230°C. The electron energy for electron impact and chemical ionization investigations was 70 and 150 eV, respectively.

Figure 2.23 shows the reconstructed gas chromatogram for the chemical ionization run. Figures 2.24 through 2.30 represent the mass spectra obtained for the seven peaks shown in Figure 2.23.

All of the cannabinoids give M+1, M+29, and M+41 peaks which are characteristic of most compounds when methane is used as reagent gas. They also show a fragmentation pattern which is characteristic of the individual ring structures.

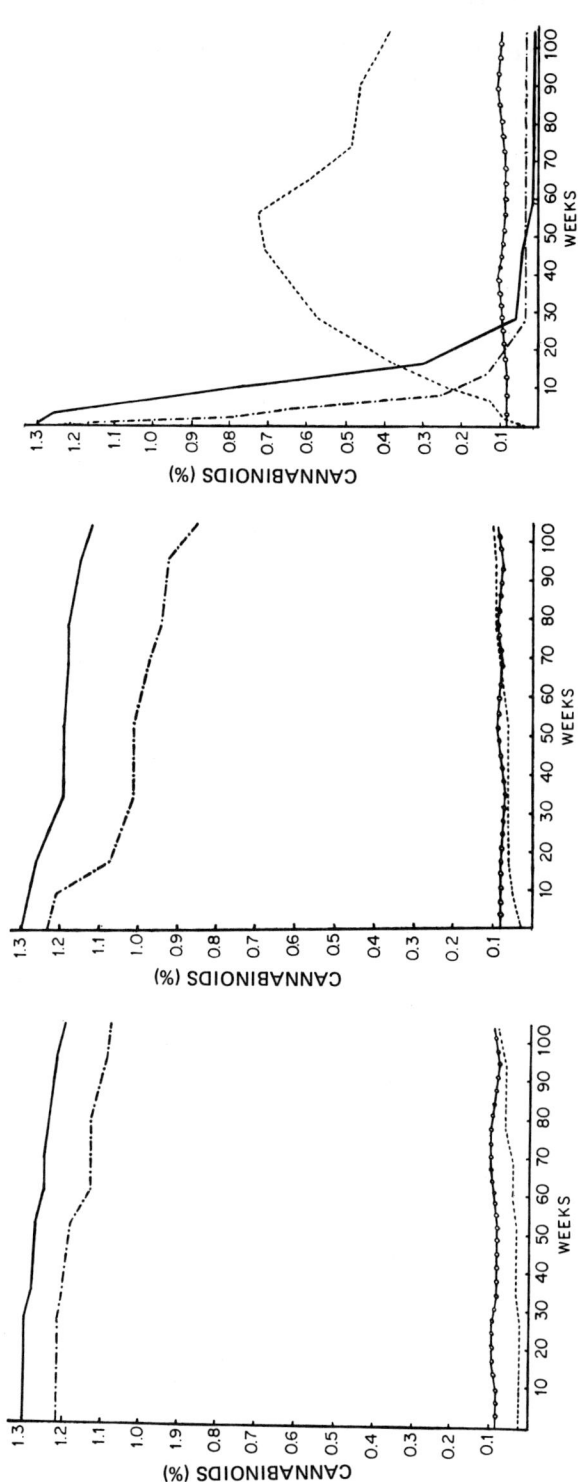

Figure 2.22. Cannabinoid concentrations in C. sativa L. material stored at −18°C, ambient temperature (22 ± 1°C), and 50°C. (a) At −18°C, ———, (−)-Δ^9-trans-THC; -·-·, (−)-Δ^9-trans-tetrahydrocannabinolic acid; o———o, cannabidiol; and ----, cannabinol. (b) At 22 ± 1°C, same component identification as in (a). (c) At 50°C, same component identification as in (a). Adapted from Turner et al. [39].

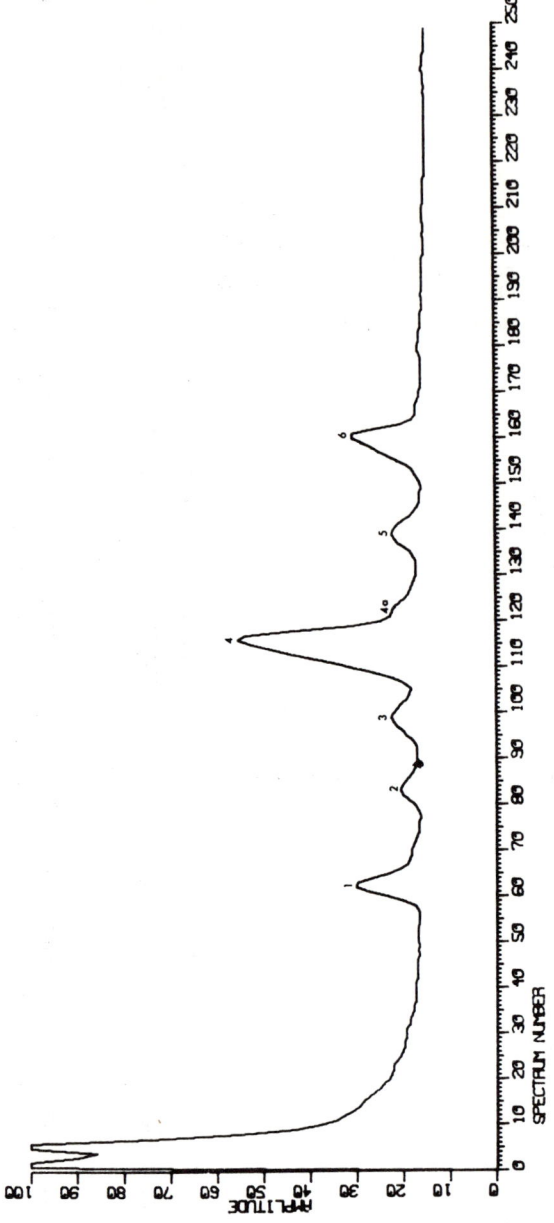

Figure 2.23. Reconstructed gas chromatogram (CI mode) of seven components observed in Nepalese hashish. From Skinner et al. [129], courtesy of Finnigan Corporation.

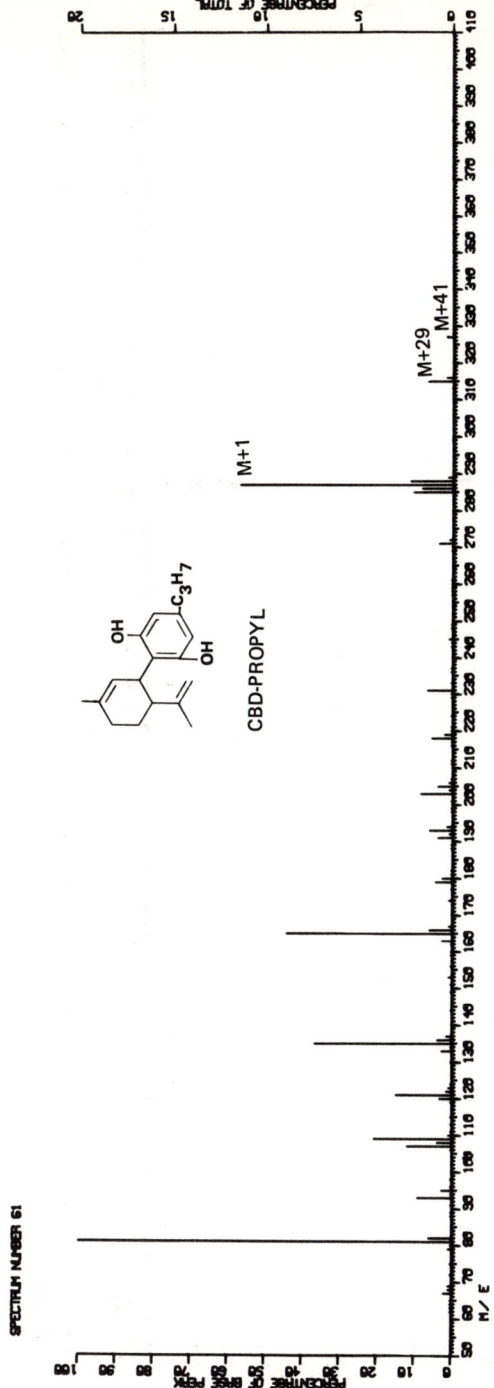

Figure 2.24. CI mass spectrum of CBD-C_3 (peak 1 in Fig. 2.23). From Skinner et al. [129], courtesy of Finnigan Corporation.

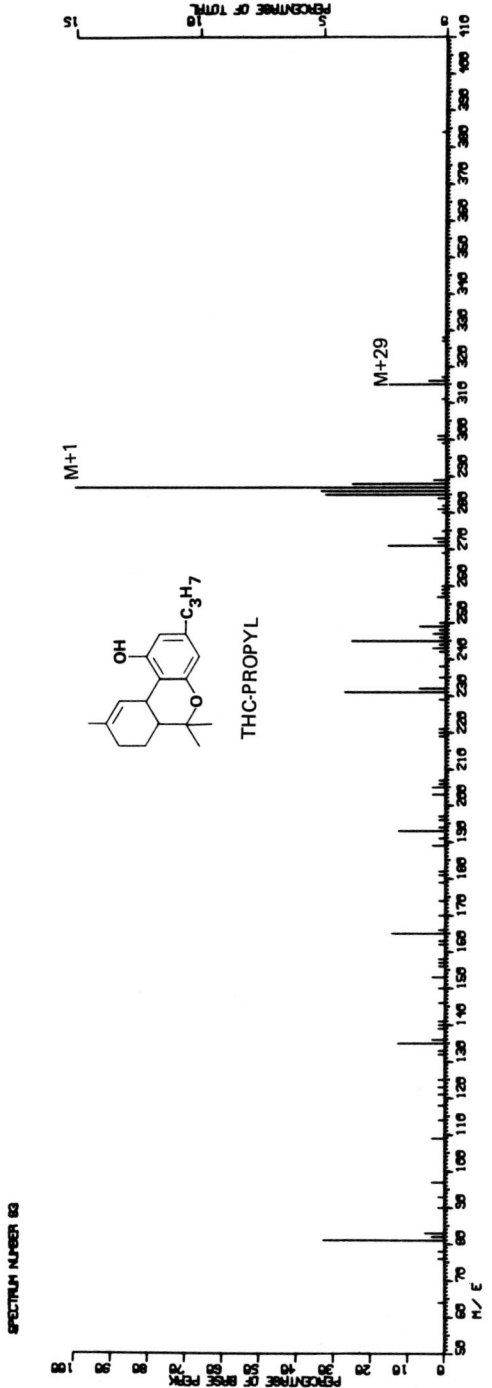

Figure 2.25. CI mass spectrum of THC-C_3 (peak 2 in Fig. 2.23). From Skinner et al. [129], courtesy of Finnigan Corporation.

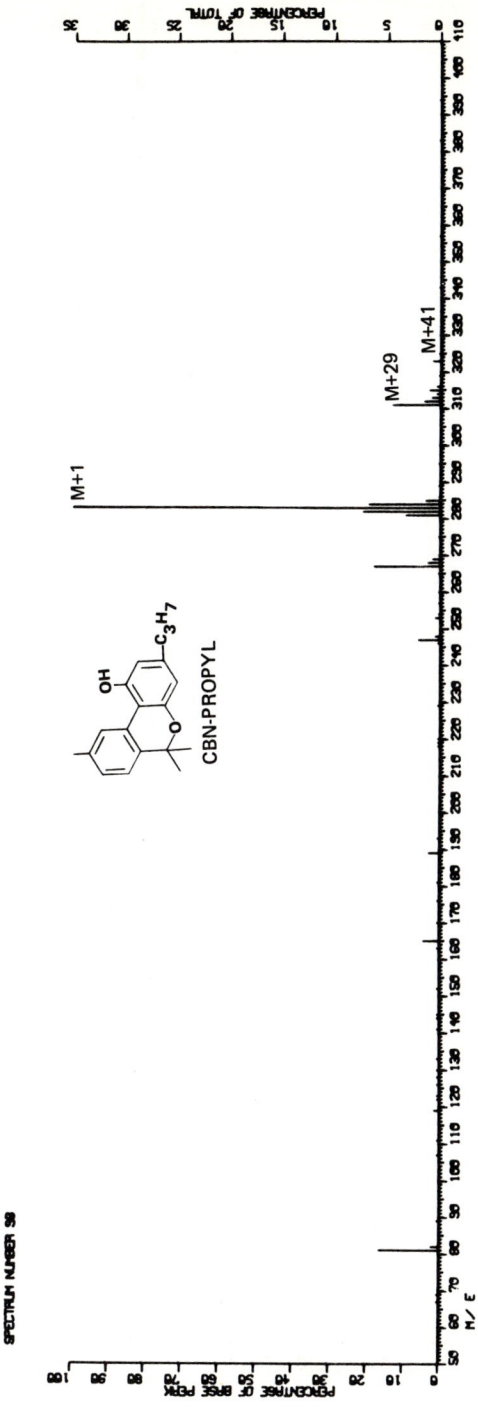

Figure 2.26. CI mass spectrum of CBN-C$_3$ (peak 3 in Fig. 2.23). From Skinner et al. [129], courtesy of Finnigan Corporation.

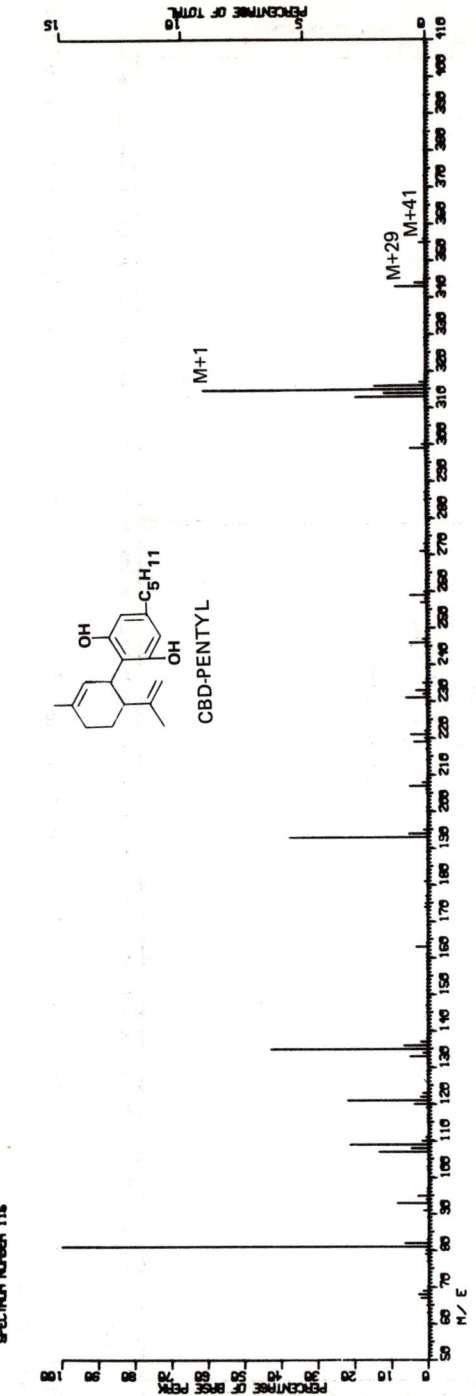

Figure 2.27. CI mass spectrum of CBD-C$_5$ (peak 4 in Fig. 2.23). From Skinner et al. [129], courtesy of Finnigan Corporation.

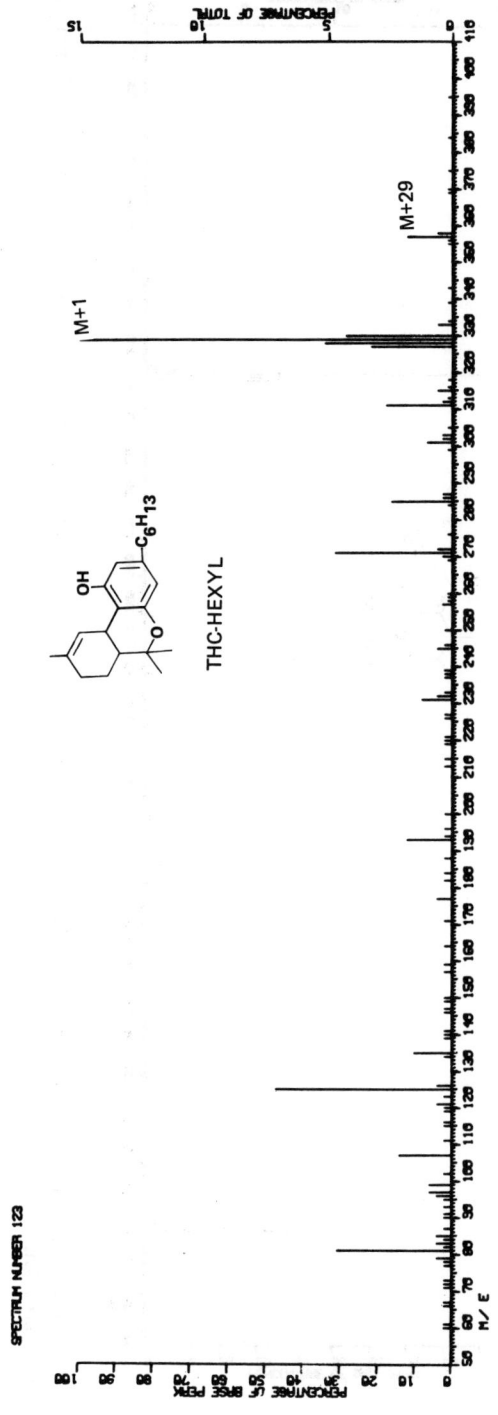

Figure 2.28. CI mass spectrum of THC-C$_6$ (peak 4a in Fig. 2.23). From Skinner et al. [129], courtesy of Finnigan Corporation.

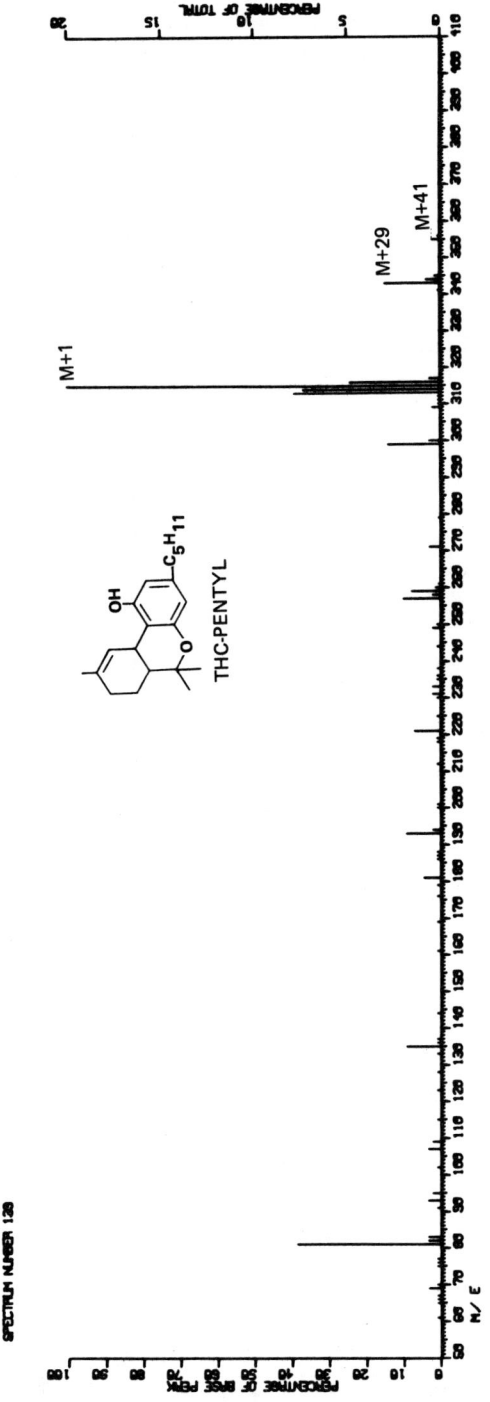

Figure 2.29. CI mass spectrum of THC-C$_5$ (peak 5 in Fig. 2.23). From Skinner et al. [129], courtesy of Finnigan Corporation.

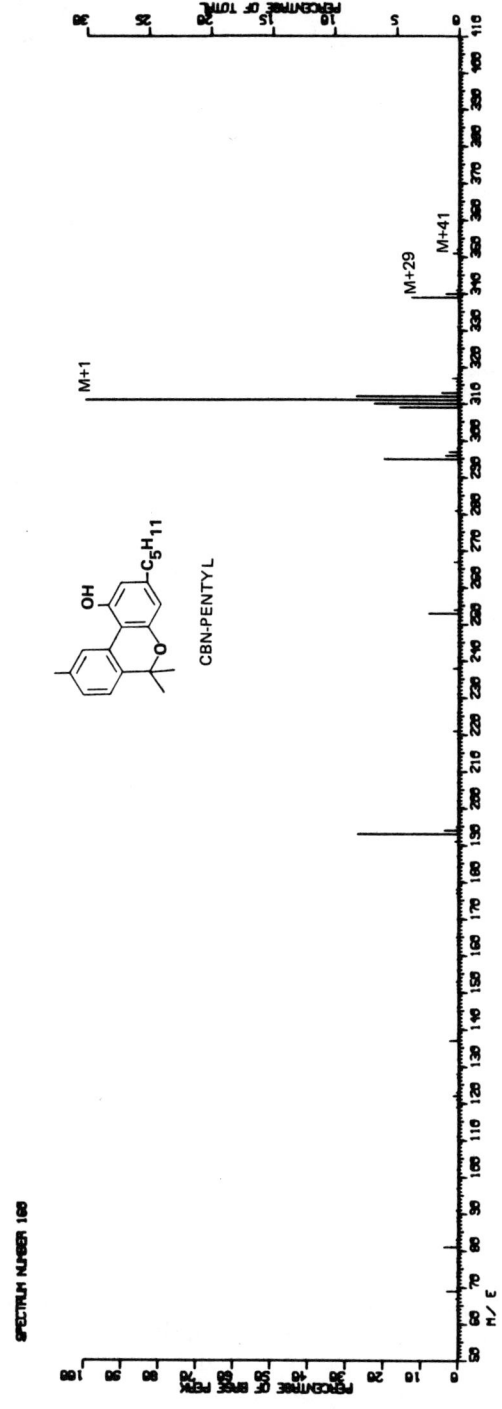

Figure 2.30. CI mass spectrum of CBN-C_5 (peak 6 in Fig. 2.23). From Skinner et al. [129], courtesy of Finnigan Corporation.

De Zeeuw et al. [25] in 1973 reported that leaves of young marijuana plants (Cannabis sativa L.) were found by GC analysis to contain appreciable amounts of two long-chain alkanes, n-heptacosane and n-nonacosane. These alkanes, together with other straight-chain alkanes ranging from C_{19} to C_{32}, could also be detected as minor components in a variety of other marijuana and hashish samples. Depending on the polarity of the GC column used, the alkanes may have retention times similar to those of the major cannabinoids and thus interfere with qualitative and quantitative analyses of the latter. The findings indicate that GC alone cannot be used as an accurate and reliable technique for cannabinoid analysis, unless the alkanes are previously removed. However, the alkane composition may be of additional advantage in determining the origin of seized cannabis samples.

For this analysis, the instrumentation, conditions, and columns used can be found in Section I of this chapter; the retention times of major alkanes and cannabinoids on three different GC column systems are listed in Table 2.8.

Paris et al. [131] also used gas chromatography for the determination of various constituents of Cannabis sativa L. In their study, an Aerograph model 200 gas chromatograph equipped with a flame ionization detector and a 6-ft by 1/8-in. column packed with 3% OV-1 on 80-100 mesh Gas Chrom Q was used. The other GC parameters recommended were injector temperature, 260°C; column temperature, 207°C; detector temperature, 235°C; nitrogen carrier-gas flow rate, 30 ml/min. Because acidic cannabinoids were thermally decomposed to their respective parent compounds in the GC unit, the cannabis extracts were silylated with bis-(trimethylsilyl)trifluoroacetamide after methyl oleate had been added as internal standard. The retention times of unmodified and silylated cannabinoids examined in this investigation are shown in Table 2.30.

In 1974, Turner et al. [56], using silylation, quantitated on a routine basis naturally occurring cannabinoids previously impossible to separate and analyze. As noted, relative retention times of many silylated cannabinoids were reported for the first time.

Sample preparation for GC analysis consisted of the following:

Three 1-g samples were extracted simultaneously with 40 ml of spectrograde chloroform. Resulting solutions were allowed to stand at room temperature for 1 hr. During the hour, each sample was shaken for approximately 15 sec at 20-min intervals. The plant material was then removed by filtration and the mother liquor was concentrated in vacuo at ambient temperature to a greenish paste void of solvent.

At this point, 1.5 ml of an ethanolic solution containing 10 mg/ml of androst-4-ene-3,17-dione was added as the internal standard. Continuous vibration from an ultrasonic vibrator was then carried out until all of the resin was in solution. At this point, 0.5 ml of bis-(trimethylsilyl)-trifluoroacetamide with 1% trimethylchlorosilane was

TABLE 2.30

Retention Times of Cannabinoids Before and After Silylation[a]

Compound	Ret. time (min)	
	Unmodified	Silylated
Methyl oleate	3.0	
Cannabidiol	7.0	5.2
Tetrahydrocannabinol	9.2	6.8
Cannabinol	11.0	8.8
CBD acid	(7.0)	11.8
THC acid	(9.2)	17.0
CBN acid	(11.0)	21.0

[a] Adapted from Paris and Paris [131].

added. The reaction mixture was heated, using a heating mantle, for approximately 10 min at 80°C. Then 0.2 µl of the reaction mixture was routinely injected into the gas chromatograph.

The Beckman instrument used to perform these analyses and separations was equipped with hydrogen flame ionization detectors and 8-ft by 2-mm-i.d. glass columns packed with 2% OV-17 coated on 100-120 mesh Chromosorb G-HP. With the specified GC conditions (injector temperature, 240°C; column temperature, 210°C; detector temperature, 260°C; nitrogen carrier-gas flow rate, 10 to 30 ml/min), the retention times of underivatized and silylated cannabinoids and other components found in Cannabis sativa L. relative to androst-4-ene-3,17-dione (internal standard) are listed in Table 2.31.

In Figure 2.31 is shown a typical OV-17 chromatogram of female Afghanistan Cannabis; Figure 2.32 is a chromatogram of the same material after silylation.

Turner et al. noted that the silyl procedure can be used as a routine method for quantitating cannabinoids not quantitated by other means. Accuracy was between 93 and 95% and cannabinoids, whether free phenols or the carboxylic acid derivatives, could be determined by this method. Furthermore, data generated by this procedure indicate that cannabichromene, previously thought to be a minor component in cannabis, is more abundant than cannabidiol in many variants.

In two articles published in 1974, Stromberg [137,138] reported mass spectrometric data and gas chromatographic retention times of terpenic

TABLE 2.31

Relative Retention Times of Underivatized and Silylated Cannabinoids and Other Components of Cannabis[a]

Compound	Rel. ret. time	U or S[b]
Olivetol	0.04	U
Cannabidivarin disilylated	0.07	S
Cannabidivarin monosilylated	0.11	S
Cannabidiol disilylated	0.11	S
(-)-Δ^9-trans-Tetrahydrocannabivarin	0.12	S
Hexahydrocannabinol (C_9 methyl axial)	0.16	S
Cannabichromene	0.17	S
Cannabidiol monosilylated	0.18	S
Cannabidivarinic acid	0.18	S
Cannabidivarin	0.18	U
(-)-Δ^8-trans-tetrahydrocannabinol	0.20	S
Cannabicyclol	0.21	S
(-)-Δ^9-trans-Tetrahydrocannabinol	0.22	S
Hexahydrocannabinol (C_9 methyl equatorial)	0.22	S
Cannabigerol disilylated	0.26	S
Tetrahydrocannabivarin	0.26	U
Cannabicyclol	0.26	U
Cannabidiolic acid	0.28	S
Cannabinol	0.31	S
Cannabichromene	0.34	U
Cannabivarin	0.34	U
Cannabidiol	0.34	U
Hexahydrocannabinol	0.37	U
Cannabigerol monomethyl ether	0.38	U
(-)-Δ^9-trans-Tetrahydrocannabivarinic acid	0.38	S
$\Delta^{9,11}$-Tetrahydrocannabinol (exocyclic)	0.41	U

TABLE 2.31 (continued)

Compound	Rel. ret. time	U or S[b]
(-)-Δ^8-trans-Tetrahydrocannabinol	0.44	U
Cannabielsoin	0.48	U
(-)-Δ^9-trans-Tetrahydrocannabinol	0.49	U
Cannabigerol	0.57	U
Cannabinol	0.63	U
(-)-Δ^9-trans-Tetrahydrocannabinolic acid A	0.64	S
C_{29}-Hydrocarbon	0.67	U
(-)-Δ^9-trans-Tetrahydrocannabinolic acid B	0.68	S
Androst-4-ene-3,17-dione	1.00	U

[a] Adapted from Turner et al. [56].
[b] U = underivatized; S = silylated.

components [137] as well as cannabinoid components [138] with retention times shorter than that of cannabidiol. Whereas 11 terpenic components of cannabis resin were tentatively identified by comparing the mass spectra with the known spectra of 134 terpenic compounds, two of the cannabinoid components were found to be cannabidivarol and cannabicyclol, whereas the other two were isomers of cannabidiol but were eluted before cannabicyclol, in contrast to hitherto described natural isomers of cannabidiol. Also, the relative retention times of eight cannabinoids, including the major ones, were determined on three different liquid stationary phases.

In both studies, the retention times were obtained with a Perkin-Elmer F-11 gas chromatograph equipped with a flame ionization detector and a 1.9-m by 2-mm-i.d. glass column with a coil diameter of 130 mm. For the terpene investigation, the column was packed with 6% OV-101 on 80-100 mesh Gas Chrom Q. With the injector temperature maintained at 220°C and the helium carrier-gas flow rate set at 60 ml/min, the retention times of identified compounds as given in Table 2.32 were obtained at three different isothermal column temperature settings, 40, 84, and 108°C.

The columns and temperatures used to separate the cannabinoids were (1) 3% OV-101 at 183°C, (2) 3% OV-17 at 215°C, and (3) 3% Dexsil 300 at 183°C. Like the terpene study, the injector temperature was 220°C but the nitrogen carrier-gas flow rate was maintained at about 30 ml/min.

For the GC-MS analysis of minor terpenic compounds, the experimental conditions were previously described [19]. The MS data were obtained by

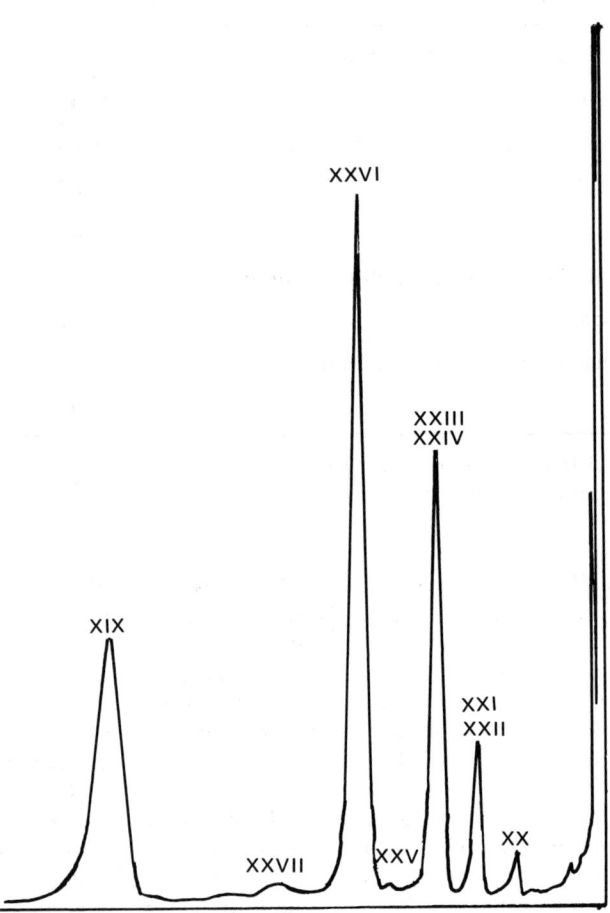

Figure 2.31. An OV-17 chromatogram of underivatized female Afghanistan Cannabis [AF-B(1) C-71] showing cannabidivarin (XX), (-)-Δ^9-trans-tetrahydrocannabivarin (XXI), cannabicyclol (XXII), cannabichromene (XXIII), cannabidiol (XXIV), (-)-Δ^8-trans-tetrahydrocannabinol (XXV), (-)-Δ^9-trans-tetrahydrocannabinol (XXVI), cannabinol (XXVII), and androst-4-ene-3,17-dione (XIX). From Turner et al. [56], courtesy of the Journal of Pharmaceutical Sciences.

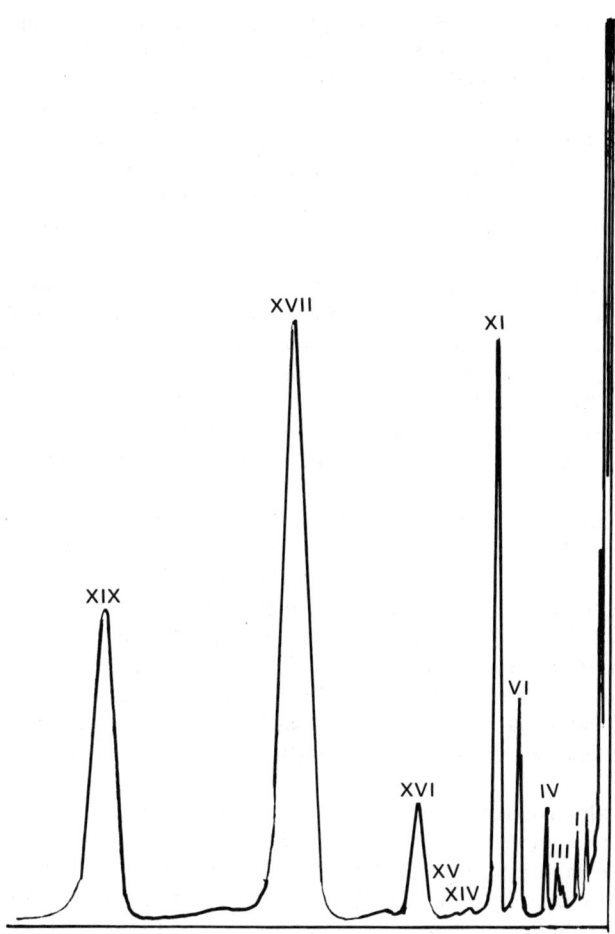

Figure 2.32. Chromatogram of silylated female Afghanistan Cannabis [AF-B(1)/C-71] showing cannabidivarin disilylated (I), cannabidiol disilylated (III), (-)-Δ^9-trans-tetrahydrocannabivarin (IV), cannabichromene (VI), cannabicyclol (X), (-)-Δ^9-trans-tetrahydrocannabinol (XI), cannabidiolic acid (XIV), cannabinol (XV), (-)-Δ^9-trans-tetrahydrocannabinolic acid A (XVII), and androst-4-ene-3,17-dione (XIX). From Turner et al. [56], courtesy of the Journal of Pharmaceutical Sciences.

scanning the eluted peaks. For the 11 components identified, their 10 major m/e ions and intensities in order of decreasing magnitude are included in Table 2.32 as well as their GC relative retention times.

With regard to the identification of the cannabinoid compounds, an LKB 9000 integrated GC-MS instrument was employed with a 4.5-m by 2-mm-i.d. glass column packed with 3% OV-101. With the helium carrier-gas flow rate kept at 30 ml/min and the column programmed from 130 to 200°C at 4°C/min, the other GC-MS settings were separator temperature, 280°C; ion source temperature, 300°C; electron energy, 70 eV.

As in his previous study, Stromberg recorded mass spectral data for the four fractions as well as their retention times on the three columns relative to Δ^9-THC as shown in Table 2.33. Included in Table 2.33 are retention data for cannabidiol, cannabichromene, and cannabinol.

Parker et al. [139] observed what appeared to be an unusually large loss of cannabidiol when stored in chloroform solution, noting a marked decomposition of CBD in chloroform as compared to its relative stability in ethanol as shown in Table 2.34. In Table 2.34, the consistency of the CBD-to-steroid (4-androstene-3,17-dione) ratio in ethanol is very evident, whereas the CBD in chloroform after 8 days in a chloroform solution is totally decomposed.

The GC analyses were carried out with a Perkin-Elmer model 990 gas chromatograph equipped with a digital integrator, dual flame ionization detectors, and 6-ft by 0.6-cm glass columns packed with 3% OV-17 on 100-120 mesh Gas Chrom Q. Using the experimental conditions given (injector temperature, 232°C; column temperature, 218°C; detector temperature, 240°C; nitrogen carrier-gas flow rate, 50 ml/min), the retention times for CBD and internal standard were 6.1 and 17.1 min, respectively.

Garrett and Tsau [140] also investigated the stability of tetrahydrocannabinols in 1974, summarizing their analytical findings as follows:

Δ^9-Tetrahydrocannabinol, as monitored by flame-ionization GC at various temperatures, degrades by a biphasic semilogarithmic curve with time in acidic aqueous solutions (<1 mg/liter) below pH 4 to GC-observable products with separate retention times and the degradations are specific hydrogen-ion catalyzed. The products are considered as Δ^8-tetrahydrocannabinol, P_1, P_2, and P_3 and can be observed and isolated by TLC. These products do not appear above pH 4 in the neutral region, and these degradations are primarily first order, are not biphasic, and are pH independent. The half-life of Δ^9-THC is about 15 min at 37°C and pH 1, typical stomach conditions. The product P_1 may give rise to cannabinol by the GC and TLC procedures since the IR, UV, TLC, NMR, and GC of thin-layer chromatographed P_1 and CBN are coincident, but chloroform extracts do not show the higher absorbances expected if the product that forms in solution to give P_1 is cannabinol. The products P_2 and P_3, isolated by TLC, are

TABLE 2.32

Relative Retention Time and Mass Spectral Data for 11 Terpenic Components[a]

Compound		m/e and RI data										Rel. ret. times		
												40°C	84°C	108°C
α-Pinene	m/e	93	92	91	79	77	136	80	41	94	69	0.46		
	RI	100	30	27	18	16	11	10	10	8	0			
Linalool	m/e	71	93	41	55	43	80	69	121	68	67	1.84	0.13	
	RI	100	62	48	40	38	33	28	16	13	5			
Limonene												1.00[b]		
Fenchyl alcohol	m/e	81	80	82	72	71	84	43	93	69	111		0.14	
	RI	100	71	22	21	20	19	19	17	17	12			
Borneol	m/e	95	43	110	71	69	139	82	121	93	136		0.20	
	RI	100	38	27	13	11	8	7	6	6	4			
α-Terpineol	m/e	59	43	136	93	121	81	68	92	67	79		0.23	
	RI	100	63	49	49	39	36	23	21	18	15			
Piperitenone	m/e	150	107	135	82	109	91	151	121	108	122		0.62	
	RI	100	55	42	25	19	19	16	15	15	7			
β-Caryophyllene	m/e	93	69	133	41	91	79	55	107	81	105		1.00[c]	1.00[c]
	RI	100	98	75	75	53	50	42	40	39	35			
trans-α-Bergamotene	m/e	93	119	69	41	91	107	55	79	105	77			1.12
	RI	100	86	47	42	27	26	26	25	22	15			
Humulene	m/e	93	80	121	107	41	147	92	94	109	79			1.19
	RI	100	39	27	22	22	19	18	17	16	13			
Alloaromadendrene	m/e	41	93	161	69	91	81	105	107	79	133			1.20
	RI	100	99	91	89	69	63	60	57	55	52			
α-Gurjunene	m/e	204	105	189	161	41	93	133	119	91	55			1.41
	RI	100	97	75	74	72	71	65	62	62	45			

[a] Adapted from Stromberg [137].
[b] Limonene used as reference material at 40°C.
[c] β-Caryophyllene used as reference material at 84 and 108°C.

TABLE 2.33

Relative Retention Time and Mass Spectral Data for Several Cannabinoid Compounds[a]

Compound		m/e and RI data										Rel. ret. times		
												Col. A	Col. B	Col. C
Fraction I[b]	m/e	203	218	204	41	174	165	43	121	91	286	0.34	0.36	0.28
	RI	100	20	15	12	11	11	9	8	7	6			
Fraction II[c]	m/e	231	314	232	233	41	299	271	315	258	174	0.42	0.39	0.41
	RI	100	22	19	11	10	8	8	6	6	6			
Fraction III[c]	m/e	231	232	43	41	217	174	314	245	243	230	0.51	0.52	0.46
	RI	100	21	21	18	12	10	8	6	5	5			
Fraction IV[d]	m/e	231	41	232	174	43	69	55	81	314	299	0.60	0.54	0.52
	RI	100	31	16	15	9	8	7	6	3	3			
Cannabidiol												0.72	0.69	0.62
Cannabichromene												0.75	0.66	0.65
Δ^9-THC												1.00	1.00	1.00
Cannabinol												1.26	1.29	1.28

[a] Adapted from Stromberg [138].
[b] Postulated to be a mixture of CBD-C_3 and CBD-C_5.
[c] Structurally related to CBD and THC.
[d] Cannabicyclol.

Col. A = 3% OV-101 at 183°C.
Col. B = 3% OV-17 at 215°C.
Col. C = 3% Dexsil 300 at 183°C.

TABLE 2.34

Stability with Time of CBD in Ethanol and Chloroform Solvents[a]

Day	CBD/steroid ratio	
	Ethanol	Chloroform
0	1.00	1.00
1	1.02	0.88
2	0.97	0.68
3	0.96	0.56
4	0.89	0.31
5	0.97	0.19
6	0.96	0.09
7	0.97	0.04
8	0.96	0.01

[a] Adapted from Parker et al. [139].

Δ^9-9-Hydroxycannabidiol

Δ^9-9-Hydroxytetrahydrocannabinol

consistent with the expected properties of Δ^9-9-hydroxycannabidiol and Δ^9-9-hydroxytetrahydrocannabinol, respectively, by IR, UV, NMR, and mass spectrometry. The final amounts of Δ^8-tetrahydrocannabinol, P_1, P_2, and P_3 are in a constant ratio independent of pH below pH 4. Δ^9-THC as monitored by GC degrades solely by a first-order process to an equilibrium with P_2 and P_3 at acidic pH values and the process is specific hydrogen-ion catalyzed. The equilibrium appears to be independent of pH below pH 4 and is the same when TLC-isolated P_2 or P_3 is used as the starting material. It follows that the acid-catalyzed isolated double-bond migration favors Δ^8-THC over

Δ^9-THC compound, and it is most probable that the equilibrating P_2 and P_3 are the results of water addition to the isolated double bond and ether solvolysis. The product that gives rise to the P_1 retention time that ultimately gives cannabinol is structurally indeterminate at present.

For this THC stability investigation, Garrett and Tsau used a Varian 2100 gas chromatograph equipped with a flame ionization detector and a glass column previously described [166] and packed with 3% OV-225 on 100-120 mesh Gas Chrom Q. Using the conditions specified (injector temperature, 260°C; column temperature, 245°C; detector temperature, 260°C; nitrogen carrier-gas flow rate, 50 ml/min; hydrogen flow rate, 26 ml/min), the retention times obtained for Δ^8-THC, Δ^9-THC, cannabinol (P_1), Δ^9-9-hydroxycannabidiol (P_2), and Δ^9-9-hydroxytetrahydrocannabinol (P_3) were approximately 2.10, 2.34, 3.50, 4.20, and 5.22 min, respectively.

Garrett and Tsau noted that "in general, the rates of appearance of P_1 and the Δ^8-THC compound paralleled the initial fast phase (k_1) of loss of the peak for Δ^9-THC and seemed to be greater than the rates of appearance of P_2 and P_3. In several cases, there may have been an anomalous discontinuity in the rate of appearance of the P_1 peak [dashed line in Fig. 2.33] which could only be explained by the fact that two products have the same retention time where the more rapidly appearing compound is further degraded or equilibrated to some other at that or some other retention time."

Kirchgessner et al. [141] identified illicit drugs using an integrated CVC Products MA-2/015/2500 GC-MS system. In order to eliminate the need for a restrictive interface system and to ensure 100% efficiency, a direct effluent coupling between the GC-MS was achieved by the use of a differential vacuum system which permitted the ion source to accept the entire effluent from the gas chromatograph. In this manner, all the sample vapor flows directly into the ionization region, resulting in optimum sensitivity. For the analyses, a 6-ft by 1/4-in.-o.d. glass column packed with 3% OV-17 coated on 80-100 mesh Chromosorb was used in conjunction with the total ion monitor of the mass spectrometer as the detector. The instrumental conditions used with the GC-MS unit were as follows:

1. Gas Chromatography

    Column: 6-ft by 4-mm-i.d., 3% OV-17 on Chromosorb, 80-100 mesh

    Column temp: see Table 1.15, Volume 2.

    Injection port temp: about 30°C above column temperature

    Detector: total output monitor on mass spectrometer

    Carrier gas: helium

    Flow rate: 20 ml/min

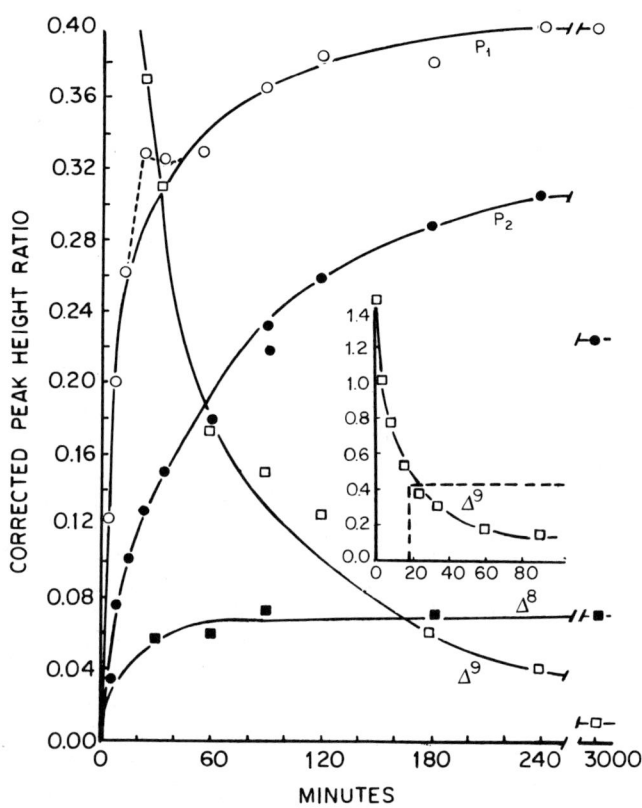

Figure 2.33. Plots of the peak height ratio to that of the internal standard tetraphenylethylene against time of Δ^9-tetrahydrocannabinol reacted at 60.8°C in aqueous solutions at pH 1.40 and the corrected peak height ratios of the products of Δ^9-tetrahydrocannabinol, P_1, P_2, and P_3 observed by GLC. The observed peak height ratios of these latter products were multiplied by 0.73, 1.6, 2.0, and 2.6, respectively, on the presumption that the resultant values would be equivalent to the peak height ratio of Δ^9-tetrahydrocannabinol. The insets are condensed plots for Δ^9-tetrahydrocannabinol, and the dashed lines encompass those values plotted in the larger graphs. From Garrett and Tsau [140], courtesy of the Journal of Pharmaceutical Sciences.

2. Mass Spectrometry

   Predynode gating: set at 25 amu

   Filament current: 2.5 A

   Electron energy: 70 eV

   Multiplier range: 3×10^{-7} A

   Total output monitor range: 10^{-9} A

   Scan rate: 7

   Gate width: minimum

   Accelerating voltage: 2.7 kV

The 1/4-in.-o.d. glass transfer tube connecting both units was maintained from 250 to 300°C.

In Table 1.15 of Volume 2 are listed the GC temperatures and major mass spectra peaks (six most intense fragment ions) used to identify CBN, CBD, and THC.

In 1974, Finkle et al. [142] published supplementary data to an earlier 1972 report (see Chapter 1, Volume 2 for details), which gives revised GC-MS reference data for toxicological and biomedical purposes including chemical ionization data for 450 drugs and metabolites and the facility of an interactive minicomputer by which the library can be manipulated. A master alphabetical list of library compounds is cross-indexed with an electron-impact base peak table (for EI spectral data) and a molecular weight table (for CI spectral data) which are rapid and simple to use. The data can be used manually with a time-share computer or through a dedicated interactive data system. The CI spectra are presented as the protonated molecular ion and most abundant fragment ions, and the EI spectra as a digital code unique to each compound with matching GC relative retention times.

This latter digital code system, which also includes molecular weight and base peak data, is illustrated below for various cannabinoids examined in their study.

In 1975, Friedrich-Fiechtl and Spiteller [144] examined by integrated GC-MS and preparative GC extracts of cannabis and found them to contain a number of so far unknown cannabinoids: 2,2-dimethyl-5-hydroxy-3-(3-oxobutyl)-7-pentyl-4-chromanon or cannabichromanon, 1-hydroxy-9-isopropyl-6-methyl-3-pentyl-dibenzofuran or cannabifuran, 1-hydroxy-9-isopropenyl-6-methyl-3-pentyl-dibenzofuran or dehydrocannabifuran, and 2-oxo-Δ^3-tetrahydrocannabinol or 2-oxo-Δ^3-THC.

CANNABIDIOL
MOL. WT. 314   BASE PEAK 231
 41 55 67 77 91 107 121 145 147 173 174 193 207 229 231
246 271 272 299 312 314 0   0   0   0   0   0   0   0   0

CANNABINOL
MOL. WT. 310   BASE PEAK 295
 41 55 65 77 91 115 128 141 152 165 178 195 209 223 238
251 265   0 295 310 0   0   0   0   0   0   0   0   0   0

CANNABINOL TMS DERIV
MOL. WT. 382   BASE PEAK 367
 45 55 73 77 91 115 128 135 152 165 178 197 209 223 238
251 265 279 295 310 323 337 351 367 382 384 0   0   0   0

CANNABIVAROL
MOL. WT. 282   BASE PEAK 267
 41 53 65 89 91 115 128 141 152 165 174 195 209 223 238
251 267 282   0   0   0   0   0   0   0   0   0   0   0   0

CANNABIDIVARIN (CANNABIDIVAROL)
MOL. WT. 286   BASE PEAK 203
 41 53 67 77 91 115 121 132 147 165 187 201 203 216 243
256 269 284 286   0   0   0   0   0   0   0   0   0   0   0

TETRAHYDROCANNABIVAROL
MOL. WT. 286   BASE PEAK 271
 41 55 67 81 91 115 128 141 159 165 187 189 203 229 243
244 271 272 286   0   0   0   0   0   0   0   0   0   0   0

TETRAHYDROCANNABINOL (DELTA-8)
MOL. WT. 314   BASE PEAK 314
 43 55 69 77 91 107 119 134 147 173 174 193 214 229 231
246 271 272 299 313 314 0   0   0   0   0   0   0   0   0

TETRAHYDROCANNABINOL (DELTA-8) TMS DERIV.
MOL. WT. 386   BASE PEAK 386
 43 55 73 79 93 105 119 134   0 167 177 195 214 221 231
246 265 273 289 303 318 330 343   0 371 386   0   0   0   0

TETRAHYDROCANNABINOL (DELTA-9)
MOL. WT. 314   BASE PEAK 299
 41 55 69 81 91 115 121 141 147 165 174 201 213 217 231
257 271 272 299 300 314   0   0   0   0   0   0   0   0   0

TETRAHYDROCANNABINOL (DELTA-9) TMS DERIV
MOL. WT. 386   BASE PEAK 303
 45 55 73 77 91 107 119 134 147 165 177 196 208 221 231
246 265 273 287 303 315 330 343 369 371 386   0   0   0   0

Structures

Cannabichroman: CH₃CO—(CH₂)₂— substituted chromanone with gem-dimethyl, (CH₂)₄—CH₃ side chain, and OH.

Cannabifuran: dibenzofuran-type structure with CH₃, (H₃C)(CH₃)CH—, OH, and (CH₂)₄—CH₃ substituents.

Dehydrocannabifuran: dibenzofuran with CH₃, H₂C=C(CH₃)—, OH, and (CH₂)₄—CH₃ substituents.

2-Oxo-Δ³-THC: CH₃, H₃C, CH₃, OH, and (CH₂)₄—CH₃ substituents on the THC skeleton with a ketone.

The above compounds were obtained in pure state by micropreparative gas chromatography followed by thin-layer chromatography, and their structures were determined by mass spectra, NMR spectra, and microchemical reactions (formation of acetyl and TMSi derivatives) followed by an investigation of the reaction products by integrated GC-MS.

The GC-MS unit consisted of a Varian 1700 gas chromatograph interfaced via a Watson-Biemann separator to a Varian CH 7 mass spectrometer. GC separations were performed with a 1.8-m by 1.8-mm-i.d. column packed with 3% OV-17 coated on 80-100 mesh, acid-washed, DMCS-treated Chromosorb W. With the injector temperature held at 270°C, the column was programmed from 230°C at 4°C/min.

Micropreparative separations were carried out with a Varian 1700 gas chromatograph equipped with a flame ionization detector and a 1.60-m by 4-mm silanized glass column packed with 3% OV-17 on 80-100 mesh, AW-DMCS, Chromosorb G. Other GC operating conditions were injector temperature, 270°C; column temperature, programmed from 200 to 300°C at 4°C/min; detector temperature, 320°C; nitrogen carrier-gas flow rate, 40 ml/min. In turn, the components trapped from the preparative GC instrument were analyzed with a Varian 1400 gas chromatograph equipped with a flame ionization detector and a 36-m by 0.3-mm-i.d. capillary column coated with 3% OV-101. This GC unit was operated as follows: injector temperature, 280°C; column temperature, programmed from 190 to 290°C at 6°C/min; detector temperature, 290°C; helium carrier-gas flow rate, 1.5 to 2.5 ml/min. Typical chromatograms obtained with the micropreparative and capillary column GC instruments are shown in Figure 2.34, chromatograms A and B, respectively.

Chemical Composition of the Marijuana Plant

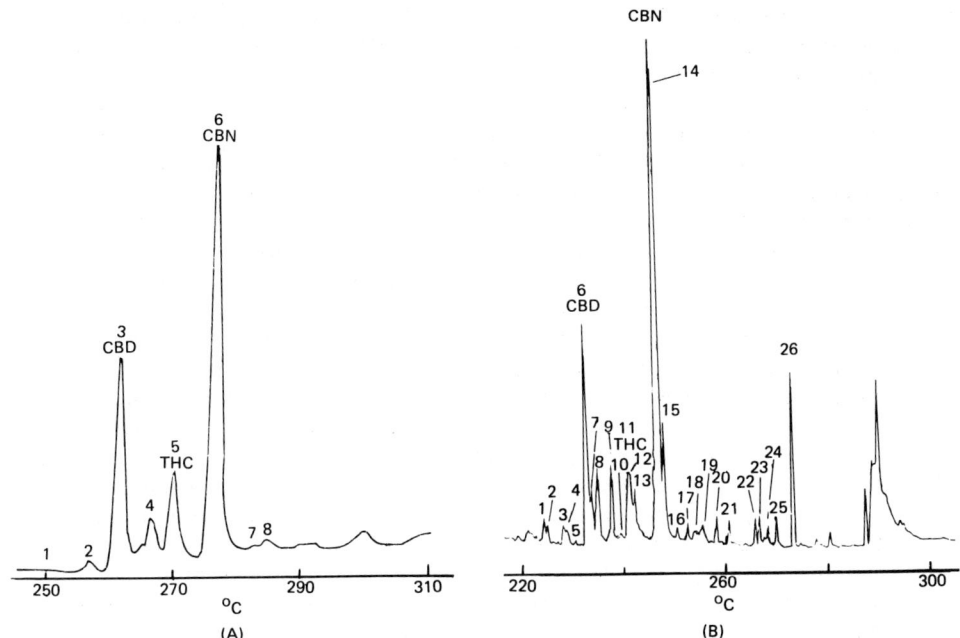

Figure 2.34. Typical chromatograms obtained with the micropreparative (A) and capillary column (B) gas chromatographs. Adapted from Friedrich-Fiechtl and Spiteller [144].

In 1975, Rasmussen [145] described a method for the formation of TMSi derivatives of the cannabinoids after injection of solid plant material and cold trapping. The compounds specifically examined were THC, CBD, and CBN, which were converted to their TMSi ethers with bis-(trimethylsilyl)-acetamide. The procedure adopted for the analysis of plant material was as follows:

> A 1-mg amount of plant material was placed in the basket of the injector as described elsewhere [105,106]. The injector was screwed on to the injection port inlet of the gas chromatograph and when equilibrium had been attained the plant material was inserted into the flash heater of the gas chromatograph and retained there for 1 min. Two minutes after the injection of the plant material the injector was removed and the septum placed on the injection port inlet. On re-attainment of equilibrium, 6 µl of BSA were injected and the column temperature was immediately increased.

All analyses were carried out with a Fractovap 2300 gas chromatograph equipped with a flame ionization detector and a 2-m by 4-mm-i.d. coiled glass column packed with 3% SE-30 on 80-100 mesh Supelcoport. With nitrogen as carrier gas (flow rate of 30 ml/min), the flash heater temperature was 300°C and the compounds were trapped in the column by use of a column temperature of 40°C during the injection. After the injection, the column temperature was increased to 230°C in 3 min and held isothermally. With these GC conditions, the retention times of underivatized THC and CBN were approximately 13.65 and 16.10 min, respectively, whereas their respective TMSi derivatives were eluted in 10.50 and 12.67 min.

For the separation of synthetic and naturally occurring cannabidiol and cannabichromene, Turner et al. [146] resolved these cannabinoids without derivatization by gas chromatography using a 6% OV-1 column. Furthermore, an artifact of cannabichromene, cannabicyclol, was also separated from (-)-Δ^9-trans-tetrahydrocannabivarin. Applicable for the quantitative analysis of cannabis containing both cannabidiol and cannabichromene, they noted that biological interaction among (-)-Δ^9-trans-THC, cannabichromene, and other cannabinoids in natural cannabis preparations could be studied. Of special interest was their observations that, in the phenyl methyl silicone series, CBD precedes cannabichromene on columns containing less than a 50% phenyl-to-methyl ratio, whereas columns containing a 50:50 or greater ratio of phenyl to methyl reverse the separation order, with cannabichromene preceding CBD.

In their investigation, Turner et al. performed all analyses with a Beckman gas chromatograph equipped with a hydrogen flame ionization detector and 8-ft by 2-mm-i.d. glass columns packed with 6% OV-1 on 100-120 mesh Gas Chrom Q. Androst-4-ene-3,17-dione was added as internal standard and the GC operating conditions were injector temperature, 240°C; column temperature, 180°C; detector temperature, 260°C; nitrogen carrier-gas flow rate, 10 and 30 ml/min, depending on instrument and column requirements. At 180°C with androst-4-ene-3,17-dione added as internal standard, the retention times of cannabicyclol, cannabidiol, and cannabichromene relative to the steroid were 0.480, 0.605, and 0.654, respectively. At 180°C, the retention times of CBD and cannabichromene were approximately 19.50 and 21.25 min, respectively.

Harvey and Paton [147] discussed the use of TMSi and other homologous trialkylsilyl derivatives for the separation and characterization of mono- and dihydroxycannabinoids using integrated GC-MS instrumentation. They noted that TMSi derivatives produced a group of peaks containing both sets of compounds, sometimes poorly resolved, whereas by increasing the alkyl chain length to n-butyl, complete fractionation into two groups was achieved. Furthermore, the mass spectra of these derivatives resembled those of the TMSi derivatives with the addition of a set of ions resulting from estimation of the silyl-alkyl chains as olefins. In this study, the following tri-n-alkylsilanes were investigated: triethyl-, tri-n-propyl-, tri-n-butyl-, and tri-n-hexylsilanes.

TABLE 2.35

Retention Indices of Trialkylsilyl Derivatives of Cannabinoids[a]

Compound	Retention index			
	TMSi[b]	TESi[b]	TPSi[b]	TBSi[b]
Propylcannabidiol	2110	2635	2835	3180
Propyl-Δ^1-THC	2170	2465	2590	2795
CBD	2270	2780	2980	3325
Cannabicyclol	2280	2550	2680	2875
Δ^6-THC	2335	2615	2740	2935
$\Delta^{1(7)}$-THC	2335	2625	2750	2950
Δ^1-THC	2350	2635	2760	2955
CBN	2430	2725	2855	3055
Cannabigerol	2440	2965	3175	3510
Heptyl-Δ^1-THC	2545	2820	2940	3130
7-Hydroxy-Δ^6-THC	2650	3220	3440	

[a] Adapted from Harvey and Paton [147].
[b] TMSi = trimethylsilyl, TESi = triethylsilyl, TPSi = tripropylsilyl, TBSi = tributylsilyl.

To obtain the retention indices of the trialkylsilyl derivatives of the various cannabinoids listed in Table 2.35, Harvey and Paton used a Varian model 2400 gas chromatograph fitted with dual flame ionization detectors and 6-ft by 2-mm-i.d. coiled glass columns packed with 3% SE-30 on 100-120 mesh Gas Chrom Q. Nitrogen was used as carrier gas (flow rate, 30 ml/min) and the column was programmed from 100 to 320°C at 4°C/min.

Spectral data were recorded with a VG Micromass 12B mass spectrometer at 70 eV, interfaced via a jet separator to a Varian 2400 gas chromatograph fitted with a SE-30 column as described above. Using a 3-sec scan and an accelerating voltage of 2.5 kV, mass spectra were recorded for each chromatographic peak.

Bailey and Gagne [149] analyzed synthetic cannabidiol, cannabichromene, and cannabivarin by gas chromatography using on-column flash methylation with trimethylanilinium hydroxide as reagent. In their method, for methylation, 10 μl of a 0.5% cannabinoid solution in methanol were treated with 6 to 10 μl of 0.2 M TMAnH in methanol, and approximately 0.2 μl of this mixture were injected into a Varian model A-700 gas chromatograph equipped with a flame ionization detector and a 6-ft by 1/4-in. glass column packed with 5% OV-7 on 80-100 mesh Chromosorb W. The retention times of various cannabinoids and their methylated derivatives shown in Table 2.36 were obtained using the following GC conditions: injector temperature, 275°C; column temperature, 225 and 250°C; detector temperature, 275°C; nitrogen carrier-gas flow rate, 30 ml/min.

Confirmation of the methylation of CBD, cannabichromene, and cannabivarin was verified by MS analyses at an ionization potential of 70 eV and a probe temperature of 200°C. Mass spectra were obtained on eluted materials from the GC unit which were collected by extinguishing the detector

TABLE 2.36

Retention Times of Cannabinoids and Their Methylated Derivatives[a]

Compound	Ret. time (min)	
	225°C	250°C
Cannabivarin	15.7	6.2
Cannabichromene	15.7	6.2
Cannabidiol	15.7	6.2
Δ^8-THC	19.8	7.8
Δ^9-THC	21.5	8.2
Cannabigerol	24.9	9.3
Cannabinol	26.3	10.3
Cannabidiol dimethyl ether	8.5	3.8
Cannabivarin methyl ether	9.9	4.3
Cannabichromene monomethyl ether	11.1	4.6
Cannabidiol monomethyl ether	11.2	4.6
Δ^8-THC methyl ether	12.6	5.3
Cannabigerol dimethyl ether	14.1	5.6
Δ^9-THC methyl ether	14.4	5.8
Cannabigerol monomethyl ether	16.8	6.9
Cannabinol methyl ether	17.6	7.1

[a] Adapted from Bailey and Gagne [149].

flame and placing an open-ended glass capillary tube of the melting-point type over the jet at the predetermined time of peak emergence.

From mass spectral data, it was shown that the methylated species had the following predominant m/e ions:

1. Cannabidiol dimethyl ether (relative percent in parentheses): 342 (14), 275 (36), 274 (100), 259 (11), 244 (21), 243 (91), 235 (19), 234 (17), 221 (67), 174 (13), 173 (57), 158 (16).
2. Cannabichromene methyl ether: 328 (6), 313 (5), 246 (21), 245 (100), 231 (6), 188 (8).
3. Cannabivarin methyl ether: 296 (6), 295 (21), 281 (35), 280 (100), 266 (9), 252 (9), 245 (9), 238 (21), 209 (11).
4. Cannabidiol monomethyl ether: 328 (5), 327 (16), 313 (8), 286 (6), 261 (10), 260 (41), 246 (27), 245 (100), 243 (9), 229 (8), 220 (9), 207 (30).

Smith [151] used integrated GC-MS to confirm the identifications of Δ^8-THC, Δ^9-THC, Δ^9-tetrahydrocannabinolic acid, CBD, cannabidiolic acid, cannabinol, cannabinolic acid, cannabichromene, and cannabichromenic acid located in the high-pressure liquid chromatogram of cannabis. Following the silylation of these components with a 4:1 mixture of bis-(trimethylsilyl)acetamide and trimethylchlorosilane, GC retention time and MS spectral studies were performed with a Varian Aerograph 2700 gas chromatograph equipped with a 50:50 effluent splitter, a 2-m by 0.4-cm glass column packed with 3% OV-17 on 80-100 mesh Gas Chrom Q, and a glass jet separator coupled to a V.G. Micromass 12F single-focusing mass spectrometer. Using the specified GC-MS operating conditions [helium carrier-gas flow rate, not specified; column temperature, about 230°C; detector (GC-FID) temperature, 250°C; accelerating voltage, 3 or 4 kV; emission current, 100 μA; electron energy, 70 eV; ion source temperature, 240°C; scan time, 3 sec; resolution of the spectra, 600-800 for 10% valley], the retention times of underivatized and TMSi derivatives of various cannabinoids relative to Δ^9-THC and Δ^9-THC-TMSi, respectively, are given in Table 2.37 as well as their predominant m/e ions (relative abundance listed as percent of base peak).

In 1976, Harvey [152] identified in several cannabis samples the butyl homologs of Δ^9-THC, Δ^9-THC acid A, CBN, and CBD. The separation was carried out by gas chromatography, whereas identification by integrated GC-MS was achieved by the preparation of trimethylsilyl, d_9-trimethylsilyl, triethylsilyl, and tri-n-propylsilyl derivatives.

Prior to derivatization, the cannabinoids were extracted from cannabis samples with ethyl acetate as follows:

TABLE 2.37

Relative Retention Time and Mass Spectral Data for Underivatized and Silylated Cannabinoids[a]

Compound	RRT As is	RRT TMSi	m/e Ions
Δ^9-THC	1.00		314 (76), 299 (100), 271 (53), 258 (32), 246 (14), 243 (33), 232 (18), 231 (90), 193 (14)
Δ^9-THC monosilylated		1.00	386 (100), 371 (80), 343 (25), 330 (15), 315 (45), 304 (10), 303 (40), 265 (5)
Δ^9-THC acid disilylated		2.96	502 (9), 487 (100), 431 (3), 419 (17)
Δ^8-THC	0.90		
Δ^8-THC monosilylated		0.93	
CBD	0.71		314 (9), 299 (6), 271 (5), 267 (9), 246 (19), 232 (18), 231 (100), 193 (7), 174 (14)
CBD disilylated		0.50	458 (11), 443 (6), 390 (100), 375 (6), 377 (26)
CBD acid trisilylated		1.36	506 (5), 491 (100), 453 (9)
CBN	1.27		310 (13), 295 (100), 239 (5), 238 (18)
CBN monosilylated		1.38	382 (14), 367 (100), 311 (2), 310 (7)
CBN acid disilylated		3.75	498 (trace), 483 (100)
Cannabichromene	0.70		314 (4), 299 (4), 232 (17), 231 (100), 174 (12)
Cannabichromenic acid disilylated		2.78	502 (3), 487 (22), 420 (43), 419 (100)

[a] Adapted from Smith [151].

Samples (about 50 mg) were crushed in a mortar and stood in ethyl acetate (10 ml) for about 1 hr. The solid material was filtered off, washed with ethyl acetate, and the solvent was removed from the combined ethyl acetate extract and washings under reduced pressure followed by a stream of nitrogen. The residue was converted into silyl derivatives for analysis by GC-MS.

The various silyl derivatives were prepared using the following reagents: (1) bis-(trimethylsilyl)-trifluoroacetamide and trimethylchlorosilane, (2) d_{18}-bis-(trimethylsilyl)-trifluoroacetamide and trimethylchlorosilane, (3) triethylchlorosilane, (4) tri-n-propylchlorosilane.

Methylene unit values and mass spectral data for the derivatives of Δ^9-THC, CBN, and CBD were obtained with the instruments and conditions given below:

1. Gas Chromatography

Chromatograph	Varian 2400 equipped with dual flame ionization detectors
Column	2-m by 2-mm-i.d. glass packed with 3% SE-30 on 100-120 mesh Gas Chrom Q
Injector temperature (°C)	280
Column temperature (°C)	Programmed from 100 to 330°C at 4°C/min
Detector temperature (°C)	300
Nitrogen carrier-gas flow rate (ml/min)	30

2. Gas Chromatography-Mass Spectrometry

GC-MS	Varian 2400 GC interfaced with a VG Micromass 12B electron-impact mass spectrometer via a glass jet separator
Column	Same as above
Injector temperature (°C)	280
Column temperature (°C)	Programmed from 170 to 280°C at 2°C/min
Helium carrier-gas flow rate (ml/min)	30
Inlet line temperature (°C)	230

Separator temperature (°C)	230
Ion source temperature (°C)	260
Ionizing voltage (eV)	25
Ionizing current (μA)	100
Accelerating voltage (kV)	2.5
Scan rate	3 sec per decade from high to low mass over the mass rnage m/e 600 to 40 with an inter-scan delay of 2 sec

Using the GC and GC-MS operating parameters, Table 2.38 lists the methylene unit values for Δ^9-THC, Δ^9-THC acid A, CBD, and CBN, and their silyl derivatives, as well as their pertinent m/e ions.

Furthermore, using the various silyl derivatives, Harvey was able to identify the following cannabis components: propyl-Δ^9-THC, propyl-cannabinol, cannabidiol, cannabichromene, Δ^9-THC, cannabinol, propyl-Δ^9-tetrahydrocannabinolic acid, cannabidiolic acid, butyl-Δ^9-tetrahydrocannabinolic acid, Δ^9-tetrahydrocannabinolic acid, cannabichromenic acid, cannabinolic acid, n-nonacosane, methyl-Δ^9-THC, methyl-cannabinol, propyl-cannabichromene, butyl-Δ^9-THC, cannabinol, palmitic acid, and butyl-cannabinol.

As noted by Harvey:

The identification of three butyl cannabinoids in three unrelated samples suggested that these homologs might occur more widely. Seven other samples with high THC-acid content were examined and butyl-Δ^9-THC acid was found in low concentrations in all of them. Methyl and propyl homologs were also present, the methyl series in low relative concentration, and the propyl series varying from about the concentration of the methyl homologs to nearly as high as that of the pentyl-cannabinoids. Ethyl and higher homologs were not detected in any of the samples using the GC-MS-COMP system.

In 1976, Knaus et al. [153] developed methods for the separation, identification, and quantitation of cannabinoids and their t-butyldimethylsilyl ether, trimethylsilylacetate, and diethylphosphate derivatives using gas chromatography and mass spectrometry.

The GC and GC-MS investigations were carried out with a Hewlett-Packard 5700 gas chromatograph equipped with a flame ionization detector and two column systems: system A for t-butyldimethylsilyl (TBDMS) derivatives, 4-ft by 1/4-in. glass column packed with 5% OV-101 on acid-washed, DMCS-treated, 80-100 mesh Chrom 750 and operated isothermally at 210°C with a helium carrier-gas flow rate of 60 ml/min; system B for trimethylsilylacetate (TMSA) derivatives, 6-ft by 1/4-in. glass column packed with

Methylene Unit Values and m/e Data for Silyl Derivatives of the Homologous Δ^9-THC, CBN, and CBD Cannabinoids[a]

Compound	Methylene value				Dominant m/e ion			
	TMSi	d_9-TMSi	TESi	TPSi	TMSi	d_9-TMSi	TESi	TPSi
Methyl-Δ^9-THC			23.21	24.58			372	414
Propyl-Δ^9-THC	21.70		24.65	25.90	358	367	400	442
Butyl-Δ^9-THC			25.48	26.80			414	456
Δ^9-THC	23.50		26.35	27.60	386	395	428	470
Methyl-Δ^9-THC acid A					431	449		
Propyl-Δ^9-THC acid A	24.99				459	474		
Butyl-Δ^9-THC acid A	25.72				473	488		
Δ^9-THC acid A	26.52				487	502		
Methyl-CBN	20.95		23.95	25.39	311		353	395
Propyl-CBN	22.42		25.30	26.76	339		381	423
Butyl-CBN	23.36		26.27	27.65	353		395	437
CBN	24.30		27.25	28.55	367		409	451
Methyl-CBD	21.10				334			
Propyl-CBD					362			
Butyl-CBD	21.90				376			
CBD	22.70				390			

[a] Adapted from Harvey [152].
TMSi = trimethylsilyl, d_9-TMSi = deuterated trimethylsilyl, TESi = triethylsilyl, TPSi = tri-n-propylsilyl.

0.5% OV-101 on acid-washed, DMCS-treated, 80-100 mesh Chrom 750 and operated isothermally at 200°C with a helium carrier-gas flow rate of 60 ml/min.

For the mass spectral analyses of the TBDMS and GC-MS analyses of the TMSA and diethylphosphate (DEP) derivatives, Knaus et al. used a Hewlett-Packard 5980A mass spectrometer interfaced with a Hewlett-Packard 5710A gas chromatograph, whereas accurate mass determinations were performed with an AEI MS-9 or MS-50 mass spectrometer. With all MS instruments, mass spectra were obtained using an ionization voltage of 70 eV.

The retention times obtained for TBDMS and TMSA derivatives of several cannabinoids using both GC columns are shown in Table 2.39; mass spectral data for these and the DEP derivatives are listed in Table 2.40.

In 1976, Fonseka et al. [154] reported chromatographic separations of Δ^1-THC, Δ^6-THC, CBN, CBD, and several of their monooxygenated derivatives from each other by a combination of liquid, thin-layer, and gas chromatography.

The GC studies were performed with a Varian Aerograph model 2100 gas chromatograph equipped with a flame ionization detector and a 6-ft by 1/8-in. glass column packed with 2% SE-30 on 125-150 mesh Gas Chrom Q. To obtain the retention data listed in Tables 2.41 and 2.42 for Δ^1-THC/Δ^6-THC and CBN/CBD compounds, respectively, the following GC conditions were used: injector temperature, 270°C; column temperature, 250°C;

TABLE 2.39

Retention Times of TBDMS and TMSA Derivatives of Some Cannabinoids[a]

Compound	Ret. time (min)	
	System A (TBDMS)	System B (TMSA)
Codeine (I.S.)	10.1[b]	
Δ^8-THC	19.8	4.1
Δ^9-THC	21.1	5.0
CBN	26.8	6.3
CBD	29.5[c]	3.6

[a] Adapted from Knaus et al. [153].
[b] Nonsilylated.
[c] Disilylated.

TABLE 2.40

Mass Spectral Data [m/e (Percent Relative Abundance)][a]

	Diagnostic Peaks of Appreciable Mass
	tert-Butyldimethylsilyl ethers (TBDMS derivatives)
Δ^9-THC	428 (M^+, 88), 413 (28), 372 (37), 371 (100), 357 (25), 345 (12), 303 (3)
Δ^8-THC	428 (M^+, 100), 413 (7), 372 (49), 371 (24), 357 (4), 345 (54), 303 (37)
Cannabinol	424 (M^+, 82), 409 (100), 367 (46)
Cannabidiol	542 (M^+, 41), 527 (5), 485 (10), 475 (100), 436 (23), 422 (51), 418 (59), 409 (41)
	Trimethylsilylacetates (TMSA derivatives)
Δ^9-THC	313 (81), 298 (93), 270 (51), 257 (33), 245 (3), 242 (42), 230 (100), 192 (21), 173 (21)
Δ^8-THC	313 (52), 298 (7), 270 (30), 257 (35), 245 (6), 242 (6), 230 (100), 192 (27), 173 (14)
Cannabinol	309 (13), 294 (100), 237 (16), 222 (6)
Cannabidiol	313 (8), 245 (17), 230 (100), 192 (11), 173 (10), 120 (14)
	Diethylphosphates (DEP derivatives)
Δ^9-THC	450 (M^+, 6), 435 (2), 421 (5), 379 (4), 367 (9), 351 (5), 297 (71), 296 (53), 281 (35), 253 (27), 228 (100)
Δ^8-THC	450 (M^+, 11), 435 (4), 421 (5), 407 (8), 379 (11), 367 (100), 357 (21), 351 (24), 297 (65), 296 (66), 281 (68), 273 (13), 253 (11), 228 (46)
Cannabinol	446 (M^+, 13), 431 (100), 415 (3), 403 (10), 375 (7) 357 (7), 300 (5), 295 (8), 238 (9)
Cannabidiol	i) di-DEP derivative: 586 (M^+, 1), 571 (2), 558 (2), 544 (30), 543 (100), 527 (14), 515 (21), 487 (11), 459 (11), 431 (11), 364 (9), 308 (6) ii) mono-DEP derivative: 450 (M^+, 6), 435 (5), 422 (3), 408 (24), 407 (100), 391 (7), 379 (19), 367 (48), 351 (18), 339 (7), 311 (7), 281 (8), 231 (10), 228 (14)

[a] From Knaus et al. [153], reproduced from the Journal of Chromatographic Science, by permission of Preston Publications, Inc.

TABLE 2.41

Retention Times of Δ^1-THC, Δ^6-THC, and Some of Their Monooxygenated Derivatives[a]

Compound	Ret. time (min)	
	Δ^1-THC	Δ^6-THC
1.		
Underivatized	4.8	4.4
2"-O-		8.7
2"-OH		8.9
1"-OH		9.3
3"-O-		10.0
6α-OH	10.1	
6β-OH	10.4	
3"-OH	11.0	10.1
4"-OH	11.1	
6-O-		
7-OH	14.2	14.1
5"-OH		14.3
2.		
TMSi ethers		
3"-OH	3.7	
4"-OH	4.1	
6-O-	4.4	
7-OH		4.7
5"-OH		5.1
6α,7-diOH	4.5	

[a] Adapted from Fonseka et al. [154].

TABLE 2.42

Retention Times of CBN, CBD, and Some of Their Monooxygenated Derivatives[a]

Compound	Ret. time (min)	
	CBN	CBD
1.		
Underivatized	6.0	3.5
2"-O-	11.8	
2"-OH	12.2	7.1
1"-OH		7.4
6α-OH		7.9
6β-OH		8.0
3"-OH	14.1	8.2
4"-OH	14.3	8.3
6-O-		8.6
7-OH	18.2	10.7
5"-OH	19.3	11.0
2.		
TMSi ethers		
3"-OH	5.3	2.1
4"-OH	5.7	2.4
6-O-		2.8

[a] Adapted from Fonseka et al. [154].

detector temperature, 270°C; nitrogen carrier-gas flow rate, 25 ml/min; hydrogen flow rate, 25 ml/min; oxygen flow rate, 200 ml/min.

In their discussion of their GC findings, Fonseka et al. stated that:

The GC patterns of the cannabinoid series were very similar. In the CBD series, the side-chain hydroxylated metabolites had retention times in the sequence 2"-, 1"-, 3"-, 4"-, and 5"-hydroxy compound. The same order was indicated for side-chain hydroxylated compounds in the other series, although some compounds were lacking.

Compounds with the same or similar retention times could be separated as their TMSi derivatives. In GC, the 3"- and 4"-hydroxy compounds showed similar properties, but could be separated as their TMSi derivatives. In the Δ^6-THC series, the 6-oxo compound decomposed in GC unless it was silylated. Also, it was possible to obtain a chromatogram of $6\alpha,7$-dihydroxy-Δ^1-THC only after silylation.

In 1976, Novotny et al. [155] developed a high-resolution GC-MS procedure for the determination of the sterol fraction of marijuana based on the separation of their TMSi derivatives with GC glass capillary columns. In their study, five phytosterols in marijuana were identified by comparisons of their retention with authentic compounds on three different stationary phases and through mass spectral data.

For the determination of retention data and recording chromatograms of the silylated derivatives, a Perkin-Elmer model 900 gas chromatograph equipped with modified injector and detector systems and 11 to 15-m by 0.25-mm-i.d. glass columns treated according to the methods established in their laboratory [167] and statically coated [168] with SE-30, SE-52, and Poly I-110 stationary phases was used. With the PE-900, samples were introduced by either a splitless injection [169] or a precolumn technique [170].

To obtain spectral data for identification purposes, Novotny et al. employed a Hewlett-Packard model 5980A integrated GC-MS (dodecapole) instrument, the GC unit provided with a modified sampling system (to use a precolumn) and the capillary column interfaced to the MS via a single-stage, all-glass jet separator. To obtain EI mass spectra, an electron energy of 70 eV was used.

Measured under temperature programming conditions (1°C/min, from 220 to 260°C), retention data for several standard TMSi sterols on the three stationary phases are shown in Table 1.31. Although present in somewhat different amounts, several phytosterols of tobacco were also found in marijuana (stigmasterol, campesterol, β-sitosterol, and 5α-stigmasta-7,24(28)-dien-3β-ol). Whereas $5,7,22$-ergostatrien-3β-ol (ergosterol) is one of the four major components of marijuana, it does not appear in tobacco.

McCallum and Cairns [157] in 1977 described a simple device for GC separations of cannabinoids using a surface-coated open tubular column without stream splitting.

The design of the inlet port (see Fig. 2.35) was such that "the capillary (A) was chosen so that its inside diameter was a 'snug' fit for the injection needle and its end rested against the septum (B) to eliminate back-diffusion of solvent vapor. The carrier gas entered the system by two holes (C) at the base of the capillary. The internal diameter of the exit (D) was 1.5 mm, providing an almost exact fit for the glass-lined steel tubing (E) leading to the column. The final seal at the exit was a cut-down, slightly compressed septum (F). Areas G, H, and I are the oven, oven wall, and inlet heater, respectively."

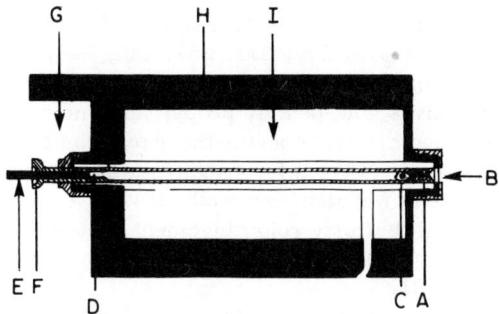

Figure 2.35. Inlet port design. From McCallum and Cairns [157], courtesy of the Journal of Pharmaceutical Sciences.

The gas chromatograph used in conjunction with this inlet port was a Tracor model 550 instrument equipped with a flame ionization detector and a 70-m by 0.5-mm-i.d. glass open-tubular column coated with SE-30. Using the specified GC conditions (injector temperature, 250°C; column temperature, 230°C; helium carrier-gas flow rate, 4 ml/min; nitrogen makeup gas flow, 25 ml/min), the retention times obtained for CBD, Δ^9-THC, and CBN were approximately 33.7, 39.6, and 45.9 min, respectively. Whereas the effective theoretical plates (N_{eff}) for n-alkanes was about 40,000, the N_{eff} for cannabinoids was only 63% that achieved for alkanes. It was noted that this reduction in efficiency apparently was due, at least in part, to support effects.

Smith et al. [158] reported the identification of hexadecanamide in a sample of cannabis resin thought to be of Pakistani origin. The parent acid, palmitic acid, had been previously identified in cannabis smoke. Its identification was confirmed in the following manner:

> When the fraction from the CBD-CBN region of the HPLC chromatogram was examined by GC on OV-17 at 240°C, a peak with a retention of 0.29 relative to Δ^9-THC was detected. This was further examined by GC-MS. The EI spectrum, obtained under conditions described previously by Smith [151], had a base peak of 59 and a second most abundant peak of 72, features characteristic of an amide. The molecular weight was found to be 255 by chemical ionization using a source temperature of 160°C, an electron energy of 100 eV, an emission current of 200 μA, an accelerating voltage of 4 kV, and isobutane as the reactant gas. The electron impact and chemical ionization spectra corresponded to literature data for hexadecanamide and the identity of the compound was confirmed by synthesizing a sample of hexadecanamide and examining it by GC-MS. The GC retention time and mass spectral data matched those of the compound found in cannabis resin.

In 1977, Vree [159] reported the mechanism of fragmentation of cannabinoids to fragments m/e 314, 299, 271, 258, 246, 243, and 231. Cannabidiol, cannabinodiol, cannabinol, Δ^8-THC, Δ^9-THC, cannabichromene, cannabicyclol, derivatives with pentyl, propyl, and methyl side chains, their methyl ethers, and cis-trans and ortho-para isomers were analyzed by GC-MS using different energies for fragmentation during GC elution. The following mechanism was distinguished: loss of a methyl radical, ring closure and rotation, McLafferty rearrangement, retro Diels-Alder, internal protonation, isomerization and internal bond formation, and one-step fragmentation to m/e 231.

The instrumentation used to obtain this mass spectral data was a LKB 9000 GC-MS unit equipped with a 1.80-m by 4-mm-i.d. glass column packed with 3% OV-17 on 60-80 mesh Gas Chrom Q. The temperatures were column, 200°C; separator, 220°C; ion source, 250°C. The helium carrier-gas flow rate was maintained at 20 ml/min, and repetitive mass spectra were taken at 10, 12, 14, 16, 18, and 20 eV during the elution of a GC peak. Other MS settings were accelerating voltage, 3.5 kV; trap current, 60 µA; analyzer tube pressure, 2×10^{-6} torr. The mass spectra obtained were normalized, and the relative abundance of a particular mass fragment was plotted against the electron voltage.

In 1978, Rosenfeld and Crocco [161] described conditions under which phenols (Δ^9-THC was included among the compounds studied) are pentafluorobenzylated in a biphasic methylene chloride/sodium hydroxide system, but in the absence of phase-transfer catalysts such as crown ethers or quaternary ammonium hydroxides.

As outlined, the general conditions for pentafluorobenzylation of Δ^9-THC were as follows:

To 40 µg of Δ^9-THC dissolved in 1 ml of methylene chloride and 0.5 ml of NaOH, pentafluorobenzyl bromide (PFBB) was added, and the mixture was shaken for 1 hr. (Note: This procedure was studied by varying the amount of PFBB added from 0 to 100 µl and the concentration of NaOH from 0 to 10 N). After shaking for 1 hr, the aqueous phase was diluted with 0.5 M phosphate buffer at pH 7.4. After the aqueous phase was aspirated, the organic phase was further diluted with methylene chloride, and dried with sodium sulfate. Δ^9-THC methyl ether was added as external standard to the dry organic phase; the solution was decanted into a second tube and taken to dryness in a sand bath at 70°C under nitrogen. After dissolving the residue with 50 µl of heptane, 1 to 2 µl was withdrawn and injected into the gas chromatograph.

To obtain maximum conversion of the Δ^9-THC to its PFBB derivative, the concentration of NaOH must be 5 N or higher.

GC analyses were performed with a Hewlett-Packard 402 dual column gas chromatograph or a Varian 2100 dual column gas chromatograph, both with flame ionization detection.

For Δ^9-THC analysis, the GC conditions and retention times were as follows: The analysis of the reaction of the Δ^9-THC was performed on a 1.52-m by 4-mm-i.d. glass column packed with 3% SE-30 on 80-100 mesh Chromosorb W and maintained at a temperature of 270°C. The retention times were 4.00 min for Δ^9-THC, 3.08 min for Δ^9-THC-O-methyl ether, and 6.85 min for pentafluorobenzyl-Δ^9-THC.

Garrett et al. [162] in 1978 studied the stability of tetrahydrocannabinols and found that the biphasic degradation of Δ^9-THC, as monitored by flame ionization gas chromatography, produced Δ^8-THC, cannabidiol, 9-hydroxyhexahydrocannabinol (or Δ^9-9-hydroxytetrahydrocannabinol), 9,10-dihydro-9-hydroxyisocannabidiol, and 6,12-dihydro-6-hydroxycannabidiol in acidic solutions. Further identification was made by GC-MS and comparison with authentic samples. They noted that only Δ^8-THC and 9-hydroxyhexahydrocannabinol were produced above pH 4 in the neutral region by first-order kinetics. The acidic degradation of CBD yielded Δ^9-THC and the products of acidic degradation of Δ^9-THC. The initial phase of acidic Δ^9-THC degradation was assigned to the development of solvolytic equilibria among Δ^9-THC, 6,12-dihydro-6-hydroxycannabidiol, CBD, and, possibly, isocannabidiol, with the concomitant production of Δ^8-THC and 9-hydroxyhexahydrocannabinol. Compounds 6,12-dihydro-6-hydroxycannabidiol, isocannabidiol, and CBD did not appear in the neutral region, since ether cleavage occurred only in strong mineral acids. Hydration of the Δ^9-double bond resulted only

9,10-Dihydro-9-hydroxyisocannabidiol

6,12-Dihydro-6-hydroxycannabidiol

Isocannabidiol

in acid-catalyzed equilibria of cleaved ethers with the Δ^8-configurations and characterized the second phase of acid degradation of Δ^9-THC. Cannabinol and hexahydrocannabinol were found together in several cases due to the disproportionation of Δ^9-THC as catalyzed by silicic acid, silica gel, and chloroform.

In this study of THC stability, Garrett et al. analyzed the various cannabinoid components with a Varian Aerograph 2400 gas chromatograph equipped with a flame ionization detector and 1.8-m glass column packed with 3% OV-17 or 3% OV-225, both coated on 100-120 mesh Gas Chrom Q. The retention times for several cannabinoids obtained with each column as shown in Table 2.43 were obtained using similar GC conditions: injector temperature, 260°C; column temperature, 235°C; detector temperature, 260°C; helium carrier-gas flow rate, 35 ml/min; hydrogen flow rate, 30 ml/min; air flow rate, 300 ml/min.

In subsequent studies for the GC characterization of degradation products of CBD and Δ^9-THC at different pH values, the following GC operating conditions were used with the OV-17 column: injector temperature, 290°C; column temperature, 240°C; detector temperature, 295°C; helium flow rate, 40 ml/min; hydrogen flow rate, 30 ml/min; air flow rate, 300 ml/min. Using these GC settings and 4-androstene-3,17-dione as internal standard, the retention times of CBD, Δ^9-THC, and the various degradation products monitored are listed in Table 2.44.

TABLE 2.43

Comparison of Retention Times of Cannabinoids on OV-17 and OV-225 Columns[a]

Compound	Ret. time (min)	
	OV-17	OV-225
Δ^8-THC	4.65	4.93
Δ^9-THC	5.19	5.53
CBN	6.43	8.45
9,10-Dihydro-9-hydroxyisocannabidiol	7.35	10.62
9-Hydroxyhexahydrocannabinol	8.07	13.14

[a] Adapted from Garrett et al. [162].

TABLE 2.44

Retention Times of CBD, Δ^9-THC, and Various Degradation Products[a]

Compound	Ret. time (min)
Cannabidiol	3.19
Hexahydrocannabinol	3.23
Δ^8-Tetrahydrocannabinol	3.97
Δ^9-Tetrahydrocannabinol	4.34
Cannabinol	5.48
9,10-Dihydro-9-hydroxyisocannabidiol	6.10
9-Hydroxyhexahydrocannabinol	6.95
9,10-Dihydro-9-hydroxycannabidiol (or Δ^9-9-hydroxycannabidiol)	7.27
6,12-Dihydro-6-hydroxycannabidiol	7.68
4-Androstene-3,17-dione (I.S.)	8.73

[a] Adapted from Garrett et al. [162].

III. SOME CHEMICAL CONSTITUENTS OF MARIJUANA SMOKE

In addition to determining the major constituents of marijuana and related materials, various constituents of marijuana smoke have been separated and identified by gas chromatography and integrated gas chromatography-mass spectrometry [99, 120, 133, 171-187].

For example, Stromberg [99] in 1971 used gas chromatography to study some of the protolytic properties for about 40 minor components of hashish whereas for the marijuana-type material the behavior of some 50 minor components under smoking conditions was investigated. In the latter case, marijuana material was rolled into cigarettes and then placed into a smoking apparatus. The condensate was collected on a glass fiber. After extracting the filter in a Soxhlet apparatus with petroleum ether for 6 hr, the extract solution was evaporated to about 200 µl. From this concentrated "smoke concentrate" extract solution, a 1-µl aliquot was withdrawn and injected into into the gas chromatograph.

For this analysis, the instrument and its operating conditions are given below.

Instrument	Perkin-Elmer F-11 equipped with flame ionization detection
Column	1.9-m by 2-mm-i.d. glass with 130-mm coil diameter, packed with 3% OV-17 on 80-100 mesh Gas Chrom Q
Injector temperature (°C)	275
Column temperature (°C)	Programmed, 0 to 2 min isothermally at 100°C; 2 to 18 min at 8°C/min; 18 to 30 min, isothermal at 228°C; 30 to 39 min at 8°C/min; 39 to 45 min, isothermally at 300°C
Detector temperature (°C)	Not specified
Nitrogen carrier-gas flow rate (ml/min)	30
Hydrogen inlet pressure (atm)	1.3
Air inlet pressure (atm)	2.0

Using these conditions, about 50 components were observed in a gas chromatogram of a petroleum ether extract of a smoke condensate; the three major components, CBD, THC and CBN, had retention times of nearly 23.55, 27.95, and 30.80 min, respectively.

Mikes and Waser [177] impregnated cigarettes with various preparations of Δ^9-THC and CBD which were then artificially smoked, and the smoke was analyzed by means of gas chromatography and integrated GC-MS. CBD and Δ^9-THC were identified by their retention times (2.9 and 3.6 min, respectively) and typical m/e fragments (M+ = 314, M+ -CH$_3$ = 299, M+ -C$_6$H$_{11}$ = 231) before and after smoking.

As noted by Mikes and Waser, GC determinations were performed with a Fractovap GI-450 gas chromatograph equipped with a flame ionization detector and a 2.7-m by 3-mm-i.d. glass spiral column packed with 1% SE-30 on silanized Chromosorb W. With this GC instrument, the nitrogen carrier-gas flow rate and the column temperature were maintained at 30 ml/min and 225°C, respectively. For GC-MS studies, a LKB 9000 GC-MS unit was used. Conditions for gas chromatography remained as above except that helium was used as carrier gas, and a total ion current detector was employed. Of the GC effluent, 20% went to the detector and 80% to the mass spectrometer. Conditions for mass spectrometry were ionization potential, 70 eV; separator temperature, 270°C; ion source temperature, 270°C; injector temperature, 250°C; multiplier voltage, 3.5 kV; trap current, 65 µA.

Their analytical data showed that retention of the two test substances (CBD and Δ^9-THC) in smoke was 21 to 23% of the starting amount. Also, the ratio of Δ^9-THC to CBD in smoke was different from that in the starting material. Hence, it was suggested that these differences were the result of partial cyclization of the cannabidiol to Δ^9-THC. An increase in the percentage of CBN in smoke was the result of a partial dehydrogenation of

Δ^9-THC or CBD. Furthermore, no evidence for the isomerization of Δ^9-THC to Δ^8-THC or for the formation of new pyrolyzed products was found.

In 1972, Fehr and Kalant [178] investigated the recovery of cannabinoids by combustion of marijuana cigarettes in a specially constructed smoking machine (see Fig. 2.36), with air flow parameters varying within potential human physiological limits. Combustion temperature was found to be about 600°C with puffing, but the average burning temperature in both tobacco and marijuana cigarettes on continuous draw was 600 ± 40°C. Percentage recovery of THC in the smoke increased slightly but significantly with increasing air flow rate, but was unaffected by continuous versus intermittent flow. However, the greatest variations in THC recovery occurred at the lowest (100 ml/min) and highest (1400 ml/min) air flow speeds. Maximum recovery after complete combustion was about 60%, with negligible change in cannabinoid ratios. Synthetic THC applied to tobacco or alfalfa cigarettes was less completely recovered, and their results indicated that total dry residue (tar content) of the cannabis smoke was somewhat greater than that of tobacco smoke.

For analysis of cannabinoids by gas chromatography, plant extracts and smoke condensate were analyzed on a column packed with 3% OV-17 on 80-100 mesh Gas Chrom Q [100] with cholestane as the internal standard [102]. The standard curve was linear over a THC sample range of 0.1 to 1.0 µg. Samples of smoke condensate were diluted so as to fall within this range.

In 1973, Novotny and Lee [133] applied gas chromatography to the detection of marijuana smoke in the atmosphere of a room. The smoke in the room atmosphere was sampled and concentrated as follows:

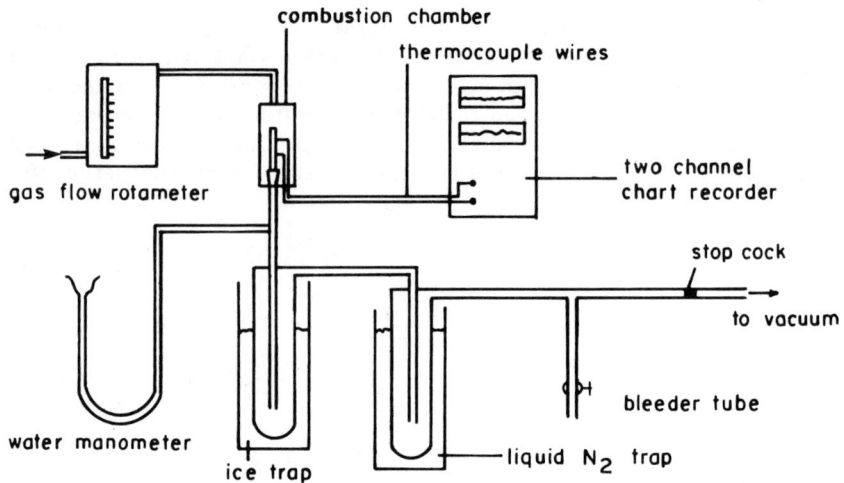

Schematic Diagram of Smoking Machine

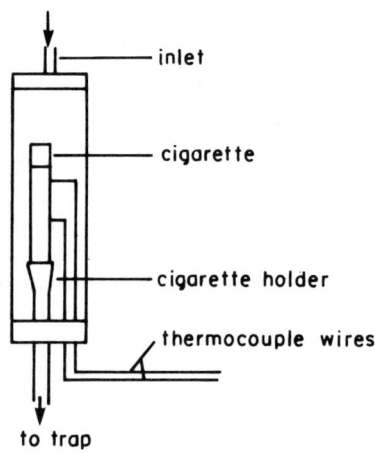

Detail of Combustion Chamber

Figure 2.36. Schematic representation of the smoking machine used for marijuana and tobacco cigarettes. From Fehr and Kalant [178], courtesy of the Canadian Journal of Physiology and Pharmacology.

A small glass tube (95 mm by 1.0 mm i.d.) was packed with a 15-mm length of thermostable porous poly-p-2,6-diphenylene oxide polymer particles and conditioned at 350°C for several hours to remove volatile materials from the polymer surface. The tube was then connected to a vacuum line and room air pumped through for 1 hr (flow rate, 400 ml/min). Trace amounts of organic compounds present in the atmosphere of the room (presumably at less than ppb levels) are effectively concentrated in the tube which is later used as part of the sample port in a commercial gas chromatograph (similar in concept to holder and GC unit shown in [Figs. 2.24 and 2.25 of Vol. 1]. A high-efficiency capillary column (60 m by 0.4 mm i.d. coated with SF-96 silicone oil; injector temperature, 300°C for 5 min during trapping; column temperature, programmed from 35 to 210°C at 2°C/min; detector temperature, 230°C) then resolves the concentrated, trapped components from the room atmosphere into individual fractions detected by the flame ionization detector.

Using this analytical approach, based on retention times, CBD, Δ^9-THC, and CBN from marijuana were tentatively identified; these having retention times of nearly 128, 146, and 159 min, respectively. It was noted that in the marijuana chromatogram, about 100 peaks could be seen which were presumably due to trace volatiles (paint constituents, plasticizers, etc.) from the furniture, books, and other objects in the room. Similar chromatograms were obtained with different commercial tobacco cigarettes. In these chromatograms, nicotine was a predominant component, having a retention time of about 51.7 min.

Kuppers et al. [179] noted that GC analysis of the products obtained by pyrolysis of cannabidiol in air at 700°C revealed the formation of several components which are not only the result of a mere cracking process. A peak with a retention time corresponding to the one of Δ^9-THC was shown by mass spectral analysis to contain the THC compound in question in addition to two components with molecular weights of 314 ($C_{21}H_{30}O_2$) and 330 ($C_{21}H_{30}O_3$). The structure of this major 330 oxidation product of CBD was identified as the decarboxylated product of naturally occurring cannabielsoic acid A by comparing its MS fragmentation pattern with one of the two decarboxylated cannabielsoic acid A C-1-stereoisomers, obtained by photochemical oxidation of cannabidiolic acid.

Cannabielsoic Acid A C-I Stereoisomers

For comparative purposes, the 70-eV mass spectrum of the GC peak having a retention time similar to Δ^9-THC is shown here, as well as that obtained for compound 330.

Upon silylation, compound 330 yielded only a monosubstituted product (parent m/e ion = 402), from which it could be concluded that only one phenolic group remained available for silylation.

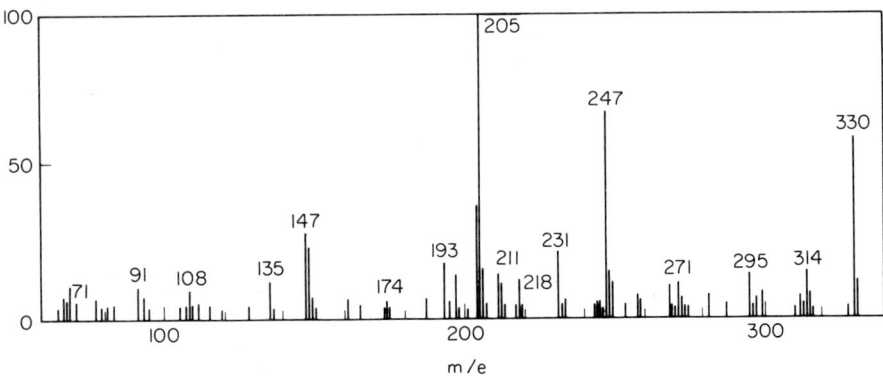

Mass Spectrum of Δ^9 — THC GC Peak (70 eV)

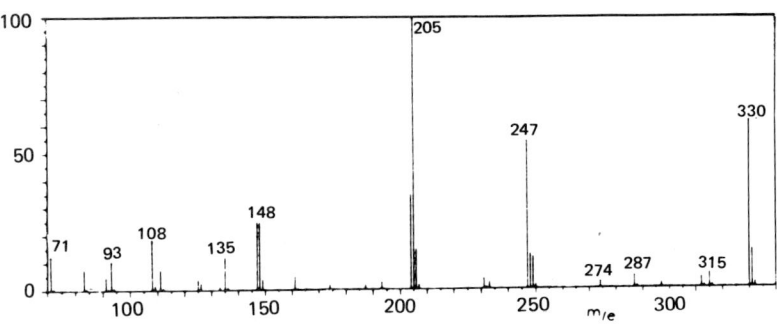

Mass Spectrum of Compound 330 (70eV)

In this study, all MS measurements were made with an AEI MS 902 instrument at an ion source temperature of about 50 to 70°C and an electron energy of 70 eV. GC analyses were performed with a Becker 409 chromatograph equipped with a flame ionization detector and a 2-m by 3-mm glass column packed with 2% OV-17 on AW-DMCS, 100-120 mesh Chromosorb W and operated at a column temperature of 250°C. Preparative GC was carried out with a Varian 1800 Autoprep instrument, whose operating conditions were injector temperature, 270°C; column temperature, 265°C (3-m by 6-mm glass column packed with 20% SE-30 on 30-50 mesh Chromosorb W); collector temperature, 270°C; detector temperature, 320°C.

With the analytical unit (carrier gas and its flow rate not specified), the retention times of peaks 1, 2, and 3 (corresponding to Δ^9-THC) relative to CBD (as determined from a reproduced chromatogram) were 0.45, 0.52, and 1.22, respectively.

With regard to the silylation of CBD, Δ^9-THC and cannabielsoin A (isolated from the pyrolysate and cannabis sample) with bis-(trimethylsilyl)trifluoroacetamide plus 1% trimethylchlorosilane, complete silylation of CBD (di-TMSi) and Δ^9-THC (mono-TMSi) occurred after 1 hr at 60°C, whereas cannabielsoin A yielded a mono-TMSi product. However, with a 50:50 BSTFA/pyridine mixture, and a 5-hr reaction time at 100°C, cannabielsoin yielded by GC a mixture composed of the following products: di-TMSi-cannabielsoin (50%), mono-TMSi-cannabielsoin (45%), and cannabielsoin (5%). In turn, cannabielsoin can also possess two C-1 stereoisomers as shown below:

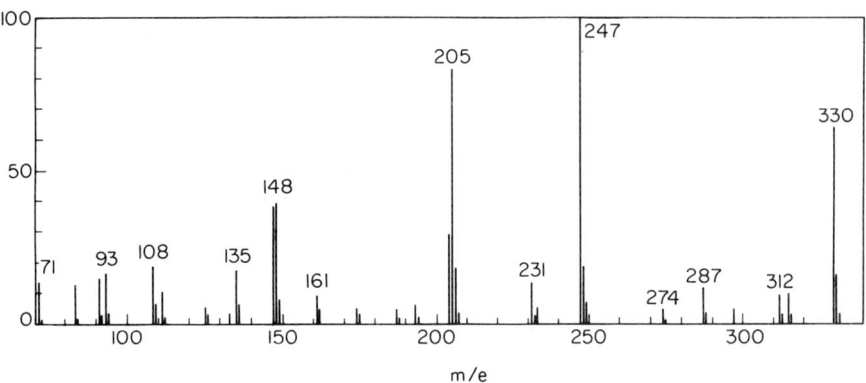

Cannabielsoin C−1 Stereoisomers I and II

In a subsequent report [182], mass spectra of isomers I and II were obtained at an electron energy of 70 eV; that of isomer I, given below, is extremely similar to isomer II.

Cannabielsoin Isomer I (Mass Spectrum at 70 eV)

In their study by mass spectrometry for the elucidation of the structures of pyrolytic products of CBD, Heerma et al. [182] assigned tentative structures to the two compounds with molecular weight 314 in peak 3:

Unknown Products A and B Having MW = 314

In 1973, Fentiman et al. [180] used GC-MS and the chemical ionization (CI) mass spectra thus generated for the identification of the noncannabinoid phenols in marijuana smoke condensate.

Following silylation with bis-(trimethylsilyl)trifluoroacetamide of the residue obtained from the smoke condensate by solvent extraction, the silyl derivatives were analyzed by conventional GC using a 3-m stainless steel column packed with 1% OV-17 on 100-120 mesh Gas Chrom Q and flame-ionization detection. The column was programmed from 75 to 300°C at 6°C/min.

GC-MS studies were performed in the following manner:

After examination of the derivatized extracts by conventional GC, the column was placed in a Varian Aerograph model 1740 gas chromatograph connected directly to a Finnigan model 1015 quadrupole mass spectrometer equipped with a chemical ionization source. No separator or splitter was used. The flow of carrier gas was adjusted to give ion source pressures which were previously found to permit maximum mass spectral sensitivity. For methane, this pressure was approximately 0.4 torr, and for helium, approximately 0.3 torr. The resulting flow rates corresponded to 30 ml/min for helium and 24 ml/min for methane as measured by a mass flow meter inserted in the carrier-gas inlet line. The ion source was differentially pumped by a 1200 liter/sec diffusion pump. The ion source temperature was maintained at 200°C and the ionization voltage at 100 eV. Figure 2.37 shows the mass spectrum of the TMSi derivative of p-hydroxyacetophenone (a component of marijuana smoke) determined using three modes of ionization.

Figure 2.38 is a computer-reconstructed gas chromatogram of the TMSi derivatives from the sodium bicarbonate extract of female Mexican marijuana

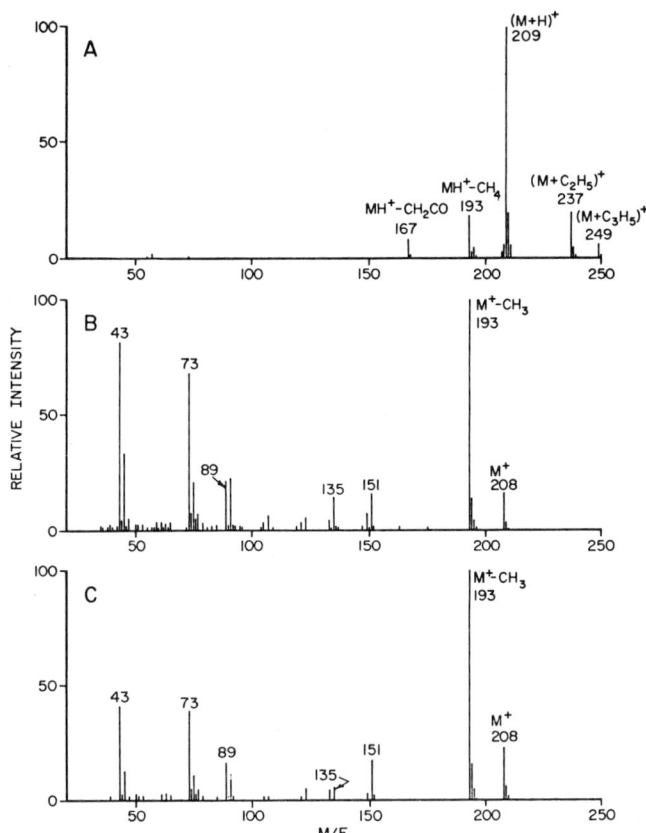

Figure 2.37. Mass spectra of the TMS derivative of p-hydroxyacetophenone. A. Methane CI-MS. B. High-pressure helium MS. C. EI-MS. From Fentiman et al. [180], courtesy of <u>Analytical Chemistry</u>.

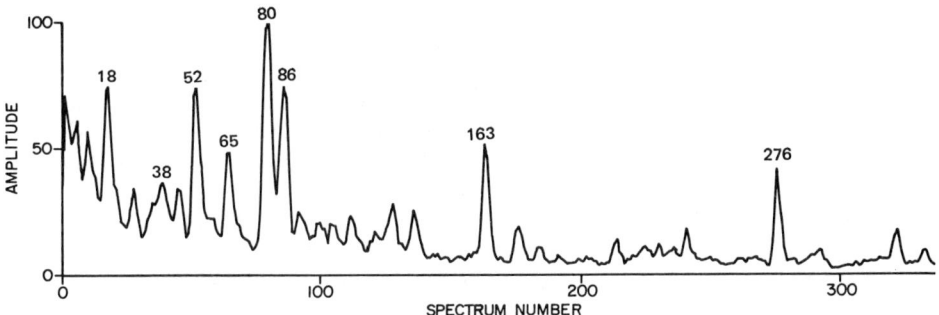

Figure 2.38. Computer-reconstructed gas chromatogram of the TMS derivatives from the sodium bicarbonate extract of female Mexican marijuana smoke condensate. From Fentiman et al. [180], courtesy of <u>Analytical Chemistry</u>.

TABLE 2.45

Composition of the Sodium Bicarbonate Extract[a]

Compound (as TMSi derivative)	Spectrum number (Fig. 2.38)	Rel. ret. time	Extract (%)
Hexanoic acid	18	0.11	5.7
Heptanoic acid	38	0.23	3.8
Furoic acid	52	0.32	3.1
Octanoic acid	65	0.40	4.0
Benzoic acid	80	0.49	9.3
Catechol	86	0.53	12.1
p-Hydroxyacetophenone	163	1.00	2.6

[a] Adapted from Fentiman et al. [180].

smoke condensate whose composition, identification, and retention times relative to p-hydroxyacetophenone are noted in Table 2.45.

A similar computer-reconstructed gas chromatogram of the TMSi derivatives from the NaOH extract of female Mexican marijuana smoke condensate is illustrated in Figure 2.39; its composition, identification, and retention times relative to linolenic acid are listed in Table 2.46.

Haq et al. [181] identified and quantitatively determined carbazole, indole, and skatole in marijuana smoke condensates using gas chromatography, mass, and liquid-scintillation spectrometry. Methods for the synthesis of carbazole-5,6,7,8,12,13-^{14}C and skatole-3a,4,5,6,7,7a-^{14}C were described which, with indole-2-^{14}C, served as internal standards. The procedure developed involves partitioning of the condensate and the respective internal standard between hexane and methanol-water (4:1). The ratio of methanol-water is changed from 4:1 to 2:1 and the resulting solution is shaken with hexane. The hexane extract is first washed with 2 N acetic acid and then with water, dried (Na_2SO_4), and chromatographed on a Florisil column. The concentrate of the appropriate fractions is analyzed for carbazole, indole, and skatole. Their analytical data showed that the average amount of carbazole, indole, and skatole per gram of fresh dry condensate is, respectively, 89 ± 3, 826 ± 4, and 597 ± 7 μg.

For the detection and quantitation of the N-heterocyclics, Haq et al. used a Beckman GC-65 gas chromatograph equipped with a flame ionization detector and a 6-ft by 2-mm-i.d. glass column packed with 3% Silar 5CP on 80-100 mesh Gas Chrom Q. For the determination of carbazole, the

Some Chemical Constituents of Marijuana Smoke

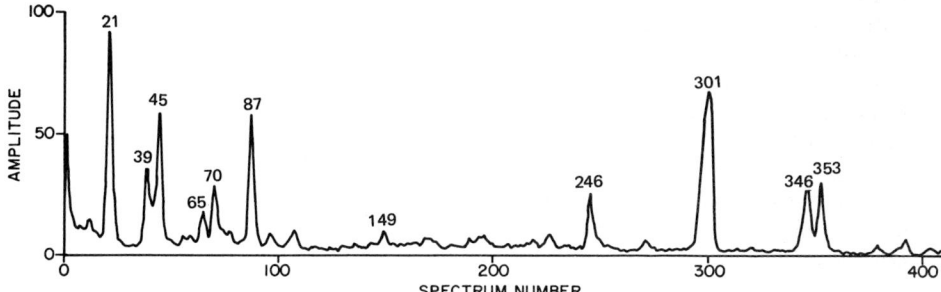

Figure 2.39. Computer-reconstructed gas chromatogram of the TMS derivatives from the sodium hydroxide extract of female Mexican marijuana smoke condensate. From Fentiman et al. [180], courtesy of <u>Analytical Chemistry</u>.

TABLE 2.46

Composition of Sodium Hydroxide Extract[a]

Compound (as TMSi derivative)	Spectrum number (Fig. 2.39)	Rel. ret. time	Extract (%)
Phenol	21	0.06	7.6
o-Cresol	39	0.11	} 9.2
p-Cresol	45	0.13	
p-Ethylphenol	70	0.20	1.9
p-Vinylphenol	87	0.25	2.1
4-Hydroxy-3-methoxystyrene	149	0.42	0.5
Myristic acid	246	0.70	4.6
Palmitic acid	301	0.85	35.2
Stearic acid	346	0.98	10.8
Linolenic acid	353	1.00	4.9

[a] Adapted from Fentiman et al. [180].

following conditions were used: injector temperature, 250°C; column temperature, 220°C; detector temperature, 300°C; helium carrier-gas flow rate, 20 ml/min. Under these conditions, carbazole was eluted in 5.2 min.

For the determination of indole and skatole, all GC settings were the same as those used for carbazole with one exception: the column was maintained at 160°C, which yielded retention times of 4.1 and 5.0 min for indole and skatole, respectively.

The identity of carbazole, indole, and skatole was conclusively established by a comparison of the mass spectra of the peaks of authentic reference compounds with those of their respective counterparts in marijuana smoke extract, these spectra were obtained on a LKB 9000 GC-MS instrument. In Figure 2.40 are shown the mass spectra for authentic carbazole (a), indole (b), and skatole (c).

In 1975, Jones and Foote [183] developed analytical methods for the identification and separation of some acids, bases, and phenols from the smoke condensate of 2638 marijuana cigarettes. Semiquantitative data were obtained by gas chromatography and integrated GC-MS. As noted in their study, the following compounds were identified: (1) acids: hexanoic, heptanoic, octanoic, benzoic, salicylic, hexadecanoic, heptadecanoic, and octadecanoic; (2) bases: dimethylamine, piperidine, pyridine, 2-methylpyridine, pyrrole, 3- and/or 4-methylpyridine, and dimethylpyridine; (3) phenols: phenol, cresols, Guaiacol, catechol, hydroquinone, p-hydroxyacetophenone, scopoletin, and/or esculetin.

The bases (extracted from smoke condensate) as their hydrochloride salts were dissolved in distilled water and analyzed with a Beckman GC-45 gas chromatograph equipped with a flame ionization detector and a 10-ft by 1/8-in. stainless steel column packed with 28% Pennwalt 223 and 4% KOH on 80-100 mesh Chromosorb R. The other GC conditions employed were injector temperature, 150°C; column temperature, 125°C; detector temperature, 250°C; helium carrier-gas flow rate, 27 ml/min.

Mass spectral data were obtained with a Beckman GC-5 gas chromatograph interfaced via a Watson-Biemann molecular separator to an AEI MS-12 mass spectrometer. As noted by the authors, 10-μl samples of the aqueous solution of hydrochloride salts were injected into a 7-in. glass forecolumn of which the first 2 in. were packed with powdered soda lime for liberating the amines and the remaining 5 in. with ascarite for the absorption of water. Following the forecolumn for the separation of the free bases was a 7-ft by 1/8-in. stainless steel column packed with 28% Pennwalt and 4% KOH on 80-100 mesh Chromosorb R. The column was held isothermally at 100°C, whereas the injector and thermal conductivity detector temperatures were maintained at 115 and 200°C, respectively. With this integrated GC-MS instrument, the helium carrier-gas flow rate was split 10:1; ten parts were diverted to the mass spectrometer whereas one part entered the GC thermal conductivity detector.

The oil from the phenol extract was silylated with bis-(trimethylsilyl)-trifluoroacetamide and its reaction products analyzed with a Beckman GC-5

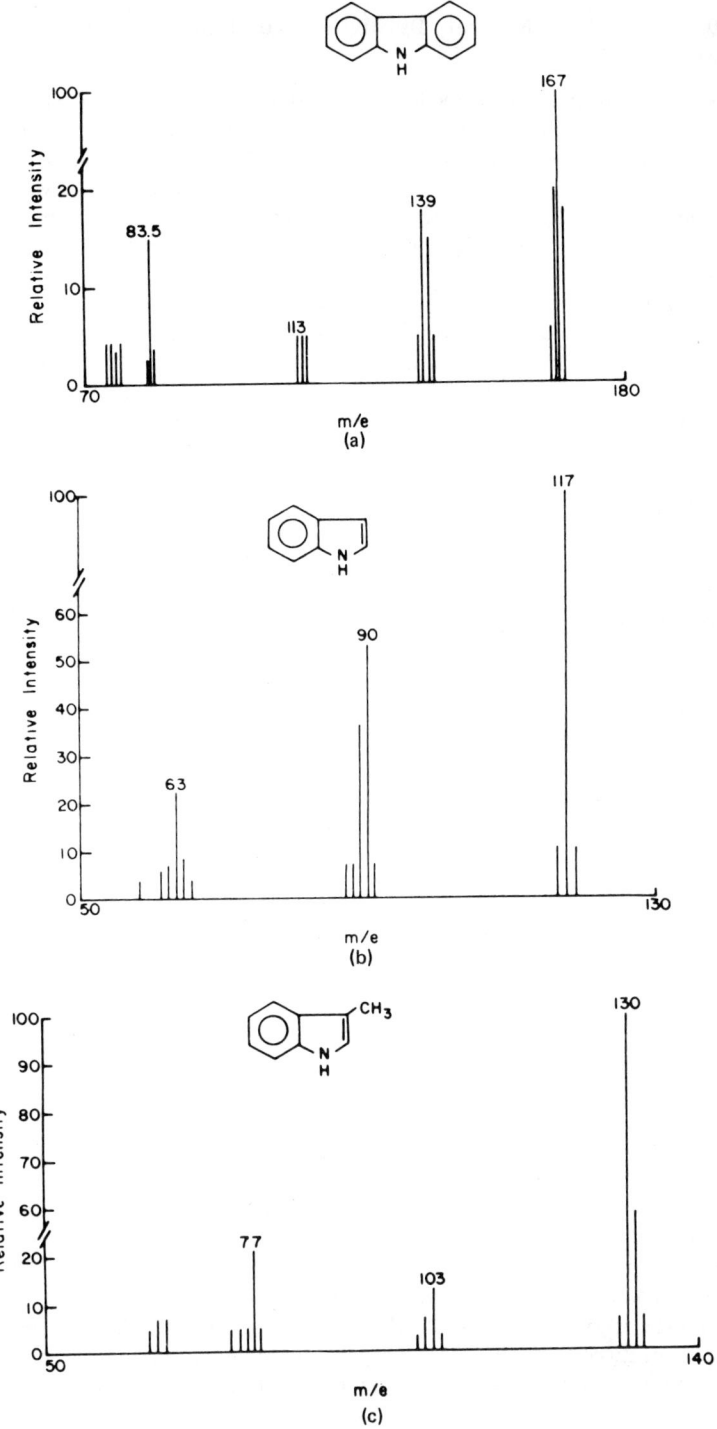

Figure 2.40. Mass spectra of (a) authentic carbazole, (b) indole, and (c) skatole. Adapted from Haq et al. [181].

gas chromatograph equipped with a thermal conductivity detector and a 10-ft by 1/8-in. stainless steel column packed with 5% OV-17 on 60-80 mesh Diatoport S. The other GC conditions used to separate the silyl derivatives of the phenolic components of marijuana smoke condensate were injector temperature, 315°C; column temperature, programmed from 60 to 300°C at 7.5°C/min; detector temperature, 325°C; helium carrier-gas flow rate, 28 ml/min.

Finally, the acidic products in the residue obtained from an extraction of the smoke condensate were esterified by refluxing with 5 ml of BF_3-methanol (14%, v/v) for 5 min and, after cooling, the mixture was poured into 25 ml of ether and washed three times with 25 ml of 5% aqueous $NaHCO_3$. The combined aqueous washes were back-extracted with ether (3 × 25 ml) and the combined ether layers were dried ($MgSO_4$), filtered, and evaporated. After the residue was dissolved in 1 ml of ether, aliquots were injected into the Beckman GC-45 gas chromatograph equipped with a flame ionization detector and a 10-ft by 1/8-in. stainless steel column packed with 2% OV-17 on 80-100 mesh Gas Chrom Q. The column was programmed from 60 to 240°C at 5.6°C/min; injector temperature, 275°C; detector temperature, 325°C; helium carrier-gas flow rate, 25 ml/min.

With the various columns and conditions cited above, the retention times for the basic, phenolic, and acidic components identified in cannabis smoke condensate are listed in Table 2.47.

Jones and Foote noted that:

1. At least 33 components were evident in the chromatogram of the bases; the compounds identified in Table 2.47 accounting for nearly 85% of the total area of all of the peaks observed.

2. Sixty-three peaks were counted in the chromatogram of the silylated phenols, of which the compounds listed in Table 2.47 account for approximately 25% of the total peak area.

3. The chromatogram of the methyl esters of smoke condensate acids contained approximately 65 peaks of which the acids shown in Table 2.47 accounted for nearly 57% of the total area.

Adams and Jones [184] in 1975 identified and quantitated in its smoke the 3-β-hydroxysterols present in American-grown (MS-13) cannabis. Their data showed that the free sterol fraction of the smoke contained campesterol, stigmasterol, and β-sitosterol in essentially the same ratio as that found in the plant material.

Following the isolation of the sterols from the condensate with 2% digitonide in 80% ethanol, the sterols were isolated for GC analysis by decomposing the digitonide with 20 ml of hot dimethylsulfoxide and extracting the cooled mixture with three 25-ml portions of hexane. The combined hexane fraction was dried ($MgSO_4$) and concentrated. The resulting solid was then

TABLE 2.47

Retention Times of Basic, Phenolic, and Acidic Components
Identified in Cannabis Smoke Condensate[a]

Compound	Ret. time (min)
A. Basic Components[b]	
Dimethylamine	1.6
Piperidine	14.7
Pyridine	16.5
2-Methylpyridine	24.5
Pyrrole	26.8
3- and/or 4-Methylpyridine and dimethylpyridine	35.3
B. Phenolic Components[c]	
Phenol	5.3
Cresols	6.8
Guaiacol	9.5
Catechol	10.1
Hydroquinone	11.7
p-Hydroxyacetophenone	14.8
Scopoletin and/or esculetin	25.6
C. Acidic Components[d]	
Hexanoic	2.7
Heptanoic	4.2
Octanoic	6.0
Benzoic	7.0
Salicylic	9.2
Hexadecanoic	22.2
Heptadecanoic	23.8
Octadecanoic	25.4

[a] Adapted from Jones and Foote [183].
[b] Hydrochloride salts; 10-ft by 1/8-in., 28% Pennwalt 223 + 4% KOH on Chromosorb R, 80-100 mesh; temperature, 125°C; detector temperature, 250°C; injector temperature, 150°C; helium flow rate, 27 ml/min.
[c] Silyl derivatives; 10-ft by 1/8-in., 5% OV-17 on Diatoport S, 60-80 mesh; temperature programmed, 60 to 300°C at 7.5°C/min; injector temperature, 315°C; detector temperature, 325°C; helium flow rate, 28 ml/min.
[d] Methyl esters; 10-ft by 1/8-in., 2% OV-17 on Gas Chrom Q, 80-100 mesh; temperature programmed, 60 to 240°C at 5.6°C/min; injector temperature, 275°C; detector temperature, 325°C; helium flow rate, 25 ml/min.

made to volume in tetrahydrofuran and aliquots were subjected to GC analysis [188,189]. Samples were injected into a Beckman GC-45 gas chromatograph equipped with a flame ionization detector and a 10-ft by 4-mm glass column packed with 5% OV-101 on 80-100 mesh Gas Chrom Q with a column temperature of 275°C and helium carrier-gas flow rate of 75 ml/min.

In 1976, Maskarinec et al. [185] also analyzed the acidic fraction of marijuana smoke condensate by capillary GC-MS. Comparative analyses of standard tobacco, Mexican, and Turkish marijuana smoke condensate carried out by means of capillary gas chromatography indicated both qualitative and quantitative changes in the constituents of chromatographic profiles. Samples were converted to volatile derivatives by methylation and trimethylsilylation, whereupon 49 aliphatic acids, aromatic acids, and phenolic compounds were identified by means of capillary GC-MS.

For this investigation, a Varian model 1400 gas chromatograph equipped with a modified splitting injector, linear temperature programmer, flame ionization detector, and a 20-m by 0.25-mm-i.d. glass capillary column prepared by etching with dry hydrogen chloride and coated dynamically with a 15% solution of FFAP in methylene chloride via the mercury plug method was used.

Methyl derivatives of the acidic fractions of marijuana smoke condensate were prepared by on-column injection with trimethylanilinium hydroxide as reagent. Using the FFAP capillary column and the specified GC conditions (injector temperature, 260°C; column temperature, from about 30 to 240°C at 2°C/min; detector temperature, 210°C), a typical chromatogram (a) of smoke condensate from Mexican marijuana is shown in Figure 2.41, where the 38 peaks identified numerically were the following compounds: (1) hexanoic acid, (2) phenol, (3) o-cresol, (4) p-cresol, (5) m-cresol, (6) furoic acid, (7) nonanoic acid, (8) decanoic acid, (9) benzoic acid, (10), o,p-divinylphenol, (11) glutaric acid, (12) m-hydroxy-p-methoxystyrene, (13) dodecanoic acid, (14) 2,4-dihydroxy anisole, (15) o-hydroxyacetophenone, (16) tetradecanoic acid, (17) olivetol, (18) 2,4-dihydroxybenzaldehyde, (19) palmitic acid, (20) palmitoleic acid, (21) palmitolenic acid, (22) stearic acid, (23) oleic acid, (24) linoleic acid, (25) linolenic acid, (26) arachidic acid, (27) eicosenoic acid, (28) eicosadienoic acid, (29) behenic acid, (30) erucic acid, (31) tricosanic acid, (32) 2-ethyl-3-hydroxy-5-pentylbenzoic acid, (33) 2-vinyl-3-hydroxy-5-pentylbenzoic acid, (34) lignoceric acid, (35) tetracosatetraenoic acid, (36) hexacosanoic acid, (37) hexacosadienoic acid, and (38) octacosanoic acid.

Using the same instrumentation and conditions, the methylated aromatic fraction isolated via a DEAE-Sephadex column contained the following compounds as indicated in chromatogram (b) in Figure 2.41:
(1) phenol, (2) o-cresol, (3) p-cresol, (4) m-cresol, (5) benzoic acid, (6) o,p-divinylphenol, (7) catechol, (8) phenylacetic acid, (9) o-isopropenylphenol, (10) m-hydroxy-p-methoxystyrene, (11) 2,4-dihydroxy anisole, (12) o-hydroxybenzaldehyde, (13) phenylpropionic acid, (14) phenylisopropionic acid, (15) o-hydroxyacetophenone, (16) olivetol, (17) 3-isopropyl-5-hydroxybenzaldehyde, (18) 2,4-dihydroxybenzaldehyde,

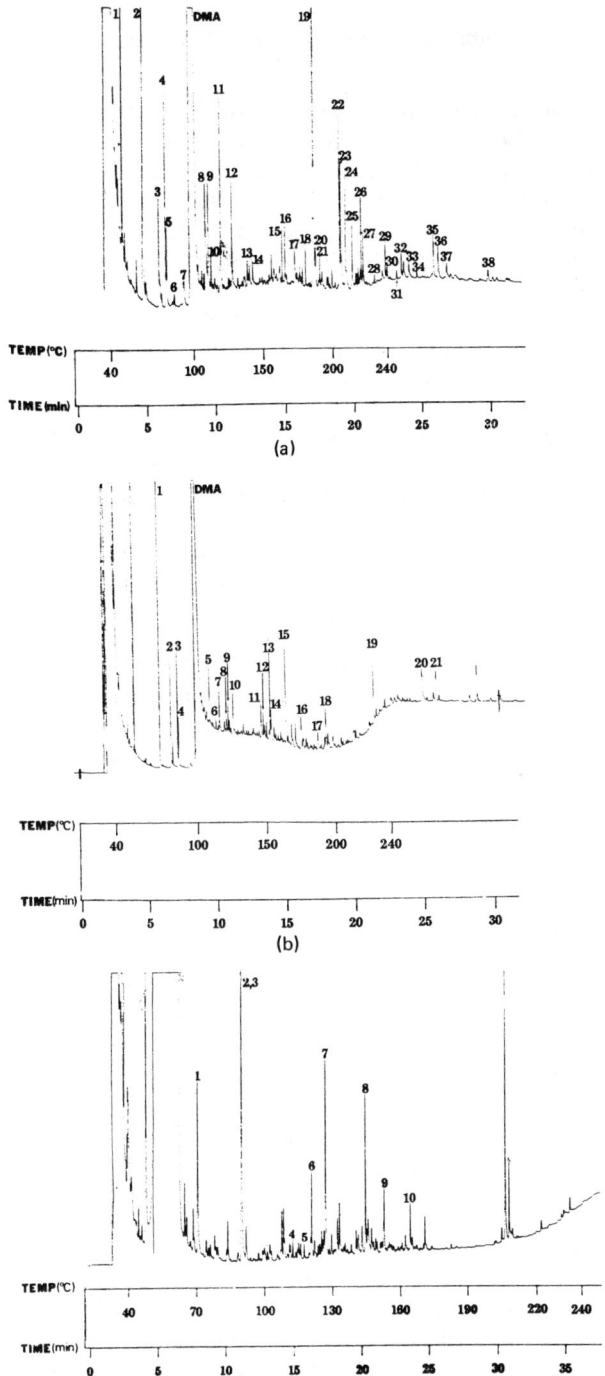

Figure 2.41. Typical chromatograms of (a) methylated total acidic extracts, (b) methylated aromatic fraction through DEAE-Sephadex column, and (c) silylated aromatic fraction from Mexican marijuana. Adapted from Maskarinec et al. [185].

(19) p-hydroxybenzyl-2-butenyl ketone, (20) 2-ethyl-3-hydroxy-5-pentylbenzoic acid, and (21) 2-vinyl-3-hydroxy-5-pentylbenzoic acid.

At a somewhat different column temperature programming rate, but with the same FFAP capillary column and GC conditions, chromatogram (c) in Figure 2.41 shows the separation of the aromatic fraction which had been reacted with N-methyl-N-trimethylsilyltrifluoroacetamide at 80°C for 2-hr; the identity of the peaks is (1) catechol, (2,3) o,p-divinylphenol, (4) o-hydroxybenzaldehyde, (5) o-hydroxyacetophenone, (6) o-isopropenylphenol, (7) m-hydroxy-p-methoxystyrene, (8) 2,4-dihydroxy anisole, (9) 2,4-dihydroxybenzaldehyde, and (10) olivetol.

To obtain mass spectra of the methylated and silylated derivatives, the FFAP capillary column was connected to the ion source of a Hewlett-Packard model 5980A dodecapole mass spectrometer via a glass jet separator. EI mass spectra were obtained with an electron energy of 20 or 70 eV with the GC peaks scanned at the rate of 100 amu/sec.

In 1976, Lee et al. [186], using a combination of chromatographic and spectral methods, carried out analyses of the polynuclear aromatic hydrocarbon fraction of marijuana smoke condensates. These PAH fraction components comprise the largest known group of chemical carcinogens which, as a class of compounds, is credited with the major carcinogenic activity of smoke condensates. The constituents of selectively enriched extracts, further purified and fractionated by a combination of column chromatography and high-resolution liquid chromatography, were analyzed with a Hewlett-Packard model 5980A integrated GC-MS (dodecapole) instrument equipped with a 11.0-m by 0.26-mm-i.d. glass capillary column coated with SE-52 liquid stationary phase. A typical chromatogram of the PAH fraction of smoke condensate from 100 g of marijuana is shown in Figure 2.42; the compounds identified by integrated GC-MS, proton NMR, and standard compound retention time data are listed in Table 1.63.

High-pressure liquid chromatograms of the Sephadex LH-20 fractions are shown in Figures 2.43, where the shaded portions represent the fractions collected for identification purposes. Capillary GC of these final, more refined, fractions are shown in Figures 2.44 through 2.47 for marijuana. In Figures 2.44 through 2.47, the peaks designated by numbers are identified in Table 1.63.

In 1977, Kettenes-van den Bosch and Salemink [187] reported a number of constituents that were identified during the preliminary stages of their research on marijuana smoke using GC and GC-MS analytical methods.

Gas chromatographic studies of acidic, phenolic, neutral, and basic fractions of marijuana smoke condensate were performed with a Becker 417 gas chromatograph equipped with a flame ionization detector and either a 200-cm by 0.3-cm-i.d. glass column packed with 3% OV-17 on 80-100 mesh Chrompack SA or a LKB 2101-2-2 glass capillary, 25-m by 0.22-mm-i.d. column coated with OV-101. Nitrogen was used as carrier gas in addition to the following: a programmed column temperature, a detector temperature of 300°C, and an injector temperature of 220°C.

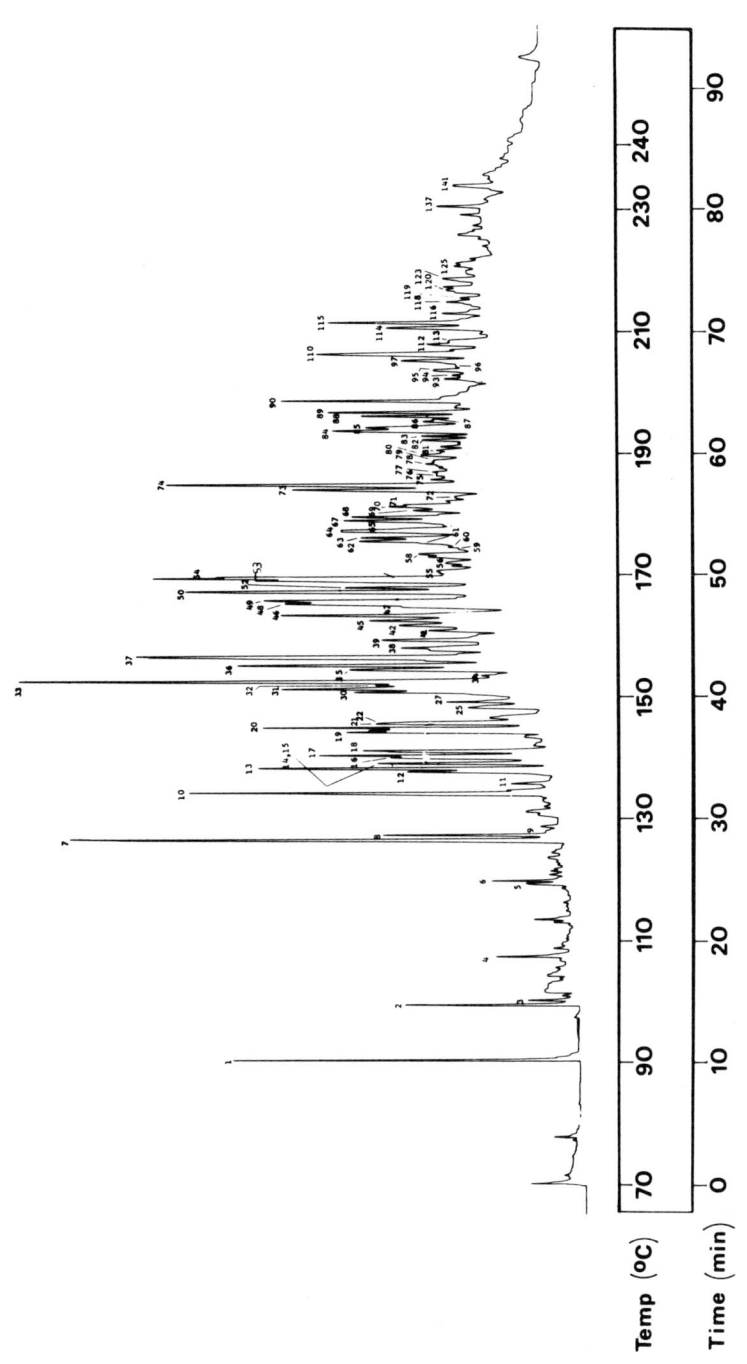

Figure 2.42. Capillary-column gas chromatogram of the polynuclear aromatic hydrocarbon fraction of smoke condensate from 100 g of marijuana. From Lee et al. [186], courtesy of Analytical Chemistry.

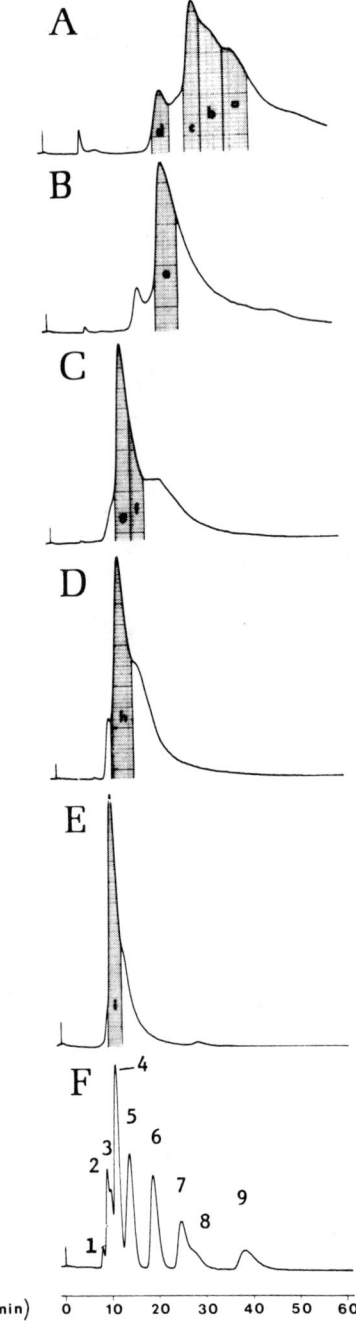

Figure 2.43. High-pressure liquid chromatograms of fractions resulting from Sephadex LH-20 chromatography of marijuana smoke condensate. Key: (1) benzene; (2) biphenyl; (3) fluorene; (4) anthracene; (5) benzo[a]-fluorene; (6) triphenylene; (7) benzo[a]pyrene; (8) perylene; (9) dibenz[a,c]-anthracene. From Lee et al. [186], courtesy of Analytical Chemistry.

Some Chemical Constituents of Marijuana Smoke   427

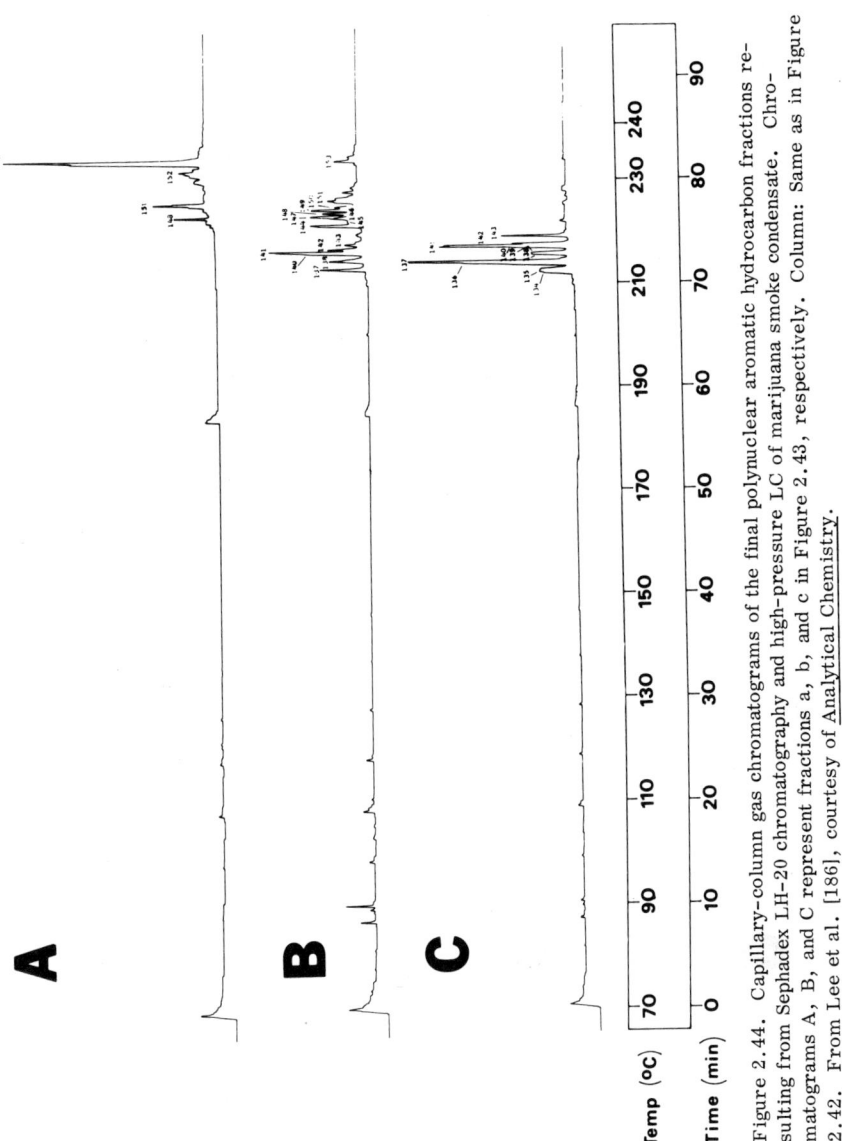

Figure 2.44. Capillary-column gas chromatograms of the final polynuclear aromatic hydrocarbon fractions resulting from Sephadex LH-20 chromatography and high-pressure LC of marijuana smoke condensate. Chromatograms A, B, and C represent fractions a, b, and c in Figure 2.43, respectively. Column: Same as in Figure 2.42. From Lee et al. [186], courtesy of Analytical Chemistry.

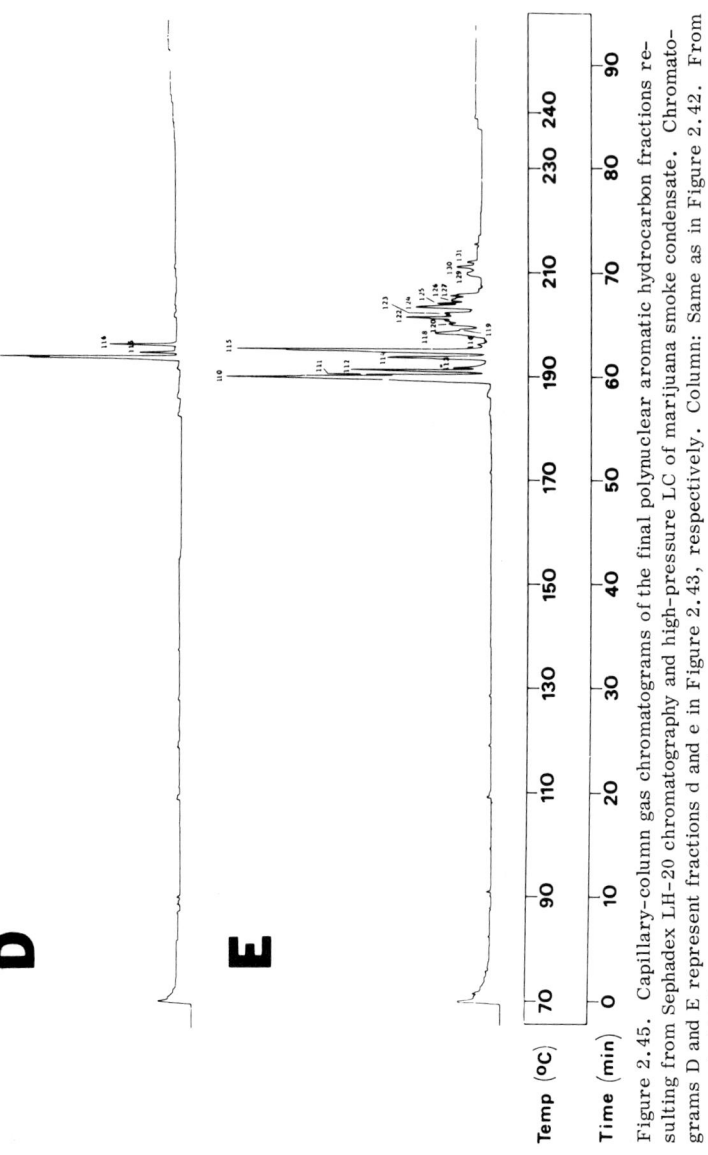

Figure 2.45. Capillary-column gas chromatograms of the final polynuclear aromatic hydrocarbon fractions resulting from Sephadex LH-20 chromatography and high-pressure LC of marijuana smoke condensate. Chromatograms D and E represent fractions d and e in Figure 2.43, respectively. Column: Same as in Figure 2.42. From Lee et al. [186], courtesy of Analytical Chemistry.

Some Chemical Constituents of Marijuana Smoke 429

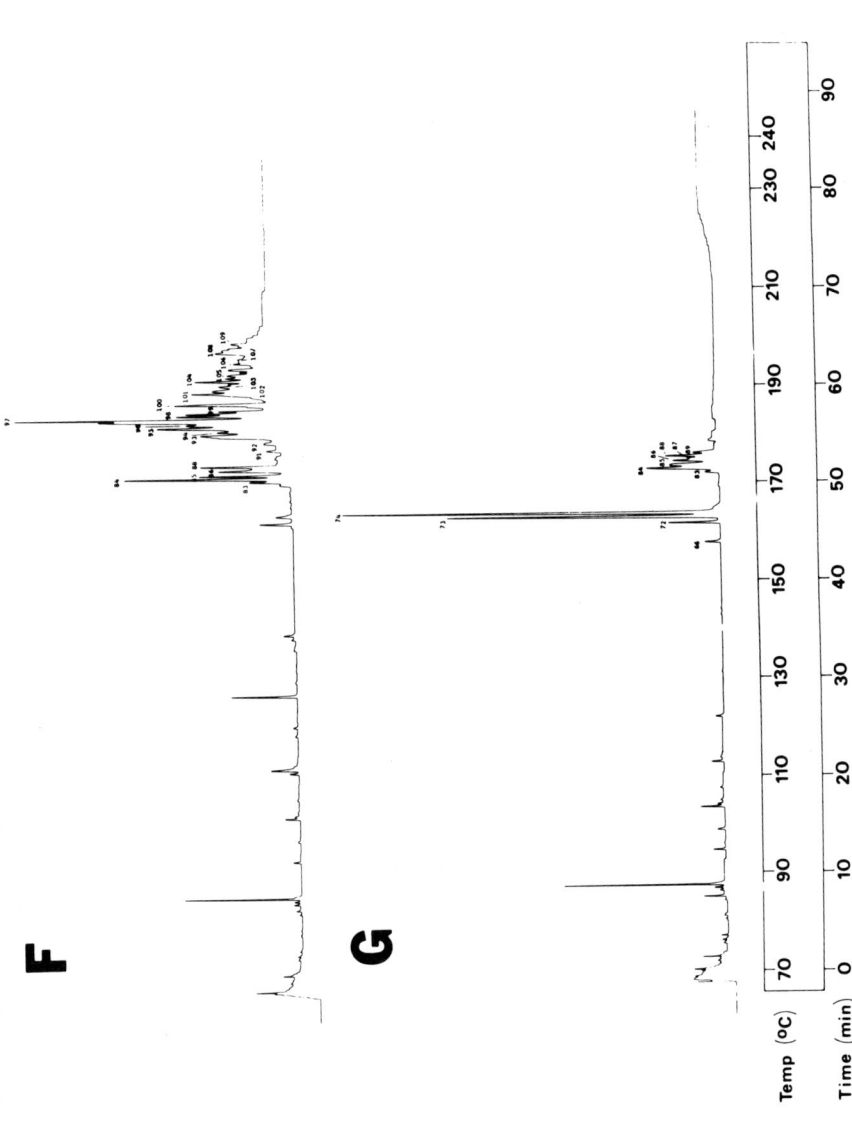

Figure 2.46. Capillary-column gas chromatograms of the final polynuclear aromatic hydrocarbon fractions resulting from Sephadex LH-20 chromatography and high-pressure LC of marijuana smoke condensate. Chromatograms F and G represent fractions f and g in Figure 2.43, respectively. Column: Same as in Figure 2.42. From Lee et al. [186], courtesy of Analytical Chemistry.

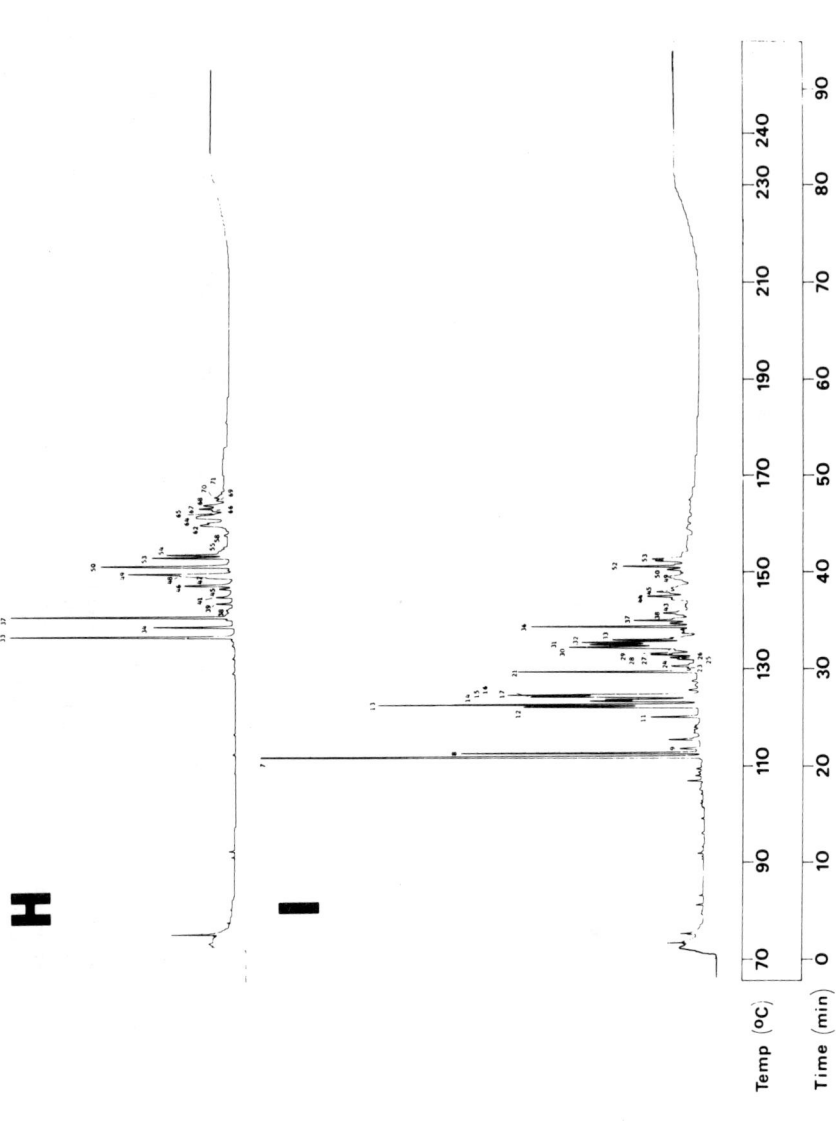

Figure 2.47. Capillary-column gas chromatograms of the final polynuclear aromatic hydrocarbon fractions resulting from Sephadex LH-20 chromatography and high-pressure LC of marijuana smoke condensate. Chromatograms H and I represent fractions h and i in Figure 2.43, respectively. Column: Same as in Figure 2.42. From Lee et al. [186], courtesy of Analytical Chemistry.

Mass spectral studies were carried out with a modified JEOL JMS-07 mass spectrometer with a double-stage separator and a 200 by 0.3-cm glass column packed with 3% OV-17 on 80-100 mesh Chrompack SA. Other important operating parameters were helium as carrier gas (flow rate not specified); injector temperature, 210°C; separator temperature, 220°C; column temperature, programmed (rate and temperature range not given); ion source temperature, 230°C; accelerating voltage, 3 kV; trap current, 300 µA.

Using these instruments, compounds identified for the first time in marijuana condensate are listed in Table 2.48.

TABLE 2.48

Compounds Identified in Marijuana Smoke Condensate[a]

---

1. Acidic fraction:

Phenylacetic acid, β-phenylpropionic acid, p-hydroxybenzaldehyde, vanillin, 2-hydroxy-3-methyl-2-cyclopenten-1-one

2. Phenolic fraction:

α-Dimethylphenol, β-naphthol, 4-methylguaiacol

3. Neutral fraction:

Benzaldehyde, acetophenone, propiophenone, benzonitrile, tolunitrile, benzyl cyanide, β-phenylethyl cyanide, three dimethyl- or ethylindoles, one trimethyl-, methylethyl-, or propylindole, three methylcarbazoles, one dimethyl- or ethylcarbazole, furfural, 5-methylfurfural, 2-acetylfuran, 5-methyl-2-acetylfuran, 4-hydroxy-6-n-pentylbenzofuran, 5-hydroxy-7-n-pentyl-2H-chromene or 4-hydroxy-2-methyl-6-n-pentylbenzofuran, 2,2-dimethyl-5-hydroxy-7-n-pentylchromene, cannabifuran, dehydrocannabifuran, 2-oxo-$\Delta^{3(4)}$-THC, cannabichromanone, Δ^9-THC methyl ether, CBN methyl ether

4. Basic fraction:

Two dimethyl- or ethylpyridines, one trimethyl-, methylethyl-, or propylpyridine, quinoline, methylpyrazine, 2,5-dimethylpyrazine, 2,6-dimethylpyrazine, methylethylpyrazine, one dimethylethyl-, diethyl, methylpropyl-, or butylpyrazine (not tetramethylpyrazine), norharman, harman

---

[a] Adapted from Kettenes-van den Bosch and Salemink [187].

IV. CANNABINOIDS AND METABOLIC PRODUCTS IN BIOLOGICAL MEDIA

In additions to the many investigations conducted to determine the composition of the marijuana plant and its pyrolytic products, other analysts have been concerned with the determination of cannabinoids and their metabolic products in biological media (human and animal) [104,166,190-236].

In 1969, Stone and Stevens [190] developed a TLC/GC procedure for the detection of cannabis components in the mouth and on the fingers of smokers. Whereas the fingers were dipped in chloroform, the mouth was rinsed with aqueous 10% ethanol containing sodium chloride and this ethanolic solution was then extracted with benzene. The fats were eluted from the extracts on an alumina column (pH 10) with a petroleum ether-benzene solution. The cannabinoid components were removed from the column with a 1:1 mixture of chloroform-benzene and spotted on acetone-washed silica gel thin-layer chromatographic plates. By spraying with tetrazotized toluidine, the CBN band or area was identified. An equivalent unsprayed spot was eluted from the TLC plate and its CBN content determined by GC using a 5-ft by 4-mm-i.d. glass column packed with neopentyl glycol plus trimer acid on silanized, 80-100 mesh Chromosorb W.

Also in 1969, Agurell et al. [191] described a method for the preparation of tritium-labeled Δ^9-THC and other cannabinols. They undertook this investigation to determine the distribution and elimination of cannabinols and Δ^9-THC in the rat as well as the presence and structures of metabolites of Δ^9-THC in tissues and urine. Intravenously injected Δ^9-THC-H^3 was eliminated very slowly by the rat, half of the administered dose still remaining in the body after 1 week. About 80% of the drug was excreted in metabolized form via feces, the remainder being eliminated as metabolites in the urine. During the first 24 hr, 2 to 6% of the injected activity appeared in the urine, but less than 0.006% of the dose, if any, was excreted unchanged. A considerable amount of the activity was extractable readily with ether and possibly an identification method for cannabis users might be based on the occurrence of this metabolite. Δ^9-THC was apparently not excreted as a glucuronide.

Labeled materials were prepared in the following manner:

Cannabinols were tritiated by heating 150 mg of cannabinols, 0.5 ml of 100% phosphoric acid, 2.5 ml of tetrahydrofuran, and 0.25 ml of tritiated water (50 mCi) to 80°C for 2 hr in a sealed ampoule under nitrogen. The reaction mixture was then dissolved in ether (25 ml), washed once with an equal volume of saturated NaHCO$_3$ solution and twice with an equal volume of water. After drying (Na$_2$SO$_4$), the ether was evaporated in vacuo. Exchangeable hydrogens were removed by dissolving the residue in 2 ml of methanol and evaporating the methanol under a stream of nitrogen. This was repeated twice.

Gas chromatographic separations were performed with an Aerograph 204 gas chromatograph equipped with a flame ionization detector and a 6-ft by 1/8-in. glass column packed with 5% SE-30 on AW-DMCS treated Gas Chrom P. With the nitrogen carrier-gas flow rate held at 25 ml/min, the column was maintained isothermally at 230°C. With these GC conditions, the retention times obtained for THC and CBN were approximately 10.35 and 12.92 min, respectively. Estimating a lower detection limit of 2 ng for the major component (Δ^9-THC) in the gas chromatograph and assuming a 70% recovery in the purification process, it was estimated that a maximum of about 300 ng of the major component, Δ^9-THC, could remain undetected in the urine.

Following the report of Stone and Stevens [190], Robinson [193] identified cannabis constituents on the hands of autopsy cases. Each hand was swabbed with chloroform-soaked cotton wool. The swabs were eluted with chloroform which was evaporated to yield a concentrated extract. This extract was examined with an F&M model 400 gas chromatograph equipped with a flame ionization detector and a 6-ft column packed with 3.8% W-98 on Diatoport S. With the column operated at 200°C, the extracts were compared with tinctures of cannabis and peaks were found that corresponded to CBD and CBN.

Wall [195] performed studies of the in vitro and in vivo metabolism of THC, the in vitro studies of the Δ^1-THC metabolism being conducted with liver homogenates prepared from mouse, rat, rabbit, guinea pig, and man (autopsy specimen).

With regard to the rat liver experiments, "500 g of liver was obtained from 350-g male rats, pretreated with phenobarbital prior to sacrifice. (The livers from animals not pretreated with phenobarbital gave similar results but lower yields of metabolites.) A homogenate was prepared with use of 0.1 M potassium phosphate (5 liters, pH 7.4), containing 0.013 M magnesium chloride. The ice-cold homogenate was centrifuged at 10,000 G. To the supernatant thus obtained was added 1.0 g of Δ^1-THC plus 394 µCi of tritium-labeled Δ^1-THC and cofactors, NADP (6.55 g, 1.6×10^{-3} M), G6P (14.3 g, 8.0×10^{-3} M), and G6P-dehydrogenase, 1000 units. The mixture was incubated aerobically and shaken for 2 hr at 37°C in 10 3-liter Fernbach flasks. The reaction was quenched by extraction with ethyl acetate."

Thin-layer chromatography of the ethylacetate extract revealed the presence of four metabolites: 7-acetoxy-Δ^1-THC, 7-hydroxy-Δ^1-THC, $6\alpha,7$-dihydroxy-Δ^1-THC, and 6β-hydroxy-Δ^1-THC.

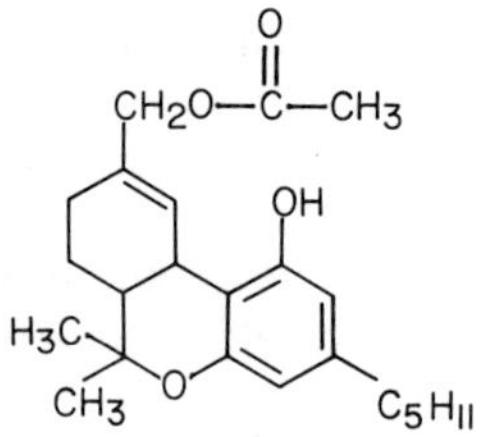

7-Acetoxy-Δ¹-THC

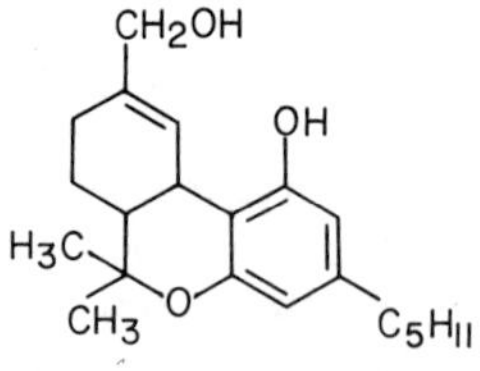

7-Hydroxy-Δ¹-THC

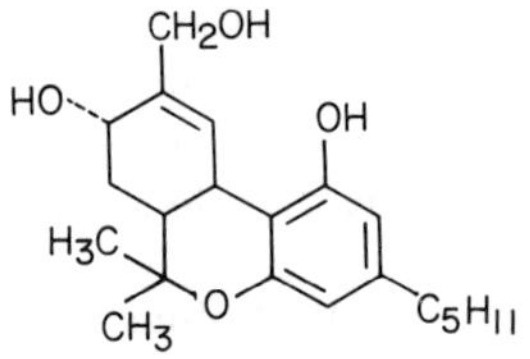

6α,7-Dihydroxy-Δ¹-THC

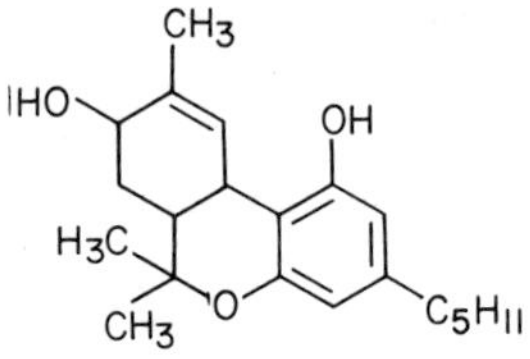

6β-Hydroxy-Δ¹-THC

In like manner, compounds isolated from large-scale rat liver incubation of Δ^6-THC yielded 7-acetoxy-Δ^6-THC, 7-hydroxy-Δ^6-THC, 5α,7-dihydroxy-Δ^6-THC, and 5β,7-dihydroxy-Δ^6-THC. Using a low-resolution mass spectrometer, the mass spectra of Δ^6-THC, Δ^1-THC and their metabolites are noted in Table 2.49. The two Δ^6-THC dihydroxy metabolites were very difficult to separate but were initially resolved by GC and eventually on a preparative scale with use of preparative TLC.

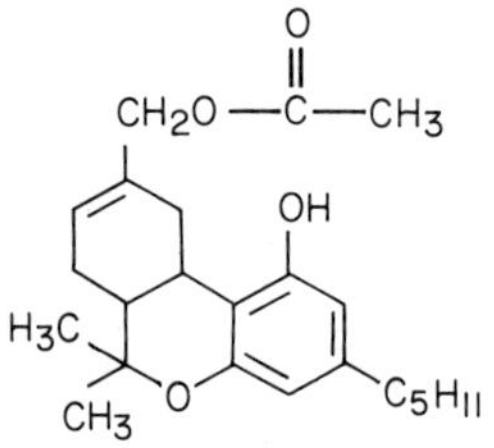

7-Acetoxy-Δ⁶-THC

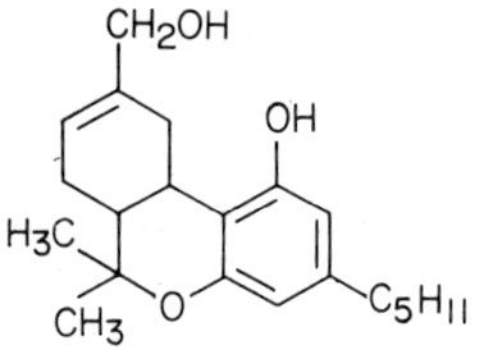

7-Hydroxy-Δ⁶-THC

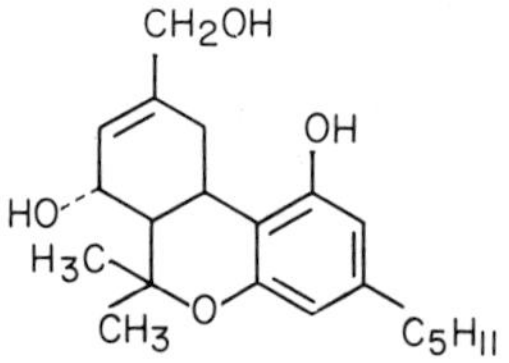

5α,7-Dihydroxy-Δ⁶-THC

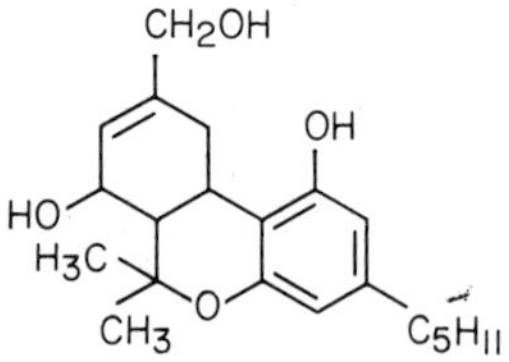

5β,7-Dihydroxy-Δ⁶-THC

TABLE 2.49

Prominent m/e Ions of Δ^6-THC, Δ^1-THC, and Their Metabolic Products[a]

Assignment	Δ^6-THC	7-Acetoxy-Δ^6-THC	7-HO-Δ^6-THC	5α,7-diHO-Δ^6-THC	5β,7-diHO-Δ^6-THC	Δ^1-THC	7-HO-Δ^1-THC	6α,7-diHO-Δ^1-THC	7-Acetoxy-Δ^1-THC	6β-HO-Δ^1-THC
M	314	372	330	346	346	314	330	346	372	330
M-CH$_3$	299		315	328		299	315	328		315
M-H$_2$O			312				312			312
M-H$_2$O-H$_2$O				310	310					
M-CH$_2$OH			299				299	315		
M-H$_2$O-CH$_3$			297	313	313		297	313		297
M-CH$_2$OH-H$_2$O				297	297			297		
M-CH$_3$COOH		312							312	
M-C$_3$H$_7$	271					271		285		
M-H$_2$O-C$_3$H$_7$			269							
[structure with OH, C$_5$H$_{11}$]	231	231	231	231	231	231	231	231	231	231
M-C$_4$H$_8$	258	316	274	290	290					
M-C$_4$H$_8$-CH$_3$	243									
[structure with OH, C$_5$H$_{11}$]	246	246	246							

[a] Adapted from Wall [195].

Similar in vitro metabolic studies were performed with CBN. Three metabolites were observed (structures given below): 7-hydroxycannabinol, 2"-hydroxycannabinol (tentative), and 2",7-dihydroxycannabinol (tentative).

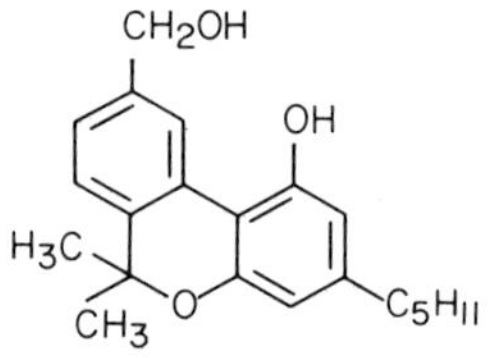

7-Hydroxy-CBN

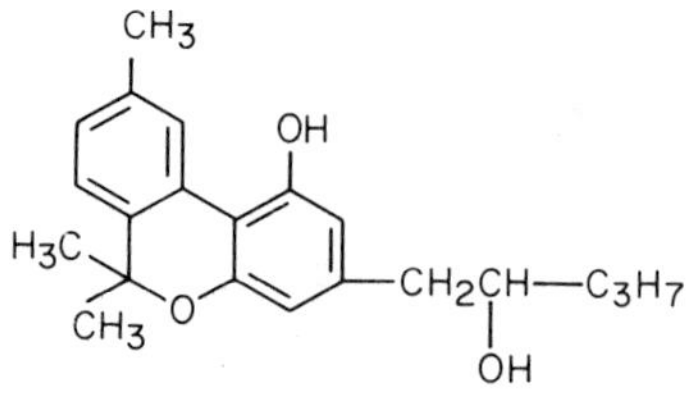

2"-Hydroxy-CBN

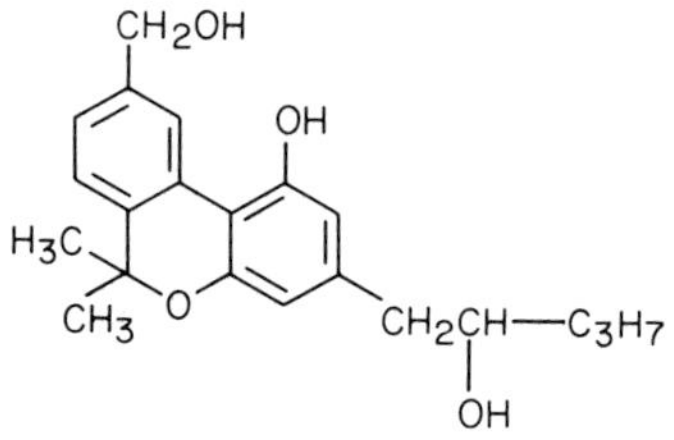

2",7-Dihydroxy-CBN

CBN and its metabolic products were silylated and analyzed by integrated GC-MS with a 6-ft column packed with 1.35% OV-17 on Chromosorb W-HP. With a carrier-gas flow rate of 35 ml/min and a column temperature of 180°C (no other conditions specified), the retention times 2"-OH-CBN (di-TMSi), 7-OH-CBN (di-TMSi), and 2",7-diOH-CBN (tri-TMSi) relative to CBN (mono-TMSi) were about 1.78, 2.67, and 3.53, respectively, whereas the major m/e ions of these silylated metabolites were 2"-OH-CBN (di-TMSi), 470 (M^+), 455, 367, 365, 363, and 145; 7-OH-CBN (di-TMSi), 470 (M^+), 455, 367, 365, and 350; 2",7-diOH-CBN (tri-TMSi), 558 (M^+), 543, 455, 315, and 145.

Wall summarized his findings as follows:

The in vitro metabolism of Δ^6-THC and Δ^1-THC by the postmitochondrial fraction obtained from the liver homogenates of various species, including man, proceeds by allylic hydroxylation. Initially, the carbon-7 methyl group, which is allylic in both the Δ^6-THC and Δ^1-THC, is hydroxylated. A second hydroxyl group is then also introduced at the allylic position, 5α- or 5β- in the case of Δ^6-THC and 6α- in the case of Δ^1-THC. In addition, a 6β-monohydroxy metabolite has been found with rabbit liver. CBN may also be hydroxylated, the predominant metabolite being 7-hydroxy-CBN. The 7-hydroxy metabolites in both Δ^6-THC and Δ^1-THC are highly active; further hydroxylation leads, depending upon the route of administration, to inactivation or reduced activity. In the case of Δ^1-THC, the same metabolites are found with human autopsy liver. These compounds are probably also formed in man in vivo, but the structures of the metabolites are not as yet rigidly established. Because 7-hydroxy-Δ^1-THC has been shown to be 15 to 20 times more active in mice by intracerebral administration, there is a possibility that this compound is indeed the active form of Δ^1-THC. However, this hypothesis requires demonstration.

Widman et al. [196] also showed 7-hydroxycannabinol to be a major, primary metabolite of cannabinol using the in vitro system described by Tagg et al. [237]. The extracted, purified phenolic metabolite reacted with on-column methylation with trimethylanilinium hydroxide was analyzed with a Varian Aerograph model 2100 gas chromatograph equipped with a flame ionization detector and a 6-ft by 1/8-in.-i.d. glass column packed with 3% JXR on 100-120 mesh Gas Chrom Q. Using a column temperature of 220°C (no other conditions specified), the retention times of unmodified CBN and 7-OH-CBN were 6.6 and 13.7 min, respectively, whereas their O-methyl derivatives, O-methyl-CBN and 7-methoxy-O-methyl-CBN, were eluted in 4.8 and 9.3 min, respectively.

Mass spectral data for 7-methoxy-O-methyl-CBN showed the following predominant m/e ions: m/e 354 (M$^+$), 340, 339 (base peak), 325, 309, 294, 284, and 268. These MS data were obtained with a LKB 9000 integrated GC-MS instrument equipped and operated as follows: ion source, 270°C; electron energy, 70 eV; ionization current, 60 µA; 4-ft by 1/8-in.-i.d. glass column packed with 5% SE-30 on Gas Chrom Q; column temperature, 248°C.

In 1972, Burstein et al. [197] isolated two of the major metabolites which appear in rabbit urine after the administration of Δ^1-THC; they were tentatively identified as 7-carboxy-Δ^1-tetracannabinols with an additional hydroxyl group on the side chain at either the 1" or 2" position. Chromatographed on a DEAE-Sephadex column with a NaCl gradient, the major peak observed in urine (fraction 150 to 165) was methylated with a mixture of CH$_3$I and K$_2$CO$_3$ in dimethylformamide, which selectively methylates acids and phenols but not aliphatic alcohols. By thin-layer chromatography, this methylated fraction gave two separated zones, each of which, upon examination by low-resolution mass spectrometry, gave a molecular ion peak at 388. This suggested isomeric methyl ester-methyl ether derivatives with a formula of C$_{23}$H$_{32}$O$_5$. At the high-mass end of the spectra of these methylated species, the principal ions of metabolite I were 388, 373, 370, 355, 329 (base peak), 316, and 311, whereas metabolite II had m/e ions at 388, 373, 370, 329, 316 (base peak), and 311. The structures of Δ^1-THC methyl ether as well as those of metabolite I (2"-hydroxy-7-carboxy-Δ^1-THC) and metabolite II (1"-hydroxy-7-carboxy-Δ^1-THC) as their respective methyl ester-methyl ether derivatives are shown below:

Metabolite I: structure with COOCH$_3$, OCH$_3$, H$_3$C, CH$_3$, and CH$_2$CH(OH)–C$_3$H$_7$ side chain.

Metabolite II: structure with COOCH$_3$, OCH$_3$, H$_3$C, CH$_3$, and CH(OH)–C$_4$H$_9$ side chain.

Δ^1-THC methyl ether: structure with CH$_3$, OCH$_3$, H$_3$C, CH$_3$, and C$_5$H$_{11}$ side chain.

As noted by Burstein et al., "the presence of a 7-carboxyl function in both metabolites is not entirely unexpected since hydroxylation of the 7 position is a well-established process. The further transformation to an acid likely involved an aldehyde (with or without the side chain hydroxyl) as an intermediate. Such an aldehyde may well have an important role in the biological activity of Δ^1-THC."

In 1972, Wall et al. [198] presented for the first time rigid analytical identification of Δ^1-THC, 7-hydroxy-Δ^1-THC, and 6α,7-dihydroxy-Δ^1-THC in the blood plasma of human volunteers. In addition to these findings, because of the availability of appropriate standards, 6α- and 6β-hydroxy-Δ^1-THC have been tentatively identified for the first time in man.

Following a solvent extract of human plasma and subsequent purification, a major component was identified as Δ^1-THC by comparison of its retention time and mass spectral fragmentation pattern with that of authentic sample. In similar fashion, the 7-hydroxy and 6α,7-dihydroxy metabolites were positively identified as their TMSi derivatives by GC-MS and comparison with known standard materials. Retention time and mass spectral data were obtained with a LKB 9000 integrated GC-MS instrument equipped with an OV-17 column operated at 210°C with a nitrogen carrier-gas flow rate of 35 ml/min.

McCallum [200] developed a method for the measurement of CBN and Δ^9-THC in the blood involving GC separation of the compounds as their phosphate esters followed by flame photometric detection.

With regard to sample preparation, the residue obtained from an extraction of blood or plasma and dissolved in 0.3 ml of dry benzene and tert-butanol (9:1) was added to about 300 mg of sodium azide pellets contained

Cannabinoids and Metabolic Products in Biological Media 439

Δ⁹-THC Phosphate Ester   CBN Phosphate Ester

in a special reaction cell for derivatization of blood extract and allowed to stand stoppered for 2 min. The solution and subsequent washings were dropped directly onto two drops of diethyl phosphorochloridate (see Fig. 2.48), shaken, and kept in the 2-ml stoppered flask at 50°C. After 30 min, the benzene was completely removed by a stream of nitrogen at a temperature not exceeding 45°C and then aqueous $NaHCO_3$ (5%; 1 ml at 45°C) was added; this solution was then mixed vigorously and permitted to stand for 1 min at 45°C. Hexane (about 1 ml) was then added, shaken vigorously with the $NaHCO_3$ solution, removed, and concentrated to a fixed volume (0.05 ml). From this final solution, 5- to 10-μl aliquots were withdrawn and injected into the gas chromatograph.

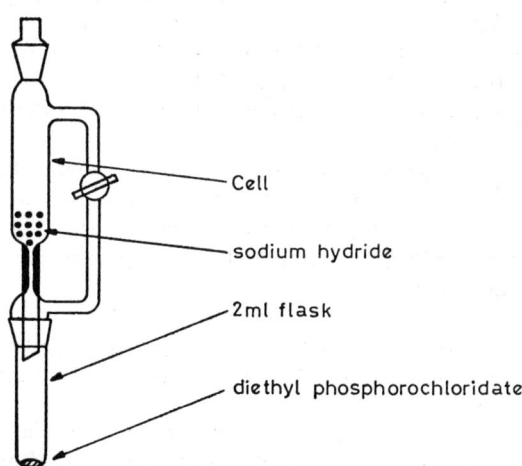

Figure 2.48. Reaction cell for derivatization of blood extract. From McCallum [200], reproduced from the Journal of Chromatographic Science, by permission of Preston Publications, Inc.

The phosphate ester derivatives were determined using a Tracor gas chromatograph equipped with a Melpar flame photometric detector and a 2-m by 6-mm-i.d. glass column packed with 3% OV-1 on 100-120 mesh Chromosorb W and operated isothermally at 240°C with a nitrogen carrier-gas flow rate of 25 ml/min.

For analyses requiring flame ionization detection, an Aerograph Hy-Fi 600 D gas chromatograph equipped with a 2-m by 3-mm-i.d. glass column packed with 1% SE-52 on 100-120 mesh Diatoport S was used; the column temperature and nitrogen flow rate were maintained at 210°C and 15 ml/min, respectively.

With the flame photometric GC unit, the retention time of CBN relative to Δ^9-THC (RRT = 1.00) was approximately 1.09. McCallum noted that, using the phosphate ester method, Δ^9-THC could be detected to less than 500 pg per injection and the minimum amount of Δ^9-THC assayable practically could be considerably less than 1 ng/ml if more than 10 ml of plasma were used.

Agurell et al. [201] developed in 1973 a method to identify and accurately measure nonlabeled Δ^9-THC in blood of cannabis smokers. As discussed, the procedure consisted of the following steps: To a 5-ml plasma sample is added deuterated Δ^9-THC (Δ^9-THC-d_2) as internal standard. After extraction with light petroleum and evaporation, the Δ^9-THC containing fraction is separated by chromatography on Sephadex LH-20 (1 × 40 cm) using light petroleum-chloroform-ethanol (10:10:1) as eluant. A fraction containing Δ^9-THC is collected and subjected to mass fragmentography; the mass spectrometer is adjusted to record the intensities of m/e 299 and 314 of Δ^9-THC and m/e 301 and 316 of Δ^9-THC-d_2. The standard curve is prepared by plotting peak height ratio of Δ^9-THC (m/e 299)/peak height Δ^9-THC-d_2 (m/e 301) against known amounts of added Δ^9-THC in nanograms per milliliter of plasma.

In this study, each volunteer smoked a cigarette containing 10 mg of Δ^9-THC during 5 min. A blood sample was taken before smoking and 10, 30, 60, and 120 min after the smoking was terminated. The heparinized blood samples were centrifuged immediately and plasma stored at -20°C in glass tubes until analyzed.

Using a LKB 9000 integrated GC-MS equipped with a 3% OV-17/100-120 mesh Gas Chrom Q column operated at 230°C (no other GC or MS parameters specified), the retention time of Δ^9-THC as indicated in a typical chromatogram was about 2.9 min.

Based on human sample analyses, it was shown that peak plasma levels of 19 to 26 ng/ml of plasma were reached within 10 min after smoking a cigarette containing 10 mg of Δ^9-THC; this was followed by a rapid decline to 5 ng/ml or less within 2 hr.

Fenimore et al. [202] described an electron capture-GC determination of Δ^9-THC in blood serum, Δ^9-THC being detected as the heptafluorobutyrate on a dual-column, dual-oven gas chromatograph utilizing a capillary column as the final resolving component. As noted by Fenimore et al., the limit of detection is less than 100 pg/ml with excellent reproducibility using hexahydrocannabinol as an internal standard. Blood serum concentrations in experimental animals injected with 0.1 mg of Δ^9-THC/kg were determined with levels below 1 ng/ml at 4 hr after administration.

A diagram of this uniquely designed gas chromatograph is shown in Figure 2.49; the packed column has linear temperature capability, whereas the capillary column oven (used in conjunction with the electron-capture detector) is maintained isothermally. As described by Fenimore et al.:

> The packed column is coiled, 6-ft by 2-mm-i.d. glass arranged for on-column injection and containing 5% SE-30 on 100-120 mesh Gas Chrom Q. The effluent from this column leads to an eight-port, high-temperature, low-dead-volume switching valve. As shown in [Fig. 2.50], the gas flow from the first column leads to a flame ionization detector when in the normal operating position. In this position, carrier gas to the capillary column flows through the trap. When the valve is in the trapping position, the effluent from the packed column flows through the 2-m length of 0.32-in.-o.d. by 0.02-i.d. nickel-200 capillary and then to the flame ionization detector while the capillary-column gas supply (argon/methane) flows directly to that column. The trapped material is swept into the capillary by returning the valve to

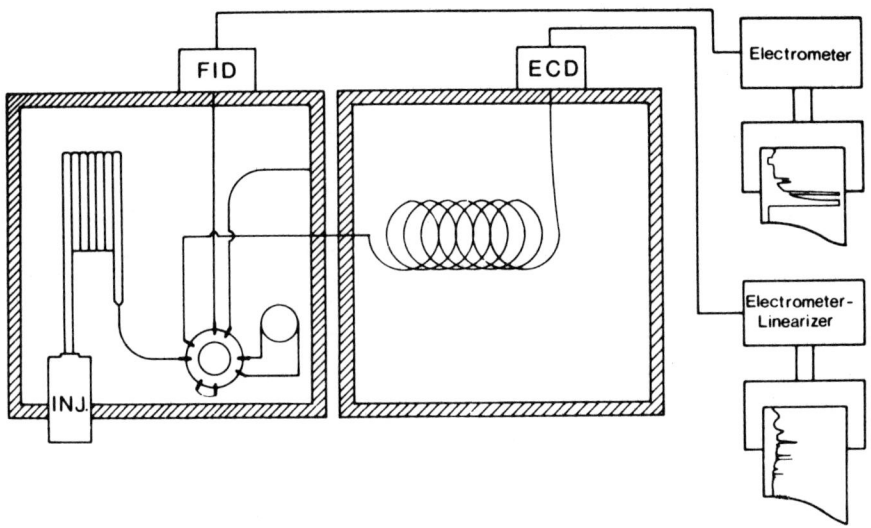

Figure 2.49. Dual oven, dual column gas chromatograph. From Fenimore et al. [202], courtesy of <u>Analytical Chemistry</u>.

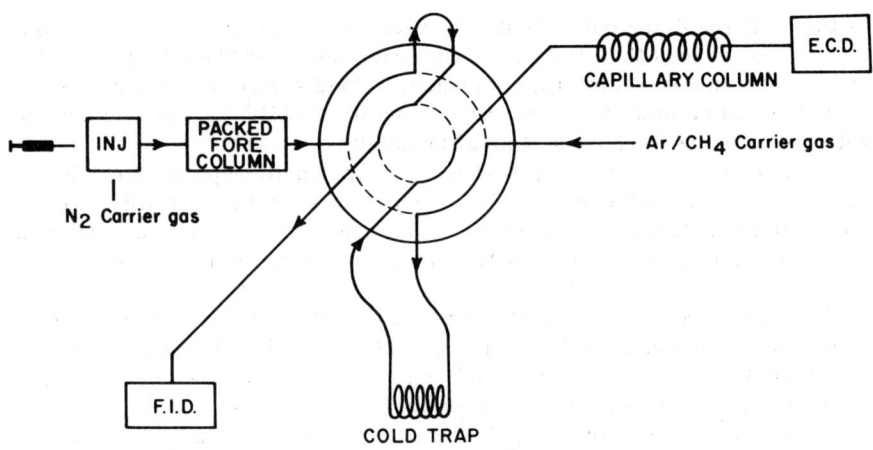

Figure 2.50. Flow diagram of eight-port valve. Normal operating flow shown by solid lines; flow during trapping shown by broken lines. From Fenimore et al. [202], courtesy of Analytical Chemistry.

the original position. Oven heat is sufficient to reheat the trap rapidly (cooled with water during the trapping cycle) to the packed column oven temperature. A heated (about 240°C) nickel transfer line leads to the 280-ft by 0.02-in.-i.d. nickel-200 needle stock capillary column; the 280-ft column is coated with a 10% OV-17 solution in chloroform after lightly etching the interior of the nickel tubing with 20% nitric acid solution and thoroughly washing with water, acetone, and chloroform.

Operating the packed 6-ft OV-17 column isothermally at 180°C with a carrier-gas flow rate of 20 ml/min, a number of electron-capturing derivatives of Δ^9-THC were compared on the basis of sensitivity, retention, ease of formation, and so on (Table 2.50). Based on this study, the heptafluorobutyrate was selected for use with blood serum extracts.

In order to determine the time interval required to trap the effluent of the packed column, a sample of Δ^9-THC-HFB was injected with the column programmed from 180 to 260°C at 2°C/min and the effluent monitored with the flame ionization detector. Using these GC conditions, the retention time of the HFB derivative of Δ^9-THC was about 13 min.

To obtain retention times of approximately 27.50 and 29.15 min for the HFB derivatives of hexahydrocannabinol (I.S.) and Δ^9-THC, the capillary column was operated isothermally at 210°C with a 3.5 ml/min argon/methane carrier-gas flow rate (total flow through ECD with purge gas: 12 ml/min).

As noted by the authors, studies of a series of human serum samples to which known amounts of Δ^9-THC had been added showed recoveries relative

TABLE 2.50

Comparison of Electron Capturing Derivatives of Δ^9-THC

R	Reagent	Sensitivity	Rel. ret. time
-COCF$_2$Cl	Chlorodifluoroacetic anhydride	1 pg	2.66
-COCF$_3$	Trifluoroacetic anhydride	5 pg	1.04
	N-Trifluoroacetyl-imidazole		
-COC$_3$F$_7$	Heptafluorobutyric anhydride	2 pg	1.00
	N-Heptafluorobutyryl-imidazole		
-COC$_6$F$_5$	Pentafluorobenzoyl chloride	1 pg	11.91
	Pentafluorobenzoyl-imidazole		
-COC$_2$F$_5$	Pentafluoropropionic anhydride	2 pg	1.00
-COC$_6$H$_4$CF$_3$	m-Trifluoromethylbenzoyl chloride	10 pg	25.52
-Si(CH$_3$)$_2$(CH$_2$Cl)	Chloromethyldimethyl-chlorosilane	100 pg	8.11
-CH$_2$C$_6$F$_5$	α-Bromopentafluoro-toluene	2 pg	11.22

From Fenimore et al. [202], courtesy of <u>Analytical Chemistry</u>.

to the internal standard to be 91.4% with a standard deviation of 0.05. The absolute recovery of Δ^9-THC through the entire procedure was 70%.

Using New Zealand white male rabbits injected with Δ^9-THC, Figure 2.51 shows the gas chromatograms obtained from blood drawn 1 and 4 hr after drug administration. It was noted that the Δ^9-THC level falls rapidly during the first 30 min and then decreases much more slowly.

In 1973, Garrett and Hunt [166] determined picogram amounts of Δ^8-THC and Δ^9-THC with CBN as internal standard in biological fluids by preparing the pentafluorobenzoate ester for analysis by gas chromatography with electron-capture detection. Following their extraction from blood and derivatization, these compounds were determined using a Varian model 2100 gas chromatograph fitted with both flame and ^{63}Ni detectors and 1.8-m by 2-mm-i.d. glass columns packed with 3% OV-225 on 100-120 mesh Gas Chrom Q. Using either detector, the GC operating conditions were detector temperature, 285°C; injector temperature, 245°C; column temperature, 210 to 230°C; nitrogen carrier-gas flow rate, 35 ml/min; standing current (^{63}Ni), 60 to 90% at 2×10^{-9}; background noise (^{63}Ni), 2 to 3% at 1×10^{-10}.

As noted by Garrett and Hunt, the procedures neither isomerized Δ^9-THC to Δ^8-THC nor the converse based on data obtained with the flame detector and a 1.8-m column when analyzing nonderivatized material (GC system and conditions the same as given for EC detection). With the 1.8-m column, Δ^8-THC, Δ^9-THC, and CBN were eluted in approximately 6.28, 7.17, and 11.25 min, respectively.

On the other hand, using their conditions and the EC detector, the chromatogram of the pentafluorobenzoyl derivatives appeared as one peak but was well separated from the peak assigned to the internal standard, CBN pentafluorobenzoate. The retention times of the PFB derivatives of Δ^8-THC/Δ^9-THC and CBN were nearly 4.05 and 6.32 min, respectively. For THC-PFB, the minimum detectable amount was 1.65×10^{-14} mole at a sensitivity setting of 1×10^{-10}.

From analyses of THC blood levels in a dog as a function of time, the resulting pharmacokinetic profile showed an initial rapid distribution and metabolism phase (apparent half-life of 7.5 min) followed by a slow loss of THC from blood (apparent half-life of 8.0 hr) and was consistent with similar results reported for humans based on radiotracer studies performed by Lemberger et al. [238].

Rosenfeld et al. [203] developed a mass fragmentographic assay for Δ^9-THC in plasma based on the chemistry of the phenolic group which is common to all cannabinoids, cannabinoidlike drugs, and their metabolites. The internal standard is Δ^9-THC perdeuteriomethyl ether (Δ^9-THC-OCD$_3$), which is easily synthesized [239]. The GC method as proposed was based on the analysis of Δ^9-THC as its O-methyl ether (Δ^9-THC-OCH$_3$), which is prepared by on-column methylation using trimethylanilinium hydroxide as reagent. The basis of the purification technique was the selective extractability of the lipid-soluble phenols from hexane by Claisen's alkali, from

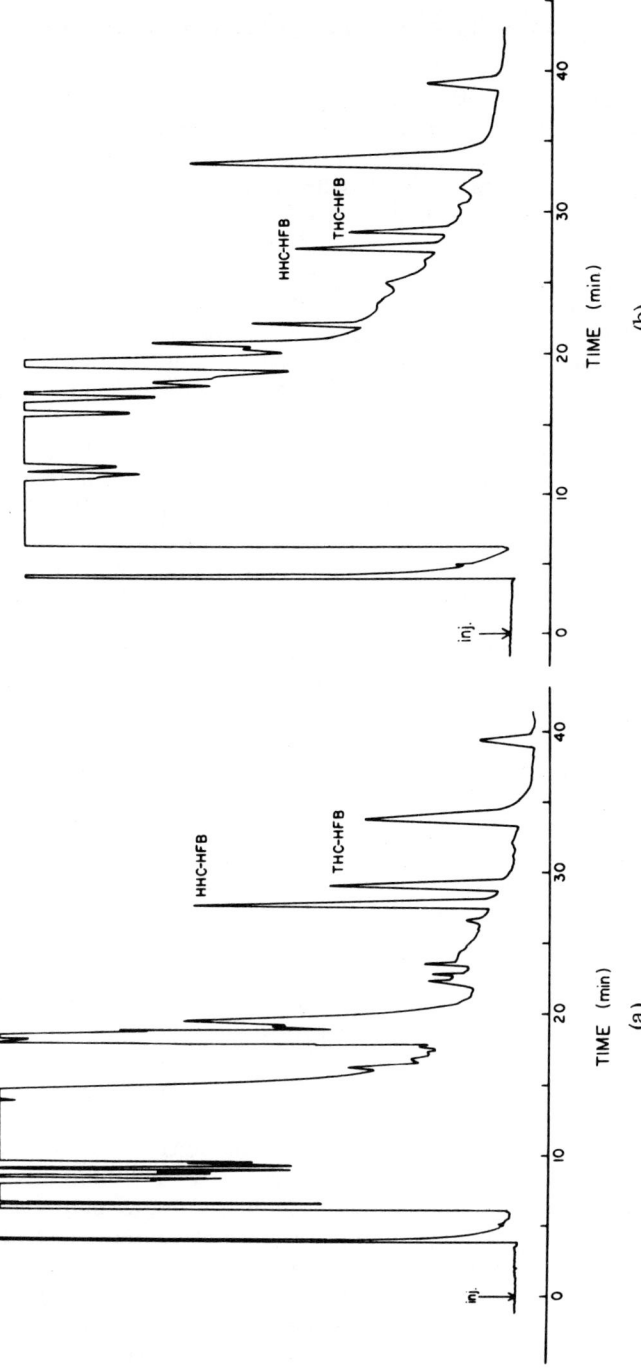

Figure 2.51. Rabbit blood serum extract (a) 1 hr and (b) 4 hr after injection of 0.1 mg/kg of Δ^9-THC with calculated Δ^9-THC serum levels in (a) and (b) being 6.5 ng/ml and 0.77 ng/ml, respectively. Adapted from Fenimore et al. [202].

which, upon reacidification, the aqueous phase was reextracted with hexane. The hexane was evaporated to dryness under a stream of nitrogen and 50 µl of solution containing 1.15 µg of internal standard was added. This was again evaporated and the residue reconstituted with 50 µl of 0.2 M TMAnH in methanol. A 1-µl aliquot of this solution was injected into the integrated GC-MS instrument for analysis.

For this study, Rosenfeld et al. performed all assays with a Varian CH 7 GC-MS instrument equipped with a 6-ft column packed with 1.5% SE-30 on 100-120 mesh Chromosorb W. The other conditions specified were column temperature, 230°C; ionizing voltage, 70 eV; emission current, 300 µA; ion source temperature, 290°C; injector temperature, 300°C; separator temperature, 300°C; carrier gas not specified. The ions monitored were the molecular ions, 328 for Δ^9-THC-OCH$_3$ and 331 for Δ^9-THC-OCD$_3$, and peak heights were used for quantitation. Chromatograms illustrating the elution of both species indicated that their retention times were approximately 2.02 min.

Using the extraction procedure outlined, recoveries of Δ^9-THC added to plasma over the range 1 to 200 ng/ml were 65% with a relative standard deviation of 5.5%.

When applied to the analysis of plasma levels of Δ^9-THC in human volunteers, data similar to that reported by Agurell et al. [201] were obtained.

In 1974, Ben-Zvi and co-workers identified 6α-hydroxytetrahydrocannabinol as a metabolic product of Δ^1-THC metabolism by mouse hepatic microsomes [205], 7-oxo-Δ^1-THC, a novel Δ^1-THC metabolite formed by rat liver microsomes [206], and cannabinol-7-oic acid in the rhesus monkey [207].

With regard to the metabolism of Δ^1-THC by mouse hepatic microsomes, the residue obtained after extraction of the incubation mixture and evaporation of the solvent was acetylated and separated by silica gel thin-layer chromatography. TLC zones 2 and 3 were identified by GC-MS based on retention time and mass spectral data as 6α-hydroxy-Δ^1-THC diacetate and 7-hydroxy-Δ^1-THC diacetate, respectively. Using the Finnigan 1015 instrument equipped with a 2-ft column packed with 2% OV-1 and the specified operating conditions (column temperature, programmed from 180 to 240°C at 8°C/min; ionizing voltage, 70 eV; injector temperature, 225°C; helium carrier gas, flow not specified), the retention times for the acetate derivatives of 6α-hydroxy-Δ^1-THC and 7-hydroxy-Δ^1-THC were listed as 7.0 and 7.5 min, respectively, whereas their respective principal m/e ions were as follows: 6α-OH-Δ^1-THC, 372, 354 (base peak), 339, 312, 297, and 295; 7-OH-Δ^1-THC, 372, 354 (base peak), 312, 297, and 259.

As for the 7-oxo-Δ^1-THC metabolite, it had been previously speculated that an intermediate aldehyde as noted below should be formed. A careful analysis of the incubation products of Δ^1-THC with rat liver microsomes revealed the presence of a hitherto unidentified metabolite. Upon further investigation, mass spectral data suggested the 7-oxo derivative which, upon reduction with LiAlH$_4$, yielded 7-OH-Δ^1-THC (thus substantiating this assignment). Such an aldehyde is a likely intermediate in the detoxification of Δ^1-THC which leads to acidic products.

Using a Finnigan GC-MS instrument, a comparison of the acetylated derivatives of 7-oxo-Δ^1-THC and 7-oxo-Δ^6-THC gave retention times of 5.7 and 5.3 min, respectively, with the 2-ft OV-1 column programmed from 180 to 235°C at 8°C/min. Using an ionizing voltage of 70 eV, the predominant m/e ions of the acetate derivatives were 7-oxo-Δ^1-THC, 370, 328, 287, 272, and 231; 7-oxo-Δ^6-THC, 370, 328, 286, 272, and 231.

Δ^1- THC Metabolism at the 7- Position

In their determination of cannabinol-7-oic acid in the rhesus monkey [207], injection of ^{14}C-Δ^1-THC into male rhesus monkeys gave rise to a complex mixture of urinary metabolites.

As noted by Ben-Zvi et al., "thin-layer chromatography on silica gel resolved the urine extract into several radioactive zones which were located by autoradiography. The most mobile zone, when analyzed by GC-MS, showed several components one of which corresponded to cannabinol-7-oic acid methyl ester-methyl ether. The mass spectrum of an authentic sample as well as its GC retention time and TLC mobility compared very favorably with those of the metabolite obtained from the urine extract. The establishment of cannabinol-7-oic acid as a transformation product of Δ^1-THC suggests the existence of a novel metabolic pathway for this cannabinoid.

Cannabinol − 7 − oic Acid Methyl Ester-Methyl Ether

Hollister et al. [208] positively identified unchanged Δ^9-THC in the urine of man following single oral doses of 30 mg in four subjects. Unchanged THC was found only in small amounts, approximately 0.010 to 0.005% of the amount administered, and for only a few hours following drug administration.

Hexane extracts of urine specimens were subjected to TLC analysis using a solvent system of petroleum ether:diethyl ether (4:1) after hydrolysis with β-glucuronidase/aryl sulfatase at pH 5.5. The sample for GC-MS assay was prepared in like manner using an aliquot of urine containing 100 mg of creatinine and the silica gel area on the TLC plate of R_f value corresponding to Δ^9-THC was eluted with ethanol. The eluate was then examined by GC-MS for the presence of m/e ions 314 and 231 using a Hewlett-Packard model 5930A instrument. Although no GC-MS operating conditions were given, superimposed single-ion chromatograms (m/e 314 and 231) indicated that Δ^9-THC had a retention time of about 17 min.

In 1975, Burstein and Varanelli [210] studied the in vivo transformations of cannabinol (CBN) in the mouse and established a pattern of metabolism.

Following the administration of ^{14}C-CBN subcutaneously to three male mice and the extraction of the metabolites from the urine and feces by procedures similar to those described by Ben-Zvi et al. [240], thin-layer chromatography was employed to resolve the crude fractions into semipurified products. The neutral fractions were acetylated with acetic anhydride in pyridine prior to chromatography, whereas the acidic fractions were methylated via conditions which permitted both the carboxyl and phenolic groups to be derivatized. For GC-MS analysis, individual TLC zones were eluted with ethanol. GC-MS analyses were performed with a Finnigan model 1015 at 70 eV using a 2-ft OV-1 column, column temperature, 180 to 240°C at 8°C/min, with helium as the carrier gas.

In Tables 2.51 and 2.52 are listed the principal m/e ions for the acetylated (Table 2.51) and methyl ester-methyl ether (Table 2.52) derivatives of the fecal metabolites of CBN formed by the metabolic pathways illustrated in Figure 2.52. It was noted that metabolites with structures such as compound F had been found in the rat and rabbit.

Widman et al. [211] studied the metabolism of Δ^1-THC in the isolated perfused dog lung. After intravascular administration of ^{3}H-Δ^1-THC, there was an overall biotransformation of 12%. Two major metabolites were isolated and identified as 3"-hydroxy-Δ^1-THC and 4"-hydroxy-Δ^1-THC

TABLE 2.51

Prominent m/e Ions of Acetylated Components in Neutral
Fraction of Fecal Extract[a]

Metabolite[b]	m/e Ions[c]
A (diacetate)	410 (s), 395 (m), 380 (h), 353 (h), 337 (h), 307 (h), 295 (m)
C (triacetate)	468 (s), 453 (m), 411 (m), 351 (m), 306 (s), 291 (m)
D (triacetate)	468 (s), 453 (m), 411 (m), 351 (m), 306 (s), 291 (m)

[a] Adapted from Burstein and Varanelli [210].
[b] Metabolites identified in Figure 2.52.
[c] s = small intensity, m = medium intensity, h = high intensity.

TABLE 2.52

Prominent m/e Ions of Methyl Ester-Methyl Ether Components
in Acidic Fraction of Fecal Extract[a]

Metabolite[b]	m/e Ions[c]
B (methyl ester-methyl ether)	368 (h), 297 (s), 281 (m), 253 (s)
E (methyl ester-methyl ether)	384 (s), 369 (h), 296 (m)
E (methyl ester-methyl ether-acetate)	426 (s), 411 (h), 395 (s), 367 (s), 351 (m), 309 (s), 296 (m), 295 (m)
G (methyl ester-methyl ether)	312 (h), 297 (s), 281 (m), 253 (s), 180 (m)

[a] Adapted from Burstein and Varanelli [210].
[b] Metabolites identified in Figure 2.52.
[c] s = small intensity, m = medium intensity, h = high intensity.

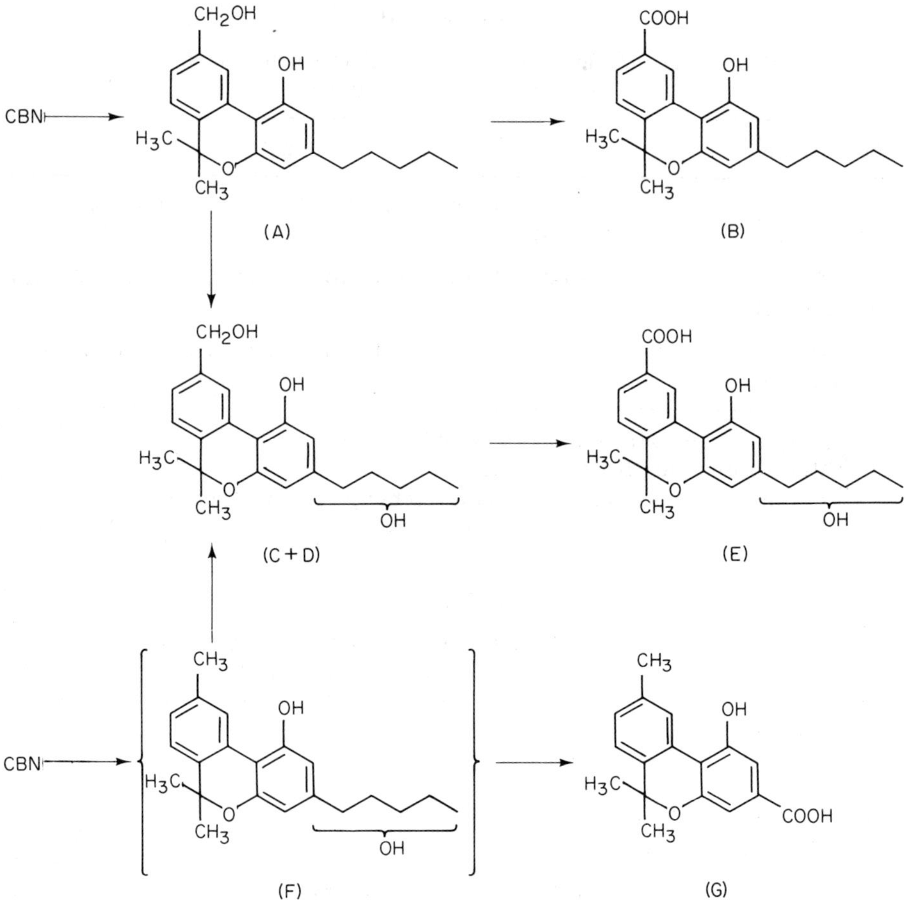

Figure 2.52. Metabolic transformations of CBN in the mouse.

(structures given in Table 2.1). 7-Hydroxy-Δ^1-THC was also present together with small amounts of 6α-OH-Δ^1-THC and 6β-OH-Δ^1-THC. An in vitro experiment using a dog liver microsomal preparation was also carried out and showed that the major metabolites were 6α-OH-Δ^1-THC and 6β-OH-Δ^1-THC. However, 7-OH-Δ^1-THC and 1,2-epoxyhexahydrocannabinol (or 9,10-epoxyhexahydrocannabinol, structure given in Table 2.1) were also isolated together with small amounts of 3"-OH-Δ^1-THC and 4"-OH-Δ^1-THC. The side-chain hydroxylated compounds were hitherto undescribed metabolites of Δ^1-THC.

Describing detailed procedures for the metabolism of Δ^1-THC by isolated perfused dog lung and dog liver supernatant and their extraction and subsequent purification prior to analysis by thin-layer chromatography, liquid scintillation counting, and integrated gas chromatography-mass spectrometry, conventional gas chromatography was performed on a column packed with 2% SE-30 on Gas Chrom Q at 230 and 250°C, whereas GC-MS studies were carried out with a LKB 9000 equipped with the above column held isothermally at 210°C. Mass spectra were obtained of the side-chain hydroxylated metabolites as their TMSi derivatives using an ionizing voltage of 70 eV.

The analytical data obtained for lung plasma metabolites pertaining to the 3"-OH-Δ^1-THC and 4"-OH-Δ^1-THC compounds were as follows:

1. 3"-OH-Δ^1-THC: GC, retention time of 0.78 relative to 7-OH-Δ^1-THC; MS, m/e 330 (M^+), 315, 258 (base peak), 247, 243, and 190.

2. 4"-OH-Δ^1-THC: GC, retention time of 0.80 relative to 7-OH-Δ^1-THC; MS, m/e 330 (M^+), 315, 287, 258, 257, 247 (base peak) and 243; MS of TMSi derivative, m/e 474 (M^+, base peak), 459, 391, 330, 315, and 117.

Liver metabolites isolated in various fractions via Sephadex LH-20 chromatography were identified as follows:

1. Fraction 9: 1,2-epoxy-Δ^1-THC; GC, retention time of 0.61 relative to 7-OH-Δ^1-THC (silylated compounds); MS, m/e 330 (M^+), 315, 312, 297, 287, 274, 259 (base peak), 246, 231, and 193.

2. Fraction 10: 6β-OH-Δ^1-THC; GC, retention time of 0.71 relative to 7-OH-Δ^1-THC; MS, m/e 330 (M^+), 312, 297 (100), 295, 271, 257, 231, and 214.

3. Fraction 13: 3"-OH-Δ^1-THC and 4"-OH-Δ^1-THC

4. Fraction 14: 6α-OH-Δ^1-THC, 6β-OH-Δ^1-THC, and 7-OH-Δ^1-THC; all identified by TLC, GC, and MS comparison; 6α-OH-Δ^1-THC had a retention time of 0.73 relative to 7-OH-Δ^1-THC.

With regard to the side-chain hydroxylated metabolites, 1"-OH-Δ^1-THC, 2"-OH-Δ^1-THC, 3"-OH-Δ^1-THC, and 5"-OH-Δ^1-THC were synthesized which, when subjected to GC and GC-MS analysis, gave retention times of 10.30, 9.85, 11.45, and 15.21 min, respectively.

In 1976, Rosenfeld and Taguchi [212] presented a mass fragmentographic assay for 7-OH-Δ^1-THC from plasma, the procedure reportedly being sensitive to 3 ng/ml of plasma with a relative standard deviation of 4%. The technique relies on the derivatization of the phenol moiety by extractive alkylation, this reaction being novel to the cannabinoid series. With this technique, it was possible to monitor 7-OH-Δ^1-THC in dog plasma for 2 hr after its administration. However, no 7-OH-Δ^1-THC was detected in dog plasma after oral or intravenous administration of Δ^1-THC.

The procedure employed for the preparation of the dog plasma for subsequent analysis by GC-MS was as follows:

The plasma assay was initiated by adding 10 ml CH_2Cl_2 to 1 ml of plasma, shaking vigorously for 5 min, and then centrifuging for 15 min at 2500 rpm. Three phases resulted: the bottom organic phase, the top aqueous phase, and a protein cake at the interface. The tubes were removed carefully from the centrifuge to prevent rupture of the protein cake. The top layer was discarded. In order to coagulate the protein cake, a large spatula of sodium chloride was added and the tube was shaken for 5 sec. After the precipitate settled, the CH_2Cl_2 was transferred to a new tube containing 20 ng of the deuterated internal (1-OC_2D_5-7-OH-Δ^1THC) solution. At this stage, the sample could be stored overnight in a refrigerator if necessary. Then, sufficient C_2H_5I (0.4 ml) to make the CH_2Cl_2 phase 0.5 M in C_2H_5I was dissolved in the extract. After the addition of 5 ml of 0.1 N NaOH and 100 µl of 0.1 M tetrahexylammonium hydroxide/methanol, the tube was shaken for 5 min. The Florisil columns were prepared and the chromatography of the organic phase was performed as before. The ether wash from the column, which could be stored overnight, was concentrated under nitrogen and then transferred to a silanized 5-ml Reacti-Vial. The samples were evaporated to dryness under nitrogen and then stored under argon in a refrigerator until analysis. At that time, each sample was dissolved in 20 µl of a 90% bis-(trimethylsilyl)trifluoroacetamide/10% trimethylchlorosilane solution. At room temperature, the reaction was complete within 5 min. A 1- to 3-µl aliquot was injected into the GC-MS instrument for analysis.

The determination of 7-OH-Δ^1-THC was carried out with a Varian CH-7 or CH-5 GC-MS instrument equipped with an accelerating voltage alternator for multiple ion detection and a coiled 6-ft column packed with 1.5% SE-30 on 80-100 mesh Chromosorb W-HP or 1.5% SE-52 on 80-100 mesh Chromosorb W-HP. The operating conditions given were ionizing voltage, 70 eV; emission current, 300 µA; ion source temperature, 290°C; injection temperature, 300°C; separator temperature, 300°C; column temperature and carrier gas not specified. Ions monitored were 327 for 1-OC_2H_5-7-TMSi-Δ^1-THC and 332 for 1-OC_2D_5-7-TMSi-Δ^1-THC.

Also in 1976, Harvey and Paton [213] characterized 7-carboxy-Δ^1-THC, together with its 2"-, 3"-, and 6α-monohydroxy and 2",6α-dihydroxy derivatives using integrated GC-MS in organic extracts of mouse liver following large doses of Δ^1-THC; their structures are shown below:

7-Carboxy-Δ^1-THC Related Compounds

7-Carboxy-Δ^1-THC Related Compounds

Compound	R_1	R_2	R_3
7-Carboxy-Δ^1-THC	H	H	H
6α-HO-7-carboxy-Δ^1-THC	OH	H	H
2"-Hydroxy-7-carboxy-Δ^1-THC	H	OH	H
3"-Hydroxy-7-carboxy-Δ^1-THC	H	H	OH
6α,2"-Dihydroxy-7-carboxy-Δ^1-THC	OH	OH	H
6α,3"-Dihydroxy-7-carboxy-Δ^1-THC	OH	H	OH

Following the extraction of the metabolite content of the liver and its separation into four fractions using a Sephadex column, the solvent was removed from each fraction and the residues were dissolved in ethyl acetate. Aliquots (0.1 ml) of these solutions were converted into derivatives for analysis by GC-MS. Silyl derivatives were prepared with either bis-(trimethylsilyl)trifluoroacetamide or d_{18}-bis-(trimethylsilyl)acetamide, each with a trace of trimethylchlorosilane. The deuterated reagent was used solely for mass spectral studies, as were the methyl ester (diazomethane)-TMSi and ethyl ester (diazoethane)-TMSi derivatives.

Conventional GC data was obtained with a Varian 2400 gas chromatograph equipped with a flame ionization detector and 2-m by 2-mm-i.d. glass columns packed with 3% SE-30 on 100-120 mesh Gas Chrom Q. Methylene unit values were determined with the following GC operating conditions: nitrogen carrier-gas flow rate, 30 ml/min; column temperature, programmed from 170 to 300°C at 4°C/min.

Mass spectra were recorded with a Varian 2400 gas chromatograph interfaced via a jet separator to a VG Micromass 12B mass spectrometer. Using the same column as that for conventional gas chromatography, the GC-MS conditions employed were column temperature, programmed from 170 to 280°C at 2°C/min; accelerating voltage, 2.5 kV; electron energy,

25 eV; ion source temperature, 260°C; scan rate, 3 sec/decade with a 2-sec interscan delay; ionizing current, 100 µA.

In Table 2.53 are listed the methylene unit values for the metabolites identified in the liver extracts as well as some of the predominant m/e ions observed in the mass spectra of these TMSi derivatives.

Martin et al. [215] reported the identification of new in vivo side-chain acid metabolites of Δ^1-THC formed by different species (guinea pig, mice, and rabbit). Following their extraction from liver homogenates and subsequent fractionation on a Sephadex LH-20 column, samples and reference standards were converted into trimethylsilyl (TMSi), deuterated-TMSi (d_9-TMSi), and methyl ester-phenolic TMSi (Me/TMSi) derivatives for GC-MS analysis as described by Harvey and Paton [214]. GC and mass spectral data were obtained with a VG Micromass 12B mass spectrometer interfaced with a Varian 2400 gas chromatograph, equipped and operated as previously noted above [213].

Martin et al. noted that analysis of the TMSi derivatives of the 10% methanol-chloroform fraction from the guinea pig showed the presence of three unidentified acid metabolites whose retention times were much shorter than that of the TMSi derivative of Δ^1-THC-7-oic acid (RT = 25.67 min).

1. Metabolite 1: This had a retention time of 20.17 min and a molecular ion at m/e 460 (d_9-TMSi, 478, which indicated two TMSi groups). Methylation with diazomethane followed by TMSi formation reduced the M⁺ to m/e 402 (indicative of the presence of a carboxylic acid group). To this metabolite the structure of 4",5"-bis-nor-Δ^1-THC-3"-oic acid (3"-acid) was assigned.

4",5"– Bisnor –Δ^1– THC – 3"– oic Acid
(3"–Acid)

2. Metabolite 2: The TMSi derivative had a retention time of 16.42 min and a M⁺ ion at 446. The Me/TMSi derivative had a M⁺ at m/e 388. The mass spectra of its TMSi and d_9-TMSi derivatives were similar to those of the 3"-acid, but containing one less methylene group. This compound was identified as 3",4",5"-trisnor-Δ^1-THC-2"-oic acid (2"-acid).

TABLE 2.53

Methylene Unit Values and Prominent m/e Ions of Silylated Metabolites[a]

Compound	MU value	m/e Ions[b]
6α-OH-Δ^1-THC	25.80	474 (0) (M$^+$), 459 (5), 384 (100)
7-OH-Δ^1-THC	26.20	474 (6) (M$^+$), 459 (5), 371 (100)
6α,7-Dihydroxy-Δ^1-THC	29.10	562 (0) (M$^+$), 547 (3), 472 (100), 459 (18)
7-Carboxy-Δ^1-THC	29.55	488 (43) (M$^+$), 473 (40), 371 (100)
6α-Hydroxy-7-carboxy-Δ^1-THC	30.25	576 (0) (M$^+$), 561 (3), 486 (100), 459 (0)
2″-Hydroxy-7-carboxy-Δ^1-THC	30.95	576 (10) (M$^+$), 561 (8), 459 (5), 504 (31), 145 (100)
3″-Hydroxy-7-carboxy-Δ^1-THC	c	576 (19) (M$^+$), 561 (21), 459 (39), 432 (100)
6α,2″-Dihydroxy-7-carboxy-Δ^1-THC	31.60	664 (0) (M$^+$), 649 (5), 574 (6), 547 (6), 502 (21), 145 (100)
6α,3″-Dihydroxy-7-carboxy-Δ^1-THC	32.65	664 (0) (M$^+$), 649 (7), 574 (100), 549 (2), 430 (32)

[a] Adapted from Harvey and Paton [213].
[b] Relative intensities given in parentheses.
[c] Component of multiple peak.

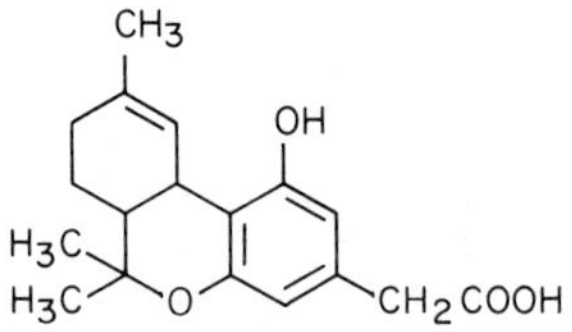

3″,4″,5″−Trisnor−Δ¹−THC−2″−oic Acid
(2″−Acid)

3. Metabolite 3: Mass spectral data of the three derivatives (TMSi, d₉-TMSi, and Me/TMSi) of the third metabolite indicated that it contained a side-chain acid group but that it contained one more methylene group than the 3″-acid. The TMSi derivative had a retention time of 23.10 min and a M⁺ ion at 474. When treated with deuterated TMSi reagent, the M⁺ ion appeared at m/e 492, whereas its Me/TMSi derivative yielded a M⁺ ion at 416. Hence, it was assigned the structure of 5″-nor-Δ^1-THC-4″-oic acid (4″-acid).

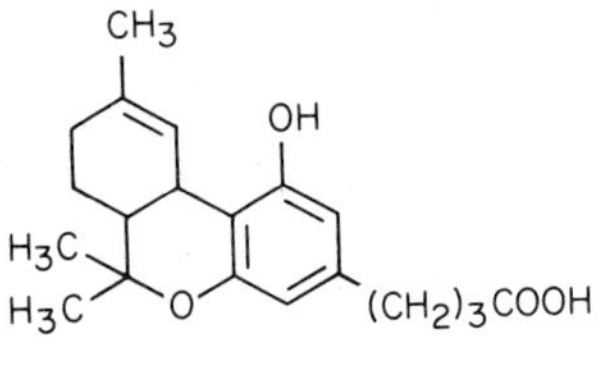

5″−Nor−Δ¹−THC−4″−oic Acid
(4″−Acid)

In 1976, Leighty et al. [216] identified long-retained metabolites of Δ^1-THC and Δ^6-THC previously detected in the liver, spleen, fat, and bone marrow of the rat after I.V. or I.P. injections of ^{14}C-Δ^1-THC or ^{14}C-Δ^6-THC as fatty acid conjugates of 7-OH-Δ^1-THC and 7-OH-Δ^6-THC. Characterization of the metabolites was facilitated by their large-scale in vitro production from 7-OH-Δ^1-THC and 7-OH-Δ^6-THC with a rat-liver microsomal enzyme system. The nonpolar metabolic fraction isolated from the in vitro system by TLC and HPLC was shown by spectrometric analysis to be the 7-acyloxy derivative of 7-OH-Δ^1-THC and 7-OH-Δ^6-THC; these structures are shown below.

7 − Acyloxy − Δ^1 − THC   7 − Acyloxy − Δ^6 − THC

In this investigation, conventional GC retention data were obtained as well as EI and CI mass spectra. To separate the 7-acyloxy metabolites of Δ^1-THC and Δ^6-THC, TMSi derivatives were prepared by reacting the purified metabolite mixture with an excess of bis-(trimethylsilyl)trifluoroacetamide at 70°C for 1 hr. These TMSi derivatives were chromatographed with a Finnigan 9500 GC unit equipped with a 0.6-m by 2-mm glass column packed with 1% Dexsil on 80-100 mesh Gas Chrom Q and a flame ionization detector. With the specified operating conditions [injector temperature, 300°C; column temperature, programmed from 250 to 320°C (heating rate not specified); detector temperature, 320°C; helium carrier-gas flow rate, 30 ml/min], the TMSi metabolites gave two peaks corresponding to retention times of 7.0 and 8.3 min.

For GC-MS investigations, the above GC column was connected to either the EI mass spectrometer via a glass jet separator, or to the CI mass spectrometer via a 1/8-in.-o.d. stainless steel tube.

The CI mass spectra were obtained with a Finnigan 3200 quadrupole mass spectrometer using a reagent gas consisting of a mixture of methane and ammonia (8:1) at a total ion source pressure of approximately 0.5 torr. Typical settings were ion source temperature, 160°C; ion energy, 15 V; lens voltage, -70 V; ion repeller, 0 V; filament emission, 1 μA; electron energy, 200 eV; electron voltage multiplier, 1.8 kV; GC and MS connecting lines, 300°C. For CI$_{NH_3}$ mass spectrometry, methane was used as the carrier gas and ammonia was bled into the ion source via a make-up gas inlet.

On the other hand, EI mass spectra were obtained from a Finnigan 1015 instrument operated under the same conditions except that the EI ion source was not heated and the ion repeller voltage was set at 30 V and the electron energy maintained at 70 eV.

The results of an EI-MS study of underivatized conjugates isolated from the in vitro system using 7-OH-Δ^6-THC indicated that "the most abundant molecular ion in the EI mass spectrum occurs at m/e 568 and has an elemental composition of $C_{37}H_{60}O_4$ based on mass measurement at high resolution (AEI MS-9 mass spectrometer; measured, 568.4490; calculated for $C_{37}H_{60}O_4$, 568.4492). This composition is consistent with esterification of 7-OH-Δ^6-THC with palmitic acid. The presence of additional molecular ions at m/e 596, 594, and 592 suggests that the enzymatic esterification is nonspecific, involving fatty acids including stearic and both mono- and diunsaturated C_{18}-fatty acids."

The CI$_{NH_3}$ spectra of the 7.0-min eluting peak showed a single protonated molecular ion at m/e 641 consistent with the structure of 7-palmitoyloxy-Δ^6-THC. They further noted that "in the EI mass spectrum obtained on the same GC peak all of the prominent fragment ions are shifted 72 amu from those of the corresponding unsilylated metabolite, consistent with attachment of a TMSi group to the phenolic hydroxyl of the cannabinoid nucleus. The second GC peak (RT = 8.3 min) is slightly broader and consists of an unresolved mixture of TMSi metabolites (molecular weights of 664, 666, and 668) containing C_{18}-fatty acids."

The structure of 7-palmitoyloxy-Δ^6-THC is shown below:

7 – Palmitoyloxy – Δ^6 – THC

With Δ^1-THC, similar palmitic/stearic conjugates were found.

In two related papers, Martin and co-workers [217,218] identified the monohydroxylated [217] and dioxygenated [218] metabolites of cannabidiol (CBD) formed by rat liver.

With regard to the identification of the monohydroxylated metabolites of cannabidiol whose structures are illustrated in Figure 2.53, cannabidiol was metabolized in vitro by rat liver enzymes. Following detailed procedures for their isolation from the incubation mixtures, the extracts were separated into fractions using Sephadex LH-20 columns from which the metabolites were resolved by thin-layer chromatography. In turn, the isolated metabolites were eluted from the TLC plate and examined by GC and GC-MS. Mass spectral data of both silylated and nonsilylated compounds were essential in determining the exact position or location of hydroxylation.

Retention time data were obtained with a Varian Aerograph model 2100 gas chromatograph equipped with a flame ionization detector and 1.8-m by 2-mm glass columns packed with 2% SE-30 on Gas Chrom Q. The other

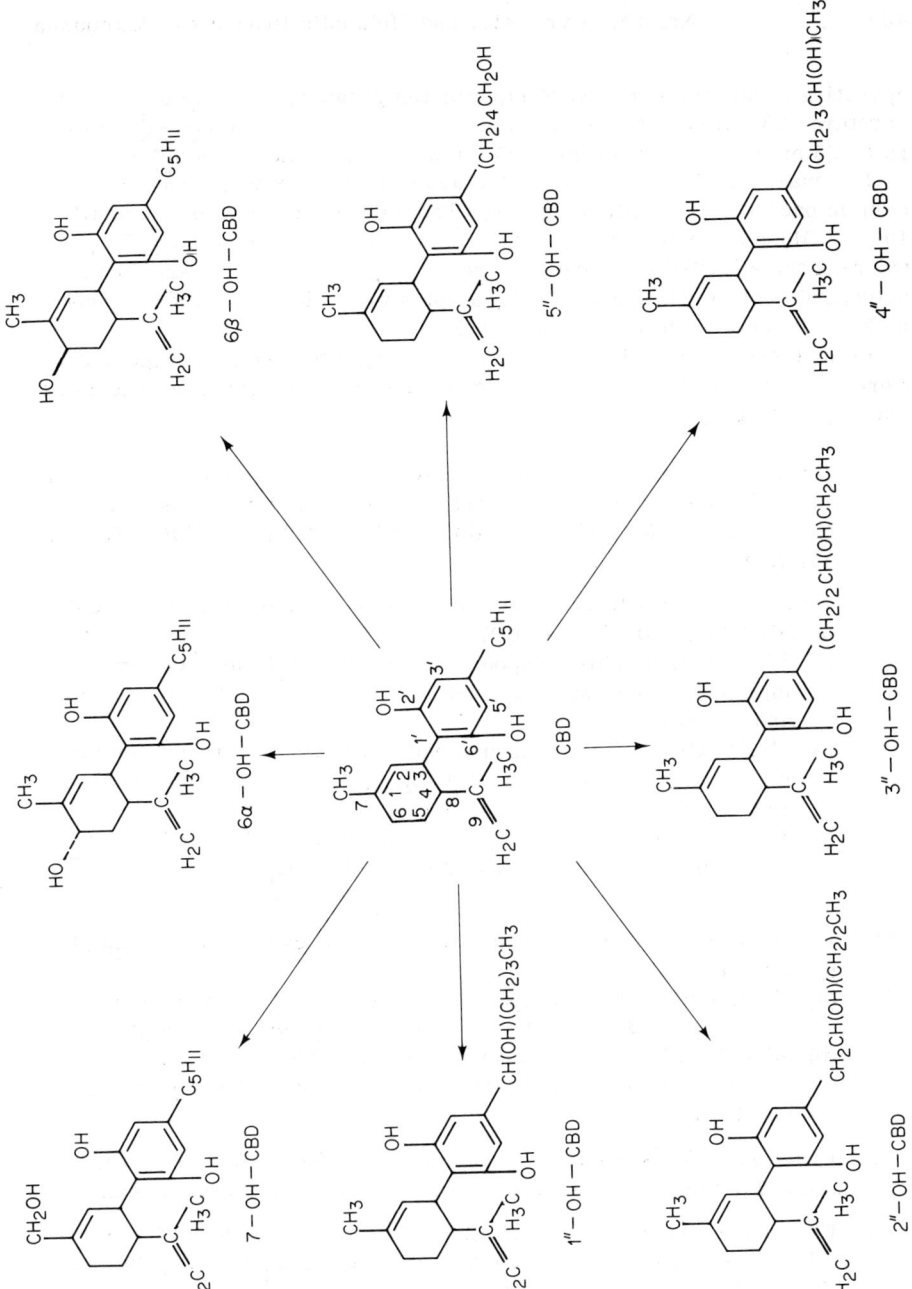

Figure 2.53. In vitro monohydroxylated metabolites of CBD. Adapted from Martin et al. [217].

operating conditions listed were injector temperature, 270°C; column temperature, 250°C; detector temperature, 270°C; nitrogen carrier-gas flow rate, 25 ml/min; hydrogen flow, 25 ml/min; oxygen flow, 200 ml/min.

For mass spectral studies, a LKB 9000 GC-MS was employed with a column packed with 3% SE-30 on Gas Chrom Q and operated isothermally at 190°C. Mass spectra of nonderivatized samples were recorded at 70 eV, whereas the silylated derivatives were recorded at 20 eV. It should be noted that, before GC analysis, all samples were silylated with bis-(trimethylsilyl)acetamide in dry acetonitrile.

Using these analytical techniques, the metabolites shown in Figure 2.53 were identified in five of the ten fractions obtained with Sephadex LH-20 column chromatography.

1. Fraction 3: Predominantly 2"-OH-CBD with a retention time of 7.10 min. Its mass spectrum showed m/e ions (relative intensity in parenthesis) at 330 (M^+, 33), 262 (45), 258 (18), 247 (100), 190 (27), and 175 (27).

2. Fraction 4: This fraction contained three metabolites, 1"-OH-CBD, 3"-OH-CBD, and 4"-OH-CBD.
   (a) 1"-OH-CBD: This compound had a RT of 7.40 min with major m/e ions at 330 (M^+, 12), 262 (32), 247 (100), 206 (8), 205 (19), and 159 (10).
   (b) 3"-OH-CBD: This compound had a RT of 8.10 min with major m/e ions at 330 (M^+, 27), 301 (5), 258 (48), 247 (100), 242 (16), and 190 (55).
   (c) 4"-OH-CBD: This compound had a RT of 8.25 min with major m/e ions at 330 (M^+, 24), 262 (45), 247 (100), 209 (18), and 122 (25).

3. Fraction 5: This fraction contained metabolites 5"-OH-CBD and 6β-OH-CBD.
   (a) 5"-OH-CBD: This compound had a RT of 11.05 min with major m/e ions at 330 (M^+, 29), 262 (45), 247 (100), and 209 (20).
   (b) 6β-OH-CBD: This compound had a RT of 7.80 min with major m/e ions at 330 (M^+, 0), 312 (77), 297 (25), 257 (100), 244 (21), and 193 (85).

4. Fraction 7: This fraction contained 7-OH-CBD with a RT of 10.75 min. Its mass spectral pattern was similar to an authentic sample.

5. Fraction 8: This fraction contained 6α-OH-CBD with a RT of 8.00 min; its major m/e ions were 330 (M^+, 4), 312 (45), 262 (100), 257 (19), 233 (39), and 193 (55).

On the other hand, mass spectral fragments characteristic of the TMSi derivatives of these metabolites were as follows:

1. 1"-OH-CBD-TMSi: 546 (15), 478 (62), 425 (47), 421 (100), 334 (2), 268 (2), 244 (2), 159 (10), 117 (4)

2. 2"-OH-CBD-TMSi: 546 (4), 478 (9), 425 (7), 334 (4), 145 (100)

3. 3"-OH-CBD-TMSi: 546 (21), 478 (100), 425 (53), 402 (16), 334 (67), 268 (48), 244 (35), 117 (14)

4. 4"-OH-CBD-TMSi: 546 (14), 478 (100), 425 (40), 334 (2), 268 (3), 244 (7), 117 (25)

5. 5"-OH-CBD-TMSi: 546 (13), 478 (100), 425 (51), 334 (2), 268 (5), 244 (8), 117 (2)

6. 6β-OH-CBD-TMSi: 546 (1), 478 (100), 337 (5)

7. 6α-OH-CBD-TMSi: 546 (1), 478 (100), 338 (3), 337 (10), 268 (1)

8. 7-OH-CBD-TMSi: 546 (11), 478 (77), 443 (100), 337 (45)

As for the dioxygenated CBD metabolites, the metabolism of CBD was studied in vitro by Martin et al. [218] using a 10,000-g supernatant from rat liver. After removal of unchanged CBD and its monohydroxylated metabolites, a polar fraction remained from which 10 dioxygenated metabolites were isolated. Mass spectrometry, integrated GC-MS, TLC, and nuclear magnetic resonance spectroscopy were used to identify the following metabolites (structures of all shown in Table 2.1): 6,7-dihydroxy-CBD, 1",7-dihydroxy-CBD, 3",7-dihydroxy-CBD, 4",7-dihydroxy-CBD, 5",7-dihydroxy-CBD, 2",6-dihydroxy-CBD, 3"-6β-dihydroxy-CBD, 4",6β-dihydroxy-CBD (tentative), 3"-hydroxy-6-oxo-CBD, and 4"-hydroxy-6-oxo-CBD.

In this investigation, the GC and GC-MS instrumentation, columns, and operating conditions were the same as those used to identify the monohydroxylated metabolites [217] of CBD. In Table 2.54 are listed the retention times of the TMSi derivatives on the 2% SE-30 column (250°C), as well as their molecular ions, base peaks (m/e), and major fragmentation ions formed with an electron energy of 20 eV.

In 1977, Rosenfeld [220] developed a GC-MS method for the simultaneous determination of Δ^9-THC and 11-hydroxy-Δ^9-THC in plasma. For the determination of both compounds, Rosenfeld used a gas chromatograph interfaced via a Watson-Biemann separator with a Varian-Mat CH-7 equipped with a two-channel accelerating-voltage alternator. The GC separations were performed on a 1.83-m by 2-mm-i.d. column packed with 3% OV-225 on 80-100 mesh Chromosorb W-HP. The other parameters for GC and MS were injector temperature, 300°C; column temperature, 220°C; separator temperature, 300°C; helium carrier-gas flow rate, 12 ml/min; ionizing voltage, 70 eV; emission current, 300 μA; ion source temperature, 300°C.

TABLE 2.54

Retention Times and Major Fragmentation Ions of TMSi Derivatives of Dioxygenated Metabolites of CBD[a]

		m/e Ions		
Metabolite	Ret. time (min)	M^+	Base peak	Major ions
6,7-diOH-CBD	3.8	634	566	478, 337, 103
1″,7-diOH-CBD	3.0	634	531	509, 425, 159
3″,7-diOH-CBD	5.6	634	531	490, 425, 268
4″,7-diOH-CBD	6.1	634	531	425, 117
5″,7-diOH-CBD	7.2	634	531	425
2″,6-diOH-CBD	2.7	634	566	425, 145
3″,6β-diOH-CBD	5.7	634	566	425, 268
4″,6β-diOH-CBD	6.0	634	566	425, 117
3″-OH-6-oxo-CBD	4.7	560	425	492, 416, 268
4″-OH-6-oxo-CBD	5.1	560	425	492, 117

[a] Adapted from Martin et al. [218].

For the determination of optimum derivatization conditions, the following studies were carried out:

A series of derivatizations were carried out with varying concentrations of base. In all cases, 100 ng of cannabinoid were dissolved in 5 ml of 0.5 M ethyl iodode in methylene chloride. Ten micromoles of tetrahexylammonium hydroxide were added followed by (except in one case) 5 ml of aqueous phase. The composition of the aqueous phases in the series was water, 0.1 N NaOH, 1 N NaOH, and 5 N NaOH. The resulting mixtures were shaken (15 min at high speed on an Eberbach shaker) and centrifuged (at 600 × G on a MSE minor centrifuge). The aqueous layers were aspirated and 82 ng of 1-0-perdeuterioethyl-Δ^9-THC and 100 ng of 1-0-perdeuterioethyl-11-hydroxy-Δ^9-THC were added to the organic phases as external standards. The organic solutions were dried with Na_2SO_4 and the excess THAH and tetrahexylammonium iodide by-product were removed by adsorption on a spatula full of Florisil, which was added directly to the tube. The solution was decanted and the methylene chloride evaporated to dryness under a stream of nitrogen accompanied by warming (50°C) in a sandbath. The residue was dissolved in 30 to 35 µl of bis-(trimethylsilyl)trifluoroacetamide/trimethylchlorosilane (9:1) and this solution was used directly for GC-MS analysis. The mass spectrometer was focused at m/e 327 for the detection of derivatized cannabinoids and m/e 332 for detection of deuterated standards. With the GC-MS conditions specified, the retention times for 1-0-ethyl-Δ^9-THC and 1-0-ethyl-11-TMSi-Δ^9-THC were approximately 1.82 and 3.73 min, respectively, as shown in Figure 2.54.

For the determination of cannabinoids in plasma, the recommended procedure consisted of the following:

Five milliliters of toluene were added to 1 ml of plasma and the mixture was shaken for 15 min on an Eberbach shaker at high speed and centrifuged at 600 × G for 5 min. The toluene was transferred to a tube containing 1 ml of Claisen's alkali and the mixture was shaken

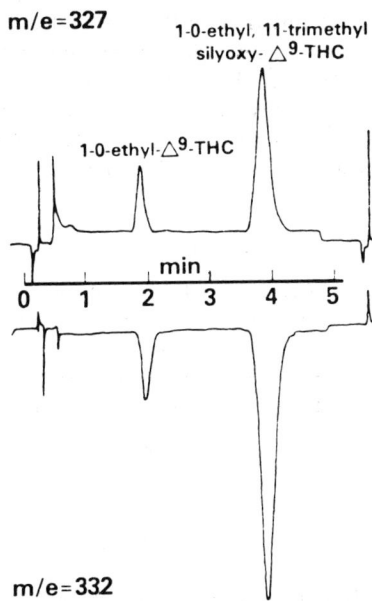

Figure 2.54. Selected ion monitor traces for detection of 1-0-ethyl-Δ^9-THC and 1-0-ethyl, 11-trimethylsilyloxy-Δ^9-THC. From Rosenfeld [220], courtesy of Analytical Letters.

and centrifuged as described above. The toluene was aspirated and the Claisen's alkali was diluted with 4 ml of water. Five milliliters of 0.5 M ethyl iodide in methylene chloride and 10 μmoles of THAH were added and the biphasic mixture subjected to the procedure described for the determination of optimum derivatization conditions.

The extraction efficiencies of Δ^9-THC from water and plasma and 11-OH-Δ^9-THC from plasma using methylene chloride, diethyl ether, and toluene are listed in Table 2.55; the data in Table 2.56 show that the reaction requires the presence of water, but for complete derivatization to occur, it is necessary that base be present. The requirement for the base indicates that the derivatization to some extent involves an alkylative extraction.

As noted by Rosenfeld, the standard curves were linear from 0 to 200 ng/ml of cannabinoid in plasma. There was an interference corresponding to 3 ng/ml of Δ^9-THC to the determination of that compound. There was no interference to the determination of the metabolite. At a concentration of 7 ng/ml of cannabinoid, the relative standard deviation of the determination of Δ^9-THC was 20%, and for 11-OH-Δ^9-THC, it was 4%. Recovery of Δ^9-THC from plasma was 42%, and the recovery of 11-OH-Δ^9-THC was 83%.

The analytical method was tested in an animal study and it was shown that, after IV injection of 30 μg/kg of 11-OH-Δ^9-THC, this compound could be monitored for 3 hr, at which time the limit of detection was approached (Fig. 2.55). Furthermore, it was demonstrated that after injection of Δ^9-THC in rabbit, 11-OH-Δ^9-THC is formed and is present in the circulation. However, despite the relatively high concentration of Δ^9-THC that was observed, the maximum concentration of the metabolite obtained was less than 10 ng/ml, and the metabolite rapidly disappeared from the plasma.

TABLE 2.55

Extraction Efficiencies[a]

Solvent	Water	Plasma
Δ^9-THC from Plasma (%)		
CH_2Cl_2	100	40
Et_2O	100	48
C_7H_8	83	88
11-Hydroxy-Δ^9-Tetrahydrocannabinol		
C_7H_8		96

[a] From Rosenfeld [220], courtesy of <u>Analytical Letters</u>.

TABLE 2.56

Percent Completion of Reaction[a]

Aqueous phase	Δ^9-THC	11-Hydroxy-Δ^9-THC
None	0	0
H$_2$O	46	76
0.1 N NaOH	100	100
1 N NaOH	100	100
5 N NaOH	100	100

[a] From Rosenfeld [220], courtesy of <u>Analytical Letters</u>.

Harvey et al. [225] examined with a GC-MS-COMP system in vivo liver metabolites of Δ^1-THC as trimethylsilyl (TMSi), d$_9$-TMSi, and methyloxime-TMSi derivatives. In addition to the reported monohydroxy, acid, and hydroxyacid metabolites, the following multiple-substituted metabolites were identified: 2",7-, 3",7-, and 6β,7-dihydroxy-Δ^1-THC; 2",6α,7-, and 3",6α,7-trihydroxy-Δ^1-THC; 2"-, 3"-, and 7-hydroxy-6-oxo-Δ^1-THC, and 2",7- and 3",7-dihydroxy-6-oxo-Δ^1-THC. The ketones and hydroxy acids were reduced to common alcohols with lithium aluminum deuteride and the number of deuterium atoms in the product was used to distinguish the metabolic alcohols from those produced by reduction. Whereas the TMSi derivatives were formed using bis-(trimethylsilyl)trifluoroacetamide: trimethylchlorosilane:acetonitrile (2:1:2) as reagent, the MO-TMSi derivatives were obtained by first dissolving the metabolites in pyridine (20 μl) and then adding an excess of methoxyamine hydrochloride. The mixture was heated for 1 hr at 60°C and then the residue was converted to TMSi derivatives.

To obtain the GC-MS data shown in Table 2.57, the following instrumentation and conditions were used. The GC retention and MS spectral data were obtained with a Varian 2400 gas chromatograph interfaced via a glass jet separator to a VG Micromass 12B mass spectrometer and equipped with a single 2-m by 2-mm-i.d. glass column packed with 3% SE-30 on 100-120 mesh Gas Chrom Q. The operating parameters specified for both analytical units were injector temperature, 270°C; column temperature, programmed from 170 to 280°C at 2°C/min; transfer-line temperature, 230°C; ion source temperature, 260°C; helium carrier-gas flow rate, 30 ml/min; accelerating voltage, 2.5 kV; electron energy, 25 eV; scan speed, 3 sec/decade with an interscan delay of 2 sec; data acquisition started when the GC column reached 190°C.

The structures of the metabolites of Δ^1-THC found in mouse liver that are included in Table 2.57 are shown below:

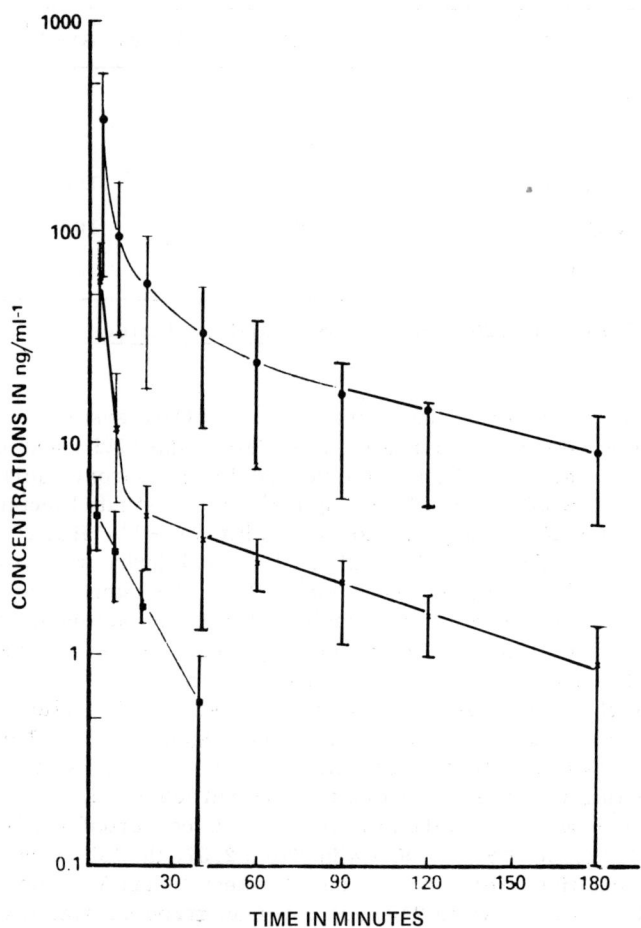

Figure 2.55. ●, Plasma concentration of Δ^9-THC after injection of Δ^9-THC; ×, plasma concentration of 11-OH-Δ^9-THC after injection of 11-OH-Δ^9-THC; ■, plasma concentration of 11-OH-Δ^9-THC after injection of Δ^9-THC. Each point is the average concentrations in three animals. The range of concentrations, I, is reported. From Rosenfeld [220], courtesy of <u>Analytical Letters</u>.

Compound	R_1	R_2	2"	3"
6α-OH-Δ^1-THC	CH_3	α-OH	H	H
3"-OH-Δ^1-THC	CH_3	H	H	OH
6β-OH-Δ^1-THC	CH_3	β-OH	H	H
7-OH-Δ^1-THC	CH_2OH	H	H	H
2"-OH-6-oxo-Δ^1-THC	CH_3	=O	OH	H
7-OH-6-oxo-Δ^1-THC	CH_2OH	=O	H	H
3"-OH-6-oxo-Δ^1-THC	CH_3	=O	H	OH
6α,7-diOH-Δ^1-THC	CH_2OH	α-OH	H	H
Δ^1-THC-7-oic acid	COOH	H	H	H
2",7-diOH-Δ^1-THC	CH_2OH	H	OH	H
6β,7-diOH-Δ^1-THC	CH_2OH	β-OH	H	H
6α-OH-Δ^1-THC-7-oic acid	COOH	α-OH	H	H
3",7-diOH-Δ^1-THC	CH_2OH	H	H	OH
2",6α,7-triOH-Δ^1-THC	CH_2OH	α-OH	OH	H
2"-OH-Δ^1-THC-7-oic acid	COOH	H	OH	H
3",6α,7-triOH-Δ^1-THC	CH_2OH	α-OH	H	OH
2",6α-diOH-Δ^1-THC-7-oic acid	COOH	α-OH	OH	H
2",7-diOH-6-oxo-Δ^1-THC	CH_2OH	=O	OH	H
3"-OH-Δ^1-THC-7-oic acid	COOH	H	H	OH
3",7-diOH-6-oxo-Δ^1-THC	CH_2OH	=O	H	OH
3",6α-diOH-Δ^1-THC-7-oic acid	COOH	α-OH	H	OH

TABLE 2.57

Retention Time and Mass Spectral Data for Derivatives of Δ^1-THC Metabolites[a]

Metabolite	Derivative	RT (min)	Diagnostic m/e ions
6α-OH-Δ^1-THC	TMSi	21.40	
3″-OH-Δ^1-THC	TMSi	21.46	
6β-OH-Δ^1-THC	TMSi	21.67	
7-OH-Δ^1-THC	TMSi	22.19	
2″-OH-6-oxo-Δ^1-THC	TMSi	23.80	488 (19), 145 (100), 473 (18)
	MO-TMSi	25.30	517 (16), 145 (100)
7-OH-6-oxo-Δ^1-THC	TMSi	26.10	488 (57), 385 (100)
	MO-TMSi	26.90	517 (5), 414 (100), 427 (41)
3″-OH-6-oxo-Δ^1-THC	TMSi	26.60	488 (49), 344 (100)
	MO-TMSi	27.80	517 (27), 373 (100)
6α,7-diOH-Δ^1-THC	TMSi	24.20	562 (0), 472 (100), 459 (8)
Δ^1-THC-7-oic acid	TMSi	25.41	

2″,7-diOH-Δ^1-THC	TMSi	26.30	562 (4), 459 (76), 145 (100)
6,7-diOH-Δ^1-THC	TMSi	27.30	562 (35), 472 (28), 459 (80), 369 (100)
6α-OH-Δ^1-THC-7-oic acid	TMSi	27.23	
3″,7-diOH-Δ^1-THC	TMSi	28.20	562 (0), 459 (100), 418 (60)
2″,6α,7-triOH-Δ^1-THC	TMSi	28.40	650 (0), 560 (57), 547 (8), 145 (100), 635 (2)
2″-OH-Δ^1-THC-7-oic acid	TMSi	29.50	
3″,6α,7-triOH-Δ^1-THC	TMSi	30.80	650 (1), 560 (100), 547 (42), 416 (39), 635 (2)
2″,6α-diOH-Δ^1-THC-7-oic acid	TMSi	30.64	
2″,7-diOH-6-oxo-Δ^1-THC	TMSi	30.20	145 (100)
	MO–TMSi	31.50	605 (4), 145 (100)
3″-OH-Δ^1-THC-7-oic acid	TMSi	31.50	
3″,7-diOH-6-oxo-Δ^1-THC	TMSi	31.85	432 (100)
	MO–TMSi	33.60	605 (14), 502 (100), 461 (35), 371 (39)
3″,6α-diOH-Δ^1-THC-7-oic acid	TMSi	33.08	

[a] Adapted from Harvey et al. [225].

Harvey et al. summarized their findings as follows:

Thus, four main sites of biotransformation have so far been identified in the mouse: the allylic positions 6 and 7 on the terpene moiety and positions 2" and 3" on the side chain. Position 7 was the most readily substituted; most of the metabolites reported to date have contained substitution at this position. Dihydroxy metabolites with one hydroxyl group at C_7 and the second group at one of the other preferred positions, 6α, 2", or 3" (6β substitution is regarded as a minor biotransformation step in this strain of mouse), were abundant as were the two possible triols. In addition to the 7-hydroxy series, the corresponding series containing the fully oxidized 7-group (7-COOH) and the oxidized 6-group (6-oxo) were also present, the former in high abundance. The observation that the positions substituted are the same in each series parallels results obtained with other cannabinoids.

In 1977, Harvey and co-workers [226, 227] identified the metabolites of Δ^1-THC and Δ^6-THC containing a reduced double bond [226] and in vivo metabolites of Δ^6-THC produced by the mouse via the epoxide-diol pathway [227].

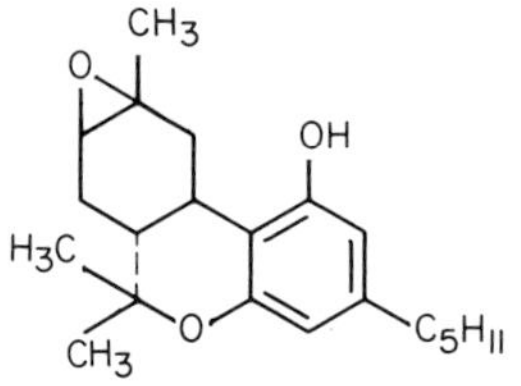

$1\beta, 6\beta$- Epoxyhexahydrocannabinol

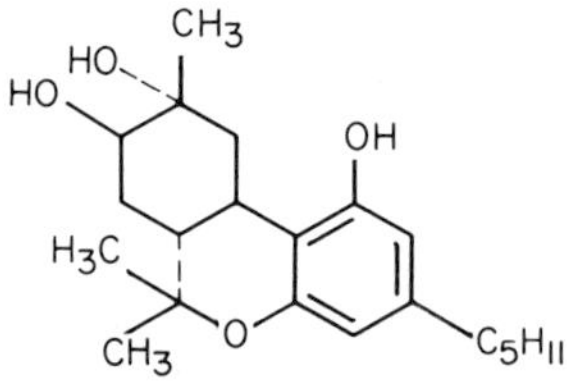

$1\alpha, 6\beta$ - Dihydroxyhexahydrocannabinol

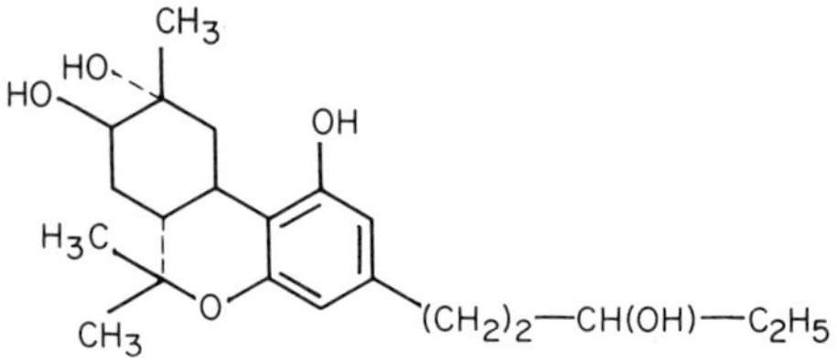

3", 1α, 6β -Trihydroxyhexahydro-cannabinol

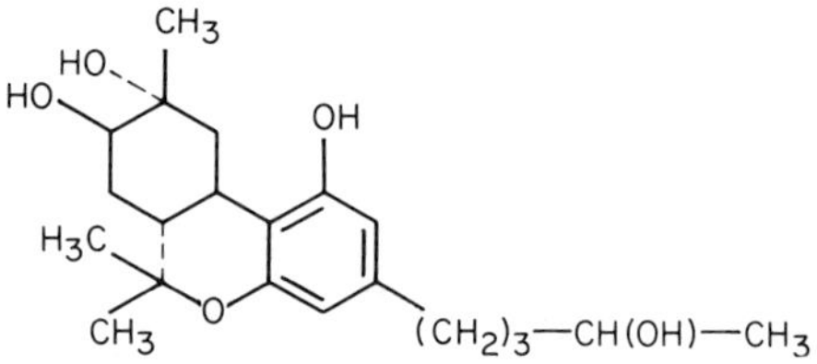

4", 1α, 6β -Trihydroxyhexahydro-cannabinol

Figure 2.56. Structures of hexahydrocannabinol metabolites.

Using isolation and derivatization techniques as well as GC-MS conditions described previously [225], the methylene unit value, retention time, and major diagnostic m/e ions of the derivatives prepared for the various 7-substituted hexahydrocannabinols (HHC) identified in their investigation are listed in Table 2.58.

The determination by Harvey and Paton of the in vivo metabolites of Δ^6-THC produced by the mouse via the epoxide-diol pathway (structures given in Fig. 2.56) were carried out with a GC-MS instrument as described previously [213]; however, the SE-30 column was programmed from 190 to 290°C at 2°C/min with the mass spectra recorded at 25 eV.

In Table 2.59 are listed some of the GC-MS data obtained for the TMSi derivatives of these hexahydrocannabinols.

Fonseka and Widman [228] identified by NMR and integrated GC-MS four dihydroxylated metabolites of cannabinol (CBN) formed by rat liver in vitro: 1",7-dihydroxy-CBN, 2",7-dihydroxy-CBN, 3",7-dihydroxy-CBN and 4",7-dihydroxy-CBN.

The silylated derivatives of these metabolites [bis-(trimethylsilyl)acetamide in dry acetonitrile as reagent] were examined chromatographically with a Varian Aerograph 2100 gas chromatograph equipped with a flame ionization detector and a 1.8-m by 2-mm glass column packed with 2% SE-30 on 125-150 mesh Gas Chrom Q and operated isothermally at 250°C. For mass spectral data, a LKB 9000 GC-MS instrument at 20 eV with the SE-30 column maintained at 200°C was used.

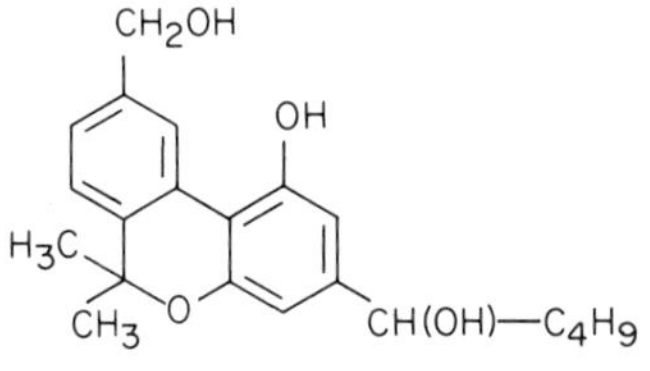

1",7 − Dihydroxy − CBN

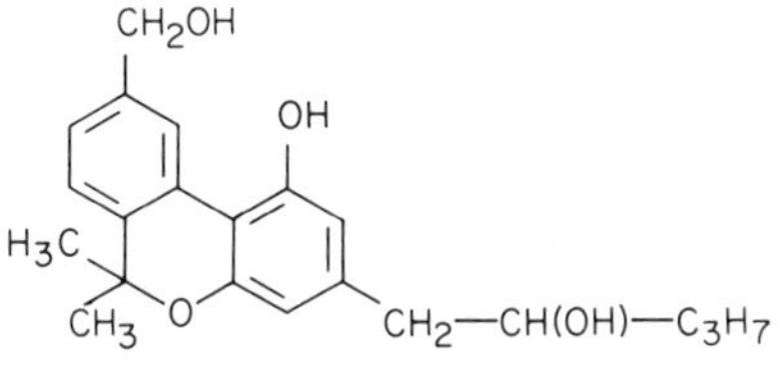

2",7 − Dihydroxy − CBN

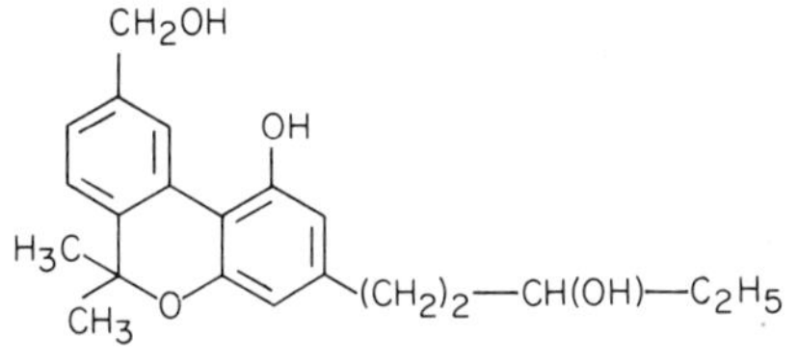

3",7 − Dihydroxy − CBN

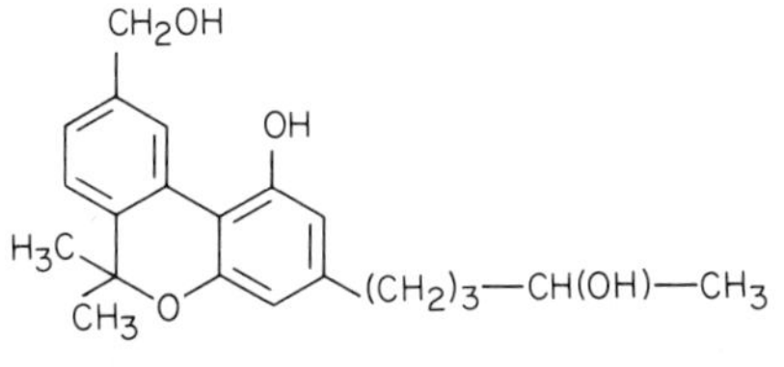

4",7 − Dihydroxy − CBN

TABLE 2.58

Methylene Unit Values, Retention Times, and Diagnostic m/e Ions of Derivatives of Several 7-Substituted Hexahydrocannabinols[a]

Compound	Derivative	MU value	RT (min)	Diagnostic m/e ions
HHC-7-oic acid (axial)	TMSi	26.62	24.50	490 (100), 475 (26), 434 (30), 372 (29)
	Me–TMSi		22.55	432 (100), 417 (7), 376 (73), 372 (18)
	Me–OAc	26.49		402 (28), 360 (100), 342 (3)
HHC-7-oic acid (equatorial)	TMSi	27.00	25.50	490 (100), 475 (12), 434 (39), 372 (4)
	Me–TMSi		23.20	432 (100), 376 (80)
	Me–OAc	26.68		402 (27), 360 (100), 346 (2), 342 (1)
7-OH-HHC (axial)	TMSi	26.46	24.35	476 (100), 420 (20), 372 (10)
7-OH-HHC (equatorial)	TMSi	26.28	23.10	476 (100), 461 (5), 420 (31)

[a] Adapted from Harvey et al. [226].

TABLE 2.59

Retention Times and Diagnostic m/e Ions for HHC-TMSi Derivatives[a]

Compound	Derivative	RT (min)	Diagnostic m/e ions
1α-6β-Epoxy-HHC	TMSi	6.30	387 (38), 384 (22), 369 (14), 359 (16), 343 (100)
1α-6β-diOH-HHC	a. bis-TMSi	13.30	492 (62), 474 (13), 459 (33), 343 (100)
	b. tris-TMSi	15.25	564 (11), 474 (39), 459 (39), 384 (9), 343 (100)
3″,1α-6β-TriOH-HHC	TMSi	22.00	508, 436, 431, 391
4″,1α-6β-TriOH-HHC	TMSi	22.20	508, 436, 431, 391, 117

[a] Adapted from Harvey et al. [227].

TABLE 2.60

Retention Times and Mass Spectral Data for TMSi Derivatives
of CBN Metabolites (Dihydroxylated)[a]

Compound	RT (min)	m/e Ions
1",7-diOH-CBN	7.9	558 (51), 543 (100), 528 (7), 501 (79), 414 (1)
2",7-diOH-CBN	8.3	558 (7), 543 (8), 528 (1), 414 (1), 145 (100)
3".7-diOH-CBN	9.7	558 (39), 543 (73), 528 (5), 414 (100), 131 (3), 117 (2)
4",7-diOH-CBN	10.5	558 (31), 543 (100), 528 (7), 414 (7), 117 (5)

[a] Adapted from Fonseka and Widman [228].

In Table 2.60, retention times (SE-30 at 250°C) and prominent m/e ions of the silylated derivatives are given.

In 1977, Yisak et al. [229] studied the in vivo transformation of cannabinol. Isolated from rat feces, they noted that "unchanged CBN and nine neutral mono-oxygenated and dioxygenated CBN metabolites [see Fig. 2.57] have been identified. In the mono-oxygenated series, the metabolites occurred in decreasing order of prominence as follows: 7-hydroxy-CBN, 4"-hydroxy-CBN, 1"-hydroxy-CBN, 2"-hydroxy-CBN, 3"-hydroxy-CBN, 5"-hydroxy-CBN, and CBN-7-aldehyde. In the dihydroxylated metabolite series, only 1",7-dihydroxy-CBN and 4",7-dihydroxy-CBN were found with the former as the more prominent metabolite."

As in the previous study, the TMSi derivatives were determined by both GC and GC-MS, using the same instrumentation and operating conditions. For these neutral in vivo CBN metabolites isolated from rat feces, the major diagnostic peaks and retention times of some of their TMSi derivatives are given in Table 2.61.

In 1978, Yisak et al. [233] identified in vivo metabolites of CBN as fatty acid conjugates and shortly thereafter the acidic in vivo metabolites of CBN isolated from rat feces [236]. In the former study, they reported the isolation and characterization of fatty acid (palmitic and oleic) conjugates of 4"-hydroxy-, 5"-hydroxy-, and 7-hydroxy-CBN. Using a Varian 2100 gas chromatograph equipped with a 1.06-m by 2-mm-i.d. glass column packed with 3% SE-30 on 100-120 mesh Gas Chrom Q and operated isothermally at 300°C, the silylated derivatives of the fatty acid conjugates yielded retention times as shown in Table 2.62, which includes mass spectral data obtained at 20 eV with a LKB 9000 GC-MS instrument.

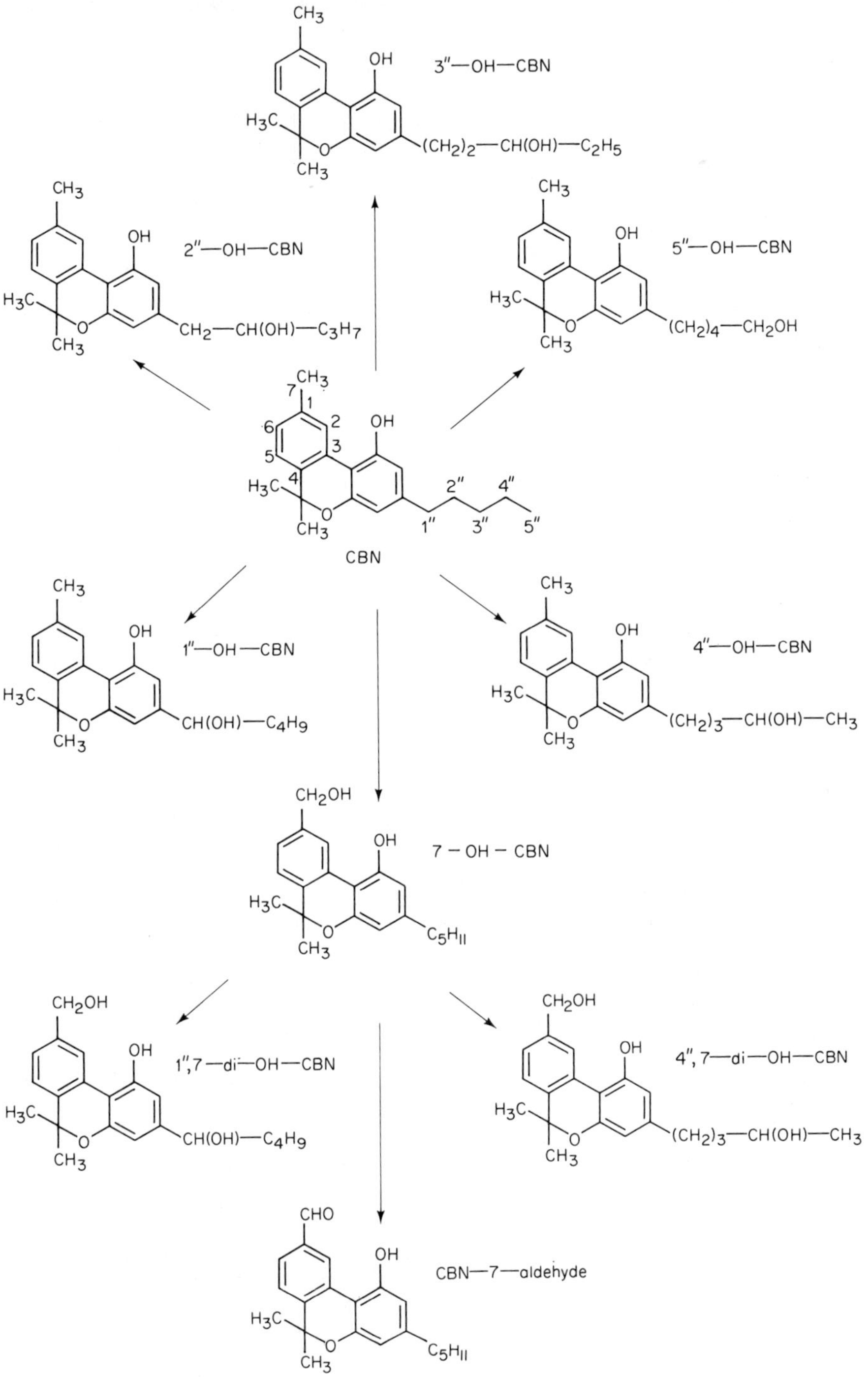

Figure 2.57. CBN metabolic pathways.

TABLE 2.61

Retention Times and Major m/e Ions of Some TMSi Derivatives of in Vivo CBN Metabolites Isolated from Feces (Rat)[a]

Compound	RT (min)	m/e Ions
CBN-7-aldehyde	10.1[b]	396 (30), 381 (100), 340 (6)
1"-OH-CBN	8.1[b]	470 (26), 455 (100), 413 (77)
2"-OH-CBN	7.8[b]	470 (8), 455 (17), 145 (100)
3"-OH-CBN	8.9[b]	470 (25), 455 (100), 326 (90)
4"-OH-CBN	9.0[b]	470 (18), 455 (100), 117 (10)
5"-OH-CBN	12.2[b]	470 (17), 455 (100)
7-OH-CBN	11.4[b]	470 (17), 455 (100), 103 (3)
1",7-diOH-CBN	6.1[c]	558 (50), 543 (100), 501 (75), 103 (5)
4",7-diOH-CBN	8.5[c]	558 (35), 543 (100), 414 (10), 117 (20)

[a] Adapted from Yisak et al. [229].
[b] For underivatized metabolite on 2% SE-30 (250°C).
[c] For TMSi derivative on 2% SE-30 (250°C).

TABLE 2.62

Retention Times and Mass Spectral Data of TMSi Derivatives of CBN Metabolites Identified as Fatty Acid Conjugates[a]

Compound	RT (min)	Diagnostic m/e ions
7-OH-CBN palmitate	3.2	636 (50), 621 (100), 381 (23), 366 (51)
4"-OH-CBN palmitate	3.0	636 (39), 621 (100), 381 (30), 366 (83)
5"-OH-CBN palmitate	3.9	636 (36), 621 (100), 381 (9), 366 (32)
7-OH-CBN oleate	4.2	662 (41), 647 (100), 381 (13), 366 (44)
4"-OH-CBN oleate	4.0	662 (34), 647 (100), 381 (40), 366 (88)
5"-OH-CBN oleate	5.3	662 (33), 647 (100), 381 (8), 366 (30)

[a] Adapted from Yisak et al. [233].

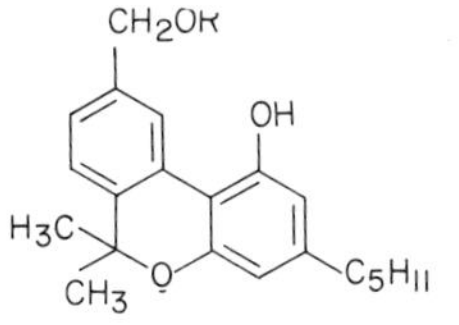

[Structures: 7-OH-CBN, 4"-OH-CBN, 5"-OH-CBN]

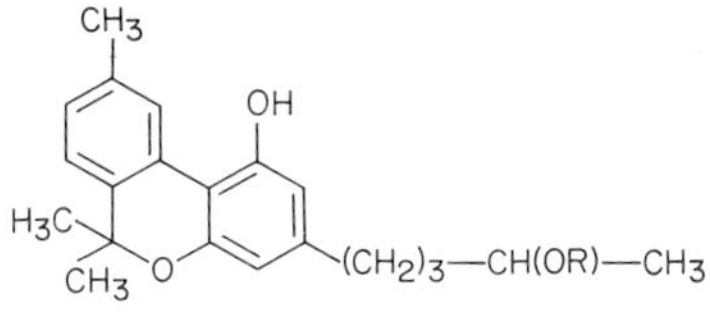

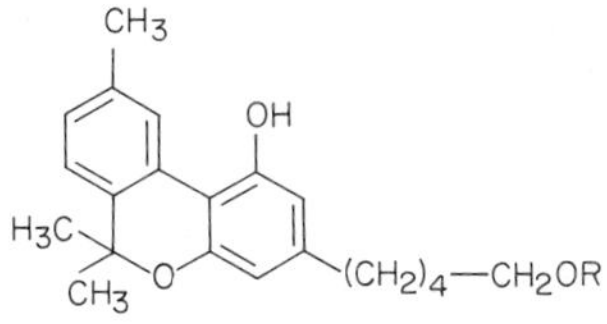

where

R = —C(=O)—(CH₂)₁₄—CH₃ (Palmitate) or —C(=O)—(CH₂)₇—CH=CH—(CH₂)₇—CH₃ (Oleate)

Hydrolysis of the conjugated metabolites with methanolic KOH yielded 7-OH-CBN, 4"-OH-CBN, and 5"-OH-CBN. These aglycones were then identified by comparing their TLC, GC, and MS properties with those of authentic, reference synthetic samples.

In a subsequent investigation [236], six acidic metabolites of CBN (formed in vivo) were isolated from rat feces and identified by GC-MS and proton magnetic resonance. Cannabinol-7-oic acid was the most abundant metabolite present; the others present in decreasing order of prominence were 1"-hydroxy-CBN-7-oic acid, 4"-hydroxy-CBN-7-oic acid, 3"-hydroxy-CBN-7-oic acid, CBN-3"-one-7-oic acid, and 2"-hydroxy-CBN-7-oic acid. Their structures are illustrated below:

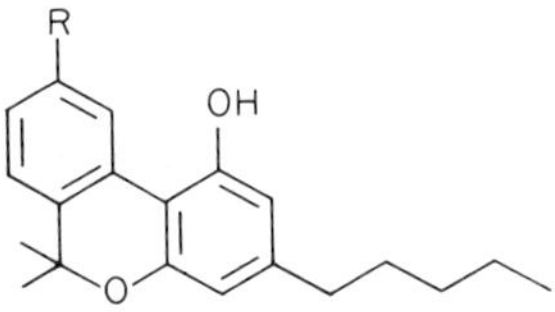

R = CH₃ CBN
R = COOH CBN-7-oic Acid

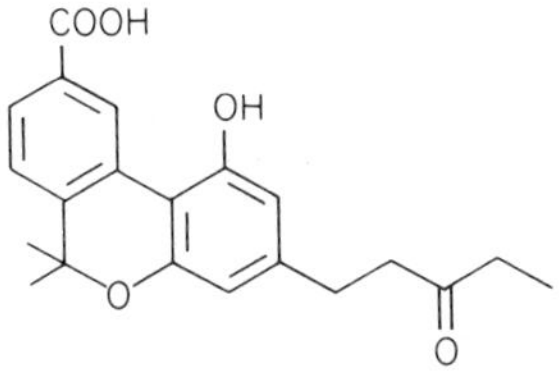

3"-O-CBN-7-oic Acid

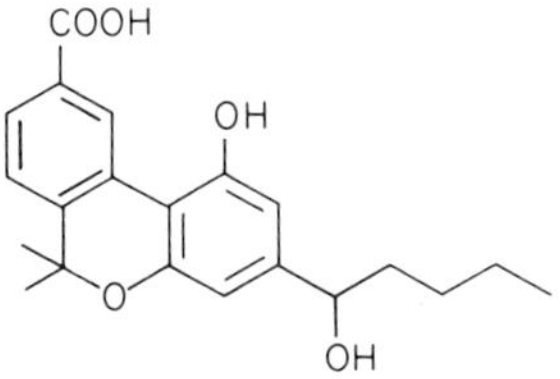

1"-OH-CBN-7-oic Acid

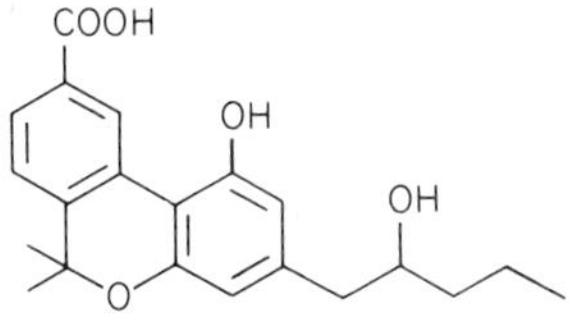

2"-OH-CBN-7-oic Acid

3"−OH−CBN−7−oic Acid   4"−OH−CBN−7−oic Acid

The methyl ester (diazomethane)-TMSi [bis-(trimethylsilyl)acetamide] derivatives were analyzed by GC and GC-MS. For gas chromatography, a Varian Aerograph 2100 gas chromatograph equipped with a flame ionization detector and a 1.8-m by 2-mm-i.d. column packed with 2% SE-30 on 125-150 mesh Gas Chrom Q was used to obtain retention time data in conjunction with the following operating conditions: injector temperature, 270°C; column temperature, 250°C; detector temperature, 270°C; nitrogen carrier-gas flow rate, 25 ml/min; hydrogen flow, 25 ml/min; oxygen flow, 200 ml/min.

To record mass spectral data, a LKB 9000 GC-MS instrument, operated at 20 eV, was employed. With this integrated analytical system, a 1.4-m by 2-mm-i.d. glass column packed with 3% SE-30 on 100-120 mesh Gas Chrom Q and operated at 200°C was employed.

With the above instruments and settings, the retention times and diagnostic m/e ions of the methyl ester-TMSi derivatives of acidic in vivo metabolites of CBN are listed in Table 2.63.

TABLE 2.63

Retention Times and Diagnostic m/e Ions of Methyl Ester-TMSi Derivatives of in Vivo Metabolites of CBN[a]

Compound	RT (min)	Diagnostic m/e ions
CBN-7-oic acid	7.1	426 (26), 411 (100), 370 (3)
CBN-3"-one-7-oic acid	23.1	440 (30), 425 (100), 383 (3)
1"-OH-CBN-7-oic acid	8.9	514 (21), 499 (27), 457 (100)
2"-OH-CBN-7-oic acid	9.6	514 (5), 499 (6), 370 (21), 145 (100)
3"-OH-CBN-7-oic acid	13.0	514 (26), 499 (79), 370 (100)
4"-OH-CBN-7-oic acid	13.2	514 (44), 499 (100), 370 (28), 117 (9)

[a] Adapted from Yisak et al. [236].

In 1978, Levy et al. [235] identified a C-glucuronide of Δ^6-THC as a mouse liver conjugate in vivo. Following an extraction of the liver homogenate with a solvent, its subsequent evaporation to dryness, and the reaction of the residue with diazomethane-acetic anhydride, the resulting Δ^6-THC-C-4'-glucuronide methyl ester tetraacetate (A) was resolved by thin-layer chromatography. The TLC zone corresponding to this compound was

(A)

extracted with ether and the solution was injected into a LKB 2091 GC-MS instrument equipped with a 6-ft by 1/8-in. glass column packed with 3% SE-30 on 100-120 mesh Gas Chrom Q. The operating conditions used were column temperature, programmed from 250 to 300°C at 5°C/min; helium carrier-gas flow rate, 25 ml/min; separator temperature, 250°C; ion source temperature, 240°C; electron energy, 70 eV; accelerating voltage, 3.5 kV. Levy et al. noted that a major peak with a retention time equivalent to that of compound A (also synthesized) was obtained, and the mass spectrum of this peak contained all the significant mass ions exhibited by the synthesized reference material under identical conditions [m/e 672 (22), 630 (100), 612 (11), 547 (11), 511 (37), 493 (14), 451 (30), 450 (28), 449 (11), 409 (22), 367 (78), and 349 (60)].

REFERENCES

1. Mechoulam, R., Science, 168, 1159 (1970).
2. Neumeyer, J. L., and Shagoury, R. A., J. Pharm. Sci., 60, 1433 (1971).
3. Caldwell, J., and Sever, P. S., J. Pharmacol. Ther., 16, 989 (1974).
4. Casarett, M. G., in L. J. Caserett and J. Doull (Eds.), Toxicology: The Basic Science of Poisons, Macmillan, New York, 1975.
5. Mechoulam, R., and Gaoni, Y., Fortschr. Chem. Org. Natur., 25, 175 (1967).
6. Martin, L., Smith, D. M., and Farmilo, C. G., Nature (London), 191, 774 (1961).

7. Farmilo, C. G., Davis, T. W. M., Vandenheuvel, F. A., and Lane, R., Scientific Research on Cannabis No. 7, UN Secretariat, Geneva, 1962.
8. Lerner, M., Science, 140, 175 (1963).
9. Davis, T. W. M., Farmilo, C. G., and Osadchik, M., Anal. Chem., 35, 751 (1963).
10. Grlic, L., Bull. Narcot., 20, 25 (1968).
11. Aramaki, H., Tomiyasu, N., Yoshimura, H., and Tsukamoto, H., Chem. Pharm. Bull. (Tokyo), 16, 822 (1968).
12. Toffoli, F., Avico, U., and Ceranni, E. S., Bull. Narcot., 20, 55 (1968).
13. Phillips, R., Turk, R., Manno, J., Jain, N. C., and Forney, R., J. Forensic Sci., 15, 191 (1970).
14. Ohlsson, A., Abou-Chaar, C. I., Agurell, S., Nilsson, I. M., and Olofsson, K., Bull. Narcot., 23, 29 (1971).
15. Fetterman, P. S., Keith, E. S., Waller, C. W., Guerrero, O., Doorenbos, N. J., and Quimby, M. W., J. Pharm. Sci., 60, 1246 (1971).
16. Fetterman, P. S., Doorenbos, N. J., Keith, E. S., and Quimby, M. W., Experientia, 27, 988 (1971).
17. Vree, T. B., Breimer, D. D., van Ginneken, C. A. M., and van Rossum, J. M., Identification of Hashish Constituents by a New Method of Combined Gas Chromatography-Mass Spectrometry, LKB-Produkter, Bromma, Sweden, 1971.
18. De Zeeuw, R. A., Wijsbeek, J., Breimer, D. D., Vree, T. B., van Ginneken, C. A. M., and van Rossum, J. M., Science, 175, 778 (1972).
19. Stromberg, L., J. Chromatogr., 68, 248 (1972).
20. Stromberg, L., J. Chromatogr., 68, 253 (1972).
21. Stromberg, L., J. Chromatogr., 68, 381 (1972).
22. Vree, T. B., Breimer, D. D., van Ginneken, C. A. M., and van Rossum, J. M., J. Pharm. Pharmacol., 24, 7 (1972).
23. Sperling, A., J. Chromatogr. Sci., 10, 268 (1972).
24. Fetterman, P. S., and Turner, C. E., J. Pharm. Sci., 61, 1476 (1972).
25. De Zeeuw, R. A., Wijsbeek, J., and Malingre, T. M., J. Pharm. Pharmacol., 25, 21 (1973).
26. Turner, C. E., and Hadley, K., J. Pharm. Sci., 62, 251 (1973).
27. Turner, C. E., and Hadley, K., J. Pharm. Sci., 62, 1083 (1973).
28. Turner, C. E., Hadley, K., and Fetterman, P. S., J. Pharm. Sci., 62, 1739 (1973).
29. Masoud, A. N., and Doorenbos, N. J., J. Pharm. Sci., 62, 313 (1973).
30. Jenkins, R. W., and Patterson, D. A., J. Forensic Sci., 2, 59 (1973).
31. Holley, J. H., Hadley, K. W., and Turner, C. E., J. Pharm. Sci., 64, 892 (1975).

32. Rasmussen, K. E., and Herweijer, J. J., Pharm. Weekblad., 110, 91 (1975).
33. Novotny, M., Lee, M. L., Low, C. E., and Raymond, A., Anal. Chem., 48, 24 (1976).
34. Rowan, M. G., and Fairbairn, J. W., J. Pharm. Pharmacol., 29, 491 (1977).
35. Segelman, A. B., J. Chromatogr., 82, 151 (1973).
36. Fairbairn, J. W., and Liebmann, J. A., J. Pharm. Pharmacol., 26, 413 (1974).
37. Waller, C. W., Pharmacol. Rev., 23, 265 (1971).
38. Turk, R. F., Manno, J. E., Jain, N. C., and Forney, R. B., J. Pharm. Pharmacol., 23, 190 (1971).
39. Turner, C. E., Hadley, K., Fetterman, P. S., Doorenbos, N. J., Quimby, M. W., and Waller, C., J. Pharm. Sci., 62, 1601 (1973).
40. Smith, D. M., and Levi, L., paper No. 17 presented at 138th National American Chemical Society Meeting, September 1960.
41. Smith, D. M., and Levi, L., Chem. Can., 12, 53 (1960).
42. Schultz, O. E., and Haffner, G., Arch. Pharm., 291 (1958).
43. Schultz, O. E., and Haffner, G., Arch. Pharm., 293, 1 (1960).
44. Kabelik, K., Krejci, K., and Santavy, F., Bull. Narcot. U.N. Social Affairs, 12, 8 (1960).
45. Grlic, L., and Andrec, A., Experientia, 17, 325 (1961).
46. Scaringelli, F., J. Assoc. Offic. Agr. Chem., 44, 296 (1961).
47. Kingston, C. R., and Kirk, P. L., Anal. Chem., 33, 1794 (1961).
48. Loewe, S. J., Arch. Exp. Pathol. Pharmacol., 211, 175 (1950).
49. Wollner, H. J., Matchett, J. R., Levine, J., and Loewe, S., J. Amer. Chem. Soc., 64, 26 (1942).
50. Leaf, G., Todd, A. R., and Wilkinson, S., J. Chem. Soc., 1942, 185.
51. Taylor, E. C., and Strojny, E. J., J. Amer. Chem. Soc., 82, 5198 (1960).
52. Todd, A. R., Experientia, 2, 55 (1946).
53. Korte, F., Hackel, E., and Sieper, H., Lieb. Ann. Chem., 685, 123 (1965).
54. Lerner, P., Bull. Narcot., 21, 39 (1969).
55. Mechoulam, R., and Gaoni, Y., Tetrahedron, 21, 1223 (1965).
56. Turner, C. E., Hadley, K. W., Henry, J. T., and Mole, M. L., J. Pharm. Sci., 63, 1872 (1974).
57. Novotny, M., and Farlow, R., J. Chromatogr., 103, 1 (1975).
58. Small, E., and Beckstead, H. D., Lloydia, 36, 144 (1973).
59. Farmilo, C. G., and Davis, T. W. M., J. Pharm. Pharmacol., 13, 767 (1961).
60. Kingston, C. R., and Kirk, P. L., Anal. Chem., 33, 1794 (1961).
61. Lerner, M., Mills, A. L., and Mount, S. F., J. Forensic Sci., 8, 126 (1963).

62. Nigam, M. C., Handa, K. L., Nigam, I. C., and Levi, L., Can. J. Chem., 43, 3372 (1965).
63. Claussen, U., Borger, W., and Korte, F., Justus Liebigs. Ann. Chem., 693, 158 (1966).
64. Taylor, E. C., Lenard, K., and Shvo, Y., J. Amer. Chem. Soc., 88, 367 (1966).
65. Claussen, U., Fehlhaber, H. W., and Korte, F., Tetrahedron, 22, 3535 (1966).
66. Lerner, M., and Zeffert, J. T., Chem. Eng. News, 44(53), 14 (1966).
67. Heaysman, L. T., Walker, E. A., and Lewis, D. T., Analyst, 92, 450 (1967).
68. Caddy, B., Fish, F., and Wilson, W. D. C., J. Pharm. Pharmacol., 19, 851 (1967).
69. Betts, T. J., and Holloway, P. J., J. Pharm. Pharmacol., 19, Suppl., 97S (1967).
70. Yamauchi, T., Shoyama, Y., Aramaki, H., Azuma, T., and Nishioka, I., Chem. Pharm. Bull. (Tokyo), 15(7), 1075 (1967).
71. Holloway, P. J., Ph.D. thesis, University of London, 1967.
72. Lerner, M., and Zeffert, J. T., Bull. Narcot., 20, 53 (1968).
73. Parker, K. D., Wright, J. A., Halpern, A. F., and Hines, C. H., Bull. Narcot., 20, 9 (1968).
74. Shoyama, Y., Fujita, T., Yamauchi, T., and Nishioka, I., Chem. Pharm. Bull. (Tokyo), 16, 1157 (1968).
75. Yamauchi, T., Shoyama, Y., Matsuo, Y., and Nishioka, I., Chem. Pharm. Bull. (Tokyo), 16, 1164 (1968).
76. Machata, G., Arch. Toxikol., 25, 19 (1969).
77. Stone, H. M., United Nations Secretariat, ST/SOA/SER S/18, August 1969.
78. Turk, R. F., Forney, R. B., King, L. J., and Ramachandran, S., J. Forensic Sci., 14, 385 (1969).
79. Backer, R. C., Jensen, W. N., Beck, A. G., and Barnett, R. J., J. Forensic Sci., 15, 287 (1970).
80. Patterson, D. A., and Stevens, H. M., J. Pharm. Pharmacol., 22, 391 (1970).
81. Song, C. H., Kanter, S. L., and Hollister, L. E., Res. Commun. Chem. Pathol. Pharmacol., 3, 375 (1970).
82. Joyce, C. R. B., Kay, H., and Curry, S. H. (Eds.), Symposium on Botany and Chemistry of Cannabis and Its Derivatives, Churchill, London, 1970.
83. Phillips, R., Turk, R., Manno, J., Jain, N., and Forney, R., J. Forensic Sci., 15, 191 (1970).
84. Street, H. V., J. Chromatogr., 48, 291 (1970).
85. Steinigen, M., Pharm. Z., 50, 1939 (1970).
86. Shoyama, Y., Yamauchi, T., and Nishioka, I., Chem. Pharm. Bull. (Tokyo), 18(7), 1327 (1970).

87. Wilson, W. D. C., Ph.D. thesis, University of Strathclyde, 1970.
88. Davis, K. H., Jr., Martin, N. H., Pitt, C. G., Wildes, J. W.. and Wall, M. E., Lloydia, 33, 453 (1970).
89. Green, D. E., Intra-Science Chem. Rep., 4, 211 (1970).
90. Archer, R. A., Boyd, D. B., Demarco, D. V., Tgminski, I. J., and Allinger, N. L., J. Amer. Chem. Soc., 92, 5200 (1970).
91. Turk, R. F., The Identification, Isolation, Toxicity, and Tissue Distribution of Δ^9-Tetrahydrocannabinol, Ph. D. thesis, Indiana University Medical Center, Dept. of Toxicology, 1970.
92. Sobotka, A. J., and Sperling, A., paper presented at 23rd Annual Meeting of the American Academy of Forensic Sciences, Phoenix, Arizona, February 1971.
93. Knight, J. B., Finnigan Applic. Tips No. 36, August 1971.
94. Gaoni, Y., and Mechoulam, R., J. Amer. Chem. Soc., 93, 217 (1971).
95. Vree, T. B., Breimer, D. D., van Ginneken, C. A. M., van Rossum, J. M., De Zeeuw, R. A., and Witte, A. H., Clin. Chim. Acta, 34, 365 (1971).
96. Merkus, F. W. H. M., Nature, 232, 579 (1971).
97. Hunneman, D. H., Beitr. Gerichtl. Med., Band 28 (1971).
98. Skinner, R. F., Proc. West. Pharmacol. Soc., 14, 4 (1971).
99. Stromberg, L. F., J. Chromatogr., 63, 391 (1971).
100. Agurell, S., and Leander, K., Acta Pharm. Suecica, 8, 391 (1971).
101. Bercht, C. A. L., Kuppers, F. J. E. M., Lousberg, R. J. C., Salemink, C. A., Svendsen, A. B., and Karlsen, J., U.N. Secretariat No. 29, July 1971.
102. Willinsky, M. D., Ph. D. thesis, University of Toronto, 1971.
103. Merkus, F. W. H. M., Pharm. Weekbl., 106, 69 (1971).
104. Schou, J., Steentoft, A., Worm, K., Anderson, J. M., and Neetsen, E., Acta Pharmacol. Toxicol., 30, 480 (1971).
105. Rasmussen, K. E., Rasmussen, S., and Svendsen, A. B., J. Chromatogr., 69, 381 (1972).
106. Rasmussen, K. E., Rasmussen, S., and Svendsen, A. B., Acta Pharm. Suecica, 9, 457 (1972).
107. Vree, T. B., Breimer, D. D., van Ginneken, C. A. M., and van Rossum, J. M., J. Chromatogr., 74, 209 (1972).
108. Merkus, F. W. H. M., Jaspers-van Wouw, M. G. J., and Bollen-Roovers, J. F. C., Pharm. Weekbl., 107, 98 (1972).
109. Verwey, A. M. A., and Witte, A. H., Pharm. Weekbl., 107, 153 (1972).
110. van Ginneken, C. A. M., Vree, T. B., Breimer, D. D., Thijssen, H. H., and van Rossum, J. M., in A. Frigerio (Ed.), Proceedings of International Symposium on GC-MS, Isle of Elba, Italy, Tamborini Editore, Milan, 1972.

111. Vree, T. B., Breimer, D. D., van Ginneken, C. A. M., and van Rossum, J. M., Chem. Weekbl., 68, H1 (1972).
112. Vree, T. B., van Ginneken, C. A. M., and van Rossum, J. M., J. Chromatogr., 74, 124 (1972).
113. Hoffman, N. E., and Yang, R. K., Anal. Lett., 5, 7 (1972).
114. Fish, F., Pharm. J., 209, 343 (1972).
115. Ek, N. A., Lonberg, E., Maehly, A. C., and Stromberg, L., J. Forensic Sci., 17, 456 (1972).
116. De Zeeuw, R. A., Malingre, T. M., and Merkus, F. W. H. M., J. Pharm. Pharmacol., 24, 1 (1972).
117. Breimer, D. D., Vree, T. B., van Ginneken, C. A. M., Henderson, P. T., and van Rossum, J. M., in A. Frigerio (Ed.), Proceedings of International Symposium on GC-MS, Isle of Elba, Italy, Tamborini Editore, Milan, 1972.
118. Kubena, R. K., Barry, H., III, Segelman, A. B., Theiner, M., and Farnsworth, N. R., J. Pharm. Sci., 61, 144 (1972).
119. Turner, C. E., Hadley, K. W., and Davis, K. H., Jr., Acta Pharm. Jugoslav., 23, 89 (1973).
120. Adams, T. C., Jr., and Jones, L. A., J. Agric. Food Chem., 21, 1129 (1973).
121. Fairbairn, J. W., and Liebmann, J. A., J. Pharm. Pharmacol., 25, 150 (1973).
122. Willinsky, M. D., and di Simone, L., Farmaco, Ed. Prat., 28, 441 (1973).
123. Bercht, C. A. L., Lousberg, R. J. J. C., Kuppers, F. J. E. M., Salemink, C. A., Vree, T. B., and van Rossum, J. M., J. Chromatogr., 81, 163 (1973).
124. Vree, T. B., Breimer, D. D., van Ginneken, C. A. M., van Rossum, J. M., and Nibbering, N. M. M., J. Chromatogr., 79, 81 (1973).
125. Bailey, K., Legault, D., and Verner, D., J. Chromatogr., 87, 263 (1973).
126. Hood, L. V. S., Dames, M. E., and Barry, G. T., Nature, 242, 402 (1973).
127. Eskes, D., Verwey, A. M. A., and Witte, A. H., Bull. Narcot., 25, 41 (1973).
128. De Zeeuw, R. A., Vree, T. B., Breimer, D. D., and van Ginneken, C. A. M., Experientia, 29, 260 (1973).
129. Skinner, R. F., Knight, J. B., and Morris, W. J., Finnigan Applic. Tips No. 48, July 1973.
130. Mechoulam, R. (Ed.), Marijuana: Chemistry, Pharmacology, Metabolism, and Clinical Effects, Academic Press, New York, 1973.
131. Paris, M. R., and Paris, R. R., Bull. Soc. Chim. France, 1, 118 (1973).

132. Turner, C. E., Bi-Annual Report, Cannabis sativa L., Project NIDA, Contract No. HSM 42-70-109, 1973.
133. Novotny, M., and Lee, M. L., Experientia, 29, 1038 (1973).
134. Binder, M., Agurell, S., Leander, K., and Lindgren, J. E., Helv. Chim. Acta, 57, 1626 (1974).
135. Parker, J. M., and Stembal, B. L., J. Assoc. Offic. Anal. Chem., 57, 888 (1974).
136. Fish, F., Chromatographia, 7, 311 (1974).
137. Stromberg, L., J. Chromatogr., 96, 99 (1974).
138. Stromberg, L., J. Chromatogr., 96, 179 (1974).
139. Parker, J. J., Borke, M. L., Block, L. H., and Cochran, T. G., J. Pharm. Sci., 63, 970 (1974).
140. Garrett, E. R., and Tsau, J., J. Pharm. Sci., 63, 1563 (1974).
141. Kirchgessner, W. G., DiPasqua, A. C., Anderson, W. A., and Delaney, G. V., J. Forensic Sci., 19, 313 (1974).
142. Finkle, B. S., Foltz, R. L., and Taylor, D. M., J. Chromatogr. Sci., 12, 304 (1974).
143. Jones, G., Widman, M., Agurell, S., and Lindgren, J. E., Acta Pharm. Suecica, 11, 283 (1974).
144. Friedrich-Fiechtl, J., and Spiteller, G., Tetrahedron, 31, 479 (1975).
145. Rasmussen, K. E., J. Chromatogr., 114, 250 (1975).
146. Turner, C. E., Hadley, K. W., Holley, J. H., Billets, S., and Mole, M. L., Jr., J. Pharm. Sci., 64, 810 (1975).
147. Harvey, D. J., and Paton, W. D. M., J. Chromatogr., 109, 73 (1975).
148. Rasmussen, K. E., J. Chromatogr., 109, 175 (1975).
149. Bailey, K., and Gagne, D., J. Pharm. Sci., 64, 1719 (1975).
150. Moffat, A. C., J. Chromatogr., 113, 69 (1975).
151. Smith, R. N., J. Chromatogr., 115, 101 (1975).
152. Harvey, D. J., J. Pharm. Pharmacol., 28, 280 (1976).
153. Knaus, E. E., Coutts, R. T., and Kazakoff, C. W., J. Chromatogr. Sci., 14, 525 (1976).
154. Fonseka, K., Widman, M., and Agurell, S., J. Chromatogr., 120, 343 (1976).
155. Novotny, M., Lee, M. L., Low, C. E., and Maskarinec, M. P., Steroids, 27, 665 (1976).
156. Fairlie, K., and Fox, B. L., J. Chromatogr. Sci., 14, 334 (1976).
157. McCallum, N. K., and Cairns, E. R., J. Pharm. Sci., 66, 114 (1977).
158. Smith, R. N., Jones, L. V., Brennan, J. S., and Vaughn, C. G., J. Pharm. Pharmacol., 29, 126 (1977).
159. Vree, T. B., J. Pharm. Sci., 66, 1444 (1977).
160. Smith, R. N., Amer. Lab., May 1978, pp. 53-60.

161. Rosenfeld, J. M., and Crocco, J. L., Anal. Chem., 50, 701 (1978).
162. Garrett, E. R., Gouyette, A. J., and Roseboom, J., J. Pharm. Sci., 67, 27 (1978).
163. Adams, R., Cain, C. K., and Baker, B. R., J. Amer. Chem. Soc., 62, 2201 (1940).
164. Adams, R., Smith, C. M., and Loewe, S., J. Amer. Chem. Soc., 63, 1973 (1941).
165. Schultz, O. E., and Haffner, G., Arch. Pharm., 293, 1 (1960).
166. Garrett, E. R., and Hunt, C. A., J. Pharm. Sci., 62, 1211 (1973).
168. Bouche, J., and Verzele, M., J. Gas Chromatogr., 6, 501 (1968).
169. Grob, K., and Grob, G., J. Chromatogr. Sci., 7, 584 (1969).
170. Novotny, M., and Farlow, R., J. Chromatogr., 103, 1 (1975).
171. Claussen, U., and Korte, F., Tetrahedron Lett., 22, 2067 (1967).
172. Vieira, F. J., Aguiar, M. B., Alencar, J. W., Seabra, A. P., Tursch, B. M., and Leclercq, J., Psychopharmacologia, 10, 361 (1967).
173. Seabra, A. P., Rev. Bras. Farm., 49, 13 (1968).
174. Claussen, U., and Korte, F., Justus Liebigs. Ann. Chem., 713, 162 (1968).
175. Fish, F., and Wilson, W. D. C., J. Forensic Sci., 9, 37 (1969).
176. Truitt, E. B., Pharmacol. Rev., 23, 273 (1971).
177. Mikes, F., and Waser, P. G., Science, 172, 1158 (1971).
178. Fehr, K. O., and Kalant, H., Can. J. Physiol. Pharmacol., 50, 761 (1972).
179. Kuppers, F. J. E. M., Lousberg, R. J. J. Ch., Bercht, C. A. L., Salemink, C. A., Terlouw, J. K., Heerma, W., and Laven, A., Tetrahedron, 29, 2797 (1973).
180. Fentiman, A. F., Jr., Foltz, R. L., and Kinzer, G. W., Anal. Chem., 45, 580 (1973).
181. Haq, M. Z., Rose, S. J., Deiderich, L. R., and Patel, A. R., Anal. Chem., 46, 1781 (1974).
182. Heerma, W., Terlouw, J. K., Laven, A., Dijkstra, G., Kuppers, F. J. E. M., Lousberg, R. J. J. C., and Salemink, C. A., in A. Frigerio and N. Castagnoli (Eds.), Mass Spectrometry in Biochemistry and Medicine, Raven Press, New York, 1974.
183. Jones, L. A., and Foote, R. S., J. Agric. Food Chem., 23, 1129 (1975).
184. Adams, T. C., Jr., and Jones, L. A., J. Agric. Food Chem., 23, 352 (1975).
185. Maskarinec, M. P., Alexander, G., and Novotny, M., J. Chromatogr., 126, 559 (1976).
186. Lee, M. L., Novotny, M., and Bartle, K. D., Anal. Chem., 48, 405 (1976).

187. Kettenes-van den Bosch, J. J., and Salemink, C. A., J. Chromatogr., 131, 422 (1977).
188. Foote, R. S., and Jones, L. A., J. Agric. Food Chem., 22, 534 (1974).
189. Grunwald, C., Anal. Biochem., 34, 16 (1970).
190. Stone, H. M., and Stevens, H. M., J. Forensic Sci., 9, 31 (1969).
191. Agurell, S., Nilsson, I. M., Ohlsson, A., and Sandberg, F., Biochem. Pharmacol., 18, 1195 (1969).
192. Agurell, S., in C. R. B. Joyce, H. Kay, and S. H. Curry (Eds.), Symposium on Botany and Chemistry of Cannabis and Its Derivatives, Churchill, London, 1970.
193. Robinson, A. E., Bull. Narcot., 23, 37 (1971).
194. Mikes, F., Hofmann, A., and Waser, P. G., Biochem. Pharmacol., 20, 2469 (1971).
195. Wall, M. E., Ann. N.Y. Acad. Sci., 191, 23 (1971).
196. Widman, M., Nilsson, I. M., Nilsson, J. L. G., Agurell, S., and Leander, K., Life Sci., 10, Part II, 157 (1971).
197. Burstein, S., Rosenfeld, J., and Wittstruck, T., Science, 176, 422 (1972).
198. Wall, M. E., Brine, D. R., Pitt, C. G., and Perez-Reyes, M., J. Amer. Chem. Soc., 94, 8579 (1972).
199. Skinner, R. F., Proc. West. Pharmacol. Soc., 15, 136 (1972).
200. McCallum, N. K., J. Chromatogr. Sci., 11, 509 (1973).
201. Agurell, S., Gustafsson, B., Holmstedt, B., Leander, K., Lindgren, J. E., Nilsson, I., Sandberg, F., and Asberg, M., J. Pharm. Pharmacol., 25, 554 (1973).
202. Fenimore, D. C., Freeman, R. R., and Loy, P. R., Anal. Chem., 45, 233 (1973).
203. Rosenfeld, J., Bowins, B., Roberts, J., Perkins, J., and Macpherson, A. S., Anal. Chem., 46, 2232 (1974).
204. Agurell, S., Gustafsson, B., Holmstedt, B., Leander, K., Lindgren, J. E., Nilsson, I., and Asberg, M., in A. Frigerio and N. Castagnoli (Eds.), Mass Spectrometry in Biochemistry and Medicine, Raven Press, New York, 1974.
205. Ben-Zvi, Z., Burstein, S., and Zikoloulos, J., J. Pharm. Sci., 63, 1173 (1974).
206. Ben-Zvi, Z., and Burstein, S., Res. Commun. Chem. Pathol. Pharmacol., 8, 223 (1974).
207. Ben-Zvi, Z., Bergen, J. R., and Burstein, S., Res. Commun. Chem. Pathol. Pharmacol., 9, 201 (1974).
208. Hollister, L. E., Kanter, S. L., Board, R. D., and Green, D. E., Res. Commun. Chem. Pathol. Pharmacol., 8, 579 (1974).
209. Widman, M., Dahmer, J., Leander, K., and Petersson, K., Acta Pharm. Suecica, 12, 385 (1975).

210. Burstein, S., and Varanelli, C., Res. Commun. Chem. Pathol. Pharmacol., 11, 343 (1975).
211. Widman, M., Nordqvist, M., Dollery, C. T., and Briant, R. H., J. Pharm. Pharmacol., 27, 842 (1975).
212. Rosenfeld, J. M., and Taguchi, V. Y., Anal. Chem., 48, 726 (1976).
213. Harvey, D. J., and Paton, W. D. M., Res. Commun. Chem. Pathol. Pharmacol., 13, 585 (1976).
214. Harvey, D. J., and Paton, W. D. M., in G. G. Nahas, W. D. M. Paton, and J. E. Idanpaan-Heikkila (Eds.), Marijuana: Biochemistry and Cellular Effects, Springer, New York, 1976.
215. Martin, B. R., Harvey, D. J., and Paton, W. D. M., J. Pharm. Pharmacol., 28, 773 (1976).
216. Leighty, E. G., Fentiman, A. F., Jr., and Foltz, R. L., Res. Commun. Chem. Pathol. Pharmacol., 14, 13 (1976).
217. Martin, B., Nordqvist, M., Agurell, S., Lindgren, J. E., Leander, K., and Binder, M., J. Pharm. Pharmacol., 28, 275 (1976).
218. Martin, B., Agurell, S., Nordqvist, M., and Lindgren, J. E., J. Pharm. Pharmacol., 28, 603 (1976).
219. Green, D. E., paper presented at NIDA Technical Review in Cannabinoid Quantifications, Washington, D.C., February 1976.
220. Rosenfeld, J., Anal. Lett., 10, 917 (1977).
221. Ohlsson, A., Agurell, S., Lindgren, J. E., Leander, K., and Widman, M., paper No. 70 presented at 173rd National American Chemical Society Meeting, New Orleans, March 1977.
222. Rosenfeld, J., and Taguchi, V. Y., paper No. 72 presented at 173rd National American Chemical Society Meeting, New Orleans, March 1977.
223. Clarke, P. A., Hidy, B. J., Lin, D. C. K., Graffeo, A. P., and Foltz, R. L., paper No. 73 presented at 173rd National American Chemical Society Meeting, New Orleans, March 1977.
224. Garrett, E. R., Gouyette, A. J., and Hunt, C. A., Paper No. 91 presented at 173rd National American Chemical Society Meeting, New Orleans, March 1977.
225. Harvey, D. J., Martin, B. R., and Paton, W. D. M., J. Pharm. Pharmacol., 29, 482 (1977).
226. Harvey, D. J., Martin, B. R., and Paton, W. D. M., J. Pharm. Pharmacol., 29, 495 (1977).
227. Harvey, D. J., and Paton, W. D. M., J. Pharm. Pharmacol., 29, 498 (1977).
228. Fonseka, K., and Widman, M., J. Pharm. Pharmacol., 29, 12 (1977).
229. Yisak, W. A., Widman, M., Lindgren, J. E., and Agurell, S., J. Pharm. Pharmacol., 29, 487 (1977).
230. Wall, M. E., and Brine, D. R., paper No. 71 presented at 173rd National American Chemical Society Meeting, New Orleans, March 1977.

231. Green, D. E., Loeffler, K. O., and Kanter, S. L., paper No. 89 presented at 173rd National American Chemical Society Meeting, New Orleans, March 1977.
232. Fenimore, D. C., Davis, C. M., and Shukla, P., paper No. 90 presented at 173rd National American Chemical Society Meeting, New Orleans, March 1977.
233. Yisak, W., Agurell, S., Lindgren, J. E., and Widman, M., J. Pharm. Pharmacol., 30, 462 (1978).
234. Harvey, D. J., Martin, B. R., and Paton, W. D. M., in A. Frigerio (Ed.), Recent Developments in Mass Spectrometry in Biochemistry and Medicine, Vol. 1, Plenum Publ. Corp., New York, 1978.
235. Levy, S., Yagen, B., and Mechoulam, R., Science, 200, 1391 (1978).
236. Yisak, W., Widman, M., and Agurell, S., J. Pharm. Pharmacol., 30, 554 (1978).
237. Tagg, J., Yashuda, D. M., Tanabe, M., and Mitoma, C., Biochem. Pharmacol., 16, 143 (1967).
238. Lemberger, L., Silberstein, S. D., Axelrod, J., and Kopin, I. J., Science, 170, 1320 (1970).
239. Burstein, S., Menezes, F., Williamson, E., and Mechoulam, R., Nature, 225, 88 (1970).
240. Ben-Zvi, Z., Bergen, J. R., and Burstein, S., in Proceedings of International Conference on the Pharmacology of Cannabis, University Park Press, 1975.

AUTHOR INDEX

Numbers in parentheses are reference numbers and indicate that an author's work is referred to although his name is not cited in the text. Underlined numbers give the page on which the complete reference is listed.

A

Abdine, H., 41(41), 45, 46, 47, 48, 49, 263
Abou-Chaar, C. I., 273(14), 281(14), 295(14), 480
Adamek, S., 146(133), 266
Adams, R., 326, 486
Adams, T. C., Jr., 325(120), 352, 353, 407(120), 407(184), 420, 484, 486
Aguiar, M. B., 407(172), 486
Agurell, S., 273(14), 281(14), 295(14), 325(100), 325(134), 325(143), 325(154), 398(154), 400(154), 401(154), 409(100), 432, 432(191), 432(192), 432(196), 432(201), 432(204), 432(217), 432(218), 432(221), 432(229), 432(233), 432(236), 437(196), 440, 458(217), 458(218), 459(217), 461(217), 461(218), 462(218), 474(229), 474(233), 474(236), 476(229), 476(233), 477(236), 478(236), 480, 483, 485, 487, 488, 489
Alencar, J. W., 407(172), 486
Alexander, G., 407(185), 422(185), 423(185), 486

Allinger, N. L., 325(90), 483
Alworth, W. L., 22, 22(13), 22(15), 25, 26, 262
Anastasov, A., 22(19), 29, 262
Anderson, E., 140(130), 266
Anderson, J. M., 325(104), 343(104), 432(104), 483
Anderson, W. A., 325(141), 384(141), 485
Andrec, A., 282(45), 481
Aramaki, H., 273(11), 285, 286, 287, 325(70), 332(70), 336(70), 480, 482
Archer, R. A., 325(90), 483
Arcos, J. C., 101, 264
Arrendale, R. F., 59(49), 60(49), 66(49), 180(172), 180(173), 180(176), 182(172), 186(172), 201(176), 204(176), 263, 268
Asberg, M., 432(201), 432(204), 440(201), 487
Avico, U., 273(12), 480
Axelrod, J., 444(238), 489
Ayers, C. W., 87, 87(75), 250, 250(75), 264
Ayres, C. I., 180(159), 267
Azuma, T., 325(70), 332(70), 336(70), 482

B

Backer, R. C., 325(79), 333, 334, 482
Badre, R., 22(23), 103(23), 103(97), 109(23), 122(97), 146(141), 146(143), 149(141), 152(141), 153(143), 156(143), 157(143), 262, 265, 267
Baer-Weiss, V., 103(100), 125(100), 265
Bailey, K., 325(125), 325(149), 357, 361, 391, 392, 484, 485
Baker, B. R., 326(163), 486
Baker, J. K., 22 (34), 36, 37, 40, 263
Barbezat-Debreuil, S., 146(134), 266
Barnett, R. J., 325(79), 333(79), 334(79), 482
Barry, G. T., 325(126), 360(126), 363(126), 484
Barry, H., III, 325(118), 484
Bartle, K. D., 180(169), 180(174), 185(174), 188(174), 198(174), 226(169), 226(193), 226(194), 230(169), 230(193), 230(194), 231(193), 233(194), 238(194), 268, 269, 407(186), 424(186), 425(186), 426(186), 427(186), 428(186), 429(186), 430(186), 486
Beck, A. G., 325(79), 333(79), 334(79), 482
Beckett, A. H., 103(92), 103(93), 119, 119(92), 119(93), 119(116), 120, 123, 265, 266
Beckstead, H. D., 273(58), 312, 313, 481
Beelen, T. C., 103(112), 134(112), 136(112), 266
Beeson, J. H., 207(180), 213, 217, 268
Bell, J. H., 146(140), 148(140), 151(140), 159, 160, 180(165), 267, 268

Belsky, T., 158(145), 267
Benveniste, P., 71(61), 264
Ben-Zvi, Z., 432(205), 432(206), 432(207), 446, 446(205), 446(206), 446(207), 447, 447(207), 448, 487, 489
Bercht, C. A. L., 325(101), 325(123), 354, 407(179) 411(179), 483, 484, 486
Bergen, J. R., 432(207), 446(207), 447(207), 448(240), 487, 489
Bergman, J., 71(62), 264
Bernos, J. B., 207(183), 268
Bertsch, W., 136(128), 140(128), 140(130), 232(200), 244, 245, 245(202), 246, 247, 266, 269
Betts, T. J., 325(69), 329, 482
Bhatia, K., 207(182), 219, 221, 268
Billets, S., 325(146), 390(146), 485
Binder, M., 325(134), 432(217), 458(217), 459(217), 461(217), 485, 488
Block, L. H., 325(139), 380(139), 383(139), 485
Board, R. D., 432(208), 448(208), 487
Bollen-Roovers, J. F. C., 325(108), 483
Booth, R., 22(24), 30(24), 262
Borger, W., 325(63), 482
Borke, M. L., 325(139), 380(139), 383(139), 485
Bouche, J., 71(55), 263, 402(168), 486
Bowins, B., 432(203), 444(203), 487
Bowman, E. R., 103(94), 114, 121(94), 265, 266
Boyd, D. B., 325(90), 483
Breimer, D. D., 273(17), 273(18), 273(22), 287(17), 287(22), 289(17), 290(17), 291 (17), 292(17), 293(17), 294(17), 301(18), 325(95), 325(107), 325(110), 325(111), 325(117), 325(124), 325(128), 341(95),

AUTHOR INDEX

[Breimer, D. D.]
  343(107), 344(107), 346(107),
  347(107), 355(124), 360(124),
  364(128), 480, 483, 484
Brennan, J. S., 325(158), 403(158),
  485
Briant, R. H., 432(211), 448(211),
  488
Brine, D. R., 432(198), 432(230),
  438(198), 487, 488
Brooks, C. J. W., 71(60), 264
Brown, J. B., 207(181), 218(181),
  268
Brunnemann, K. D., 103(90),
  159(90), 169, 170, 171, 173,
  174, 250(205), 252, 257, 265,
  269
Burchfield, W. H., 207(183), 268
Burdick, D., 74(65), 171(154),
  264, 267
Burlingame, A. L., 158(145), 267
Burns, D. T., 22(33), 41, 42, 263
Burrows, I. E., 103(105), 103(108),
  128, 131, 131(108), 132(108),
  265, 266
Burstein, S., 432(197), 432(205),
  432(206), 432(207), 432(210),
  437, 438, 444(239), 446(205),
  446(206), 446(207), 447(207),
  448, 448(240), 449, 487, 488,
  489
Bush, L. P., 22(21), 22(25), 29(21),
  30(21), 31, 31(21), 31(36), 71,
  262, 263
Buyske, D. A., 75(67), 264

C

Caddy, B., 22(26), 22(27), 22(28),
  22(29), 31, 31(26), 32, 32(27),
  32(28), 32(29), 33, 262, 263,
  325(68), 329, 331, 482
Cain, C. K., 326(163), 486
Cairns, E. R., 325(157), 402, 403,
  485
Caldwell, J., 271(3), 479

Calvin, M., 158(145), 267
Campbell, I. M., 79(72), 264
Cano, J. P., 22(23), 103(97), 109,
  122(97), 262, 265
Carmella, S., 59(50), 65(50),
  68(50), 69(50), 103(23), 248(50),
  249(50), 263
Caroff, J., 146(141), 146(143), 149,
  152, 153, 156, 157, 267
Carroll, D. I., 103(95), 103(98),
  121(95), 121(98), 122(98),
  123(98), 124(98), 265
Carruthers, W., 50, 50(42), 51, 59,
  59(48), 73, 73(42), 136(117),
  146, 146(42), 148, 158(42), 159,
  159(148), 247, 263, 266, 267,
  269
Carugno, N., 146(136), 146(138),
  180, 180(157), 180(160), 184,
  266, 267
Cassarett, L. J., 479
Cassarett, M. G., 271, 271(4), 479
Cassel, K., 267
Cassidy, F. J., 207(181), 218(181),
  268
Castagnoli, N., Jr., 103(112),
  134(112), 136(112), 266, 486,
  487
Catalin, J., 22(23), 103(23),
  109(23), 262
Ceranni, E. S., 273(12), 480
Chakraborty, B. B., 180(167), 268
Chambaz, E., 71, 263
Chamberlain, W. J., 171(154),
  180(170), 180(172), 181(170),
  182(170), 267, 268
Chang, R. C., 245(202), 269
Chopra, N. M., 258, 258(211), 258(212),
  259, 269
Chortyk, O. T., 79(73), 180(170),
  180(172), 180(173), 180(176),
  181(170), 182(170), 182(172),
  186(172), 201(176), 204(176),
  264, 268
Clarke, P. A., 22(32), 34(32), 263,
  432(223), 488

Claussen, U., 325(63), 325(65), 407(171), 407(174), 482, 486
Cochran, T. G., 325(139), 380(139), 383(139), 485
Cogbill, E. C., 106(113), 266
Collin, E. J., 22(33), 41, 42, 263
Commins, B. T., 159(147), 267
Cook, J. W., 180(158), 187, 267
Corcoran, A. C., 114, 266
Corn, M., 207(179), 213, 216, 217, 268
Corp, P. J., 103(105), 128(105), 131(105), 265
Coutts, R. T., 325(153), 396(153), 398(153), 399(153), 485
Craig, J. C., 20, 21(10), 22, 23, 24, 103(112), 134(112), 136(112), 262, 266
Crocco, J. L., 325(161), 404, 486
Crouse, R. H., 159(150), 162, 163, 164, 166, 267
Cundiff, R. H., 17, 18, 27, 262, 263
Curry, S. H., 325(82), 482, 487
Cuzin, J. L., 146(135), 266

D

Dahmer, J., 432(209), 487
Dames, M. E., 325(126), 360(126), 363(126), 484
Davis, C. M., 432(232), 489
Davis, D. L., 22(21), 29(21), 30(21), 31(21), 262
Davis, H. J., 180(161), 189, 268
Davis, K. H., Jr., 325(88), 325(119), 336, 483, 484
Davis, T. W. M., 273(7), 273(9), 281(9), 282, 282(7), 284, 285, 286, 325, 325(59), 480, 481
Deiderich, L. R., 407(181), 416(181), 419(181), 486
Delaney, G. V., 325(141), 384(141), 485

DeMaio, L., 207(179), 213, 216, 217, 268
Demarco, D. V., 325(90), 483
DeSelms, R. C., 22(13), 25(13), 26(13), 262
De Zeeuw, R. A., 273(18), 273(25), 281, 301, 302, 304, 307, 325(95), 325(116), 325(128), 341(95), 351, 364, 374, 480, 483, 484
Dijkstra, G., 407(182), 413(182), 414(182), 486
DiPasqua, A. C., 325(141), 384(141), 485
DiSimone, L., 325(122), 484
Dobson, V. F., 265
Dollery, C. T., 432(211), 448(211), 488
Dong, M., 207(187), 269
Donike, M., 22(30), 33, 263
Doorenbos, N. J., 273(15), 273(16), 273(29), 281(29), 281(39), 295(15), 299(15), 303(16), 306(15), 308(16), 311, 312, 325(39), 361(39), 365(39), 480, 481
Dorfman, R. I., 103(91), 114(91), 119(91), 120(91), 265
Doull, J., 479
Dow, J., 103(111), 133, 266
Drews, C. J., 22(20), 27, 262
Dumas, C., 22(23), 103(23), 103(97), 109(23), 122, 262, 265
Durand, A., 103(97), 122(97), 265
Dzidic, I., 103(95), 103(98), 121(95), 121(98), 122(98), 123(98), 124(98), 265

E

Eglinton, G., 50, 50(43), 53, 158(145), 263, 267
Ek, N. A., 325(115), 350, 484
Ellington, J. J., 50(46), 54, 59(49), 60(49), 63, 64, 66(49), 68, 70,

AUTHOR INDEX

[Ellington, J. J.]
  79, 81, 82, 84, 85, 86, 263, 264
Enzell, C. R., 103(109), 266
Eskes, D., 325(127), 484

F

Fairbairn, J. W., 273(34), 281(36), 319, 325(121), 354, 481, 484
Fairlie, K., 325(156), 485
Falkman, S. E., 103(108), 131, 132, 263, 266
Farlow, R., 71(57), 263, 319(57), 402(170), 481, 486
Farmilo, C. G., 273(6), 273(7), 273(9), 281(6), 281(9), 282(7), 282(9), 284(9), 285(9), 286(9), 325, 325(6), 325(59), 479, 480, 481
Farnsworth, N. R., 325(118), 484
Fehlhaber, H. W., 325(65), 482
Fehr, K. O, 407(178), 409, 410, 486
Fenimore, D. C., 432(202), 432(232), 440, 441, 442, 443, 445, 487, 489
Fentiman, A. F., Jr., 407(180), 414, 415, 416, 417, 432(216), 456(216), 486, 488
Ferrand, E., 207(187), 269
Fetterman, P. S., 273(15), 273(16), 273(24), 273(28), 281(28), 281(39), 295, 299, 303, 303(16), 303(24), 304, 305, 306, 306(28), 308(16), 308(28), 311(28), 325(28), 325(39), 361(39), 365(39), 480, 481
Feyerabend, C., 103(107), 129, 131, 133, 265
Finkle, B. S., 34, 263, 325(142), 386, 485
Fish, F., 22(26), 22(27), 22(28), 22(29), 31(26), 31(29), 32(27),

[Fish, F.]
  32(28), 32(29), 33(28), 262, 263, 325(68), 325(114), 325(136), 329(68), 331(68), 407(175), 482, 484, 485, 486
Fisher, P. G., 79(71), 81(71), 82(71), 264
Foltz, R. L., 22(32), 34, 35, 263, 325(142), 386(142), 407(180), 414(180), 415(180), 416(180), 417(180), 432(216), 432(223), 456(216), 485, 486, 488
Fonseka, K., 325(154), 398, 400, 401, 432(228), 471, 474, 485, 488
Foote, R. S., 407(183), 418, 421, 422(188), 486, 487
Forehand, J. B., 136(126), 266
Forney, R., 273(13), 281(13), 281(38), 325(38), 325(78), 325(83), 338(38), 480, 481, 482
Fox, B. L., 325(156), 485
Frahm, B., 103(96), 265
Freeman, R. R., 432(202), 440(202), 441(202), 442(202), 443(202), 445(202), 487
Friedrich-Fiechtl, J., 325(144), 386, 389, 485
Frigerio, A., 483, 484, 486, 487, 489
Fujita, T., 325(74), 332(74), 333(74), 336(74), 482

G

Gagne, D., 325(149), 391, 392, 485
Gaoni, Y., 273, 287(55), 325(94), 339, 341, 479, 481, 483
Garner, J. W., 159(150), 162(150), 163(150), 164(150), 166(150), 267
Garrett, E. R., 325(140), 325(162), 380, 384, 384(166), 385, 405, 406, 407, 432(166), 432(224), 444, 485, 486, 488
Gelpi, E., 146(144), 154, 158, 267

Gerber, J. N., 103(102), 126, 127, 265
Giger, W., 226(195), 232, 239, 269
Gilman, A., 262
Giovannozzi-Sermanni, G., 146(136), 266
Goldfarb, T., 103(99), 123, 265
Goldman, N. L., 20, 22(9), 23(9), 24(9), 262
Gonzalez, A. G., 50(43), 53(43), 263
Goodman, L. S. 262
Gori, G. B., 257
Gouw, T. H., 226(191), 227, 269
Gouyette, A. J., 325(162), 405(162), 406(162), 407(162), 432(224), 486, 488
Graffeo, A. P., 432(223), 488
Green, D. E., 325(89), 432(208), 432(219), 432(231), 448(208), 483, 487, 488, 489
Griest, W. H., 250, 251(204), 269
Gritz, E. R., 103(99), 103(100), 123(99), 125, 265
Grlic, L., 273(10), 282(45), 480, 481
Grob, G., 71(56), 263, 402(169), 486
Grob, K., 71(56), 103(86), 136(118), 136(119), 136(122) 136(123), 136(125), 136(127), 137, 137(118), 137(119), 138, 138(122), 140, 140(122), 141, 145, 263, 264, 266, 402(169), 486
Groenen, P. J., 232(197), 240, 269
Gruenke, L. D., 103(112), 134(112), 136(112), 266
Grundke, K., 103(103), 126, 265
Grunwald, C., 31(36), 71, 263, 422(189), 487
Gudzinowicz, B. J., 88(77), 207, 207(177), 207(178), 213, 215, 216, 264, 268

Gudzinowicz, M. J., 88(77), 207, 207(178), 264, 268
Guerin, M. R., 103(87), 111, 113, 115, 116, 117, 118, 171(156), 178, 181, 232, 232(196), 232(198), 232(199), 240, 241, 242, 243, 244, 250, 251(204), 253, 256(208), 264, 267, 269
Guerrero, O., 273(15), 295(15), 299(15), 306(15), 480
Guichon, G., 159(152), 162(152), 267
Guillerm, R., 22(23), 103(23), 103(97), 109(23), 122(97), 146(141), 146(143), 149(141), 152(141), 153(143), 156(143), 157(143), 262, 265, 267
Gustafsson, B., 432(201), 432(204), 440(201), 487

H

Hackel, E., 289(53), 481
Hadley, K. W., 273(26), 273(27), 273(28), 273(31), 281(28), 281(39), 306(26), 306(27), 306(28), 308, 308(26), 308(27), 308(28), 311(28), 313(31), 313(56), 314(31), 315(31), 318(31), 325(28), 325(39), 325(56), 325(119), 325(146), 361(39), 365(39), 374(56), 377(56), 378(56), 379(56), 390(146), 480, 481, 484, 485
Haegele, K. D., 103(98), 121(98), 122(98), 123(98), 124(98), 265
Haffner, G., 282(42), 282(43), 287(43), 332, 481, 486
Hall, K., 103(111), 133, 266
Halmer, O. M., 114, 266
Halpern, A. F., 325(73), 482
Hamilton, R. J., 50(43), 53(43), 263
Handa, K. L., 325(62), 326(62), 327(62), 482
Haq, M. Z., 407(181), 416, 419, 486

Harke, H. P., 22(20), 27, 103(96), 262, 265
Harlow, E. S., 106(113), 266
Harvey, D. J., 325(147), 325(152), 390, 391, 393, 396, 397, 432(213), 432(214), 432(215), 432(225), 432(226), 432(227), 432(234), 452, 454, 454(213), 454(215), 455, 465, 469, 470, 470(226), 490(227), 471, 471(213), 471(225), 472, 473, 485, 488, 489
Heaysman, L. T., 325(67), 328, 329, 331, 482
Hecht, S. S., 41(39), 41(40), 42, 44, 59(50), 65, 68, 69, 248, 249, 263
Heerma, W., 407(179), 407(182), 411(179), 413(182), 414, 486
Henderson, P. T., 325(117), 484
Henderson, W., 71(60), 264
Hengen, M., 103(110), 133, 266
Hengen, N., 103(110), 133, 266
Henry, J. T., 313(56), 325(56), 374(56), 377(56), 378(56), 379(56), 481
Herweijer, J. J., 273(32), 481
Hidy, B. J., 432(223), 488
Higman, H. C., 79(71), 79(73), 81(71), 82(71), 180(176), 201(176), 204(176), 264, 268
Hines, C. H., 325(73), 482
Hirth, L., 71(61), 264
Ho, C. H., 250, 251, 269
Hobbs, M. E., 75(67), 146(132), 146(133), 171, 171(153), 176, 177, 178, 179, 264, 266, 267
Hobson, G. D., 267
Hodgkins, J. E., 22(16), 27, 28, 262
Hoffman, N. E., 325(113), 348, 349, 350, 484
Hofmann, A., 432(194), 487
Hoffmann, D., 41, 41(39), 41(40), 42(40), 44(40), 59(50), 65(50), 68(50), 69(50), 79(70), 87(76),

[Hoffman, D.]
103(88), 103(90), 159(88), 159(90), 159(149), 161, 169(90), 170(90), 171(70), 171(88), 171(90), 173(90), 174(90), 177, 180(162), 180(163), 180(164), 180(166), 180(168), 180(171), 187(171), 199(171), 200, 203, 205, 206, 209, 210, 211, 212, 248(50), 249(50), 250(205), 252, 252(76), 253, 254, 255, 257, 258, 261, 263, 264, 265, 267, 268, 269
Holley, J. H., 273(31), 313, 314, 315, 318, 325(146), 390(146), 480, 485
Hollister, L. E., 325(81), 336(81), 432(208), 448, 482, 487
Holloway, P. J., 325(69), 325(71), 329, 482
Holmstedt, B., 432(201), 432(204), 440(201), 487
Holzer, G., 136(128), 140, 140(130), 266
Honeycutt, R. G., 146(142), 153(142), 155(142), 267
Hood, L. V. S., 325(126), 360(126), 363, 484
Horning, E. C., 71, 103(95), 103(98), 121, 122, 123, 124, 263, 265
Horning, M. G., 103(95), 103(98), 121(94), 121(98), 122(98), 123(98), 124(98), 265
Horton, A. D., 103(87), 111(87), 113(87), 115(87), 116(87), 117(87), 118(87), 136(129), 145, 146, 147, 232(198), 232(199), 244, 253, 256, 264, 266, 269
Hoshaku, H., 103(82), 264
Hoyland, J. R., 22(32), 34(32), 263
Hsu, F., 232(200), 244(200), 245(200), 246(200), 247(200), 269
Hunneman, D. H., 325(97), 483

Hunt, C. A., 384(166), 432(166), 432(224), 444, <u>486</u>, <u>488</u>

I

Idanpaan-Heikkila, J. E., <u>488</u>
Ikeda, R. M., 103(85), 104, 107, 108, 109, 110, <u>265</u>
Ireland, M. S., 180(165), <u>268</u>
Ireland, S., 159, 160(146), <u>267</u>
Isaac, P. F., 103(104), 103(106), 129, 130, <u>265</u>

J

Jacin, H., 22(18), 27, 29, 103(18), 104, 126, <u>262</u>
Jackson, G. C., 103(105), 128(105), 131(105), <u>265</u>
Jaffe, J. H., 1, 2, <u>262</u>
Jain, N. C., 273(13), 281(13), 281(38), 325(38), 325(83), 338(38), <u>480</u>, <u>481</u>, <u>482</u>
James, A. T., 207, <u>269</u>
Janini, G. M., 180(175), 189, 207(185), 207(186), 220, 220(186), 223, 225, <u>268</u>, <u>269</u>
Jarboe, C. H., 17, 19, 20, 103(8), <u>262</u>
Jarvik, M. E., 103(99), 103(100), 123(99), 125(100), <u>265</u>
Jaspers-von Wouw, M. G. J., 325(108), <u>483</u>
Jenden, D. J., 22(24), 30, <u>262</u>
Jensen, W. N., 325(79), 333(79), 334(79), <u>482</u>
Jenkins, R. W., Jr., 103(89), 136(24), <u>265</u>, <u>266</u>, 273(30), <u>480</u>
Jentoft, R. E., 226(191), 227(191), <u>269</u>
Johnston, K., 207(185), 220(185), 223(185), <u>269</u>
Johnston, R. L., 50(45), 54, 58, 59, 60, 61, 62, 158(45), <u>263</u>
Johnstone, R. A. W., 50, 50(42), 51, 59, 59(48), 73, 73(42), 75(66), 136, 136(117), 146,

[Johnstone, R. A. W.]
146(42), 148, 158(42), 159, 159(148), 180(158), 187(158), 247, <u>263</u>, <u>264</u>, <u>266</u>, <u>267</u>, <u>269</u>
Jones, G., 325(143), <u>485</u>
Jones, L. A., 50(45), 54, 58, 59, 60, 61, 62, 158(45), <u>263</u>, 325(120), 352, 353, 407(120), 407(183), 407(184), 418, 420, 421, 422(188), <u>484</u>, <u>486</u>, <u>487</u>
Jones, L. V., 325(158), 403(158), <u>485</u>
Jones, T. C., 79(69), 171(69), <u>264</u>
Jones, W. L., 103(89), 136(124), <u>265</u>, <u>266</u>
Joyce, C. R. B., 325(82), <u>482</u>, <u>487</u>

K

Kabelik, K., 282(44), <u>481</u>
Kalant, H., 407(178), 409, 410, <u>486</u>
Kanazawa, J., 22, 22(12), <u>262</u>
Kanter, S. L., 325(81), 336(81), 432(208), 432(231), 448(208), <u>482</u>, <u>487</u>, <u>489</u>
Karlsen, J., 325(101), <u>483</u>
Kay, H., 325(82), <u>482</u>, <u>487</u>
Kaye, W. I., 54, <u>263</u>
Kazakoff, C. W., 325(153), 396(153), 398(153), 399(153), <u>485</u>
Keith, C. H., 136(121), 146(139), <u>266</u>, <u>267</u>
Keith, E. S., 273(15), 273(16), 295(15), 299(15), 303(16), 306(15), 308(16), <u>480</u>
Keller, C. J., 31, 71, <u>263</u>
Kettenes-van den Bosch, J. J., 407(187), 424, 431, <u>487</u>
Kiefer, J. E., 103(83), <u>264</u>
Kilburn, K. D., 180(167), <u>268</u>
King, L. J., 325(78), <u>482</u>
King, W. H., 207(181), 218(181), <u>268</u>
Kingston, C. R., 282(47), 325, 325(60), 326, <u>481</u>

AUTHOR INDEX

Kinzer, G. W., 407(180), 414(180), 415(180), 416(180), 417(180), <u>486</u>
Kirchgessner, W. G., 325(141), 384, <u>485</u>
Kirk, P. L., 282(47), 325, 325(60), 326, <u>481</u>
Knaus, E. E., 325(153), 396, 398, 399, <u>485</u>
Knight, J. B., 325(93), 325(129), 339, 340, 364(129), 366(129), 367(129), 368(129), 369(129), 370(129), 371(129), 372(129), 373(129), <u>483</u>, <u>484</u>
Knights, B. A., 71(58), 71(59), <u>264</u>
Knowlton, D. A., 22(32), 34(32), <u>263</u>
Kobashi, Y., 11, 14, 15, 16, 22, 22(14), 29, 103, 103(82), 105, <u>262</u>, <u>264</u>
Kopin, I. J., 444(238), <u>489</u>
Korte, F., 289, 325(63), 325(65), 407(171), 407(174), <u>481</u>, <u>482</u>, <u>486</u>
Kostenbauder, H. B., 22(31), 33, 34, 35, 36, <u>263</u>
Kozak, A. I., 146(137), <u>267</u>
Krejci, K., 282(44), <u>481</u>
Kriz, J., 226(190), 227(190), 228(190), <u>269</u>
Kubena, R. K., 325(118), <u>484</u>
Kuhn, W. F., 136(126), <u>266</u>
Kuksis, A., 81, 83(74), <u>264</u>
Kuppers, F. J. E. M., 325(101), 325(123), 354(123), 407(179), 407(182), 411, 413(182), 414(182), <u>483</u>, <u>484</u>, <u>486</u>

L

Lane, R., 273(7), 282(7), <u>480</u>
Lao, R. C., 207(184), 219, 223, <u>268</u>
Lassiter, C. W., 146(140), 148(140), 151(140), <u>267</u>
Laurie, W., 71(59), <u>264</u>
Laven, A., 407(179), 407(182), 411(179), 413(182), 414(182), <u>486</u>
Leaf, G., 282(50), <u>481</u>
Leander, K., 325(100), 325(134), 409(100), 432(196), 432(201), 432(204), 432(209), 432(217), 432(221), 437(196), 440(201), 458(217), 459(217), 461(217), <u>483</u>, <u>485</u>, <u>487</u>, <u>488</u>
Leclercq, J., 407(172), <u>486</u>
Lee, H. C., 103(90), 159(90), 169(90), 170(90), 171(90), 173(90), 174(90), <u>265</u>
Lee, M. L., 70, 72(51), 73(51), 180(169), 180(174), 185, 188, 198, 226(169), 226(193), 226(194), 230, 230(169), 230(193), 230(194), 231, 233, 238, <u>263</u>, <u>268</u>, <u>269</u>, 273(33), 313(33), 320(33), 324(33), 325(133), 325(155), 402(155), 407(133), 407(186), 409, 424, 425, 426, 427, 428, 429, 430, <u>481</u>, <u>485</u>, <u>486</u>
Legault, D., 325(125), 357(125), 361(125), <u>484</u>
Leighty, E. G., 432(216), 456, <u>488</u>
Lemberger, L., 444, <u>489</u>
Lenard, K., 325(64), 326(64), <u>482</u>
Lerner, M., 273(8), 282, 283, 325(61), 325(66), 325(72), <u>480</u>, <u>481</u>, <u>482</u>
Lerner, P., 295, 306, 325(54), <u>481</u>
Levi, L., 281(40), 281(41), 325(62), 326(62), 327(62), <u>481</u>, <u>482</u>
Levine, J., 282(49), <u>481</u>
Levins, R. J., 103(85), 104, 107, 108, 109, 110, <u>265</u>
Levitt, T., 103(107), 129(107), 131(107), 133(107), <u>265</u>
Levy, S., 432(235), 479, <u>489</u>
Lewis, D. T., 325(67), 328(67), 329(67), 331(67), <u>482</u>

AUTHOR INDEX

Lewis, J. S., 54, 263
Lichenstein, H. A., 245(201), 269
Liebmann, J. A., 281(36), 325(121), 354, 481, 484
Lin, D. C. K., 432(223), 488
Lindgren, B. O., 71(62), 264
Lindgren, J. E., 325(134), 325(143), 432(201), 432(204), 432(217), 432(218), 432(221), 432(229), 432(233), 440(201), 458(217), 458(218), 459(217), 461(217), 461(218), 462(218), 474(229), 474(233), 476(229), 476(233), 485, 487, 488, 489
Lindsey, A. L., 159(147), 267
Liu, Y. Y., 87(76), 252, 253, 254, 264, 269
Locke, D. C., 207(187), 269
Loeffler, K. O., 432(231), 489
Loewe, S. J., 282(48), 282(49), 326(164), 481, 486
Loheac, J., 159(152), 162(152), 267
Lonberg, E., 325(115), 350(115), 484
Louis, M. C., 267
Lousberg, R. J. J. C., 325(101), 325(123), 354(123), 407(179), 407(182), 411(179), 413(182), 414(182), 483, 484, 486,
Low, C. E., 70, 72(51), 73(51), 263, 273(33), 313(33), 320(33), 324(33), 325(155), 402(155), 481, 485
Loy, P. R., 432(202), 440(202), 441(202), 442(202), 443(202), 445(202), 487
Lundgren, R. A., 103(108), 131(108), 132(108), 266
Lyerly, L. A., 103(84), 264

M

Machata, G., 325(76), 482

Macpherson, A. S., 432(203), 444(203), 487
Maehly, A. C., 325(115), 350(115), 484
Malaterre, M., 159(152), 162, 267
Malingre, T. M., 273(25), 281(25), 304(25), 307(25), 325(116), 351(116), 374(25), 480, 484
Manno, J., 273(13), 281(13), 281(38), 325(38), 325(83), 338(38), 480, 481, 482
Marai, L., 81, 83(74), 264
Markunas, P. C., 17, 18, 27, 262, 263
Martin, B. R., 432(215), 432(217), 432(218), 432(225), 432(226), 432(234), 454, 458, 458(217), 458(218), 459, 461, 461(217), 462, 465(225), 469(225), 470(226), 471(225), 472(226), 488, 489
Martin, H. F., 88(77), 207(178), 264, 268
Martin, L., 273(6), 281, 325(6), 479
Martin, N. H., 325(88), 336(88), 483
Mary, N. Y., 20, 21(10), 22(9), 23(9), 24(9), 262
Maskarinec, M. P., 70, 72(51), 73(51), 263, 325(155), 402(155), 407(185), 422, 423, 485, 486
Masoud, A. N., 273(29), 281(29), 311, 312, 480
Massingill, J. L., Jr., 22(16), 27, 28, 262
Masuda, Y., 253, 255, 269
Matchett, J. R., 282(49), 481
Matsuo, Y., 325(75), 333(75), 482
Mauch, A., 103(96), 265
McCallum, N. K., 325(157), 402, 403, 432(200), 438, 440, 485, 487
McKennis, H., Jr., 103(94), 103(109), 114, 121, 265, 266

McNiven, N. L., 103(91), 114, 119, 120, 265
Means, R. E., 50(44), 52(44), 55(44), 56(44), 57(44), 77, 78(68), 158(44), 263, 264
Mechoulam, R., 271(1), 273, 274, 287(55), 325(1), 325(94), 325(130), 335, 339, 341, 432(235), 444(239), 479(235), 479, 481, 483, 484, 489
Melville, R. S., 265
Menefee, B. S., 103(81), 159(81), 264
Menezes, F., 444(239), 489
Merkus, F. W. H. M., 325(96), 325(103), 325(108), 325(116), 342, 351(116), 483, 484
Mikes, F., 407(177), 408, 432(194), 486, 487
Miller, R. L., 74(63), 74(64), 74(65), 75(63), 76(63), 264
Mills, A. L., 325(61), 481
Mitoma, C., 437(237), 489
Moffat, A. C., 22(26), 31(26), 36, 262, 263, 325(150), 485
Mokhnachev, I. G., 22(19), 29(19), 262
Mold, J. D., 50(44), 52, 55, 56, 57, 77, 78, 158(44), 263, 264
Mole, M. L., Jr., 313(56), 325(56), 325(146), 374(56), 377(56), 378(56), 379(56), 390(146), 481, 485
Monkman, J. L., 207(184), 268
Moore, H., 146(142), 153(142), 155(142), 267
Morel, S., 146(135), 266
Morris, W. J., 325(129), 364(129), 366(129), 367(129), 368(129), 369(129), 370(129), 371(129), 372(129), 373(129), 484
Moshy, R. J., 22(18), 27(18), 103(18), 104(18), 262
Mostecky, J., 226(190), 227, 228, 269

Mount, S. F., 325(61), 481
Mumpower, R. C., 103(83), 264
Munson, J. W., 41(41), 45, 46, 47, 48, 49, 263
Murayama, T., 22(22), 262
Muschik, G. M., 207(186), 220(186), 225(186), 269
Myher, J. J., 81, 83, 264

N

Nahas, G. G., 488
Naworal, J., 79(72), 264
Neelakantan, L., 22(31), 33, 34, 35, 36, 263
Neetsen, E., 325(104), 343(104), 432(104), 483
Nesnow, S., 180(163), 200(163), 203(163), 268
Neumeyer, J. L., 271(2), 272(2), 274, 325(2), 479
Newman, R. H., 103(89), 136(124), 265, 266
Newsome, R. J., 136(121), 146(139), 266, 267
Nibbering, N. M. M., 325(124), 355(124), 360(124), 484
Nigam, I. C., 325(62), 326(62), 327(62), 482
Nigam, M. C., 325(62), 326, 327, 482
Nilsson, I. M., 273(14), 281(14), 295(14), 432(191), 432(196), 432(201), 432(204), 437(196), 440(201), 480, 487
Nilsson, J. L. G., 432(196), 437(196), 487
Nishioka, I., 325(70), 325(74), 325(75), 325(86), 332(70), 332(74), 333(74), 333(75), 336(70), 336(74), 336(86), 337(86), 482
Nordqvist, M., 432(211), 432(217), 432(218), 448(211), 458(217), 458(218), 459(217), 461(217),

[Nordqvist, M.]
    461(218), 462(218), <u>488</u>
Norman, V., 136(121), 146(139),
    <u>266</u>, <u>267</u>
Novotny, M., 70, 71, 71(54),
    71(57), 72, 73, 88(78),
    180(169), 180(74), 185(174),
    188(174), 198(174), 226(78),
    226(169), 226(193), 226(194),
    230(169), 230(193), 230(194),
    231(193), 233(194), 238(194),
    <u>263</u>, <u>264</u>, <u>268</u>, <u>269</u>, 273,
    273(33), 313, 319(57), 320,
    324, 325(133), 325(133),
    325(155), 402, 402(167),
    402(170), 407(133), 407(185),
    407(186), 409, 422(185),
    423(185), 424(187), 425(186),
    426(186), 427(186), 428(186),
    429(186), 430(186), <u>481</u>, <u>485</u>,
    <u>486</u>

O

Oakley, E. T., 171(155), 175, <u>267</u>
Ohlsson, A., 273(14), 281(14),
    295, 432(191), 432(221), <u>480</u>,
    <u>487</u>, <u>488</u>
Olerich, G., 103(87), 111(87),
    113(87), 115(87), 116(87),
    117(87), 118(87), 171(156),
    178(156), 181(156), <u>264</u>, <u>267</u>
Olerich, M. R., 232(199), <u>269</u>
Olofsson, K., 273(14), 281(14),
    295(14), <u>480</u>
Omura, I., 136(122), 138(122),
    140(122), <u>266</u>
O'Neill, H. J., 159(150), 162(150),
    163(150), 164(150), 166(150),
    <u>267</u>
Ornaf, R. M., 41(39), 41(40),
    42(40), 44(40), <u>263</u>
Oro, J., 136(128), 140(128),
    146(144), 154, 158, <u>266</u>,
    <u>267</u>

Osadchik, M., 273(9), 281(9),
    282(9), 284(9), 285(9), 286(9),
    <u>480</u>
Osborne, J. S., 146(133), <u>266</u>
Osborne, N. B., 258, 258(211), <u>269</u>
Ourisson, G., 71(61), <u>264</u>

P

Page, B. F. J., 103(105), 103(108),
    128(105), 131(105), 131(108),
    132(108), <u>265</u>, <u>266</u>
Page, I., 114, <u>266</u>
Pappas, N. A., 15, 18, 103(81),
    159(81), <u>262</u>, <u>264</u>
Parker, J. J., 325(139), 380, 383,
    <u>485</u>
Parker, J. M., 325(135), <u>485</u>
Parker, K. D., 325(73), <u>482</u>
Paris, M. R., 325(131), 374, 375,
    <u>484</u>
Paris, R. R., 325(131), 374(131),
    375, <u>484</u>
Patashnik, S., 103(91), 114(91),
    119(91), 120(91), <u>265</u>
Patel, A. R., 407(181), 416(181),
    419(181), <u>486</u>
Paton, W. D. M., 325(147), 390,
    391, 432(213), 432(214),
    432(215), 432(225), 432(226),
    432(227), 432(234), 452, 454,
    454(213), 454(215), 455,
    465(225), 469(225), 470(226),
    470(227), 471, 471(213),
    471(225), 472(226), 473(227),
    <u>485</u>, <u>488</u>, <u>489</u>
Patrianakos, C., 257, <u>269</u>
Patterson, D. A., 273(30), 325(80),
    335, <u>480</u>, <u>482</u>
Patton, H. W., 54, 146(131), <u>263</u>,
    <u>266</u>
Pecsar, R. E., 207(180), 213, 217,
    <u>268</u>
Perez-Reyes, M., 432(198),
    438(198), <u>487</u>

AUTHOR INDEX 503

Perkins, J., 432(203), 444(203), 487
Petersson, K., 432(209), 487
Petrakis, N. L., 103(112), 134, 135, 136, 266
Phillipe, R. J., 146(132), 146(142), 153, 155, 266, 267
Phillips, R., 273(13), 281, 325(83), 480, 482
Pilotti, A., 103(109), 266
Pitt, C. G., 325(88), 336(88), 432(198), 438(198), 483, 487
Plimmer, J. R., 75(66), 247, 264, 269
Popl, M., 226(190), 227(190), 228(190), 269

Q

Quan, P. M., 136(117), 180(158), 187(158), 266, 267
Quimby, M. W., 273(15), 273(16), 281(39), 295(15), 299(15), 303(16), 306(15), 308(16), 325(39), 361(39), 365(39), 480, 481
Quin, L. D., 2, 13, 15, 18, 22, 103, 103(11), 103(80), 103(81), 105, 106, 107, 159, 171, 171(153), 176, 177, 178, 179, 262, 264, 267

R

Rainey, W. T., 171(156), 178(156), 181(156), 267
Raisinghani, K. H., 103(91), 114(91), 119(91), 120(91), 265
Ramachandran, S., 325(78), 482
Rand, M. J., 103(104), 103(106), 126, 129, 130, 265
Raphael, R. A., 50(43), 53(43), 263
Rapoport, H., 22(13), 22(15), 25(13), 26, 26(13), 262
Rasmussen, K. E., 273(32), 325(105), 325(106), 325(145),

[Rasmussen, K. E.]
325(148), 389, 389(105), 389(106), 481, 483, 485
Rasmussen, S., 325(105), 325(106), 389(105), 389(106), 483
Rathkamp, G., 180(162), 180(163), 180(164), 180(166), 180(168), 200(163), 203(163), 205, 206, 209, 210, 211, 212, 257, 253, 261, 268, 269
Raymond, A., 273(33), 313(33), 320(33), 324(33), 481
Resnik, F. E., 171(155), 175(155), 267
Richter, W., 158(145), 267
Roberts, J., 432(203), 444(203), 487
Robinson, A. E., 432(193), 433, 487
Roch, M., 22(24), 30(24), 262
Rose, S. J., 407(181), 416(181), 419(181), 486
Roseboom, J., 325(162), 405(162), 406(162), 407(162), 486
Rosene, C. J., 17, 19, 20, 103(8), 262
Rosenfeld, J. M., 325(161), 404, 432(197), 432(203), 432(212), 432(220), 432(222), 437(197), 444, 446, 451, 461, 463, 464, 465, 466, 486, 487, 488
Rossi, S., 180, 180(157), 180(160), 184, 267
Rowan, M. G., 273(34), 319, 481
Rowland, M., 103(92), 119(92), 119(116), 265, 266
Rushneck, D. R., 136(120), 138, 139, 266
Russell, M. A. H., 103(107), 129(107), 131(107), 133(107), 265
Ruth, J. M., 50(44), 52(44), 55(44), 56(44), 57(44), 77, 78(68), 146(142), 153(142), 155(142), 158(44), 263, 264, 267

S

Salemink, C. A., 325(101), 325(123), 354(123), 407(179), 407(182), 407(187), 411(179), 413(182), 414(182), 424, 431, 483, 484, 486, 487
Sandberg, F., 432(191), 432(201), 440(201), 487
Santavy, F., 282(44), 481
Sato, R., 22, 22(12), 262
Sawicki, E., 267
Scaringelli, F., 282(46), 481
Schaffner, C., 226(195), 232, 239, 269
Schepartz, A. I., 50(46), 54(46), 59(49), 60(49), 63(46), 64(46), 66(49), 68(46), 70(46), 79(71), 81(71), 82(71), 84(46), 85(46), 86(46), 263, 264
Schievelbein, H., 103(103), 126, 265
Schmeltz, I., 74, 74(65), 75, 76, 79(69), 87(76), 171(69), 171(154), 180(171), 187, 199, 252(76), 264, 267, 268
Scholtzhauer, P. F., 50(46), 54(46), 59(49), 60(49), 63(46), 64(46), 66(49), 68(46), 70(46), 84(46), 85(46), 86(46), 263
Schou, J., 325(104), 343, 432(104), 483
Schroer, J. A., 207(186), 220(186), 225(186), 269
Schuller, D., 103(96), 265
Schultz, F. J., 180(165), 268
Schultz, O. E., 282(42), 282(43), 287(43), 332, 481, 486
Scott, D., 22(26), 22(27), 22(28), 22(29), 31(26), 31(29), 32(27), 32(28), 32(29), 33(28), 262, 263
Scott, P. M., 158(145), 267
Seabra, A. P., 407(172), 407(173), 486

Searl, T. D., 207(181), 218, 268
Segelman, A. B., 279(35), 325(118), 481, 484
Seibl, J., 136(122), 138(122), 140(122), 266
Sellier, N., 159(152), 162(152), 267
Sever, P. S., 271(3), 479
Severson, R. F., 59(49), 60, 61, 65, 66, 79(73), 180(170), 180(172), 180(173), 181(170), 182, 182(170), 186, 263, 264, 268
Shagoury, R. A., 271(2), 272(2), 274, 325(2), 479
Shaikl, B., 180(175), 189(175), 268
Sherman, L. R., 258(212), 259, 269
Sherstyackh, N. A., 22(19), 29(19), 262
Shoyama, Y., 325(70), 325(74), 325(75), 325(86), 332, 332(70), 333, 333(75), 336, 336(70), 336(74), 337, 482
Shukla, P., 432(232), 489
Shunbo, F., 245(202), 269
Shvo, Y., 325(64), 326(64), 482
Sieper, H., 289(53), 481
Silberstein, S. D., 444(238), 489
Simon, W., 136(122), 138(122), 140(122), 266
Skinner, R. F., 325(98), 325(129), 364, 366, 367, 368, 369, 370, 371, 372, 373, 432(199), 483, 484, 487
Slanski, J. M., 22(18), 27(18), 103(18), 104(18), 262
Small, E., 273(58), 312, 313, 481
Smith, C. M., 326(164), 486
Smith, D. M., 273(6), 281(6), 281(40), 281(41), 325(6), 479, 481
Smith, R. N., 325(151), 325(158), 325(160), 393, 394, 403, 485

Snook, M. E., 180(170), 180(172), 180(173), 180(176), 181, 182, 182(172), 186(172), 201, 204, 268
Sobotka, A. J., 325(92), 483
Song, C. H., 325(81), 336, 482
Spears, A. W., 146(140), 148, 151, 159, 159(151), 160(146), 162, 168, 180(165), 267, 268
Sperling, A., 273(23), 295, 301, 325(92), 480, 483
Spiteller, G., 325(144), 386, 389, 485
Srivastava, S. C., 103(94), 121(94), 265
Stead, A. H., 22(26), 31, 262
Stedman, R. L., 2, 11, 74(63), 74(64), 74(65), 75(63), 76(63), 101, 171(154), 262, 264, 267
Steel, G., 71(60), 264
Steentoft, A., 325(104), 343(104), 432(104), 483
Steinigen, M., 325(85), 482
Stembal, B. L., 325(135), 485
Stevens, H. M., 325(80), 335, 432, 432(190), 433, 482, 487
Stevens, R. K., 50(44), 52(44), 55(44), 56(44), 57(44), 158(44), 263
Stillwell, R. N., 103(95), 103(98), 121(95), 121(98), 122(98), 123(98), 124(98), 265
Stokely, J. R., 253, 256(208), 269
Stolerman, I. P., 103(99), 123(99), 265
Stone, H. M., 325(77), 432 432(190), 433, 482, 487
Stratmann, D., 22(30), 33, 263
Street, H. V., 325(84), 482
Strojny, E. J., 282(51), 481
Stromberg, L. F., 273(19), 273(20), 273(21), 281, 302, 302(19), 302(20), 302(21), 325(99), 325(115), 325(137), 325(138), 342, 350(115), 375, 377(19),

[Stromberg, L. F.] 377(137), 377(138), 380, 381, 382, 407, 407(99), 480, 483, 484, 485
Stuckey, C. L., 226(192), 227, 229, 269
Svahn, C. M., 71(62), 264
Svendsen, A. B., 325(101), 325(105), 325(106), 389(105), 389(106), 483
Swinehart, J. S., 146(137), 267

T

Tagg, J., 437, 489
Taguchi, V. Y., 432(212), 432(222), 451, 488
Tanabe, M., 437(237), 489
Taylor, E. C., 282(51), 325(64), 326, 328, 481, 482
Taylor, D. M., 34, 263, 325(142), 386(142), 485
Terlouw, J. K., 407(179), 407(182), 411(179), 413(182), 414(182), 486
Tgminski, I. J., 325(90), 483
Theiner, M., 325(118), 484
Thijssen, H. H., 325(110), 483
Thomas, R. S., 207(184), 219(184), 223(184), 268
Thor, L. V., 146(135), 266
Thornton, R. E., 180(159), 180(167), 267, 268
Tishbee, A., 245(201), 269
Todd, A. R., 282(50), 282(52), 481
Toffoli, F., 273(12), 480
Tomiyasu, N., 273(11), 285(11), 287(11), 480
Tosk, J., 180(171), 187(171), 199, 268
Triggs, E. J., 103(92), 103(93), 119(92), 119(93), 120, 123, 265
Truitt, E. B., 407(176), 486
Tsau, J., 325(140), 380, 384, 385, 485

Tsukamoto, H., 273(11), 285(11), 287(11), 480
Tuoey, G. P., 146(131), 266
Turk, R. F., 273(13), 281(13), 281(38), 325(38), 325(78), 325(83), 325(91), 338, 480, 481, 482, 483
Turnbull, L. B., 114, 266
Turner, C. E., 273(24), 273(26), 273(27), 273(28), 273(31), 281, 281(39), 303, 303(24), 304, 305, 306, 306(26), 306(27), 306(28), 308, 308(26), 308(27), 311, 313, 313(31), 313(56), 314(31), 315(31), 318(31), 325(28), 325(39), 325(56), 325(119), 325(132), 325(146), 361, 365, 374, 377, 378, 379, 390, 480, 481, 484, 485
Tursch, B. M., 407(172), 486

V

Vandenheuvel, F. A., 273(7), 282(7), 480
Van Gemert, L. J., 232(197), 240, 269
Van Ginneken, C. A. M., 273(17), 273(18), 273(22), 287(17), 287(22), 289(17), 290(17), 291(17), 292(17), 293(17), 294(17), 301(18), 325(95), 325(107), 325(110), 325(111), 325(112), 325(117), 325(124), 325(128), 341(95), 343(107), 344(107), 345(112), 346(107), 347(107), 355(124), 360(124), 364(128), 480, 483, 484
Van Rossum, J. M., 273(17), 273(18), 273(22), 287(17), 287(22), 289(17), 290(17), 291(17), 292(17), 293(17), 294(18), 301(18), 325(95), 325(107), 325(110), 325(111), 325(112), 325(117), 325(123),

[Van Rossum, J. M.] 325(124), 341(95), 343(107), 344(107), 345(112), 346(107), 347(107), 354(123), 355(124), 360(124), 480, 483, 484
Varanelli, C., 432(210), 448, 449, 488
Vaughn, C. G., 325(158), 403(158), 485
Veal, J. T., 103(101), 125, 265
Verner, D., 325(125), 357(125), 361(125), 484
Veron, J., 146(141), 146(143), 149(141), 152(141), 153(143), 156(143), 157(143), 267
Verwey, A. M. A., 325(109), 325(127), 483, 484
Verzele, M., 71(55), 263, 402(168), 486
Viala, A., 22(23), 103(23), 103(97), 109(23), 122(97), 262, 265
Vieira, F. J., 407(172), 486
Vollmin, J. A., 103(86), 136(122), 136(125), 138, 140, 141, 145, 265, 266
Vree, T. B., 273(17), 273(18), 273(22), 287, 287(17), 287(22), 288, 289, 290, 291, 292, 293, 294, 301(18), 325(95), 325(107), 325(110), 325(111), 325(112), 325(117), 325(123), 325(124), 325(128), 325(159), 341, 343, 344, 345, 346, 347, 354(123), 355, 356, 360, 364(128), 404, 480, 483, 484, 485

W

Walker, E. A., 325(67), 328(67), 329(67), 331(67), 482
Wall, M. E., 325(88), 336(88), 432(195), 432(198), 432(230), 433, 435, 436, 438, 483, 487, 488
Waller, C. W., 273(15), 281(37),

AUTHOR INDEX

[Waller, C. W.]
 281(39), 295(15), 299(15), 306(15), 325(39), 361(39), 365(39), 480, 481
Waltz, P., 180(157), 267
Wartman, W. B., 106(113), 266
Waser, P. G., 407(177), 408, 432(194), 486, 487
Watanabe, M., 15, 22, 22(14), 29, 103(82), 262, 264
Weeks, W. W., 22(21), 29, 30, 31, 262
Weissbecker, L., 171(155), 175(155), 267
Wheeler, R. J., 207(183), 268
Whittemore, I. M., 226(191), 227(191), 269
Widman, M., 325(143), 325(154), 398(154), 400(154), 401(154), 432(196), 432(209), 432(211), 432(221), 432(228), 432(229), 432(233), 432(236), 437, 448, 471, 474, 474(229), 474(233), 474(236), 476(229), 476(233), 477(236), 478(236), 485, 487, 488, 489
Wijsbeek, J., 273(18), 273(25), 281(25), 301(18), 304(25), 307(25), 374(25), 480
Wilder, P., Jr., 75(67), 264
Wildes, J. W., 325(88), 336(88), 483
Wilkinson, S., 282(50), 481
Williamson, E., 444(239), 489
Willinsky, M. D., 325(102), 325(122), 409(102), 483, 484
Willis, D. E., 226, 226(189), 269
Wilson, W. D. C., 325(68), 325(87), 329(68), 331(68), 407(175), 482, 483, 486
Witte, A. H., 325(95), 325(109), 325(127), 341(95), 483, 484
Wittstruck, T., 432(197), 437(197), 487
Wolf, L., 20, 21(10), 22(9), 23(9), 24(9), 262

Wollner, H. J., 282(49), 481
Wolstenholme, W. A., 103(102), 126, 127, 265
Worm, K., 325(104), 343(104), 432(104), 483
Woziwodzki, H., 79(70), 171(70), 180(162), 264, 268
Wright, J. A., 325(73), 482
Wynder, E. L., 41(39), 103(88), 159(88), 159(149), 161, 171(88), 177, 263, 265, 267

Y

Yagen, B., 432(235), 479(235), 489
Yamauchi, T., 325(70), 325(74), 325(75), 325(86), 332, 332(74), 333, 333(74), 336(70), 336(74), 336(86), 337(86), 482
Yang, R. K., 325(113), 348, 349, 350, 484
Yashuda, D. M., 437(237), 489
Yasumatsu, N., 22(17), 22(22), 29, 262
Yisak, W. A., 432(229), 432(233), 432(236), 474, 474(236), 476, 477(236), 478, 488, 489
Yoshimura, H., 273(11), 285(11), 287(11), 480

Z

Zeffert, J. T., 325(66), 325(72), 482
Zeldes, S. G., 136(129), 145, 146, 147, 266
Zielinski, W. L., Jr., 180(175), 189(175), 207(185), 207(186), 220(185), 220(186), 223(185), 225(186), 268, 269
Zikoloulos, J., 432(205), 446(205), 487
Zlatkis, A., 71(54), 232(200), 244(200), 245(200), 245(201), 245(202), 246(200), 247(200), 263, 269, 402(167), 486

SUBJECT INDEX

A

Acefluoranthylene, 193, 198
Acenaphth-(1,2-a)-acenaphthylene, 198, 238
Acenaphthene, 90, 144, 183, 185, 227, 228, 230, 239
Acenaphthylene, 90, 144, 185, 199
Aceperylene, 198
Acephenanthrylene, 185
Acepyrylene, 193, 198
Acetaldehyde, 5, 94, 116, 140, 145, 146, 147, 154, 361
Acetic acid, 6, 15, 75, 96, 175, 177, 416
  methyl derivative of, see methyl acetate
  recovery of, 177
Acetic anhydride, 326, 448, 479
Acetoin, 142
Acetone, 5, 25, 28, 94, 116, 137, 138, 140, 147, 154, 250, 361, 432, 442
Acetonitrile, 95, 114, 116, 137, 138, 140, 154, 159, 460, 465, 471
Acetophenone, 143, 431
7-Acetoxy-Δ^1-tetrahydrocannabinol, see 11-acetoxy-Δ^9-tetrahydrocannabinol
7-Acetoxy-Δ^6-tetrahydrocannabinol, see 11-acetoxy-Δ^8-tetrahydrocannabinol

11-Acetoxy-Δ^8-tetrahydrocannabinol, 434, 435
  structure of, 434
11-Acetoxy-Δ^9-tetrahydrocannabinol, 433, 434, 435
  structure of, 434
Acetylcarbromal, 38
Acetylene, 89, 102, 152, 154
  structure of, 102
2-Acetylfuran, 94, 140, 142, 431
Acetylmethylfuran, 143
Acetylnaphthalene, 183
Acridine, 222
Acrolein, 5, 94, 116, 137, 145, 146, 147, 154
Acrylonitrile, 95, 116, 137
7-Acyloxy-Δ^1-tetrahydrocannabinol, see 11-acyloxy-Δ^9-tetrahydrocannabinol
7-Acyloxy-Δ^6-tetrahydrocannabinol, see 11-acyloxy-Δ^8-tetrahydrocannabinol
11-Acyloxy-Δ^8-tetrahydrocannabinol, 456, 457
  silyl derivative of, 457
  structure of, 457
11-Acyloxy-Δ^9-tetrahydrocannabinol, 456, 457
  silyl derivative of, 457
  structure of, 457
Adenine, 8
Adipic acid, 6, 96, 178, 179
  methyl derivative of, 178, 179

α-Alanine, 10, 101
β-Alanine, 10, 101
Alkali flame ionization detector,
  see nitrogen detector
Alkylbenzo(a)pyrene, 90
Alkylchrysene, 90
1-Alkyl-3,5-dihydroxybenzene(s),
  357, 358, 361
  structure of, 357, 358
1-Alkyl-3,5-dimethoxybenzene(s),
  357, 358, 361
  structure of, 357, 358
Alkylfluoranthene, 90
Alkylpyrene, 90
Allene, 89, 153
Alloaromadendrene, 381
Allobarbital, 38
Allylamine, 98
4-Allylcatechol, 7
Allylmercaptan, 244
5-Allyl-2,3-methylenedioxyphenyl-
  methylether, see myristicin
α-Aminoadipic acid, 10
α-Aminobutyric acid, 10
γ-Aminobutyric acid, 10, 101
γ-Aminopyridine, 15, 16
β-Aminopyridine, 15, 16
Ammonia, 8, 87, 98, 250-252,
  254, 457, 458
Ammonium chloride, 120
Ammonium hydroxide, 121
Amobarbital, 38
Amphetamine, 38, 119
Amylamine, 98
sec-Amylamine, 98
n-Amylmercaptan, 244
α-Amyrin, 324
  acetyl derivative of, 324
β-Amyrin, 3, 92, 324
  acetyl derivative of, 324
γ-Amyrin, 324
  acetyl derivative of, 324
Anabasine, 8, 12, 13, 15, 16, 17,
  22, 23, 24, 26, 27, 29, 30,
  31, 41, 42, 98, 103, 105,
  107
  recovery of, 41

[Anabasine]
  structure of, 12
Anatabine, 8, 12, 15, 16, 25, 26,
  27, 29, 30, 31, 42, 98, 103,
  105, 107
  structure of, 12
4-Androstene-3,17-dione, 295, 304,
  305, 306, 308, 312, 314, 315,
  354, 362, 374, 375, 377, 378,
  379, 380, 390, 406, 407
Aniline, 98, 143
p-Anisaldehyde, 5, 7
m-Anisidine, 98
Anisole, 97, 137, 140, 142, 159,
  161, 247
Anthracene, 90, 145, 183, 185, 190,
  200, 214, 221, 222, 223, 227,
  230, 234, 239, 328, 329, 426
Anthranthrene, 184, 198, 223, 238,
  239
9,10-Anthraquinone, 5
Aprobarbital, 38
Arabinose, 9
Arabogalactan, 9
Arachidic acid, 6, 96, 422
  methyl derivative of, 422
Arachidonic acid, 6
Arginine, 10
Argon (GC), 21, 50, 114, 137, 257,
  282, 283, 325, 328, 441, 442
Argon ionization detector (GC), 50,
  114, 162, 282, 325, 328
Aryl sulfatase, 448
Ascorbic acid, 45
Asparagine, 10
Aspartic acid, 10, 101
Atmospheric pressure ionization
  (API) detector, 121, 122
Auxin acid, 6
Azelaic acid, 6
Azulene, 90, 183

B

Barbital, 38
Barium carbonate, 232

SUBJECT INDEX

Barium hydroxide, 27, 87
Behenic acid, 422
  methyl derivative of, 422
Benzacenaphthylene, 191, 234
Benz(c)acridine, 218, 219
Benzalazine, 88
Benzaldehyde, 5, 20, 94, 142, 431
Benz(a)anthracene, 90, 215, 217,
    218, 219, 222, 223, 224,
    236, 239
Benz(c)anthracene, 194, 218
Benzanthracene-7,12-dione, 219
Benz(a)anthrone, 218, 219
Benzene, 15, 17, 21, 23, 27, 29,
    41, 50, 56, 88, 90, 104, 122,
    123, 125, 132, 136, 137, 145,
    185, 201, 213, 283, 285, 326,
    338, 343, 426, 432, 438, 439
Benzimidazole, 98
Benzindene, 222
Benz(f)indene, 185
1,2-Benzoanthracene, 183, 185,
    221
Benzo(a)carbazole, 202
Benzo(b)carbazole, 202
Benzo(c)carbazole, 202
Benzo(def)dibenzothiophene, 235
Benzo(a)fluoranthene, 185
Benzo(b)fluoranthene, 90, 185, 222
Benzo(ghi)fluoranthene, 90, 183,
    185, 194, 222, 236
Benzo(j)fluoranthene, 90, 102, 185,
    196, 222, 237
  structure of, 102
Benzo(k)fluoranthene, 90, 185, 196,
    200, 217, 221, 222, 223, 224,
    237
Benzo(mno)fluoranthene, 90, 200,
    223, 224
2,3-Benzofluoranthene, 184
3,4-Benzofluoranthene, 184
11,12-Benzofluoranthrene, 184
Benzo(a)fluorene, 192, 222, 235,
    239, 426
5H-Benzo(a)fluorene, 90
11-H-Benzo(a)fluorene, 90

Benzo(b)fluorene, 90, 192, 222,
    235, 239
11H-Benzo(b)fluorene, 90
Benzo(c)fluorene, 222
7H-Benzo(c)fluorene, 90
1,2-Benzofluorene, 183, 185
2,3-Benzofluorene, 183, 185, 221
3,4-Benzofluorene, 183, 185
Benzofuran, 142
Benzoic acid, 6, 96, 416, 418, 421,
    422
  methyl derivative of, 421, 422
  silyl derivative of, 416
Benzo(a)naphthacene, 90
Benzonitrile, 19, 143, 431
Benzo(rst)pentaphene, 225
Benzo(ghi)perylene, 90, 184, 197,
    217, 223, 225, 226, 238, 239
Benzo(c)phenanthrene, 90, 222, 236
3,4-Benzophenanthrene, 183
Benzo(a)pyrene, 88, 90, 102, 185,
    189, 196, 200, 215, 217, 218,
    219, 220, 221, 223, 224, 226,
    230, 234, 239, 426
  detection limits of, 189, 200, 220
  recovery of, 189
  structure of, 102
Benzo(e)pyrene, 90, 185, 189, 196,
    200, 217, 218, 221, 223, 224,
    230, 234, 239
1,2-Benzopyrene, 184, 217, 218,
    227
3,4-Benzopyrene, 184, 227
Benzoquinoline, 222
Benzotetraphene, 184
1,12-Benzperylene, 221
Benzyl acetate, 4, 93
Benzyl alcohol, 4, 93, 143
Benzyl benzoate, 93
Benzyl butyl phthalate, 323
Benzyl cinnamate, 93
Benzyl cyanide, 431
α-Bergamotene, 326, 327
trans-α-Bergamotene, 363, 381
Betaine, 10
Binaphthyl, 195, 236, 237

1,1'-Binaphthyl, 217
2,2'-Binaphthyl, 217
Biphenyl, 90, 143, 183, 199, 227, 228, 230, 232, 239, 426
2,2'-Bipyridyl, 15, 16, 104
2,3'-Bipyridyl, 8, 15, 16, 99, 103, 105, 107
Bis-(p-chlorophenyl)-chloromethane, 298
Bis-(p-chlorophenyl)-methane, 258
Bis-(ethylhexyl)-tetrachlorophthalate, 214
4",5"-Bisnor-Δ^1-tetrahydrocannabinol-3"-oic acid, see 4",5"-bisnor-Δ^9-tetrahydrocannabinol-3"-oic acid
4",5"-Bisnor-Δ^9-tetrahydrocannabinol-3"-oic acid, 454
 methyl/silyl derivative of, 454
 structure of, 454
Bis-(trimethylsilyl)-acetamide, 60, 66, 68, 85, 389, 393, 460, 471, 478
Bis-(trimethylsilyl)-acetamide-d_{18}, 453
Bis-(trimethylsilyl)-trifluoroacetamide, 67, 178, 303, 308, 336, 348, 362, 374, 395, 413, 414, 418, 452, 453, 457, 463, 465
Bis-(trimethylsilyl)-trifluoroacetamide-d_{18}, 395
Bombiprenone, 64, 65, 66
 structure of, 65
Borax, 87
Borneol, 3, 381
Boron trifluoride, 326, 339, 420
α-Bromopentafluorotoluene, 443
Butabarbital, 38
Butalbital, 38
Butadiene, 140
1,2-Butadiene, 89
1,3-Butadiene, 89, 102, 152, 154, 155
 structure of, 102

3-(Buta-1,3-dienyl)-pyridine, 19
n-Butanal, see n-butyraldehyde
Butane, 140, 152, 154, 227
2,3-Butanedione, 94, 140
2-Butanone, 5, 94, 140
2-Butenaldehyde, 140
1-Butene, 89, 152, 154
2-Butene, 89
cis-Butene-2, 152, 154
trans-Butene-2, 152, 154
Butene-3-yne-1, see 3-Buten-1-yne
Butenone, 94, 137
1-Buten-3-one, 140
3-Buten-1-yne, 89, 155
n-Butylacetate, 93, 137
Butyl alcohol, 93, 137
sec-Butyl alcohol, 93, 137
tert-Butyl alcohol, 438
Butylamine, 98
sec-Butylamine, 98
n-Butylbenzene, 227
n-Butylmercaptan, 244
Butylpyrazine, 431
1-Butyne, 89, 155
n-Butyraldehyde, 5, 137, 140, 154
n-Butyric acid, 6, 75, 96
Butyronitrile, 95, 137, 140

C

3,8,10(15)-Cadinatriene, 321
Caffeic acid, 7, 97
Caffeine, 1, 37, 39, 124
1-O-Caffeoylglucose, 7
4-Caffeoylquinic acid, 7
Calcium chloride, 136
Calcium sulfate, 338
Campesterol, 3, 69, 70, 72, 73, 92, 402, 420
 silyl derivative of, 69, 70, 72, 73, 402
Camphene, 326, 327, 363
Cannabichromanone, 386, 388, 431
 structure of, 388
Cannabichromene, 274, 292, 306, 307, 308, 312, 313, 314, 315,

SUBJECT INDEX 513

[Cannabichromene]
  319, 323, 332, 335, 337,
  339, 341, 344, 346, 359,
  364, 375, 376, 378, 379,
  380, 382, 390, 391, 392,
  393, 394, 396, 404
  methoxy derivative of, 391, 392,
    393
  silyl derivative of, 306, 308, 313,
    315, 332, 376, 379, 393
  structure of, 274
Cannabichromene-C_3, 277, 344,
  348, 359, 364, 396
  silyl derivative of, 396
  structure of, 277
Cannabichromenic acid, 274, 332,
  336, 337, 393, 394, 396
  silyl derivative of, 332, 336,
    337, 393, 394
  structure of, 274
Cannabicyclol, 274, 306, 314, 315,
  322, 335, 337, 338, 339,
  341, 344, 345, 346, 359,
  376, 377, 378, 379, 382,
  390, 391, 404
  structure of, 274
  tributylsilyl derivative of, 391
  triethylsilyl derivative of, 391
  trimethylsilyl derivative of, 315
    376, 379, 391
  tripropylsilyl derivative of, 391
Cannabicyclol-C_3, 277, 345, 346,
  348, 359
  structure of, 277
Cannabicyclolic acid, 275, 337
  structure of, 275
Cannabidiol, 274, 277, 281, 282,
  283, 284, 285, 286, 287,
  289, 292, 293, 294, 295,
  296, 297, 298, 299, 300,
  301, 302, 303, 304, 305,
  306, 307, 308, 311, 312,
  313, 314, 315, 319, 323,
  325, 329, 330, 331, 332,
  333, 334, 335, 336, 337,
  338, 341, 342, 343, 344,
  345, 347, 350, 351, 354,

[Cannabidiol]
  355, 360, 362, 364, 365,
  370, 375, 376, 377, 378,
  379, 380, 382, 383, 387,
  389, 390, 391, 392, 393,
  394, 395, 396, 397, 398,
  399, 401, 403, 404, 405,
  406, 407, 408, 409, 411,
  413, 414, 433, 458, 459,
  461, 462
  tert-butyldimethylsilyl derivative
    of, 398, 399
  cyclization of, 409
  decomposition of, in methanol,
    325
  diethylphosphate derivative of,
    399
  dimethoxy derivative of, 347, 392,
    393
  isomerization of, 287
  metabolism of, 458, 459, 461, 462
  monomethyl ether derivative of,
    276, 344, 347, 354, 391, 392
    393
    structure of, 277
  ortho isomer of, 344, 345
    structure of, 345
  pyrolysis of, 411, 414
  side-chain hydroxy metabolites of,
    401
  stability of, 365, 380, 383, 405,
    406
    in acid solution, 380, 405, 406
    in chloroform, 380, 383
    in ethanol, 380, 383
  structure of, 274, 370, 409, 459
  tributylsilyl derivative of, 391
  triethylsilyl derivative, 391
  trifluoroacetyl derivative of, 329,
    331
  trimethylsilyl derivative of, 302,
    306, 308, 313, 315, 329, 331,
    332, 336, 337, 351, 375, 376,
    379, 389, 391, 393, 394, 397,
    413
  trimethylsilyl/acetyl derivative
    of, 398, 399

[Cannabidiol]
   trimethylsilyl-d$_9$ derivative of, 395
   tripropylsilyl derivative of, 391, 395
Cannabidiol-C$_4$, 278, 397
   silyl derivative of, 397
   structure of, 278
Cannabidiolic acid, 274, 282, 283, 285, 308, 311, 313, 315, 332, 333, 336, 337, 351, 375, 376, 379, 393, 396, 411
   decarboxylation of, 283, 285, 351, 375
   diacetyl/methyl derivative of, 283
   methyl derivative of, 282, 283
   silyl derivative of, 308, 311, 313, 315, 333, 336, 337, 351, 375, 376, 379, 393
   structure of, 274
Cannabidiorcin, 277, 287, 288, 289, 292, 293, 294, 344, 359, 397
   silyl derivative of, 397
   structure of, 277, 288, 289
Cannabidivarin, 277, 287, 288, 289, 292, 293, 294, 301, 302, 304, 306, 307, 308, 314, 315, 322, 341, 342, 344, 354, 359, 367, 376, 377, 378, 379, 382, 387, 391, 397
   structure of, 277, 288, 289, 367
   tributylsilyl derivative of, 391
   triethylsilyl derivative of, 391
   trimethylsilyl derivative of, 302, 308, 315, 376, 379, 391, 397
   tripropylsilyl derivative of, 391
Cannabidivarinic acid, 275, 376
   silyl derivative of, 376
   structure of, 275
Cannabidivarol, see cannabidivarin

Cannabielsoic acid A, 274, 411
   C-1 stereoisomers of, 411
   structure of, 411
Cannabielsoin A, 354, 377, 413
   C-1 stereoisomers of, 413
   mass spectra of, 413
   structure of, 413
   silyl derivative of, 413
Cannabifuran, 386, 388, 431
   structure of, 388
Cannabigerol, 274, 275, 292, 304, 305, 306, 307, 312, 313, 314, 315, 333, 335, 337, 339, 341, 344, 345, 347, 354, 360, 362, 376, 377, 391, 392
   dimethoxy derivative of, 333, 392
   structure of, 333
   monomethoxy derivative of, 275, 292, 306, 312, 313, 314, 333, 337, 344, 345, 347, 360, 362, 376, 392
   structure of, 275, 345
   trimethylsilyl derivative of, 362
   structure of, 274
   tributylsilyl derivative of, 391
   triethylsilyl derivative of, 391
   trimethylsilyl derivative of, 315, 333, 376, 391
   tripropylsilyl derivative, 391
Cannabigerol-C$_3$, 344, 345, 359
   methoxy derivative of, 344, 345, 359
   structure of, 345
Cannabigerolic acid, 274, 275, 336, 337
   monomethoxy derivative of, 275, 336, 337
   structure of, 275
   trimethylsilyl derivative of, 336, 337
   structure of, 274
Cannabinodiol, 404
Cannabinodiol-C$_3$, 344, 345, 359
   structure of, 345
Cannabinodiol-C$_5$, 344, 345

SUBJECT INDEX
515

[Cannabinodiol-C_5]
  structure of, 345
Cannabinoids (general)
  biogenesis of (scheme), 273, 274
  chemical composition of, in
    biological media, 432-479
    marijuana plant, 273-407
    marijuana smoke, 407-431
  mechanism of fragmentation of, 404
  metabolic products of, 431-479
  numbering systems for, 271, 272
    formal chemical, 271, 272
    monoterpenoid, 271, 272
  stability of, 361, 362
  structure-retention time relationships of, 343, 345, 346, 347, 360
  synthetic, 343, 355, 357, 358, 359, 360
  structures of some, 356
Cannabinol, 272, 274, 281, 282,
    283, 284, 285, 286, 287,
    290, 292, 293, 294, 295,
    296, 297, 298, 299, 301,
    302, 303, 304, 305, 306,
    307, 312, 314, 315, 323,
    325, 326, 329, 331, 332,
    333, 334, 335, 336, 337,
    338, 339, 341, 342, 343,
    344, 346, 347, 350, 351,
    354, 360, 362, 364, 365,
    373, 375, 376, 377, 378,
    379, 380, 382, 384, 385,
    387, 389, 390, 391, 392,
    393, 394, 395, 396, 397,
    398, 399, 401, 403, 404,
    406, 407, 408, 409, 411,
    432, 433, 436, 437, 438,
    439, 440, 444, 448, 449,
    471, 474, 475, 476, 477,
    478
  acetyl derivative of, 326
  tert-butyldimethylsilyl derivative
    of, 398, 399

[Cannabinol]
  ^{14}C-labeled, 448
  diethylphosphate derivative of,
    399, 438, 439, 440
    structure of, 439
  fatty acid conjugates of, 474
  metabolic pathways of, 448, 450, 475
  metabolism of, 436, 437, 448,
    449, 450, 471, 474, 475, 476,
    477, 478
  monomethoxy derivative of, 276,
    347, 354, 392, 431, 437
    structure of, 276
  pentafluorobenzoyl derivative of,
    444
  stability of, 362, 365
  structure of, 272, 274, 373, 477
  tributylsilyl derivative of, 391
  trichloroacetyl derivative of, 343
  triethylsilyl derivative of, 391, 397
  trifluoroacetyl derivative of, 329, 331
  trimethylsilyl derivative of, 302,
    315, 329, 331, 332, 333, 336,
    347, 362, 375, 376, 379, 387,
    389, 390, 391, 393, 394, 397,
    436
  trimethylsilyl/acetyl derivative of,
    398, 399
  tripropylsilyl derivative of, 391, 397
Cannabinol-C_4, 278, 393, 396, 397
  structure of, 278
  triethylsilyl derivative of, 397
  trimethylsilyl derivative of, 397
  tripropylsilyl derivative of, 397
Cannabinol-7-aldehyde, <u>see</u>
  7-carbonylcannabinol
Cannabinolic acid, 274, 332, 336,
  337, 351, 362, 375, 393, 396
  decarboxylation of, 351, 375
  silyl derivative of, 336, 337, 351,
    362, 375, 393

[Cannabinolic acid]
  structure of, 274
Cannabinol-7-oic acid, 276, 446,
    477, 448, 477, 478
  methyl/methoxy derivative of,
    447, 448
    structure of, 448
  methyl/silyl derivative of, 478
  structure of, 276, 477
Cannabinol-11-oic acid, see
  cannabinol-7-oic acid
Cannabiorcin, 277, 287, 288, 290,
    292, 293, 294, 344, 346,
    359, 396, 397
  structure of, 277, 288, 290
  triethylsilyl derivative of, 397
  trimethylsilyl derivative of, 397
  tripropylsilyl derivative of, 397
Cannabivarin, 287, 288, 290, 292,
    293, 294, 301, 302, 306, 307,
    311, 323, 341, 342, 344, 347,
    359, 369, 376, 387, 391, 392,
    393, 396, 397
  methoxy derivative of, 391, 392,
    393
  structure of, 288, 290, 369
  triethylsilyl derivative of, 397
  trimethylsilyl derivative of, 302,
    347, 397
  tripropylsilyl derivative of, 397
Cannabivarol, see cannabivarin
Caproaldehyde, 94
Caproic acid, 6, 75, 96
Capronaldehyde, 137
n-Capronitrile, 95, 140
n-Caprylic acid, 75
Carbazole, 98, 144, 190, 201, 202,
    221, 416, 418, 419
  structure of, 419
Carbazole-5,6,7,8,12,13-^{14}C,
  416
Carbisoprodol, 39
N'-Carbomethoxyanabasine, 42,
    43, 44
  mass spectrum of, 44
  structure of, 43, 44

N'-Carbomethoxynornicotine, 42,
    43, 44
  mass spectrum of, 44
  structure of, 43, 44
Carbon dioxide, 101, 139, 255, 256
Carbon-14 dioxide, 22, 26
Carbon disulfide, 95, 244
Carbon monoxide, 101, 139
Carbon tetrachloride, 258-259
7-Carbonylcannabinol, 474, 475,
    476
  structure of, 475
Carbonyl sulfide, 95, 238, 240, 244
11-Carboxy-Δ^9-tetrahydrocanna-
    binol, 437, 447, 452, 454,
    455, 467, 468
  silyl derivative of, 454, 455, 468
  structure of, 447, 467
7-Carboxyhexahydrocannabinol
    (axial), 472
  methyl/acetyl derivative of, 472
  methyl/silyl derivative of, 472
  silyl derivative of, 472
7-Carboxyhexahydrocannabinol
    (equatorial), 472
  methyl/acetyl derivative of, 472
  methyl/silyl derivative of, 472
  silyl derivative of, 472
Carbromal, 38
Δ^3-Carene, 363
Cariophyllene oxide, see caryo-
  phyllene oxide
α-Carotene, 3
β-Carotene, 3
α-Caryophyllene, 282, 327
β-Caryophyllene, 282, 327, 363,
    381
Caryophyllene oxide, 326
Catechol, 7, 97, 170, 171, 172,
    174, 416, 418, 421, 422, 424
  N-methyl-N-silyl derivative of,
    424
  silyl derivative of, 416, 421
Catechol-^{14}C, 169, 170
Cellulose, 9
Cerotic acid, 96

Chlordiazepoxide, 40
Chlorobenzene, 258
4-Chlorobenzyl chloride, 253
Chlorodifluoroacetic anhydride, 443
Chloroform, 15, 27, 29, 41, 45, 67, 73, 104, 145, 258, 259, 303, 305, 307, 354, 374, 380, 383, 406, 432, 433, 440, 442, 454
Chlorogenic acid, 7, 97
Chloromethyldimethylchlorosilane, 443
o-Chlorophenol, 161, 163
Chlorophyll, 11
Chlorphentermine, 120, 121, 126, 128
 recovery of, 121
Cholestane, 409
5-Cholesten-3β-ol, see cholesterol
5α-Cholest-7-en-3β-ol, 72, 73
 silyl derivative of, 72, 73
Cholesterol, 3, 69, 70, 72, 73, 85
 recovery of, 85
 silyl derivative of, 69, 70, 72, 73, 85
Cholesteryl acetate, 85
Cholesteryl palmitate, 85
Choline, 10
Chrysene, 90, 184, 185, 194, 200, 214, 217, 218, 221, 223, 224, 230, 236, 239
Chrysofluorene, 221
Cinnamonitrile, 95
Citral, 326, 328
 structure of, 328
Citric acid, 6, 45
Citrulline, 10
Cocaine, 40, 286
Codeine, 40, 126, 127, 398
 structure of, 127
Collidine, 98
Column packing (solid support)(GC)
 types of
  Aeropak-30, 126
  Alumina, 153, 155

[Column packing (solid support) (GC)]
 [types of]
  Anakrom-series, 37, 114, 159, 207
  Celite, 22, 50, 126, 159, 176, 177, 179
  Chromoport-30, 259
  Chromosorb-series, 29, 31, 33, 34, 54, 59, 67, 111, 112, 129, 130, 132, 146, 162, 169, 175, 179, 180, 181, 182, 189, 200, 201, 213, 219, 220, 223, 244, 250, 259, 282, 283, 286, 295, 303, 306, 312, 325, 326, 328, 329, 332, 333, 336, 338, 343, 348, 351, 352, 354, 358, 360, 362, 375, 384, 388, 392, 396, 398, 405, 408, 412, 413, 418, 421, 432, 436, 440, 446, 452, 461
  Chromosorb-101, 360, 361
  Chromosorb-103, 252
  Chromosorb-104, 232, 240
  Chrompack SA, 424, 431
  Diatomite, 41, 335
  Diatoport S, 54, 58, 59, 60, 121, 213, 420, 421, 433, 440
  Firebrick, 2, 17, 21, 24, 26, 106, 111, 137, 149, 153, 159, 162, 176, 177, 178, 179, 187, 247
  Gas Chrom-series, 29, 30, 43, 45, 52, 74, 75, 79, 81, 88, 105, 109, 112, 122, 125, 149, 161, 169, 200, 205, 207, 213, 220, 253, 257, 288, 295, 302, 303, 304, 312, 335, 339, 341, 342, 345, 350, 352, 354, 356, 360, 362, 364, 374, 377, 380, 384, 390, 391, 393, 395, 398, 404, 406, 408, 409, 414, 416, 420, 421, 422, 433, 437, 440, 441, 444, 451, 453, 457, 458, 460, 465, 471, 474, 478, 479
  Glass beads, 149, 219, 221

[Column packing (solid support) (GC)]
 [types of]
  Nylon-66, 162
  Phasepak Q, 87
  Porapak, 30, 232, 254, 255
  Silica, 244
  Silocel, 148, 149
  Supelcon, 69
  Supelcoport, 133, 390
  Teflon-6, 177
  Varaport-30, 129
  W-98
Coronene, 90, 184, 223, 225, 238, 239
Cotinine, 8, 9, 12, 13, 15, 16, 20, 21, 22, 23, 24, 35, 36, 98, 100, 103, 105, 107, 114, 119, 120, 121, 122, 123, 133, 134, 135
 detection limits of, 133
 deuterated, 135
 rate of excretion (urine), 123
 recovery of, 119, 121, 123, 133
 structure of, 12
Coulson electrolytic conductivity detector (GC), see nitrogen detector (GC)
p-Coumaric acid, 7, 97
Coumarin, 144
p-Coumarylquinic acid, 7
Creatinine, 448
m-Cresol, 7, 97, 113, 144, 161, 163, 418, 421, 422
 silyl derivative of, 113, 421
o-Cresol, 97, 113, 144, 161, 417, 418, 421, 422
 silyl derivative of, 113, 417, 421
p-Cresol, 97, 113, 144, 161, 163, 169, 417, 418, 421, 422
 silyl derivative of, 113, 417, 421
m-Cresyl methyl ether, 161, 247, 248
o-Cresyl methyl ether, 161, 248
p-Cresyl methyl ether, 161, 247, 248

Crotonaldehyde, 5, 94, 137, 154
Crotonic acid, 6
Crotononitrile, 95
Cryptoxanthin, 3
Cumene, 136, 145
Curcumene, 326, 327
Cyanogen, 95
2-Cyanopyridine, 19
3-Cyanopyridine, 19, 104, 143
Cyclobarbital, 39
Cyclohexane, 137, 160, 163, 187, 218, 320
Cyclohexanone, 140
Cyclohexene, 89, 145
1,3-Cyclopentadiene, 89
Cyclopenta(cd)floranthene, 238
Cyclopentamine, 38
Cyclopentane, 153, 154, 155
Cyclopentanone, 94, 140
Cyclopenta(cd)perylene, 238
4H-Cyclopenta(def)phenanthrene, 190, 234
Cyclopenta(cd)pyrene, 238
Cyclopentene, 89, 153, 155
p-Cymene, 327
Cysteic acid, 10
Cysteine, 10
Cystine, 10

D

Damascenone, 250, 251
p,p'-DDE, 258
p,p'-DDM, 259
p,p'-DDT, 258-259
 pyrolysis of, 258, 259
Decafluorobenzaldehyde azine, 88, 252, 253
 structure of, 252
n-Decane, 140, 145
Decanoic acid, 6, 96, 422
 methyl derivative of, 422
1-Decene, 89, 140
Dehydrocannabifuran, 386, 388, 431
 structure of, 388
3,4-Dehydropiperidine, 98

SUBJECT INDEX

1-Deoxy-1-L-alanino-D-fructose, 9
1-Deoxy-1-(N-γ-aminobutyric acid)-D-fructose, 9
1-Deoxy-1-L-proline-D-fructose, 9
Deoxyribose, 9
Desmethylcotinine, see norcotinine
Desmethylnicotine, see nornicotine
Dextromethorphan, 39
Diacetyl, 154
Diallyl sulfide, 244
1,7-Diazaindene, 19, 20
Diazepam, 40
Diazoethane, 453
Diazomethane, 73, 79, 171, 282, 333, 453, 454, 478, 479
Dibenz(ah)acridine, 98
Dibenz(aj)acridine, 98
Dibenzanthracene, 223, 238
Dibenz(ac)anthracene, 197, 225, 230, 426
Dibenz(ah)anthracene, 90, 197, 215, 225, 226
Dibenz(ai)anthracene, 197
1,2,3,4-Dibenzanthracene, 184, 217
1,2,5,6-Dibenzanthracene, 184, 221
1,2,7,8-Dibenzanthracene, 184
1,2,6,7-Dibenzpyrene, 225
4,5,7,8-Dibenzpyrene, 225
7H-Dibenz(cg)carbazole, 98
Dibenzo(defp)chrystene, 225
Dibenzo(bmno)fluoranthene, 198, 238
Dibenzo(emno)fluoranthene, 238
Dibenzo(fmno)fluoranthene, 238
Dibenzo(ai)fluorene, 91
Dibenzofuran, 185, 190
Dibenzo(ac)naphthacene, 91
Dibenzo(aj)naphthacene, 91
Dibenzo(bh)phenanthrene, 91
Dibenzopyrene, 198, 223, 238
Dibenzo(ah)pyrene, 91
Dibenzo(ai)pyrene, 91, 221

Dibenzo(al)pyrene, 91
Dibenzo(cdjk)pyrene, 91
1,2,3,4-Dibenzopyrene, 184
1,2,4,5-Dibenzopyrene, 184
3,4,8,9-Dibenzopyrene, 184
3,4,9,10-Dibenzopyrene, 184
Dibenzothiophene, 234
Dibenzylphthalate, 312, 328, 329, 331, 335
Dibromobutane, 257
2,3-Dibromo-n-butane, 257
Dibromobutene, 257
1,2-Dibromo-1-chloroethane, 257
Dibromopropane, 257
Dibromopropene, 257
2,6-Di-tert-butylphenol, 162, 163
Dibutylphthalate, 4
Di-n-butyl sulfide, 244
1,3-Dichlorobenzene, 145
p,p'-Dichlorobiphenyl, 258
Dichlorocarbene, 259
Dichloroethylene, 145
Dichloromethane, see methylene chloride
cis-p,p'-Dichlorostilbene, 259
trans-p,p'-Dichlorostilbene, 259
α,p-Dichlorotoluene, 258
Diethylamine, 98, 285
Diethyl disulfide, 244
Diethylene glycol, 4, 93
Diethyl ether, 41, 58, 60, 88, 120, 126, 128, 129, 133, 136, 140, 145, 170, 185, 328, 420, 432, 448, 464
Di-(2-ethylhexyl)phthalate, 4
Diethylketone, 137, 154
Diethylpentane, 145
Diethylphosphorochloridate, 439
Diethylphthalate, 145, 321
Diethylpyrazine, 431
Diethyl sulfide, 244
Dihydroanthracene, 222
9,10-Dihydroanthracene, 91
Dihydrobenz(a)anthracene, 222
Dihydrobenzo(a)fluorene, 222
Dihydrobenzo(b)fluorene, 222

Dihydrobenzo(c)fluorene, 222
Dihydrobenzofuran, 143
5,6-Dihydro-8H-benzo(a)cyclopent(h)anthracene, 91
10,11-Dihydro-9H-benzo(a)cyclopent(i)anthracene, 91
3,4-Dihydrobenzo(a)pyrene, 91
Dihydrochrysene, 222
Dihydrocodeine, 40
16,17-Dihydro-15H-cyclopent(a)phenanthrene, 91
Dihydrofluoranthene, 222
Dihydrofluorene, 222
6,12-Dihydro-6-hydroxycannabidiol, 405, 407
 structure of, 405
9,10-Dihydro-9-hydroxycannabidiol, 383, 384, 407
 structure of, 383, 384
9,10-Dihydro-9-hydroxyisocannabidiol, 405, 406, 407
 structure of, 405
Dihydrometanicotine, 15, 16, 98, 100
Dihydromethylbenz(a)anthracene, 222
Dihydromethylchrysene, 222
Dihydromethyltriphenylene, 222
Dihydrophenanthrene, 222
Dihydropyrene, 222
Dihydrotriphenylene, 222
2,4-Dihydroxyanisole, 422, 424
 N-methyl-N-silyl derivative of, 424
2,4-Dihydroxybenzaldehyde, 422, 424
 N-methyl-N-silyl derivative of, 424
6,7-Dihydroxycannabidiol, 278, 461, 462
 silyl derivative of, 461, 462
 structure of, 278
1",7-Dihydroxycannabidiol, 278, 461, 462
 silyl derivative of, 461, 462
 structure of, 278

2",6-Dihydroxycannabidiol, 279, 461, 462
 silyl derivative of, 461, 462
 structure of, 279
3",6β-Dihydroxycannabidiol, 279, 461, 462
 silyl derivative of, 461, 462
 structure of, 279
3",7-Dihydroxycannabidiol, 278, 461, 462
 silyl derivative of, 461, 462
 structure of, 278
4",6β-Dihydroxycannabidiol, 279, 461, 462
 silyl derivative of, 461, 462
 structure of, 279
4",7-Dihydroxycannabidiol, 278, 461, 462
 silyl derivative of, 461, 462
 structure of, 278
5",7-Dihydroxycannabidiol, 279, 461, 462
 silyl derivative of, 461, 462
 structure of, 279
1",7-Dihydroxycannabinol, 272, 273, 471, 474, 475, 476
 silyl derivative of, 471, 474, 476
 structure of, 273, 471, 475
2",7-Dihydroxycannabinol, 436, 471, 474
 silyl derivative of, 436, 471, 474
 structure of, 436, 471
3",7-Dihydroxycannabinol, 471, 474
 silyl derivative of, 471, 474
 structure of, 471
4",7-Dihydroxycannabinol, 272, 273, 471, 474, 475, 476
 silyl derivative of, 471, 474, 476
 structure of, 273, 471, 475
2"-6α-Dihydroxy-7-carboxy-Δ^1-tetrahydrocannabinol, see 2",8α-dihydroxy-11-carboxy-Δ^9-tetrahydrocannabinol
2",8α-Dihydroxy-11-carboxy-Δ^9-tetrahydrocannabinol, 452,

SUBJECT INDEX

[2",8α-Dihydroxy-11-carboxy-Δ^9-tetrahydrocannabinol] 453, 455, 467, 469
 silyl derivative of, 455, 469
 structure of, 454, 467
3",6α-Dihydroxy-7-carboxy-Δ^1-tetrahydrocannabinol, see 3",8α-dihydroxy-11-carboxy-Δ^9-tetrahydrocannabinol
3",8α-Dihydroxy-11-carboxy-Δ^9-tetrahydrocannabinol, 453, 455, 467, 469
 silyl derivative of, 455, 469
 structure of, 453, 467
1α,6β-Dihydroxyhexahydrocannabinol, see 8β,9α-dihydroxyhexahydrocannabinol
8β,9α-Dihydroxyhexahydrocannabinol, 470, 473
 silyl derivative of, 473
 structure of, 470
6,8-Dihydroxy-11-isopropyl-4,8-dimethyl-14-oxo-4,9-pentadecadienoic acid, 4
2",7-Dihydroxy-6-oxo-Δ^1-tetrahydrocannabinol, see 2",11-dihydroxy-8-oxo-Δ^9-tetrahydrocannabinol
2",11-Dihydroxy-8-oxo-Δ^9-tetrahydrocannabinol, 465, 467, 469
 methyl/silyl derivative of, 469
 silyl derivative of, 469
 structure of, 467
3",7-Dihydroxy-6-oxo-Δ^1-tetrahydrocannabinol, see 3",11-dihydroxy-8-oxo-Δ^9-tetrahydrocannabinol
3",11-Dihydroxy-8-oxo-Δ^9-tetrahydrocannabinol, 465, 467, 469
 methyl/silyl derivative of, 469
 silyl derivative of, 469
 structure of, 467

5α-7-Dihydroxy-Δ^6-tetrahydrocannabinol, see 7α,11-dihydroxy-Δ^9-tetrahydrocannabinol
5β,7-Dihydroxy-Δ^6-tetrahydrocannabinol, see 7β,11-dihydroxy-Δ^9-tetrahydrocannabinol
6α,7-Dihydroxy-Δ^1-tetrahydrocannabinol, see 8α,11-dihydroxy-Δ^9-tetrahydrocannabinol
6α,7-Dihydroxy-Δ^6-tetrahydrocannabinol, see 8α,11-dihydroxy-Δ^8-tetrahydrocannabinol
6β,7-Dihydroxy-Δ^1-tetrahydrocannabinol, see 8β,11-dihydroxy-Δ^9-tetrahydrocannabinol
6β,7-Dihydroxy-Δ^6-tetrahydrocannabinol, see 8β,11-dihydroxy-Δ^8-tetrahydrocannabinol
7α,11-Dihydroxy-Δ^8-tetrahydrocannabinol, 434, 435
 structure of, 434
7β,11-Dihydroxy-Δ^8-tetrahydrocannabinol, 434, 435
 structure of, 434
8α,11-Dihydroxy-Δ^9-tetrahydrocannabinol, 400, 402, 433, 434, 435, 438, 455, 467, 469
 silyl derivative of, 400, 402, 438, 455, 469
 structure of, 434, 467
8β,11-Dihydroxy-Δ^9-tetrahydrocannabinol, 280, 465, 467, 469
 silyl derivative of, 469
 structure of, 280, 467
11,16-Dihydroxy-Δ^9-tetrahydrocannabinol, 280
 structure of, 280
2",7-Dihydroxy-Δ^1-tetrahydrocannabinol, see 2",11-dihydroxy-Δ^9-tetrahydrocannabinol
2",11-Dihydroxy-Δ^9-tetrahydrocannabinol, 465, 467, 469
 silyl derivative of, 469
 structure of, 467

3",7-Dihydroxy-Δ^1-tetrahydrocannabinol, see 3",11-dihydroxy-Δ^9-tetrahydrocannabinol
3",11-Dihydroxy-Δ^9-tetrahydrocannabinol, 465, 467, 469
  silyl derivative of, 469
  structure of, 467
Diisobutyl sulfide, 244
p-Diisopropylbenzene, 229
Dimenhydrinate, 39
Dimethoxybenzene, 143
2,6-Dimethoxyphenol, 97
Dimethylacenaphthylene, 185
9,9-Dimethylacridan, 98
Dimethylamine, 99, 418, 421
2,3-Dimethylaniline, 99
2,4-Dimethylaniline, 99
2,5-Dimethylaniline, 99
2,6-Dimethylaniline, 99
3,5-Dimethylaniline, 99
7,12-Dimethyl-1,2-benzanthracene, 184
Dimethylbenz(a)anthracene, 222
9,10-Dimethylbenz(a)anthracene, 91
9,10-Dimethyl-1,2-benzanthracene, 210, 215
1,2-Dimethylbenzene, 140
1,3-Dimethylbenzene, 140
1,4-Dimethylbenzene, 140
Dimethylbenzocarbazole, 202
Dimethylbenzo(b)fluoranthene, 223
Dimethylbenzo(k)fluoranthene, 223
Dimethylbenzo(a)pyrene, 223
Dimethylbutene, 145
2,3-Dimethyl-1-butene, 89
3,3-Dimethyl-1-butene, 89, 140
Dimethylcarbazole, 202
1,9-Dimethylcarbazole, 200
2,9-Dimethylcarbazole, 200
3,9-Dimethylcarbazole, 200
4,9-Dimethylcarbazole, 200, 201, 203
  structure of, 203

Dimethylchlorosilane, 29, 31, 69, 303, 306, 328, 329, 333, 338, 343, 351, 354, 388, 396, 398, 412, 433
Dimethylchrysene, 91, 185, 222
1,1-Dimethylcyclohexane, 145
1,3-Dimethylcyclohexane, 145
1,2-Dimethylcyclopentane, 145
Dimethyl disulfide, 95, 244
1,2-Dimethyl-3-ethylbenzene, 229
1,2-Dimethyl-4-ethylbenzene, 227, 229
1,3-Dimethyl-2-ethylbenzene, 229
1,3-Dimethyl-4-ethylbenzene, 229
1,4-Dimethyl-2-ethylbenzene, 229
Dimethylethylpyrazine, 431
Dimethylfluoranthene, 91
1,9-Dimethylfluorene, 210, 212
2,3-Dimethylfluorene, 205, 210, 211, 212
  structure of, 211
9,9-Dimethylfluorene, 210, 212
Dimethylfluoren-9-one, 160
Dimethylformamide, 60, 66, 68, 85, 437
Dimethylfuran, 137
2,4-Dimethylfuran, 321
2,5-Dimethylfuran, 95, 136, 140, 154
1-(2,6-Dimethylheptyl)-p-menthene-8(9), 322
2,4-Dimethylhexane, 145
2,5-Dimethylhexane, 145
2,5-Dimethyl-1,2-hexene, 145
2,2-Dimethyl-5-hydroxy-3-(3-oxobutyl)-7-pentyl-4-chromanon, see cannabichromanone
2,2-Dimethyl-5-hydroxy-7-n-pentylchromene, 431
Dimethylindole, 201
1,2-Dimethylindole, 144, 205, 206, 208, 209
  structure of, 208
1,3-Dimethylindole, 205, 206, 208, 209

[1,3-Dimethylindole]
  structure of, 208
1,4-Dimethylindole, 205, 206, 209
  structure of, 209
1,5-Dimethylindole, 205, 206
1,6-Dimethylindole, 205, 206
1,7-Dimethylindole, 205, 206, 209
  structure of, 209
2,3-Dimethylindole, 206
2,5-Dimethylindole, 206
2,7-Dimethylindole, 206
3,4-Dimethylisopropylbenzene, 229
Dimethylnaphthalene(s), 91, 143, 144, 185, 187, 189, 199, 228
  recovery of, 189
1,2-Dimethylnaphthalene, 199, 228
1,3-Dimethylnaphthalene, 185, 199, 228
1,4-Dimethylnaphthalene, 185, 199, 228
1,5-Dimethylnaphthalene, 185, 199, 228
1,6-Dimethylnaphthalene, 91, 144, 185, 187, 199, 228
1,7-Dimethylnaphthalene, 185, 199, 228
1,8-Dimethylnaphthalene, 91, 185, 187, 199, 228
2,3-Dimethylnaphthalene, 185, 199, 228
2,6-Dimethylnaphthalene, 91, 185, 187, 199, 228
2,7-Dimethylnaphthalene, 91, 185, 187, 199, 228
1-(2,7-Dimethyloctyl)-p-menthene-8(9), 322
2,4-Dimethylpentane, 140
2,4-Dimethylpentan-3-one, 94, 140
2,5-Dimethylphenanthrene, 91
2,3-Dimethylphenol, 161, 163, 165, 167
2,4-Dimethylphenol, 144, 161, 163, 164, 166, 168

2,5-Dimethylphenol, 144, 161, 163, 168
2,6-Dimethylphenol, 143, 163, 164, 166, 168, 431
3,4-Dimethylphenol, 144, 161, 163, 168
3,5-Dimethylphenol, 144, 161, 163, 165, 167
Dimethylphentermine, 125
Dimethylphthalate, 145
2,2-Dimethylpropionaldehyde, 140
2,5-Dimethylpyrazine, 431
2,6-Dimethylpyrazine, 99, 431
Dimethylpyrazole, 142
Dimethylpyridine, 418, 421
2,3-Dimethylpyridine, 99, 142
2,4-Dimethylpyridine, 99, 142
2,5-Dimethylpyridine, 99, 142
2,6-Dimethylpyridine, 99, 142
3,4-Dimethylpyridine, 99
3,5-Dimethylpyridine, 99
Dimethyl sulfide, 95, 244
Dimethyl sulfoxide, 187, 420
2,4-Dimethylthiophene, 244
Dimethyltriphenylene, 222
Dimethyl trisulfide, 244
N,N-Dimethyltryptamine, 39
2,4-Dimethyl-4-vinylcyclohexene, 89, 187
1,3-Dimyristin, 60, 61
  silyl derivative of, 60, 61
Dipentene, 89, 136
Diphenylacenaphthalene, 238
Diphenylacenaphthylene, 198
Diphenylamine, 33, 34, 35, 99
  pentafluoropropionyl derivative of, 34, 35
Diphenylhydantoin, 40
Dipropylamine, 99
Dipropylphthalate, 4
Di-n-propyl sulfide, 244
α,α'-Dipyridyl, see 2,2'-bipyridyl
α,β'-Dipyridyl, see 2,3'-bipyridyl
2,3'-Dipyridyl, see 2,3'-bipyridyl
Disulfiram, 39

o,p-Divinylphenol, 422, 424
  N-methyl-N-silyl derivative of, 424
n-Docosane, 145
n-Docosanol, 4, 60, 93
n-Dodecane, 142, 145
Dodecanoic acid, 422
  methyl derivative, 422
n-Dotriacontane, 64, 221
Doxylamine, 39
Durene, 142, 215
γ-3,8,13-Duvatriene-1,5-diol, 3
β-3,8,13-Duvatriene-1,5-diol, 3
α-4,8,13-Duvatriene-1,3-diol, 3
β-4,8,13-Duvatriene-1,3-diol, 3

E

Eicosadienoic acid, 422
  methyl derivative of, 422
n-Eicosane, 144, 329, 332
1-Eicosanol, 4, 93
Eicosenoic acid, 422
  methyl derivative of, 422
Electron affinity detector (GC), 34, 35, 88, 121, 180, 181, 182, 189, 253, 257, 259, 343, 440, 441, 442, 444
Ephedrine, 38
1,2-Epoxyhexahydrocannabinol, see 9,10-epoxyhexahydrocannabinol
1α,6β-Epoxyhexahydrocannabinol, see 8β,9α-epoxyhexahydrocannabinol
1β,6 -Epoxyhexahydrocannabinol, see 8β,9β-epoxyhexahydrocannabinol
8β,9α-Epoxyhexahydrocannabinol, 470, 473
  silyl derivative of, 470, 473
8β,9β-Epoxyhexahydrocannabinol, 470
  structure of, 470

9,10-Epoxyhexahydrocannabinol, 276, 450, 451
  silyl derivative of, 451
  structure of, 276
5,7,22-Ergostatrien-3β-ol, see ergosterol
5-Ergosten-3β-ol, see campesterol
Ergosterol, 3, 402
  silyl derivative of, 402
Erucic acid, 422
  methyl derivative of, 422
Erythrose, 9
Esculetin, 7, 97, 418, 421
  silyl derivative of, 421
Esculetin-7-glucoside, 7
Ethambutol, 122, 124
Ethane, 152, 154
Ethinamate, 38
Ethylacetate, 4, 67, 93, 116, 137, 145, 163, 395, 433
Ethylacetylene, 153
Ethyl alcohol, 4, 50, 68, 84, 93, 116, 135, 137, 154, 174, 248, 336, 352, 354, 361, 380, 383, 420, 432, 440, 448
Ethylamine, 8, 99
2-Ethylaniline, 99
4-Ethylaniline, 99
Ethylanthracene, 191, 192, 222, 234
Ethylbenz(a)anthracene, 195, 237
Ethylbenzene, 91, 136, 137, 140, 145, 187, 226
Ethylbenzofluoranthene, 197
Ethylbenzofuran(s), 199
Ethylbenzopyrene, 197
Ethylbinaphthyl, 196
Ethylbutyrate, 4, 93
Ethylcaproate, 5, 93
Ethylcapronate, 137
Ethylcarbazole, 200, 201, 202, 203, 431
  structure of, 203
4-Ethylcatechol, 170, 171

Ethyl chloride, 140
Ethylchrysene, 195, 237
Ethyl-4H-cyclopenta(def)phenanthrene, 192, 235
Ethyldibenzothiophene, 234
Ethyldimethylindole, 202
Ethyl-3,10-dimethylundecanoate, 322
Ethylene, 89, 152, 154
Ethylene glycol, 93
Ethylfluoranthene, 193, 194, 235, 236
2-Ethylfluorene, 210, 212
Ethylfluoren-9-one, 160
Ethyl formate, 93, 116, 147
3-Ethylheptane, 145
2-Ethyl-3-hydroxy-5-pentylbenzoic acid, 422, 424
  methyl derivative of, 422, 424
Ethylindene(s), 199
1-Ethylindole, 190, 205, 206, 208, 209
  structure of, 208
3-Ethylindole, 99, 145, 201, 206
Ethyl iodide, 452, 462, 464
Ethylisovalerate, 5, 93
Ethylmercaptan, 244
Ethylmethylanthracene, 192, 193, 235
Ethylmethylbenz(a)anthracene, 196
Ethylmethylbiphenyl, 190
Ethylmethylchrysene, 196
Ethylmethyl-4H-cyclopenta(def)-phenanthrene, 235
Ethylmethylfluoranthene, 194, 236
Ethylmethylindole, 202
Ethylmethylphenanthrene, 192, 193, 235
Ethylmethylpyrene, 194, 236
Ethyl-β-methylvalerate, 5, 93
1-Ethylnaphthalene, 189, 199, 228
  recovery of, 189
2-Ethylnaphthalene, 189, 199, 228
  recovery of, 189

Ethylphenanthrene, 191, 192, 222, 234
2-Ethylphenol, 97, 144, 161, 165, 167
3-Ethylphenol, 97, 144, 161, 165, 167, 168
4-Ethylphenol, 97, 144, 161, 165, 167, 168, 417
  silyl derivative of, 417
Ethylpropionate, 5, 93
Ethyl-n-propyl sulfide, 244
Ethylpyrene, 193, 194, 235, 236
Ethylpyridine, 431
2-Ethylpyridine, 13
3-Ethylpyridine, 13, 15, 19, 20, 99, 104
4-Ethylpyridine, 13
Ethyl-3-pyridylketone, 159
m-Ethyltoluene, 91, 137, 140, 187
o-Ethyltoluene, 91, 137, 140, 226
p-Ethyltoluene, 91, 137, 140, 187
Ethylvalerate, 5
Eudesma-2,6,8-triene, 321
Eugenol, 7, 97

F

Farnesene, 89, 158
β-Farnesene, 326, 327, 363
Farnesylacetone, 92, 322
Fenchyl alcohol, 381
Fentanyl, 40
Ferulic acid, 7, 97
1-O-Feryloylglucose, 7
3-Feruloylquinic acid, 7
Flame ionization detector (GC), 27, 29, 30, 31, 32, 33, 34, 36, 37, 41, 43, 46, 54, 61, 67, 69, 74, 75, 77, 79, 81, 85, 88, 105, 109, 111, 122, 125, 126, 129, 132, 137, 138, 140, 141, 145, 149, 153, 156, 157, 158, 159, 168, 169, 179, 180, 181, 185, 189, 201, 205, 213, 218, 219, 220, 223, 227, 230,

[Flame ionization detector (GC)]
231, 232, 245, 246, 247,
250, 253, 286, 288, 295,
302, 303, 305, 312, 329,
332, 333, 335, 336, 338,
341, 342, 343, 348, 350,
351, 352, 354, 356, 358,
360, 374, 375, 377, 380,
384, 388, 390, 391, 392,
393, 395, 396, 398, 405,
406, 408, 411, 412; 414,
416, 418, 420, 422, 424,
433, 437, 440, 441, 442,
444, 453, 457, 458, 471,
478
Flavoxanthin, 3
Fluoranthene, 91, 183, 185, 191,
200, 218, 221, 222, 223,
224, 227, 230, 234, 239
Fluorene, 91, 144, 183, 185, 205,
210, 212, 222, 223, 230,
239, 426
Fluorene carbonitrile, 222
Fluorene-9-one, 159, 160
Flurzepam, 40
Formaldehyde, 5, 20, 94
Formic acid, 6, 75, 96, 175, 177
 methyl derivative of, see methyl formate
 recovery of, 177
Fructose, 9
Fucosterol, 72, 73
 silyl derivative of, 72, 73
Fumaric acid, 6
Furan, 95, 116, 137, 140, 145, 154
Furfural, 5, 94, 140, 142, 431
Furfuryl alcohol, 4, 93, 143
Furoic acid, 6, 96, 176, 179, 416, 422
 methyl derivative of, 176, 179, 422
 silyl derivative of, 416

G

Galactan, 10
Galactosamine, 10
Galactose, 10
Galacturonic acid, 10
Geranyl pyrophosphate, 274
 structure of, 274
Glucosamine, 10
Glucose, 10
β-Glucuronidase, 448
Glutamic acid, 10, 101
Glutamine, 10, 101
Glutaric acid, 96, 178, 179, 422
 methyl derivative of, 178, 179, 422
Glutathione, 10
Glutethimide, 39
D-Glyceric acid, 6
Glycerol, 4, 93, 107, 108, 113, 144
 silyl derivative of, 113
Glyceryl triacetate, 93
Glycine, 10, 101
Glycolaldehyde, 5
Glycolic acid, 6, 96, 171, 176, 177, 179
 methyl derivative of, 176, 177, 179
Glyoxal, 5, 94
Glyoxylic acid, 6, 96
Guaiacol, 7, 97, 418, 421
 silyl derivative of, 421
Guanine, 8
α-Gurjuene, 381

H

Harmane, 8, 99, 431
Helium (GC), 13, 14, 15, 17, 22,
24, 25, 27, 29, 32, 33, 37,
43, 52, 54, 56, 61, 63, 69,
74, 75, 77, 79, 81, 82, 85,
86, 87, 88, 104, 105, 106,
112, 130, 132, 133, 135, 145,

SUBJECT INDEX

[Helium (GC)]
   149, 153, 159, 161, 162,
   170, 175, 176, 177, 178,
   179, 180, 181, 185, 189,
   200, 201, 205, 216, 218,
   220, 227, 230, 232, 246,
   247, 250, 252, 253, 254,
   282, 289, 325, 326, 338,
   339, 341, 345, 346, 352,
   354, 357, 360, 377, 380,
   384, 388, 393, 396, 398,
   403, 404, 406, 408, 414,
   415, 418, 420, 421, 422,
   446, 448, 457, 461, 465,
   479
Hemimellitene, 137
1-Heneicosanol, 4, 93
n-Hentriacontane, 64, 324
Hentriacontanylhentriacontanoate, 93
n-Heptacosane, 64, 305, 323, 374
n-Heptadecane, 143
Heptadecanoic acid, 418, 421
   methyl derivative of, 421
1-Heptadecanol, 4, 93
Heptafluorobutyric anhydride, 443
N-Heptafluorobutyrylimidazole, 443
n-Heptane, 34, 129, 133, 140, 145, 404
Heptanoic acid, 6, 75, 96, 416, 418, 421
   methyl derivative of, 421
   silyl derivative of, 416
4-Heptanone, 94, 140
n-Heptylbenzene, 227
Heptyl-Δ^1-tetrahydrocannabinol, see heptyl-Δ^9-tetrahydrocannabinol
Heptyl-Δ^9-tetrahydrocannabinol, 391
   tributylsilyl derivative of, 391
   triethylsilyl derivative of, 391
   trimethylsilyl derivative of, 391
   tripropylsilyl derivative of, 391

Heroin, 40
Hexachloroethane, 258
Hexacosadienoic acid, 422
   methyl derivative of, 422
Hexacosanoic acid, 422
   methyl derivative of, 422
Hexacosanol, 68, 69
   silyl derivative of, 68, 69
Hexadecanamide, 403
n-Hexadecane, 143
Hexadecanoic acid, 418, 421
   methyl derivative of, 421
1,5-Hexadiene, 145
Hexahydrocannabinol, 276, 362, 364, 376, 406, 407, 441, 442, 445, 470, 471, 472
   C-9 axial isomer, 362, 364, 376
      silyl derivative of, 376
      structure of, 364
   C-9 equatorial isomer, 362, 364, 376
      silyl derivative of, 376
      structure of, 364
   heptafluorobutyryl derivative of, 442, 445
   metabolites of, 470
   silyl derivative of, 362
   structure of, 276, 362
   7-substituted derivatives of, 471, 472
cis-ortho-Hexahydrocannabinol, 356, 359
   structure of, 356
cis-para-Hexahydrocannabinol, 356, 359
   structure of, 356
trans-para-Hexahydrocannabinol, 356, 360
   structure of, 356
Hexahydrofarnesylacetone, 4, 92
Hexamethylbenzene, 215
Hexamethyldisilazane, 129, 132, 329
n-Hexane, 56, 58, 60, 65, 66, 68, 69, 70, 79, 84, 85, 88, 140,

[n-Hexane]
  145, 152, 153, 154, 155, 174, 248, 285, 288, 312, 328, 352, 364, 416, 420, 439, 444, 446, 448
2-Hexanone, 94, 140
3-Hexanone, 94, 140
Hexanoic acid, 416, 418, 421, 422
  methyl derivative of, 421, 422
  silyl derivative of, 416
n-Hexatriacontane, 221
1-Hexene, 89
2-Hexene, 89
Hexobarbital, 39
Hexylamine, 99
n-Hexylbenzene, 227
n-Hexylmercaptan, 244
HHC-7-oic acid (axial), see 7-carboxyhexahydrocannabinol (axial)
HHC-7-oic acid (equatorial), see 7-carboxyhexahydrocannabinol (equatorial)
Histidine, 10
Homocystine, 11
Homoserine, 11
Humulene, 363, 381
β-Humulene, 326, 327
Hydrazine, 87, 88, 252-253
  detection limits of, 88, 253
  pentafluorobenzyl derivative of, 88, 252, 253
Hydrocaffeic acid, 7
Hydrocarbons, 50-64, 89, 90, 307
  natural products of tobacco leaf, 50-58, 59, 60, 61, 62, 63, 64
  recovery of, from urea adducts, 50, 54, 61
Hydrochloric acid, 41, 45, 67, 69, 84, 114, 125, 135, 170, 175
Hydrogen (GC), 112, 169, 232
Hydrogen cyanide, 95
Hydrogen sulfide, 96, 238, 240, 244

Hydroquinone, 97, 113, 170, 171, 418, 421
  silyl derivative of, 421
6-Hydroxy-23-aceto-Δ^{12}-oleanene, 324
6-Hydroxy-23-aceto-$\Delta^{13(18)}$-oleanene, 324
2- or o-Hydroxyacetophenone, 7, 97, 162, 168, 422, 424
  N-methyl-N-silyl derivative of, 424
3- or m-Hydroxyacetophenone, 7, 97
4- or p-Hydroxyacetophenone, 7, 97, 414, 415, 416, 418, 421
  silyl derivative of, 414, 415, 416, 421
6-Hydroxy-23-aceto-Δ^{12}-ursene, 324
6-Hydroxy-23-aceto-$\Delta^{13(18)}$-ursene, 324
2- or o-Hydroxybenzaldehyde, 7, 97, 422, 424
  N-methyl-N-silyl derivative of, 424
3- or m-Hydroxybenzaldehyde, 7, 97
4- or p-Hydroxybenzaldehyde, 7, 97, 431
3-Hydroxybenzoic acid, 7, 97
4-Hydroxybenzoic acid, 7, 97
m-Hydroxybenzyl alcohol, 65, 67, 68, 248, 249
  silyl derivative of, 67, 68
o-Hydroxybenzyl alcohol, 67, 68
  silyl derivative of, 67, 68
p-Hydroxybenzyl alcohol, 67, 68, 249
  recovery of, 67
  silyl derivative of, 67, 68
p-Hydroxy-2-butenylketone, 424
2-Hydroxy-3,7-cadinadiene, 321
2-Hydroxy-3,7(11)-cadinadiene, 321
6α-Hydroxycannabidiol, 401, 459, 460, 461

[6α-Hydroxycannabidiol]
  silyl derivative of, 461
  structure of, 459
6β-Hydroxycannabidiol, 401, 459,
    460, 461
  silyl derivative of, 461
  structure of, 459
7-Hydroxycannabidiol, 401, 459,
    460, 461
  silyl derivative of, 461
  structure of, 459
Δ^9-9-Hydroxycannabidiol, see
    9,10-Dihydro-9-hydroxy-
    cannabidiol
1"-Hydroxycannabidiol, 401, 459,
    460, 461
  silyl derivative of, 461
  structure of, 459
2"-Hydroxycannabidiol, 401, 459,
    461
  silyl derivative of, 461
  structure of, 459
3"-Hydroxycannabidiol, 401, 402,
    459, 460, 461
  silyl derivative of, 401, 402, 461
  structure of, 459
4"-Hydroxycannabidiol, 401, 402,
    459, 460, 461
  silyl derivative of, 401, 402, 461
  structure of, 459
5"-Hydroxycannabidiol, 401, 459,
    460, 461
  silyl derivative of, 461
  structure of, 459
Hydroxycannabigerol, 274
  structure of, 274
7-Hydroxycannabinol, 280, 400,
    436, 437, 474, 475, 476,
    477
  fatty acid conjugates of, 474,
    476, 477
    silyl derivative of, 474, 476
    structure of, 477
  oleate derivative of, 474, 476,
    477

[7-Hydroxycannabinol]
  [oleate derivative of]
    silyl derivative of, 474, 476
    structure of, 477
  palmitate derivative of, 474, 476,
    477
    silyl derivative of, 474, 476
    structure of, 477
  silyl derivative of, 436, 476
  structure of, 280, 436, 475
1"-Hydroxycannabinol, 474, 475,
    476
  silyl derivative of, 476
  structure of, 475
2"-Hydroxycannabinol, 401, 436,
    474, 475, 476
  silyl derivative of, 436, 476
  structure of, 436, 475
3"-Hydroxycannabinol, 401, 474,
    475, 476
  silyl derivative of, 401, 476
  structure of, 475
4"-Hydroxycannabinol, 401, 474,
    475, 476, 477
  fatty acid conjugates of, 474, 476,
    477
    silyl derivative of, 474, 476
    structure of, 477
  oleate derivative of, 474, 476,
    477
    silyl derivative of, 474, 476
    structure of, 477
  palmitate derivative of, 474, 476,
    477
    silyl derivative of, 474, 476
    structure of, 477
  silyl derivative of, 401, 476
  structure of, 475
5"-Hydroxycannabinol, 401, 474,
    475, 476, 477
  fatty acid conjugates of, 474, 476,
    477
    silyl derivative of, 474, 476
    structure of, 477
  oleate derivative of, 474, 476, 477

[5"-Hydroxycannabinol]
  [oleate derivative of]
    silyl derivative of, 474, 476
    structure of, 477
  palmitate derivative of, 474, 476, 477
    silyl derivative of, 474, 476
    structure of, 477
  structure of, 475, 476
1"-Hydroxycannabinol-7-oic acid, 477, 478
  methyl/silyl derivative of, 478
  structure of, 477
2"-Hydroxycannabinol-7-oic acid, 477, 478
  methyl/silyl derivative of, 478
  structure of, 477
3"-Hydroxycannabinol-7-oic acid, 477, 478
  methyl/silyl derivative of, 478
  structure of, 478
4"-Hydroxycannabinol-7-oic acid, 477, 478
  methyl/silyl derivative of, 478
  structure of, 478
1"-Hydroxy-7-carboxy-Δ^1-tetrahydrocannabinol, see 1"-hydroxy-11-carboxy-Δ^9-tetrahydrocannabinol
2"-Hydroxy-7-carboxy-Δ^1-tetrahydrocannabinol, see 2"-hydroxy-11-carboxy-Δ^9-tetrahydrocannabinol
3"-Hydroxy-7-carboxy-Δ^1-tetrahydrocannabinol, see 3"-hydroxy-11-carboxy-Δ^9-tetrahydrocannabinol
6α-Hydroxy-7-carboxy-Δ^1-tetrahydrocannabinol, see 8α-hydroxy-11-carboxy-Δ^9-tetrahydrocannabinol
1"-Hydroxy-11-carboxy-Δ^9-tetrahydrocannabinol, 437, 438
  methyl/methoxy derivative of, 437, 438
  structure of, 438
2"-Hydroxy-11-carboxy-Δ^9-tetrahydrocannabinol, 437, 438, 452, 453, 455, 467, 469
  methyl/methoxy derivative of, 437, 438
  silyl derivative of, 455, 469
  structure of, 438, 453, 467
3"-Hydroxy-11-carboxy-Δ^9-tetrahydrocannabinol, 452, 453, 455, 467, 469
  silyl derivative of, 455, 469
  structure of, 453, 467
8α-Hydroxy-11-carboxy-Δ^9-tetrahydrocannabinol, 452, 453, 455, 467
  silyl derivative of, 455
  structure of, 453, 467
Hydroxycotinine, 12
  structure of, 12
12α-Hydroxy-13-epimanoyloxide, 3, 92
12-Hydroxy-2,8-eudesmadiene, 321
7-Hydroxyhexahydrocannabinol (axial), 472
  silyl derivative of, 472
7-Hydroxyhexahydrocannabinol (equatorial), 472
  silyl derivative of, 472
9-Hydroxyhexahydrocannabinol, 383, 384, 405, 406, 407
  structure of, 383
α-Hydroxyisocaproic acid, 6
β-Hydroxyisocaproic acid, 6
1-Hydroxy-9-isopropenyl-6-methyl-3-pentyldibenzofuran, see dehydrocannabifuran
1-Hydroxy-9-isopropyl-6-methyl-3-pentyldibenzofuran, see cannabifuran
m-Hydroxy-p-methoxystyrene, see 3-hydroxy-4-methoxystyrene
3-Hydroxy-4-methoxystyrene, 422, 424
  N-methyl-N-silyl derivative of, 424
4-Hydroxy-3-methoxystyrene, 417

[4-Hydroxy-3-methoxystyrene]
  silyl derivative of, 417
2-Hydroxy-3-methyl-2-cyclopenten-1-one, 431
5-Hydroxymethylfurfural, 5, 94
4-Hydroxy-2-methyl-6-n-pentylbenzofuran, 431
α-Hydroxy-β-methylvaleric acid, 6
β-Hydroxy-β-methylvaleric acid, 6
Hydroxynicotine, 20, 21
3"-Hydroxy-6-oxo-cannabidiol, 279, 461, 462
  silyl derivative of, 461, 462
  structure of, 279
4"-Hydroxy-6-oxo-cannabidiol, 279, 461, 462
  silyl derivative of, 461, 462
  structure of, 279
2"-Hydroxy-6-oxo-Δ^1-tetrahydrocannabinol, see 2"-hydroxy-8-oxo-Δ^9-tetrahydrocannabinol
3"-Hydroxy-6-oxo-Δ^1-tetrahydrocannabinol, see 3"-hydroxy-8-oxo-Δ^9-tetrahydrocannabinol
7-Hydroxy-6-oxo-Δ^1-tetrahydrocannabinol, see 11-hydroxy-8-oxo-Δ^9-tetrahydrocannabinol
2"-Hydroxy-8-oxo-Δ^9-tetrahydrocannabinol, 465, 467, 468
  methoxy/silyl derivative of, 468
  silyl derivative of, 468
  structure of, 467
3"-Hydroxy-8-oxo-Δ^9-tetrahydrocannabinol, 465, 467, 468
  methoxy/silyl derivative of, 468
  silyl derivative of, 468
  structure of, 467
11-Hydroxy-8-oxo-Δ^9-tetrahydrocannabinol, 465, 467, 468
  methoxy/silyl derivative of, 468

[11-Hydroxy-8-oxo-Δ^9-tetrahydrocannabinol]
  silyl derivative of, 468
  structure of, 467
4-Hydroxy-6-n-pentylbenzofuran, 431
5-Hydroxy-7-n-pentyl-2H-chromene, 431
2-Hydroxyphenylacetic acid, 7, 97
3-Hydroxyphenylacetic acid, 7, 97
4-Hydroxyphenylacetic acid, 8, 97
m-Hydroxyphenylethanol, 67, 68
  silyl derivative of, 67, 68
o-Hydroxyphenylethanol, 67, 68
  silyl derivative of, 67, 68
2-(p-Hydroxyphenyl)ethanol, 65, 67, 68, 69, 248, 249
  detection limits of, 68
  silyl derivative of, 67, 68
3-Hydroxyphenylpropionic acid, 8, 97
4-Hydroxyphenylpropionic acid, 8, 97
Hydroxyproline, 11
Hydroxypyruvic acid, 6
6-Hydroxy-Δ^1-tetrahydrocannabinol, see 8-Hydroxy-Δ^9-tetrahydrocannabinol
6α-Hydroxy-Δ^1-tetrahydrocannabinol, see 8α-hydroxy-Δ^9-tetrahydrocannabinol
6β-Hydroxy-Δ^1-tetrahydrocannabinol, see 8β-hydroxy-Δ^9-tetrahydrocannabinol
7-Hydroxy-Δ^1-tetrahydrocannabinol, see 11-hydroxy-Δ^9-tetrahydrocannabinol
1"-Hydroxy-Δ^1-tetrahydrocannabinol, see 1"-hydroxy-Δ^9-tetrahydrocannabinol
2"-Hydroxy-Δ^1-tetrahydrocannabinol, see 2"-hydroxy-Δ^9-tetrahydrocannabinol
3"-Hydroxy-Δ^1-tetrahydrocannabinol, see 3"-hydroxy-Δ^9-tetrahydrocannabinol

4"-Hydroxy-Δ^1-tetrahydrocannabinol, see 4"-hydroxy-Δ^9-tetrahydrocannabinol
5"-Hydroxy-Δ^1-tetrahydrocannabinol, see 5"-hydroxy-Δ^9-tetrahydrocannabinol
1"-Hydroxy-Δ^6-tetrahydrocannabinol, see 1"-hydroxy-Δ^8-tetrahydrocannabinol
2"-Hydroxy-Δ^6-tetrahydrocannabinol, see 2"-hydroxy-Δ^8-tetrahydrocannabinol
3"-Hydroxy-Δ^6-tetrahydrocannabinol, see 3"-hydroxy-Δ^8-tetrahydrocannabinol
5"-Hydroxy-Δ^6-tetrahydrocannabinol, see 5"-hydroxy-Δ^8-tetrahydrocannabinol
7-Hydroxy-Δ^6-tetrahydrocannabinol, see 11-hydroxy-Δ^8-tetrahydrocannabinol
1"-Hydroxy-Δ^8-tetrahydrocannabinol, 400
2"-Hydroxy-Δ^8-tetrahydrocannabinol, 400
3"-Hydroxy-Δ^8-tetrahydrocannabinol, 400
5"-Hydroxy-Δ^8-tetrahydrocannabinol, 400
  silyl derivative of, 400
11-Hydroxy-Δ^8-tetrahydrocannabinol, 391, 400, 434, 435, 436-437, 456, 457, 458
  fatty acid conjugates of, 456, 458
  structure of, 434
  triethylsilyl derivative, 391, 400
  trimethylsilyl derivative, 391
  tripropylsilyl derivative, 391
8-Hydroxy-Δ^9-tetrahydrocannabinol, 275
  structure of, 275
8α-Hydroxy-Δ^9-tetrahydrocannabinol, 400, 438, 446, 450, 451, 455, 467, 468

  acetyl derivative of, 446
  silyl derivative of, 451, 455, 468
  structure of, 467
8β-Hydroxy-Δ^9-tetrahydrocannabinol, 400, 433, 434, 435, 438, 450, 451, 467, 468
  silyl derivative of, 451, 468
  structure of, 434, 467
11-Hydroxy-Δ^9-tetrahydrocannabinol, 35, 273, 280, 400, 433, 434, 435, 436-437, 438, 446, 447, 450, 451, 452, 453, 455, 456, 458, 461, 462, 463, 464, 465, 466, 467, 468
  acetyl derivative of, 433, 434, 446
  detection limits of, 451, 464
  ethoxy derivative of, 452, 462, 463
    deuterated, 452, 462, 463
    silyl derivative of, 452, 463
  silyl derivative of, 452, 462, 463
  fatty acid conjugates of, 456, 458
  perdeuterioethoxy derivative of, 452, 462, 463
    silyl derivative of, 452, 463
  pharmacokinetics of, 466
  recovery of, 464
  structure of, 280, 434, 447, 453, 467
  trimethylsilyl derivative of, 438, 451, 452, 455, 468
1"-Hydroxy-Δ^9-tetrahydrocannabinol, 451
2"-Hydroxy-Δ^9-tetrahydrocannabinol, 451
3"-Hydroxy-Δ^9-tetrahydrocannabinol, 276, 400, 448, 450, 451, 467, 468
  silyl derivative of, 400, 451, 468
  structure of, 276, 467
4"-Hydroxy-Δ^9-tetrahydrocannabinol, 276, 400, 448, 450, 451

[4"-Hydroxy-Δ^9-tetrahydrocannabinol]
  silyl derivative of, 400, 451
  structure of, 276
5"-Hydroxy-Δ^9-tetrahydrocannabinol, 451
Δ^9-9-Hydroxytetrahydrocannabinol, see 9-hydroxyhexahydrocannabinol
2"-Hydroxy-Δ^1-tetrahydrocannabinol-7-oic acid, see 2"-hydroxy-11-carboxy-Δ^9-tetrahydrocannabinol
α-Hydroxyvaleric acid, 6

I

Indan, 142, 227
Indene, 91, 142
Indeno-(1,2,3-cd)fluoranthene, 91, 198, 238
Indeno-(1,2,3-cd)pyrene, 91, 198, 238, 239
Indole, 99, 113, 144, 201, 206, 416, 418, 419
  structure of, 419
Indole-2-^{14}C, 416
Indoleacetic acid, 6
Inositol, 4
Internal standard, 15, 19, 31, 33, 34, 37, 41, 42, 45, 46, 60, 61, 67, 68, 69, 79, 85, 119, 120, 121, 125, 126, 128, 129, 132, 133, 135, 162, 168, 169, 189, 218, 219, 249, 252, 286, 295, 304, 308, 312, 314, 315, 328, 329, 332, 349, 350, 354, 362, 374, 375, 380, 385, 390, 398, 404, 406, 407, 409, 416, 440, 441, 442, 444, 446
Ionene, 91
Isoamylamine, 8, 99
Isoamylmercaptan, 244

Isobutanal, see isobutyraldehyde
Isobutane, 152, 154, 227, 403
Isobutene, 152, 154
Isobutyl alcohol, 93, 137
Isobutylamine, 8, 99
N-Isobutylbutylamine, 99
Isobutylmercaptan, 244
Isobutyraldehyde, 5, 94, 137, 154, 361
Isobutyric acid, 6, 75, 76, 77, 96
Isobutyronitrile, 95, 137, 140
Isocannabidiol, 405
  structure of, 405
Isocaproic acid, 6, 75, 76, 96
Isocapronitrile, 95, 140
Isoeugenol, 8, 97, 144
Isoleucine, 11
Isonicotein, see 2,3'-bipyridyl
Isonicotine, 12
  structure of, 12
Isooctane, 348
Isopentane, 149, 152, 153
Isopentenone, 137
Isoprene, 89, 116, 140, 145, 146, 147, 152, 154, 207
o-Isopropenylphenol, 422, 424
  N-methyl-N-silyl derivative of, 424
4-Isopropenyltoluene, 91, 136, 140
Isopropyl alcohol, 140
Isopropylamine, 99
Isopropylbenzene, 91, 226
Isopropyl formate, 94
3-Isopropyl-5-hydroxybenzaldehyde, 422
2-Isopropylmalic acid, 6
Isopropylmercaptan, 244
N-Isopropylpropylamine, 99
3-Isopropyltoluene, 226
4-Isopropyltoluene, 91
Isoquercetrin, 8
Isoquinoline, 19, 20, 99, 104
Isotetrahydrocannabinol, 355
$\Delta^{4(8)}$-Isotetrahydrocannabinol, 322

cis-ortho-8,9-Isotetrahydrocannabinol, 355
  structure of, 355
cis-para-8,9-Isotetrahydrocannabinol, 355
  structure of, 355
trans-ortho-8,9-Isotetrahydrocannabinol, 355, 359
  α-methyl derivative of, 359
  structure of, 355
trans-para-8,9-Isotetrahydrocannabinol, 355, 359
  α-methyl derivative of, 359
  structure of, 355
Isovaleraldehyde, 5, 94, 137, 154
Isovaleric acid, 6, 75, 77, 96
Isovaleronitrile, 95, 140
Isovanillic acid, 97

K

Kaempferol-3-rhamnoglucoside, 8
5-(9-Ketodecyl)-2-furfuraldehyde, 322
α-Ketoglutaric acid, 6, 96
6-Keto-10-nor-23-aceto-Δ^{12}-oleanene, 324
6-Keto-10-nor-23-aceto-$\Delta^{13(18)}$-oleanene, 324
6-Keto-10-nor-23-aceto-Δ^{12}-ursene, 324
6-Keto-10-nor-23-aceto-$\Delta^{13(18)}$-ursene, 324
6-Keto-Δ^{12}-oleanene, 324
6-Keto-$\Delta^{13(18)}$-oleanene, 324
2-[5-(2-Keto-n-pentyl)resorcinolyl]-p-mentha-1,8(9)-diene, 322
6-Keto-Δ^{12}-ursene, 324
6-Keto-$\Delta^{13(18)}$-ursene, 324

L

Lactic acid, 6, 96, 171, 176, 177, 179
  methyl derivative of, 171, 176, 177, 179

Lauric acid, 6, 96
Leucine, 11, 101
Levallorphan, 40
α-Levantanolide, 3, 92
α_2-Levantanolide, 3
β-Levantanolide, 3, 92
Levulinic acid, 96, 176, 179
  methyl derivative of, 176, 179
Lidocaine, 120, 121, 133
  recovery of, 121
Lignin, 10
Lignocaine, see lidocaine
Lignoceric acid, 422
  methyl derivative of, 422
Limonene, 137, 138, 140, 142, 145, 250, 251, 282, 327, 363, 381
Linalool, 3, 326, 327, 363, 381
trans-Linalool oxide, 326, 327
Linoleic acid, 6, 77, 84, 86, 96, 113, 178, 179, 180, 181, 416, 422
  methyl derivative of, 422
  recovery of, 180
  silyl derivative of, 86, 113, 178, 179, 180, 181
Linolenic acid, 6, 77, 84, 86, 96, 113, 178, 179, 180, 181, 417, 422
  methyl derivative of, 422
  recovery of, 180
  silyl derivative of, 86, 113, 178, 179, 180, 181, 417
Liquid stationary phase (GC)
  types of
    Apiezon-series, 26, 27, 28, 29, 31, 33, 50, 54, 58, 59, 60, 62, 130, 137, 159, 160, 162, 163, 213, 227, 247, 282, 343
    BHxBT [N,N'-bis(p-hexyloxybenzylidene)-α,α'-bi-p-toluidine], 220, 223, 225
    m-Bis-(m-phenoxyphenoxy)benzene, 227
    BMeBT [N,N'-bis(p-methoxybenzylidene)-α,α'-bi-p-toluidine], 189, 223

SUBJECT INDEX

[Liquid stationary phase (GC)]
 [types of]
  BPhBT [N,N'-bis(p-phenylbenzylidene)-α,α'-bi-p-
   toluidine], 200, 220, 223,
   225
  BTPPC, 125
  Carbowax-series, 19, 31, 41,
   43, 104, 105, 111, 121, 126,
   129, 132, 135, 169, 175,
   177, 207, 259, 328
  CDMS (cyclohexane dimethanol
   succinate), 335
  DEGA (diethylene glycol adipate), 74
  DEGS (diethylene glycol succinate), 162, 163
  Dexsil-series, 54, 59, 60, 61,
   63, 64, 85, 86, 181, 182,
   201, 219, 220, 250, 302,
   348, 349, 350, 379, 382,
   457
  Didecylphthalate, 169, 175, 176
  Dimethyl sulfolane, 149
  Dinonylsebacate, 175, 176
  Dioctylphthalate, 13, 14, 15,
   161
  Di-n-octylsebacate, 161
  Ditridecylphthalate, 226
  4-Dodecyldiethylenetriamine
   succinimide (DDTS), 30, 31
  E-301, 137, 148, 159, 187
  EGSS-X, 81, 83
  Emulphor-series, 141, 245,
   246
  Epon-1001, 28
  FFAP, 112, 244, 422, 424
  Flexol-series, 175, 176, 178
  Hallcomid, 153
  Hexadecane, 149
  Igepal CO-880, 33
  JXR, 27, 28, 295, 302, 350, 437
  Neopentyl glycol, 432
  NGA (neopentylglycol adipate),
   29

[Liquid stationary phase (GC)]
 [types of]
  β,β'-Oxydipropionitrile, 111,
   146, 153, 155
  OV-1, 81, 154, 159, 169, 200,
   205, 210, 213, 217, 218, 220,
   338, 339, 343, 348, 349, 350,
   364, 374, 377, 390, 440, 446,
   448
  OV-7, 219, 221, 358, 392
  OV-17, 34, 37, 45, 88, 126,
   159, 160, 169, 170, 205, 210,
   245, 247, 257, 259, 288, 292,
   293, 294, 295, 303, 304, 306,
   312, 335, 336, 338, 341, 342,
   343, 345, 348, 349, 350, 352,
   354, 356, 362, 375, 377, 378,
   380, 382, 384, 388, 393, 403,
   404, 406, 408, 409, 412, 414,
   420, 421, 424, 431, 436, 438,
   440, 442
  OV-25, 306
  OV-101, 30, 31, 111, 125, 140,
   145, 179, 180, 227, 302, 360,
   361, 363, 377, 380, 382, 388,
   396, 398, 422
  OV-210, 348, 349, 350
  OV-225, 200, 257, 348, 349,
   350, 384, 406, 444, 461
  PBG (polybutylene glycol), 11,
   13, 17, 24, 105
  PEG (polyethylene glycol), 11,
   13, 15, 16, 21, 22, 29, 105,
   107, 126, 137, 159
  Pennwalt-223, 112, 418, 421
  PMPE (poly-m-phenoxylene),
   213, 217, 218
  Polyamine, 26
  Polyethyleneimine, 87
  Poly-I-100, 71, 402
  Polyethyleneglycol adipate
   (PEGA), 227
  Polysev [m-bis-m-(phenoxy-
   phenoxy)phenoxybenzene],
   154, 158

[Liquid stationary phase (GC)]
  [types of]
    PPG (polypropylene glycol), 11, 13, 14, 17, 19, 27, 105, 106, 107, 109, 122, 138, 139, 162
    QF-1, 27, 28, 88, 207, 210, 213, 214, 216, 253, 306, 338
    Reoplex, 187, 326, 360, 361, 363
    SE-52, 27, 28, 71, 73, 159, 185, 230, 232, 233, 319, 320, 332, 342, 402, 424, 440, 452
    SF-96, 138, 411
    Silar-10C, 79, 81, 82, 83, 416
    Silicone grease, 15, 16, 50
    Silicone oil (DC-series), 29, 30, 31, 227, 253
    Silicone rubber (SE-30), 27, 28, 29, 30, 36, 52, 54, 59, 62, 71, 77, 114, 149, 159, 189, 207, 210, 213, 214, 216, 219, 259, 283, 286, 295, 306, 325, 332, 333, 336, 351, 390, 391, 395, 398, 402, 403, 405, 408, 413, 433, 437, 441, 446, 451, 452, 453, 458, 460, 461, 465, 471, 474, 476, 478, 479
    SP-series, 69, 133
    Squalane, 153, 154
    Succinimide polymer, 125
    TCEP [1,2,3-tris-(cyanoethoxy)-propane], 153, 227, 244
    TCP (tricresylphosphate), 175
    Trimer acid, 177, 432
    3,3'-(Trimethylene dioxy)dipropionitrile, see β,β'-oxydipropionitrile
    Tween, 74, 75
    Versamid, 29, 30
    W-98, 43, 67, 169, 189, 343, 433

[Liquid stationary phase (GC)]
  [types of]
    XE-60, 27, 28, 54, 59, 205, 253, 295, 329, 331, 343
Lithium aluminum deuteride, 465
Lithium aluminum hydride, 60, 77, 447
Lutein, 3
2,4-Lutidine, 13
2,5-Lutidine, 13
Lysine, 11

M

Magnesium chloride, 433
Magnesium sulfate, 420
Maleic acid, 6
Maleic hydrazide, 253, 254
  bis-(4-chlorobenzyl) derivative of, 253, 254
  structure of, 254
  detection limits of, 253
Malic acid, 6, 96, 178, 179
  methyl derivative of, 178, 179
Malonic acid, 6, 96, 171, 176, 179
  methyl derivative of, 171, 176, 179
Maltose, 10
Mannose, 10
Marijuana, 2, 70, 71, 185, 190-198, 252, 271-479
  metabolic products of, in biological media, 432-479
  plant constituents, 70, 71, 273-407
  smoke constituents, 190-198, 407-431
Melilotic acid, 8
Melvalonate, 274
Menthol, 4, 93, 109, 113, 143
Meperidine, 39
Mephenteramine, 38
Mephobarbital, 39
Meprobamate, 39
Mescaline, 39
Mesitylene, 136, 137, 215
Mesoxaldialdehyde, 5

SUBJECT INDEX

Metanicotine, 13, 15, 16, 19, 20, 22, 23, 99, 101, 105
Methacrolein, 94, 137, 154
Methacrylonitrile, 95, 137
Methadone, 39
Methamphetamine, 38
Methane, 26, 34, 35, 45, 46, 47, 152, 154, 255, 364, 414, 415, 441, 442, 457
Methanethiol, 96
Methanol, 4, 41, 45, 50, 58, 88, 93, 116, 135, 136, 137, 138, 140, 154, 175, 177, 185, 325, 333, 352, 361, 391, 416, 420, 433, 446, 452
Methapyriline, 39
Methaqualone, 39
Metharbital, 38
Methionine, 11
Methionine sulfone, 11
Methohexital, 39
m-Methoxyacetophenone, 161
p-Methoxyacetophenone, 161
Methoxyamine hydrochloride, 465
7-Methoxy-O-methylcannabinol, 437
2-Methoxyphenol, 144, 165, 167, 168
3-Methoxyphenol, 97, 163, 168
4-Methoxyphenol, 97, 163, 168
m-Methoxyphenyl methyl ether, 248
o-Methoxyphenyl methyl ether, 248
1-Methylacenaphthene, 185
1-Methylacenaphthylene, 185, 190
Methyl acetate, 94, 116, 137, 140, 147, 154, 175, 177, 361
Methylacetophenone, 143
Methylacetylene, 89, 153
5-Methyl-2-acetylfuran, 431
Methylacrylate, 94, 137
Methyl allyl disulfide, 244
Methyl allyl sulfide, 244
Methylamine, 8, 99, 252

3-Methylaminopropyl-3'-pyridyl ketone, 20
3-Methylaminopyridine, 99
N-Methylanabasine, 8, 15, 16
N-Methylanatabine, 8
N-Methylaniline, 99
Methylanisole, 142
1-Methylanthracene, 185, 221
2-Methylanthracene, 91, 183, 185, 190
9-Methylanthrecene, 91, 183
Methylarachidate, 73
Methylarachidonate, 73
Methylbehenate, 60, 73
Methylbenz(a)anthracene, 194, 195, 222, 236
3-Methylbenz(a)anthracene, 91
5-Methylbenz(a)anthracene, 91
Methyl-1,2-benzanthracene, 185
7-Methyl-1,2-benzanthracene, 184
Methylbenz(a)pyrene, 222
Methylbenzocarbazole, 202
Methylbenzofluoranthene, 196, 197, 237
Methylbenzo(b)fluoranthene, 185, 223
Methylbenzo(ghi)fluoranthene, 185, 236
Methylbenzo(j)fluoranthene, 185
Methylbenzo(k)fluoranthene, 185, 223
Methylbenzofluorene, 185
11-Methyl-11H-benzo(a)fluorene, 91
Methylbenzo(ghi)perylene, 223
Methylbenzo(c)phenanthrene, 236
Methylbenzopyrene, 196, 197, 237
Methylbenzo(a)pyrene, 91, 185, 223
Methylbenzo(c)pyrene, 222
Methylbenzo(e)pyrene, 185
Methylbinaphthyl, 195, 237
2-Methylbiphenyl, 228
3-Methylbiphenyl, 228
4-Methylbiphenyl, 199, 228
Methylbutane, 154
2-Methylbutane, 145, 155
3-Methyl-2-butanone, 94

3-Methyl-3-butanone, 140
2-Methylbutene-1, 89, 152, 153, 154, 155
2-Methylbutene-2, 89, 152, 153, 154, 155
3-Methylbutene-1, 89, 152, 153, 154, 155
2-Methyl-1-buten-3-one, 140
3-Methyl-3-buten-2-one, 95
2-(3-Methyl-2-butenyl)-p-mentha-2,6-diene, 321
N-Methylbutylamine, 99
2-Methylbutylamine, 99
1-Methyl-4-butylbenzene, 229
1-Methyl-3-tert-butylbenzene, 229
1-Methyl-4-tert-butylbenzene, 229
Methyl-n-butyl sulfide, 244
2-Methylbutyraldehyde, 94, 140
3-Methylbutyraldehyde, 140
α-Methylbutyric acid, 7
Methyl caproate, 73
Methylcarbazole, 99, 191
1-Methylcarbazole, 202
2-Methylcarbazole, 202
3-Methylcarbazole, 202
4-Methylcarbazole, 202
9-Methylcarbazole, 200, 201, 203
  structure of, 203
3-Methylcatechol, 170, 171
4-Methylcatechol, 170, 171
Methyl chloride, 101, 139, 152, 258, 259
Methylcholanthrene, 214
3-Methylcholanthrene, 184, 210, 221, 224
20-Methylcholanthrene, 184
Methylchrysene, 91, 194, 195, 236
1-Methylchrysene, 185
2-Methylchrysene, 185
3-Methylchrysene, 185
4-Methylchrysene, 185
6-Methylchrysene, 185
1-(1-Methylcyclohex-3-enyl)-18-methylnonadeca-1,5,9-trien-4,5-diol, 323

Methyl-4H-cyclopenta(def)phenanthrene, 191, 234
2-Methylcyclopentane, 145
3-Methylcyclopentanone, 140
1-Methyl-1-cyclopentene, 89
3-Methyl-1-cyclopentene, 89
4-Methyl-1-cyclopentene, 90
Methyldecane, 145
2-Methyldecane, 145
5-Methyldecane, 145
Methyldibenzanthracene, 223, 238
Methyldibenzothiophene, 234
1-Methyl-3,5-diethylbenzene, 229
4-Methyl-2,6-di-tert-butylphenol, 169
2-Methyldotriacontane, 64
Methylene chloride, 30, 45, 114, 121, 125, 131, 133, 135, 145, 258, 259, 319, 350, 404, 422, 452, 462, 463, 464
4,5-Methylene phenanthrene, 183, 230, 239
4-Methyl-5α-ergosta-7,24(28)-dien-3β-ol, 71, 73
  silyl derivative of, 71, 73
  structure of, 71
Methylethylamine, 99
Methylethyl disulfide, 244
Methylethylfluoren-9-one, 160
Methylethylindole, 431
Methylethylketone, 137, 154
Methylethylpyrazine, 431
Methylethylpyridine, 431
2-Methyl-5-ethylpyridine, 13
Methylethyl sulfide, 244
Methyl(ghi)fluoranthene, 185
Methylfluoranthene, 192, 193
1-Methylfluoranthene, 185, 192, 235
2-Methylfluoranthene, 185
8-Methylfluoranthrene, 91, 185
1-Methylfluorene, 91, 183, 185, 190, 205, 210, 211, 212, 222
  structure of, 211

2-Methylfluorene, 183, 185, 190, 205, 210, 212, 222
3-Methylfluorene, 185, 205, 210, 212
4-Methylfluorene, 185, 205, 210, 211, 212, 222
  structure of, 211
9-Methylfluorene, 92, 185, 205, 210, 212, 222
1-Methylfluoren-9-one, 160
2-Methylfluoren-9-one, 160
3-Methylfluoren-9-one, 160
4-Methylfluoren-9-one, 160
Methyl formate, 94, 116, 137, 147, 175, 177
Methylfuran, 95, 137, 147
2-Methylfuran, 116, 140, 147, 154
5-Methylfurfural, 5, 94, 143, 431
Methylglyoxal, 5, 94
4-Methylguaiacol, 431
3-Methylhentriacontane, 64
3-Methylheptacosane, 64
2-Methylheptane, 145
3-Methylheptane, 145
2-Methyl-2-heptene-6-one, 361, 363
Methylhexane, 145
3-Methylhexane, 140
2-Methyl-2-hexene, 145
2-Methylhistidine, 11
1-Methylindane, 145
1-Methylindene, 143, 199
1-Methylindole, 143, 190, 205, 206, 208, 209
  structure of, 208
2-Methylindole, 144, 206
3-Methylindole, 99, 144, 206
4-Methylindole, 206
5-Methylindole, 145, 206
6-Methylindole, 206
7-Methylindole, 144, 206
Methyl iodide, 25, 26, 437
Methyl-^{14}C iodide, 26
Methyl isocyanate, 94, 101
1-Methyl-4-isopropylcyclohex-1-ene, 89, 140

Methylisopropylketone, 137, 154
1-Methyl-7-isopropylphenanthrene, see retene
Methyl laurate, 73
Methyl linoleate, 73
Methyl linolenate, 73
Methyl margarate, 73
Methyl mercaptan, 140, 244
6-Methyl-2-(1-methyl-3-cyclohex-1-enyl)hepta-1,5-diene, 321
N-Methylmyosmine, 20, 21, 22, 23, 24, 99
Methyl myristate, 73
1-Methylnaphthalene, 92, 143, 145, 185, 187, 189, 199, 228
  recovery of, 189
2-Methylnaphthalene, 92, 143, 145, 185, 187, 189, 199, 228
  recovery of, 189
Methylnaphthobenzothiophene, 236
Methylnaphthylketone, 95
N-Methylnicotinamide, 9, 15, 16, 22, 23, 105
Methyl nitrite, 94, 116
3-Methylnonacosane, 64
Methyl nonadecylate, 73
2-Methyloctacosane, 64
9-Methyloctacosane, 323
Methyloctane, 145
Methyl oleate, 73, 374, 375
Methyl palmitate, 73, 85
Methylpentane, 152
2-Methylpentane, 140, 145, 153, 154, 155
3-Methylpentane, 145, 153, 154, 155
3-Methyl-1-pentanol, 4
2-Methyl-3-pentanone, 95, 140
3-Methyl-2-pentanone, 95, 140
4-Methyl-2-pentanone, 5, 95, 140
2-Methyl-4-pentenal, 94, 140
2-Methylpenten-4-aldehyde, see 2-methyl-4-pentenal
2-Methyl-1-pentene, 90
2-Methyl-2-pentene, 90, 154
3-Methyl-1-pentene, 90

4-Methyl-1-pentene, 90
4-Methyl-2-pentene, 90, 145
1-Methylphenanthrene, 92, 183, 185, 191, 221, 234
2-Methylphenanthrene, 185, 190, 221, 234
3-Methylphenanthrene, 185, 190, 221, 234
9-Methylphenanthrene, 92, 185, 191, 230, 234
N-Methyl-2-phenethylamine, 9, 99
Methylphenidate, 38
2-Methylphenol, 163, 164, 165, 166, 168
3-Methylphenol, 163, 164, 165, 166, 168
4-Methylphenol, 163, 164, 165, 166, 168
N-Methyl-2-phenylethylamine, 9
Methylphenylindole, 202
3-Methyl-3-phenylpiperidine, 119
2-Methyl-3-phytyl-1,4-naphthoquinone, see vitamin K_1
3-Methyl-3-piperidinopyrazine, see modaline
N-Methyl-4-(3'-piperidyl)-n-butylamine, see octahydronicotine
2-Methylpropenaldehyde, 140
2-Methylpropene, 90
Methylpripionaldehyde, 140
Methylpropionate, 94
Methyl-n-propyl disulfide, 244
Methylpropylketone, 137
Methylpropylpyrazine, 431
Methyl-n-propyl sulfide, 244
Methylpyrazine, 142, 431
2-Methylpyrazine, 99
1-Methylpyrene, 92, 183, 185, 193, 217, 230, 235
2-Methylpyrene, 92, 185, 192, 235
3-Methylpyrene, 183
4-Methylpyrene, 92, 185, 193, 235

2-Methylpyridine, 9, 99, 101, 418, 421
3-Methylpyridine, 9, 100, 101, 142, 418, 421
4-Methylpyridine, 100, 418, 421
Methyl-3-pyridylketone, 159
N-Methylpyrrole, 100, 140
N-Methylpyrrolidine, 9, 101
2-Methylpyrrolidine, 9, 100
N-Methyl-3-pyrroline, 9
Methyl-α-pyrrylketone, 5
Methylquinoline, 144
2-Methylquinoline, see quinaldine
Methyl reductone, 5
Methyl salicylate, 5, 8
Methyl stearate, 73
4-Methyl-5α-stigmasta-7,24(28)-dien-3β-ol, 71, 72, 73
  silyl derivative of, 71, 72, 73
  structure of, 72
Methylstyrene, 145
m-Methylstyrene, 92, 137
o-Methylstyrene, 92, 137
1-Methyl-(1,2,3,4-tetrahydronaphthalene), 145
Methylthionitrite, 94, 96
2-Methyltriacontane, 55, 64
N-Methyl-N-trimethylsilyltrifluoroacetamide, 424
Methyltriphenylene, 185, 222
1-Methyltritriacontane, 55, 64
2-Methyltritriacontane, 55
3-Methyltritriacontane, 55, 64
4-Methyl tritriacontane, 55
5-Methyltritriacontane, 55
Methylundecanoate, 79
2-Methylvaleraldehyde, 94
α-Methylvaleric acid, 7
β-Methylvaleric acid, 7, 75, 77, 96
2-Methyl-5-vinylpyridine, 13
Methyprylon, 38
Microcoulometric detector (GC), 259
Modaline, 129, 133

Morphine, 40
Myosmine, 9, 13, 15, 16, 19, 20, 22, 23, 24, 27, 29, 30, 42, 100, 103, 104, 105, 107, 144
Myrcene, 282, 327, 363
Myristic acid, 7, 84, 86, 96, 417
  silyl derivative of, 86, 417
Myristicin, 109

N

Naphthacene, 92, 184, 223, 224
Naphthalene, 19, 20, 92, 143, 145, 183, 185, 187, 199, 227, 228, 229, 230
  recovery of, 189
Naphthalene-^{14}C, 187, 189
Naphthobenzothiophene, 236
11-H-Naphtho-(2,1-a)fluorene, 92
1-Naphthol, 97
2-Naphthol, 97, 431
Naphtho(2,3-a)pyrene, 92
1-Naphthylamine, 100, 253, 255
  detection limits of, 253
  pentafluoropropionyl derivative of, 253, 255
2-Naphthylamine, 253, 255
  detection limits of, 253
  pentafluoropropionyl derivative of, 253, 255
Naringenin, 8
Naringin, 8
Neo-β-carotene, 3
Neochlorogenic acid, 8, 97
Neophytadiene, 3, 54, 58, 59, 63, 64, 86, 89, 107, 108, 144, 250, 251
Neoxanthin, 3
Nicotelline, 9, 100
Nicotinamide, 9, 20, 100
Nicotine, 1, 2-41, 42, 100, 101, 103, 104, 105, 106, 107, 108, 109, 113, 114, 119, 120, 121, 122, 123, 124,

[Nicotine]
  125, 126, 127, 128, 129, 130, 131, 132, 133, 134, 135, 136, 143, 171, 172, 174, 411
  biosynthesis of, 22, 24, 26
  blood levels (plasma) of, 126, 128, 129, 130, 131
  catalytic hydrogenation of, 33, 34
  demethylation of, 2, 12, 20
    mechanism of, 20
  detection limits of, 33, 34, 121, 125, 129, 132, 133, 134
  deuterated, 135
  half-life of, 129
  metabolism of, 2, 12, 20, 114, 119
  N-methylation of, 2, 12
  oxidation of, 2, 12
  pentafluoropropionyl derivative of, 33, 34, 121
  pharmacokinetics of, 128, 130, 131
  pyrolysis of, 17, 19, 20, 103, 104
    mechanism of, 20
    products of, 19, 20, 104
  rate of excretion (urine) of, 119, 120, 123, 125
    effect of pH on, 119, 120
  recovery of, 17, 29, 109, 119, 121, 122, 126, 129, 130, 132, 133, 134
  silyl derivative of, 113
  specific activity of, 25, 26
  structure of, 12, 33, 127, 134
Nicotine-N-oxide, 9, 20, 21, 22, 24
  rearrangement of, 20, 21
  pathways for, 20, 21
Nicotine sulfate, 22
Nicotinic acid, 9, 20, 100
Nicotinonitrile, 100
3-Nicotinoylpropionic acid, 20
α-Nicotyrine, 9, 12, 13, 15, 16, 20, 22, 23, 24, 27, 42, 100, 105, 144

[α-Nicotyrine]
  structure of, 12
β-Nicotyrine, 12, 13, 15, 16, 20, 22, 23, 24, 27, 42, 100, 105, 144
  structure of, 12
Nikethamide, 38
Nitric oxide, 101, 253-257
  detection limits of, 257
Nitrobenzene(s), 257, 258, 260-261
  detection limits of, 257
N,p-Nitrobenzyldimethylamine oxide, 21
2-Nitrocumene, 258
4-Nitrocumene, 257, 258, 260, 261
  structure of, 260
2-Nitro-1,3-dimethylbenzene, 258
2-Nitro-1,4-dimethylbenzene, 257, 258, 261
  structure of, 261
4-Nitro-1,2-dimethylbenzene, 257, 258, 260, 261
  structure of, 260
4-Nitro-1,3-dimethylbenzene, 257, 258, 260, 261
  structure of, 260
5-Nitro-1,3-dimethylbenzene, 258
2-Nitroethylbenzene, 258
4-Nitroethylbenzene, 258
Nitrogen (GC), 29, 31, 34, 41, 45, 50, 88, 109, 111, 121, 122, 126, 129, 132, 137, 138, 139, 146, 148, 158, 159, 181, 187, 223, 228, 231, 232, 244, 247, 286, 288, 295, 302, 303, 304, 306, 312, 331, 333, 335, 336, 341, 342, 345, 348, 350, 351, 354, 356, 358, 362, 374, 375, 377, 380, 390, 391, 392, 395, 401, 408, 424, 433, 438, 440, 444, 453, 460, 478

Nitrogen ionization detector (GC), 32, 33, 36, 37, 38, 39, 40, 111, 112, 114, 118, 129, 130, 133, 134, 230, 253-254, 256
  response index of, 37, 38, 39, 40
  effect of structure on, 37, 40
Nitrogen profiling, 112, 114, 118
N'-Nitrosoanabasine, 42, 43
  structure of, 43
N'-Nitrosonornicotine, 12, 41-49
  deuterated, 45, 46, 47, 48
    mass spectrum of, 47
  effect of pH on extraction of, 45, 46, 48
  mass spectrum of, 44, 46
  recovery of, 49
  structure of, 12, 43, 44
N'-Nitroso-(2'-^{14}C)-nornicotine, 42
2-Nitrotoluene, 257, 258, 260, 261
  structure of, 260
3-Nitrotoluene, 258
4-Nitrotoluene, 257, 258, 261
  structure of, 261
2-Nitro-1,3,5-trimethylbenzene, 258
Nitrous oxide, 101
n-Nonacosane, 64, 305, 323, 354, 374, 396
n-Nonadecane, 144
1-Nonadecanol, 4, 93
n-Nonane, 140, 145
Nonanoic acid, 7, 96, 422
  methyl derivative of, 422
n-Nonyne, 145
Norcotinine, 12, 15, 17, 20
  structure of, 12
Norharmane, 9, 100, 431
Norleucine, 11
Nornicotine, 9, 12, 13, 15, 16, 18, 20, 21, 22, 23, 24, 26, 27, 29, 30, 31, 41, 42, 100, 101, 103, 104, 105, 107, 133
  recovery of, 41

[Nornicotine]
  structure of, 12
Nor-α-nicotyrine, 12, 19, 100, 105
  structure of, 12
Nor-β-nicotyrine, 12, 19, 100, 105
  structure of, 12
Norphytene, 89, 250, 251
Norpristane, 158
5"-Nor-Δ^1-tetrahydrocannabinol-4-oic acid, see 5"-nor-Δ^9-tetrahydrocannabinol-4-oic acid
5"-Nor-Δ^9-tetrahydrocannabinol-4-oic acid, 456
  methyl/silyl derivative of, 456
  structure of, 456

O

cis-Ocimene, 363
trans-Ocimene, 363
n-Octacosane, 64, 221, 323
Octacosanoic acid, 422
  methyl derivative of, 422
n-Octadecane, 144
Octadecanoic acid, 418, 421
  methyl derivative of, 421
1-Octadecanol, 4, 93
1-Octadecene, 144
Octahydroanthracene, 222
Octahydrofluoranthene, 222
Octahydronicotine, 33, 34, 35, 36
  pentafluoropropionyl derivative of, 33, 34, 35, 36
  detection limits of, 33, 34
  structure of, 33
  structure of, 33
Octahydrophenanthrene, 222
Octahydropyrene, 222
n-Octane, 140
Octanoic acid, 7, 96, 416, 418, 421

[Octanoic acid]
  methyl derivative of, 421
  silyl derivative of, 416
Oleic acid, 7, 77, 84, 86, 96, 113, 178, 179, 180, 181, 422, 474
  methyl derivative of, 422
  recovery of, 180
  silyl derivative of, 86, 113, 178, 179, 180, 181
Olivetol, 273, 274, 306, 326, 328, 376, 422, 424
  N-methyl-N-silyl derivative of, 424
  structure of, 274, 328
Olivetolic acid, 274
  structure of, 274
Orcinol, 326
Ornithine, 11, 101
Oxalacetic acid, 7
Oxalic acid, 7, 96, 176, 177, 179
  methyl derivative of, 176, 177, 179
Oxazepam, 40
α-5,8-Oxido-3,9,13-duvatrien-1-ol, 3, 92
α-5,8-Oxido-3,9(17),13-duvatrien-1-ol, 3, 92
β-5,8-Oxido-3,9(17),13-duvatrien-1-ol, 3
6-Oxo-cannabidiol, 401
  silyl derivative of, 401
2"-Oxo-cannabinol, 401
3"-Oxo-cannabinol-7-oic acid, 477, 478
  methyl/silyl derivative of, 478
  structure of, 477
6-Oxo-Δ^1-tetrahydrocannabinol, see 8-oxo-Δ^9-tetrahydrocannabinol
7-Oxo-Δ^1-tetrahydrocannabinol, see 11-oxo-Δ^9-tetrahydrocannabinol
2-Oxo-Δ^3-tetrahydrocannabinol, 386, 388, 431
  structure of, 388

2"-Oxo-Δ^6-tetrahydrocannabinol, see 2"-oxo-Δ^8-tetrahydrocannabinol
3"-Oxo-Δ^6-tetrahydrocannabinol, see 3"-oxo-Δ^8-tetrahydrocannabinol
2"-Oxo-Δ^8-tetrahydrocannabinol, 400
3"-Oxo-Δ^8-tetrahydrocannabinol, 400
11-Oxo-Δ^8-tetrahydrocannabinol, 447
  acetyl derivative of, 447
8-Oxo-Δ^9-tetrahydrocannabinol, 400
  silyl derivative of, 400
11-Oxo-Δ^9-tetrahydrocannabinol, 446, 447
  acetyl derivative of, 447
Oxycodone, 40
Oxymetazoline, 40
Oxymorphone, 40

P

Palmitic acid, 7, 77, 84, 85, 86, 96, 113, 178, 179, 180, 181, 396, 403, 417, 422, 458, 474
  methyl derivative of, 85, 422
  recovery of, 85
  silyl derivative of, 85, 86, 113, 178, 179, 180, 181, 417
Palmitic-^{14}C acid, 180
  recovery of, 180
  silyl derivative of, 180
Palmitoleic acid, 96, 422
  methyl derivative of, 422
Palmitolenic acid, 422
  methyl derivative of, 422
Palmitone, 95
7-Palmitoyloxy-Δ^6-tetrahydrocannabinol, see 11-palmitoyloxy-Δ^8-tetrahydrocannabinol
11-Palmitoyloxy-Δ^8-tetrahydrocannabinol, 458
  silyl derivative of, 458
  structure of, 458
Paraffins (marijuana), 352, 353, 354, 374
  anteiso, 352, 353, 354
  iso-, 352, 353, 354
  straight-chain, 352, 353, 354, 374
Pentacene, 225
n-Pentadecane, 143
Pentadiene, 137
1,2-Pentadiene, 90, 154
1,3-Pentadiene, 90, 140, 153, 154, 155
  cis-isomer, 153, 154, 155
  trans-isomer, 153, 154, 155
1,4-Pentadiene, 90, 140, 153, 155
2,3-Pentadione, see 2,3-Pentanedione
Pentafluorobenzaldehyde, 88, 252
Pentafluorobenzaldehyde azine, 88
Pentafluorobenzoyl chloride, 443
Pentafluorobenzoylimidazole, 443
Pentafluorobenzyl bromide, 404
Pentafluoropropionic anhydride, 33, 34, 121, 253, 443
Pentamethylbenzene, 215, 227
Pentamethylnaphthalene(s), 189
  recovery of, 189
n-Pentane, 140, 145, 148, 152, 153, 154, 155, 258
2,3-Pentanedione, 95, 140
2-Pentanone, 5, 95, 140
3-Pentanone, 95, 140
Pentatriacontane, 185
Pentazocine, 40
1-Pentene, 90, 140, 145, 152, 153, 154, 155
2-Pentene, 90, 152, 153, 154, 155
  cis isomer, 152, 154, 155
  trans isomer, 152, 153, 154, 155
1-Penten-3-one, 140
1-Penten-4-one, 140

4-Penten-2-one, 95
Pentobarbital, 38
n-Pentylbenzene, 227, 228
tert-Pentylbenzene, 229
Perhydropyrene, 217
1,8,9-Perinaphthoxanthene, 100
Perylene, 92, 184, 185, 189, 196,
 200, 221, 223, 224, 226,
 227, 230, 237, 239, 426
 detection limits of, 189
β-Phellandrene, 326, 327, 363
Phenacetin, 38
Phenanthrene, 92, 145, 183, 185,
 190, 200, 214, 221, 222,
 223, 227, 230, 234, 239
Phenanthrene-^{14}C, 187, 189
Phenanthro(10,1,2,3,cdef)fluorene,
 198, 238
Phenazocine, 40
Phencyclidine, 39
Phendimetrazine, 38
β-Phenethylacetate, 5
β-Phenethylalcohol, 4, 93
Phenmetrazine, 38
Phenobarbital, 39, 433
Phenol, 8, 97, 107, 108, 113, 143,
 161, 162, 163, 164, 165,
 166, 167, 168, 169, 417,
 418, 421, 422
 silyl derivative of, 113, 417, 421
Phenylacetic acid, 7, 96, 422, 431
 methyl derivative of, 422
Phenylacetylene, 92
Phenylalanine, 11, 101
Phenyl disulfide, 215
o-Phenylenepyrene, 223
2-Phenylethylamine, 9, 100
β-Phenylethylcyanide, 431
Phenylhexane, 145
3-Phenylindole, 202
Phenylisopropionic acid, 422
 methyl derivative of, 422
N-Phenyl-4-isopropylphenylamine,
 100

α-Phenyllactic acid, 7
Phenyl methyl ether, 248
N-Phenyl-2-naphthylamine, 100
Phenyloctane, 145
Phenylpropanolamine, 38
Phenylpropionic acid, 97, 422, 431
 methyl derivative of, 422
3-Phenylpropionitrile, 95
Phenylpyruvic acid, 7
Phenyl sulfide, 210, 215
Phenyl sulfone, 215
Phenyl sulfoxide, 215
Phosphoric acid, 74, 75, 432
Phosphorus detector (flame) (GC),
 438, 440
Phthalic acid, 97, 178, 179
 methyl derivative of, 178, 179
Phytane, 158
Phytoene, 3
Phytofluene, 3
Phytol, 92, 250, 251
Picene, 184, 225
α-Picoline, 13
β- or 3-Picoline, 13, 15, 19, 20,
 104
γ-Picoline, 13
α-Pinene, 145, 326, 327, 363, 381
β-Pinene, 89, 137, 145, 326, 327,
 363
Pipecolic acid, 11
Piperidine, 9, 13, 100, 418, 421
Piperitenone, 381
Pivaldehyde, 94, 137
Planteose, 10
Plastoquinone A, 64
Poly-p-2,6-diphenylene oxide
 polymer, 411
Polynuclear aromatic hydrocarbons,
 1, 60, 180-232, 233, 234,
 235, 236, 237, 238, 239, 424,
 425, 427, 428, 429, 430
Potassium carbonate, 135, 437
Potassium bromide, 333
Potassium chloride, 69, 84

Potassium dihydrogen phosphate, 87
Potassium hydroxide, 17, 24, 27, 31, 41, 60, 65, 66, 68, 84, 87, 109, 112, 121, 122, 129, 130, 132, 135, 418, 421, 477
Potassium phosphate, 433
Pristane, 158
Procaine, 39
Proline, 11, 101
Promethazine, 40
Propadiene, 154, 155
Propane, 152, 154
Propenaldehyde, 140
Propene, 90, 152, 154
Propionaldehyde, 5, 94, 116, 137, 140, 147, 154
Propionic acid, 7, 75, 97, 177
Propionitrile, 95, 140
Propiophenone, 143, 431
n-Propanol, 93, 137
n-Propylallyl disulfide, 244
n-Propylallyl sulfide, 244
Propylamine, 100
n-Propylbenzene, 226
n-Propyl-n-butyl sulfide, 244
4-n-Propylcatechol, 170, 171
Propyldimethylindole, 202
Propylene glycol, 4, 93, 107, 108, 109, 113, 143
Propylethylindole, 202
Propylhexidrine, 38
Propylindole, 201
n-Propylmercaptan, 244
Propylmethylindole, 202
Propylpyridine, 431
2-n-Propyltoluene, 227
4-n-Propyltoluene, 227
Protocatechuic acid, 8, 98
Protocatechuic aldehyde, 8, 98
Propyne, 154, 155
Proxyphene, 39
Pseudocumene, 137
Pulegone, 326
Pyrahexyl, 283, 284, 325

Pyrazine, 100
Pyrene, 92, 183, 185, 192, 200, 214, 217, 218, 219, 221, 222, 223, 224, 230, 235, 239
Pyridine, 9, 11, 13, 15, 19, 100, 101, 104, 142, 178, 308, 329, 348, 413, 418, 421, 465
Pyridine-3-aldehyde, 100
3-Pyridinol, 100
Pyrido(2,3-b)indole, 100
3-Pyridylacetic acid, 12
   structure of, 12
γ-3-Pyridyl-γ-hydroxybutyric acid, 20
3-Pyridylethylketone, 9, 13, 22, 23, 100, 105
γ-3-Pyridyl-γ-methylaminobutyric acid, 20, 21
β-Pyridylmethylketone, 13, 15, 16, 100, 104, 105
3-Pyridylmethylketone, see β-pyridylmethylketone
γ-3-Pyridyl-γ-oxo-N-methyl-butyramide, 20
β-Pyridyl-n-propylketone, 9, 13, 15, 16, 22, 23, 27, 100, 105
3-Pyridyl-n-propylketone, see β-pyridyl-n-propylketone
Pyrocatechol, 144
Pyrocoll, 100
Pyrrole, 9, 13, 100, 101, 104, 140, 142, 418, 421
Pyrrolidine, 9, 13, 100
Pyrrolidine-2-acetic acid, 11
3-Pyrroline, 100
Pyrrolo(2,3-b)pyridine, 100
Pyruvic acid, 7, 97

Q

Qualitative methods of analysis, by use of
   alphabetical index, 386
   base peak index, 386, 387
   digital code index, 386, 387

SUBJECT INDEX 547

[Qualitative methods of analysis, by use of]
MS/integrated GC-MS/GC-MS-COMP, 34, 35, 45, 46, 47, 48, 49, 70, 71, 72, 79, 71, 88, 113, 121, 122, 123, 124, 126, 127, 132, 133, 134, 135, 136, 138, 139, 140, 141, 142, 143, 144, 145, 148, 158, 169, 178, 179, 180, 181, 185, 187, 189, 201, 203, 205, 208, 209, 211, 212, 219, 220, 232, 234, 235, 236, 237, 238, 250, 252, 253, 254, 255, 257, 260, 261, 273, 281, 287, 288, 289, 302, 304, 308, 309, 319, 321, 322, 323, 324, 325, 339, 340, 341, 342, 343, 345, 346, 348, 352, 354, 357, 361, 364, 366, 367, 368, 369, 370, 371, 372, 377, 380, 381, 382, 384, 386, 388, 390, 391, 392, 393, 394, 395, 396, 397, 398, 399, 402, 403, 404, 405, 407, 408, 412, 413, 414, 415, 416, 417, 418, 419, 422, 424, 431, 436, 437, 438, 440, 444, 446, 447, 448, 451, 452, 453, 454, 455, 456, 457, 458, 460, 461, 462, 463, 465, 468, 469, 471, 472, 473, 474, 476, 477, 478, 479
internal standard, see internal standard
Kovats index, 31, 32, 36, 71, 72, 181, 183, 184, 391
mass fragmentography, see multiple/specific ion (m/e) detection
mass spectrum (CI/EI)/specific ions (m/e), 34, 35, 43, 49,

[Qualitative methods of analysis, by use of]
[mass spectrum (CI/EI)/specific ions (m/e)]
54, 55, 56, 59, 71, 72, 73, 77, 79, 123, 124, 126, 127, 132, 134, 142, 143, 144, 145, 158, 169, 179, 180, 201, 203, 205, 208, 209, 211, 212, 232, 252, 253, 254, 255, 257, 260, 261, 289, 290, 291, 293, 295, 308, 309, 319, 321, 322, 323, 324, 339, 340, 342, 346, 348, 354, 361, 364, 366, 367, 368, 369, 370, 371, 377, 380, 381, 382, 386, 388, 391, 392, 393, 394, 395, 397, 398, 399, 402, 403, 404, 408, 411, 412, 413, 414, 415, 416, 417, 418, 419, 424, 434, 435, 436, 437, 438, 440, 444, 446, 447, 448, 449, 451, 452, 453, 454, 455, 456, 457, 458, 460, 461, 462, 463, 465, 468, 469, 471, 472, 473, 474, 476, 478, 479
methylene unit value, 348, 349, 350, 395, 396, 397, 453, 454, 455, 472
molecular ion, 35, 46, 59, 79, 122, 123, 124, 127, 134, 158, 180, 386, 408, 412, 415, 436, 437, 446, 451, 454, 456, 457, 458, 460, 461
molecular weight, 35, 49, 60, 81, 126, 190, 191, 192, 193, 194, 195, 196, 197, 198, 207, 210, 214, 215, 216, 234, 235, 236, 237, 238, 293, 294, 295, 321, 322, 323, 324, 361, 386, 387, 403, 411, 458

[Qualitative methods of analysis, by use of]
multiple/specific ion (m/e) detection, 45, 46, 47, 48, 49, 79, 123, 124, 132, 133, 134, 135, 220, 232, 440, 444, 446, 451, 452, 462, 463

retention index, see Kovats index

relative retention time, 13, 14, 15, 16, 24, 29, 32, 33, 37, 38, 39, 40, 42, 43, 50, 68, 80, 82, 152, 153, 161, 213, 217, 220, 227, 228, 232, 239, 244, 247, 282, 283, 304, 306, 308, 325, 329, 330, 331, 332, 335, 342, 344, 345, 346, 347, 354, 355, 357, 358, 359, 360, 361, 362, 363, 364, 374, 375, 376, 377, 380, 381, 382, 386, 390, 393, 394, 403, 413, 416, 417, 436, 440, 443, 451

retention time, 2, 13, 17, 19, 21, 22, 23, 24, 25, 26, 27, 29, 30, 31, 33, 34, 35, 38, 39, 40, 42, 43, 46, 52, 58, 59, 60, 61, 63, 68, 69, 73, 75, 76, 78, 80, 81, 82, 83, 86, 87, 88, 105, 106, 107, 109, 110, 116, 117, 118, 119, 121, 122, 126, 127, 129, 130, 132, 133, 134, 135, 139, 147, 149, 150, 151, 155, 158, 160, 161, 163, 168, 169, 170, 171, 176, 177, 178, 179, 181, 183, 184, 186, 188, 189, 199, 200, 204, 205, 206, 207, 210, 213, 214, 215, 216, 217, 219, 220, 221, 223, 224, 225, 226, 227, 228, 229, 230, 231, 232, 233, 240, 243, 244, 246,

[Qualitative methods of analysis, by use of]
[retention time]
249, 250, 251, 253, 254, 256, 257, 258, 282, 283, 284, 286, 294, 295, 302, 303, 305, 307, 312, 314, 315, 320, 325, 326, 327, 328, 329, 331, 332, 333, 335, 336, 337, 338, 341, 343, 349, 351, 352, 353, 354, 355, 358, 360, 374, 375, 377, 380, 384, 390, 392, 398, 400, 401, 403, 405, 406, 407, 408, 411, 412, 418, 420, 421, 423, 424, 425, 426, 427, 428, 429, 430, 433, 437, 438, 440, 442, 444, 445, 446, 447, 448, 451, 454, 456, 457, 458, 460, 461, 462, 463, 468, 469, 471, 472, 473, 474, 476, 478, 479

specific ion detection, see multiple/specific ion (m/e) detection

Quantitative methods of analysis, by use of

calibration curve, 17, 29, 34, 36, 46, 47, 67, 125, 130, 134, 282, 286, 336, 343, 354, 409, 440, 464

peak area measurements, 15, 25, 62, 103, 282, 286, 295, 312, 336, 354

peak height measurements, 17, 29, 34, 46, 67, 125, 130, 134, 303, 343, 348, 349, 440, 446

Quaterphenyl, 198, 237
Quercetin methyl ether, 8
Quercimeritrin, 8
Quinaldine, 41, 42
Quinic acid, 8, 98
Quinic acid γ-lactone, 98

SUBJECT INDEX

Quinoline, 15, 19, 20, 31, 100, 128, 129, 130, 132, 133, 134, 143, 431
  structure of, 134
Quinone, 142
Quinoxaline, 100

R

Raffinose, 10
Reductic acid, 95
Reductone, 5
Resorcinol, 98, 170, 171
Response index, 37, 38, 39, 40
  effect of structure on, 37, 40
Retene, 210, 214
Rhamnose, 10
Ribose, 10
Rutin, 8
Rutinose, 10

S

Sabinene hydrate, 326, 327
Salicylaldehyde, 8, 98, 161
Salicylamide, 39
Salicylic acid, 98, 418, 421
  methyl derivative of, 421
Scopoletin, 8, 98, 418, 421
  silyl derivative of, 421
Scopoletin-7-glucoside, 8
Scopoletin rhamnoglucoside, 8
Secobarbital, 39
α-Selinene, 326, 327
Serine, 11, 101
Shikimic acid, 8
Silver nitrate, 338
Sinapic acid, 98
β-Sitosterol, 3, 69, 70, 72, 73, 92, 402, 420
  silyl derivative of, 69, 70, 72, 73, 402
Skatole, 201, 416, 418, 419
  structure of, 419
Skatole-3a, 4, 5, 6, 7, 7a-^{14}C, 416

Sodium azide, 438, 439
Sodium biacrbonate, 34, 120, 170, 414, 415, 416, 420, 432, 439
Sodium bisulfite, 88
Sodium chloride, 88, 131, 132, 432, 437, 452
Sodium hydroxide, 13, 15, 41, 67, 87, 114, 120, 121, 125, 128, 129, 131, 132, 135, 404, 416, 417, 452, 462, 465
Sodium phosphate, 45
Sodium sulfate, 67, 88, 170, 404, 416, 432, 462
Solanesene(s), 64, 65, 66, 89
  structures of, 65
Solanesol, 4, 59, 60, 61, 65, 66, 86, 92
  silyl derivative of, 60, 61, 66, 86
  structure of, 59, 65
Solanochromene, 4
Solanone, 4, 92
Sorbic acid, 97
Sorbitol, 10
Squalane, 214
Squalene, 89, 250, 251
Stachyose, 10
Starch, 10
Stearic acid, 7, 77, 84, 86, 97, 113, 178, 179, 180, 181, 417, 422, 458
  methyl derivative of, 422
  recovery of, 180
  silyl derivative of, 86, 113, 178, 179, 180, 181, 417
Sterols (marijuana), 402
  silyl derivatives of, 402
Sterols (tobacco), 58, 68-73, 86
  silyl derivative of, 68, 69, 70, 71, 86
5α-Stigmasta-7,24(28)-dien-3β-ol,
  silyl derivative of, 71, 72, 73, 402
  structure of, 72
5,21-Stigmastadien-3β-ol, <u>see</u> stigmasterol

5,24(28)-Stigmastadien-3β-ol, see fucosterol
5-Stigmasten-3β-ol, see β-sitosterol
5α-Stigmast-7-en-3β-ol, 72
 silyl derivative of, 72
7-Stigmasten-3β-ol, 73
 silyl derivative of, 73
Stigmasterol, 3, 69, 70, 72, 73, 92, 402, 420
 silyl derivative of, 69, 70, 72, 73, 402
trans-Stilbene, 145, 183
Styrene, 92, 136, 137, 140, 145, 187
Succinic acid, 7, 97, 171, 176, 179
 methyl derivative of, 171, 176, 179
Sucrose, 10
Sulfur detector (flame photometer) (GC), 111, 112, 114, 117, 232, 244, 245, 246, 247
Sulfur dioxide, 244
Sulfuric acid, 87, 114, 131, 132, 250, 252
Sulfur profiling, 112, 114, 117, 245
Syringaldehyde, 8, 98
Syringic acid, 8, 98

T

Taurine, 11
o,p'-TDE, 193
p,p'-TDE, 193, 258
p,p'-TDEE, 192
Temperature programming (GC column), 33, 37, 54, 56, 58, 60, 61, 63, 67, 69, 71, 72, 73, 74, 77, 79, 81, 82, 85, 86, 104, 106, 107, 109, 110, 111, 112, 117, 126, 135, 137, 138, 139, 141, 145, 146, 149, 158, 160

[Temperature programming (GC column)] 169, 180, 181, 182, 185, 188, 189, 201, 216, 217, 218, 219, 220, 221, 223, 226, 227, 230, 231, 232, 233, 243, 244, 246, 247, 250, 302, 303, 319, 320, 338, 342, 352, 360, 380, 388, 389, 390, 391, 395, 402, 408, 411, 414, 420, 421, 422, 423, 424, 425, 427, 428, 429, 430, 442, 446, 447, 453, 457, 465, 471, 479
Terephthalic acid, 7
α-Terpinene, 326, 327, 363
γ-Terpinene, 326, 327, 363
α-Terpineol, 326, 327
Terpinene-4-ol, 326, 327
Terpinolene, 363
Tetrachloroethylene, 145, 258
n-Tetracosane, 221
1-Tetracosanol, 93
Tetracosatetraenoic acid, 422
 methyl derivative of, 422
n-Tetradecane, 143
Tetradecanoic acid, 422
 methyl derivative of, 422
1-Tetradecene, 143
Tetrahexylammonium hydroxide, 452, 462, 464
Tetrahexylammonium iodide, 462
Δ^1-Tetrahydrocannabinol, see Δ^9-tetrahydrocannabinol
dl-Δ^1-cis-3,4-tetrahydrocannabinol, see dl-Δ^9-cis-3,4-tetrahydrocannabinol
dl-Δ^1-trans-3,4-tetrahydrocannabinol, see dl-Δ^9-trans-3,4-tetrahydrocannabinol
Δ^1-Tetrahydrocannabinol-7-oic acid, see 11-carboxy-Δ^9-tetrahydrocannabinol
$\Delta^{1(6)}$-Tetrahydrocannabinol, see Δ^8-tetrahydrocannabinol

SUBJECT INDEX

$\Delta^{1(7)}$-Tetrahydrocannabinol, see $\Delta^{9(11)}$-tetrahydrocannabinol
Δ^6-Tetrahydrocannabinol, see Δ^8-tetrahydrocannabinol
dl-Δ^6-cis-3,4-tetrahydrocannabinol, see dl-Δ^8-cis-3,4-tetrahydrocannabinol
dl-Δ^6-trans-3,4-tetrahydrocannabinol, see dl-Δ^8-trans-3,4-tetrahydrocannabinol
Δ^8-Tetrahydrocannabinol, 272, 275, 276, 289, 291, 292, 295, 304, 305, 306, 312, 313, 314, 326, 328, 335, 337, 338, 339, 340, 341, 343, 344, 346, 347, 356, 357, 358, 359, 360, 362, 376, 377, 378, 380, 383, 384, 385, 387, 391, 392, 393, 394, 398, 399, 400, 402, 404, 405, 406, 407, 409, 434, 435, 436, 437, 444, 456, 457, 470, 471, 479
  acid A, 276, 337
    silyl derivative of, 337
    structure of, 276
  acid B, 276
    structure of, 276
  tert-butyldimethylsilyl derivative of, 398, 399
  cis-ortho, 356, 359
    α-methyl derivative of, 356, 359
    structure of, 356
    structure of, 356
  cis-para, 356, 359
    α-methyl derivative of, 356, 359
    structure of, 356
    structure of, 356
  ^{14}C-labeled, 456
  diethylphosphate derivative of, 399
  dl-cis-3,4-, 326, 328
  dl-trans-3,4-, 326, 328

[Δ^8-Tetrahydrocannabinol]
[dl-trans-3,4-]
  structure of, 328
  glucuronide of, 479
    methyl/acetyl derivative of, 479
    structure of, 479
  metabolism of, 434, 435, 436, 437, 456, 457, 470, 471
  methoxy derivative of, 346, 347, 392
  ortho, 346, 347
    methoxy derivative of, 346, 347
  pentafluorobenzoyl derivative of, 444
  structure of, 272, 275
  synthetics with varying C-3 side-chain substituents, 357, 358, 361
    structures of, 357, 358
  trans-ortho, 356, 359
    α-methyl derivative of, 356, 359
    structure of, 356
    structure of, 356
  trans-para, 356, 360
    α-methyl derivative of, 356, 360
    structure of, 356
    structure of, 356
  tributylsilyl derivative of, 391
  trichloroacetyl derivative of, 343
  triethylsilyl derivative of, 391
  trimethylsilyl derivative of, 376, 387, 391, 393, 394
  trimethylsilyl/acetyl derivative of, 398, 399
  tripropylsilyl derivative of, 391
Δ^9-Tetrahydrocannabinol, 2, 35, 271, 272, 273, 274, 275, 280, 281, 282, 283, 284, 285, 286, 287, 289, 290, 291, 292, 293, 294, 295, 296, 298, 299, 300, 301, 302, 303, 304, 305, 306, 307, 308, 311, 312, 313, 314, 315, 319, 323, 325, 326, 328, 329, 331, 332, 333, 334, 335, 336, 337,

[Δ⁹-Tetrahydrocannabinol]
    338, 339, 341, 342, 343,
    344, 345, 346, 347, 348,
    349, 350, 351, 352, 354,
    355, 356, 357, 359, 360,
    362, 364, 365, 371, 372,
    375, 376, 377, 378, 379,
    380, 382, 383, 384, 385,
    387, 389, 390, 391, 392,
    393, 394, 395, 396, 397,
    398, 399, 400, 403, 404,
    405, 406, 407, 408, 409,
    411, 412, 413, 431, 432,
    433, 434, 435, 436, 437,
    438, 439, 440, 441, 442,
    443, 444, 445, 446, 447,
    448, 450, 451, 452, 454,
    456, 457, 458, 461, 462,
    463, 464, 465, 466, 468,
    469, 470, 471
  acid A, 274, 275, 303, 308, 311,
    312, 315, 332, 333, 336,
    337, 351, 352, 362, 365,
    375, 377, 379, 393, 394,
    396, 397
    decarboxylation of, 312, 351,
      375
    silyl derivative of, 303, 308,
      311, 315, 332, 333, 336,
      337, 351, 362, 375, 377,
      379, 393, 394, 397
    silyl-d_9 derivative of, 397
    stability of, 365
    structure of, 274, 275
  acid B, 275, 311, 312, 315, 377
    decarboxylation of, 312
    silyl derivative of, 315, 377
    structure of, 275
  biotransformation of, 280, 450,
    470
    sites of, 450, 470
  tert-butyldimethylsilyl derivative
    of, 398, 399
  ^{14}C-labeled, 447, 456
  C_4-homolog, 278, 393, 396, 397

[Δ⁹-Tetrahydrocannabinol]
  [C_4-homolog]
    acid of, 278, 393, 396, 397
      structure of, 278
      trimethylsilyl derivative of,
        397
      trimethylsilyl-d_9 derivative of,
        397
    structure of, 278
    triethylsilyl derivative of, 393,
      397
    trimethylsilyl derivative of, 393
    trimethylsilyl-d_9 derivative of,
      393
    tripropylsilyl derivative of, 393,
      397
  C_6-homolog, 371
  chlorodifluoroacetyl derivative of,
    443
  chloromethyldimethylchlorosilyl
    derivative of, 443
  cis-ortho, 356, 359
    structure of, 356
  cis-para, 356, 359
    structure of, 356
  dehydrogenation of, 409
  detection limits of, 349, 433, 440,
    441, 443, 444
  deuterated, 440
  diethylphosphate derivative of,
    399, 438, 439, 440
    structure of, 439
  dl-cis-3,4-, 326
  dl-trans-3,4-, 328
    structure of, 328
  half-life of, 380, 444
  heptafluorobutyryl derivative of,
    440, 441, 442, 443, 445
  isomerization of, 409
  mass spectrum of, 291, 412
  metabolism of, 273, 433, 434,
    435, 436, 437, 438, 446, 447,
    448, 450, 451, 452, 454, 456,
    457, 465, 468, 469, 470, 471
  methoxy derivative of, 275, 346,

SUBJECT INDEX

[Δ^9-Tetrahydrocannabinol]
  [methoxy derivative of]
    347, 355, 392, 404, 405,
    431, 437, 438, 444, 446
  deuterated, 444, 446
  structure of, 275, 438
  numbering system for, 271, 272
    formal chemical, 271, 272
    monoterpenoid, 271, 272
  ortho, 346, 347
    methoxy derivative of, 346, 347
  oxidation of, 287
  pentafluorobenzoyl derivative of,
    443, 444
  pentafluorobenzyl derivative of,
    404, 405
  pentafluoropropionyl derivative
    of, 443
  pentafluorotoluoyl derivative of,
    443
  perdeuterioethyl derivative of,
    452, 462, 463
    silyl derivative of, 452, 463
  pharmacokinetics of, 440, 444,
    445, 466
  recovery of, 444, 446, 464
  stability of, 336, 338, 362, 365,
    380, 405, 406
    in acetone, 338
    in acid solution, 380, 405, 406
    in ethyl alcohol, 336
    under nitrogen, 338, 362
  structure of, 272, 274, 280, 290,
    372, 409, 447
  trans-ortho, 356, 360
    structure of, 356
  trans-para, 356, 357, 359, 360
    structure of, 356
  tributylsilyl derivative of, 391
  trichloroacetyl derivative of, 343
  triethylsilyl derivative of, 391,
    397
  trifluoroacetyl derivative of, 329,
    331, 348, 349, 350, 443
  trifluoromethylbenzoyl derivative
    of, 443

[Δ^9-Tetrahydrocannabinol]
  trimethylsilyl derivative of, 302,
    303, 308, 313, 315, 329, 331,
    332, 333, 336, 347, 348, 349,
    350, 351, 352, 362, 376, 379,
    387, 389, 390, 391, 393, 394,
    397, 413
  trimethylsilyl/acetyl derivative of,
    398, 399
  trimethylsilyl-d_9 derivative of,
    397
  tripropylsilyl derivative of, 391,
    397
  tritium-labeled, 432, 433, 448
Δ^9-Tetrahydrocannabinol-11-
  aldehyde, 447
  structure of, 447
$\Delta^{9(11)}$-Tetrahydrocannabinol, 306,
  376, 391
  tributylsilyl derivative of, 391
  triethylsilyl derivative of, 391
  trimethylsilyl derivative of, 391
  tripropylsilyl derivative of, 391
Δ^9-Tetrahydrocannabiorcin, 277,
  287, 288, 290, 292, 293,
  294, 344, 359, 396, 397
  acid A, 397
  structure of, 277, 288, 290
    trimethylsilyl derivative of, 397
    trimethylsilyl-d_9 derivative of,
      397
  structure of, 277, 288, 290
  trans-para, 359
  triethylsilyl derivative of, 397
  trimethylsilyl derivative of, 396
  tripropylsilyl derivative of, 397
Δ^9-Tetrahydrocannabivarin, 287,
  288, 290, 292, 293, 294, 301,
  302, 304, 306, 307, 308, 311,
  314, 341, 342, 344, 346, 362,
  368, 376, 378, 379, 387, 390,
  391, 396, 397
  acid A, 277, 376, 396, 397
    structure of, 277
    trimethylsilyl derivative of, 376,
      397

[Δ^9-Tetrahydrocannabivarin]
[acid A]
  trimethylsilyl-d$_9$ derivative of, 397
  structure of, 288, 290, 368
  (-)-trans, 362
  trans-para, 359
  tributylsilyl derivative of, 391
  triethylsilyl derivative of, 391, 397
  trimethylsilyl derivative of, 302, 308, 376, 379, 391, 397
  trimethylsilyl-d$_9$ derivative of, 397
  tripropylsilyl derivative, 391, 397
Δ^9-Tetrahydrocannabivarol, see Δ^9-tetrahydrocannabivarin
Tetrahydrofuran, 95, 137, 422, 432
Tetrahydronaphthalene, see tetralin
1,2,3,4-Tetrahydronaphthalene, see tetralin
Tetrahydropyran, 95, 137
1,2,5,6-Tetrahydropyridine, 9
Tetralin, 102, 143, 227, 229
  structure of, 102
1,2,3,4-Tetramethylbenzene, 227, 229
1,2,3,5-Tetramethylbenzene, 229
1,2,4,5-Tetramethylbenzene, 227, 229
Tetramethylcarbazole, 202
Tetramethylindole, 202
Tetramethylnaphthalene(s), 189
  recovery of, 189
2,3,4,6-Tetramethylphenol, 168
Tetraphenylethylene, 385
n-Tetratriacontane, see 1-methyltritriacontane
Theophylline, 39
Thermal conductivity detector (GC), 2, 17, 52, 74, 87, 153, 162, 171, 175, 177, 252, 282, 325, 418, 420

Thioamyl, 39
Thiocyanic acid, 96
Thiocyanogen, 96
Thiopental, 39
Thiophene, 96, 137, 244
Threonine, 11, 101
Thymine, 9
Thymol, 98
Tobacco leaf, 1-88
  alcohol(s), 59-68, 69
  alkaloid(s), 2-41
    relationship between boiling point and relative retention time, 13, 14, 15
    thermal decomposition of, 13, 15
  ammonia, 87
  chemical composition of, see natural products of
  fatty acid(s)/ester(s), 73-85, 86, 87, 88
    dimethylacetals of, 83
    reduction of, 77, 78
  hydrazine, 87, 88
  hydrocarbon(s), 50-58, 59, 60, 61, 62, 63, 64, 77, 78
    recovery from urea adducts, 50, 54, 61
  natural products of, 1, 2-88
  pyrolysis, 1, 13, 15, 60
  polyaromatic hydrocarbons of, 1, 60
  products of, 13, 15
  sterols, 58, 68-73, 86
Tobacco smoke, 1, 17, 60, 65, 66, 88-261
  acidic components, 171-180, 181
  alkaloids, 103-135, 136
    in biological fluids, 114-135, 136
      breast fluid, 134-135, 136
      plasma/blood, 126-134
      urine, 114-126
    in smoke, 103-114
  ammonia, 250-252
  aryl methyl ethers, 247
  carcinogenic compounds, 1, 2, 88, 102

[Tobacco smoke]
  chemical composition, 88-261
  p,p'-DDT in, 258-259
  hydrazine, 252-253
  hydrocarbons, 102, 146-158
  hydroxybenzyl alcohols/hydroxyphenyl ethanols, 65, 67, 248-249
  ketones, 159, 160
  low-boiling volatiles, 136-146, 147
  maleic hydrazide, 253, 254
  miscellaneous compounds of, 330-361
  1- and 2-naphthylamines, 253, 255
  nitrobenzene(s), 257, 258, 260-261
  nitric oxide, 253-257
  phenolic constituents, 159-171, 172, 173, 174
  polynuclear aromatic hydrocarbons and related compounds, 1, 60, 180-232, 233, 234, 235, 236, 237, 238, 239
  sulfur-containing compounds, 232-247
  terpenes, 250, 251
  vinyl chloride in, 257
Tocopherol(s), 4, 92
m-Tolualdehyde, 5
p-Tolualdehyde, 143
Toluene, 136, 138, 140, 145, 149, 218, 463, 464
m-Toluic acid, 97
p-Toluic acid, 97
m-Toluidine, 100
o-Toluidine, 100
p-Toluidine, 100
α-Tolunitrile, 143, 431
Triacetin, 108, 109
n-Triacontane, 64, 323
Tribenz(ach)anthracene, 92
Tri-n-butylchlorosilane, 390
Trichloroacetic anhydride, 343

Trichloroethylene, 145, 258
Tricosanoic acid, 422
  methyl derivative of, 422
1-Tricosanol, 4, 93
n-Tridecane, 143, 145, 229
1-Tridecene, 143
Triethylamine, 121
Triethylbenzene, 229
Triethylene glycol, 4, 93, 108
Triethylchlorosilane, 390, 395
Trifluoroacetic acid, 348
Trifluoroacetic anhydride, 343, 348, 443
N-Trifluoroacetylimidazole, 443
N-Trifluorobutyrylimidazole, 443
m-Trifluoromethylbenzoyl chloride, 443
Trihexylchlorosilane, 390
$3'',1\alpha,6\beta$-Trihydroxyhexahydrocannabinol, see $3'',8\beta,9\alpha$-trihydroxyhexahydrocannabinol
$3'',8\beta$-9α-Trihydroxyhexahydrocannabinol, 470, 473
  silyl derivative of, 473
  structure of, 470
$4'',1\alpha,6\beta$-Trihydroxyhexahydrocannabinol, see $4'',8\beta$-9α-Trihydroxyhexahydrocannabinol
$4'',8\beta$-9α-Trihydroxyhexahydrocannabinol, 470, 473
  silyl derivative of, 473
  structure of, 470
$2'',6\alpha,7$-Trihydroxy-Δ^1-tetrahydrocannabinol, see $2'',8\alpha,11$-trihydroxy-Δ^9-tetrahydrocannabinol
$3'',6\alpha,7$-Trihydroxy-Δ^1-tetrahydrocannabinol, see $3'',8\alpha,11$-trihydroxy-Δ^9-tetrahydrocannabinol
$2'',8\alpha,11$-Trihydroxy-Δ^9-tetrahydrocannabinol, 465, 467, 469
  silyl derivative of, 469
  structure of, 467

3",8α,11-Trihydroxy-Δ⁹-tetra-
    hydrocannabinol, 465, 467,
    469
  silyl derivative of, 469
  structure of, 467
1,2,3-Trimethoxybenzene, 8
Trimethylamine, 9, 100
Trimethylamine oxide, 21
2,4,6-Trimethylaniline, 100
Trimethylanilinium hydroxide,
    345, 391, 422, 437, 444,
    446
1,2,3-Trimethylbenzene, 92, 227,
    229
1,2,4-Trimethylbenzene, 92, 137,
    140, 226
1,3,5-Trimethylbenzene, 92, 137,
    140
1,3,6-Trimethylbenzene, 92
Trimethylcarbazole, 202
Trimethylchlorosilane, 67, 178,
    303, 308, 329, 362, 374,
    393, 395, 413, 452, 453,
    463, 465
1,1,3-Trimethylcyclohexane, 145
Trimethylcyclopentane, 145
Trimethylindole, 202
Trimethylnaphthalene(s), 189, 199
  recovery of, 189
1,3,6-Trimethylnaphthalene, 187
2,3,6-Trimethyl-1,4-naphtho-
    quinone, 95
Trimethylphenol, 144
2,3,4-Trimethylphenol, 168, 169
2,2,4-Trimethylpentane, 336
2,3,5-Trimethylphenol, 98, 144,
    161, 163, 165, 167, 168
2,3,6-Trimethylphenol, 168
2,4,5-Trimethylphenol, 163, 165,
    167
2,4,6-Trimethylphenol, 98, 163,
    168
3,4,5-Trimethylphenol, 161, 163,
    168, 169
Trimethylsilylimidazole, 345

1,3,5-Trimethyl-2,4,6-tris-(3,5-
    t-butyl-4-hydroxybenzyl)-
    benzene, 207
1-[4(β,γ,γ-Trimethyl-γ-valero-
    lactonyl)]-p-menthene-8(9),
    322
Tripalmitin, 85
1,3,5-Triphenylbenzene, 217, 218,
    219
Triphenylcarbinol, 350
1,3,5-Triphenyldecane, 227
Triphenylene, 184, 185, 200, 217,
    222, 223, 224, 230, 239, 426
Tri-n-propylchlorosilane, 390, 395
3",4",5"-Trisnor-Δ¹-tetrahydro-
    cannabinol-2"-oic acid, see
    3",4",5"-trisnor-Δ⁹-tetra-
    hydrocannabinol-2"-oic acid
3",4",5"-Trisnor-Δ⁹-tetrahydro-
    cannabinol-2"-oic acid, 454,
    455
  methyl/silyl derivative of, 454
  structure of, 456
Tri-o-tolylphosphate, 210, 214
n-Tritriacontane, 64
Tri-2,4-xylylphosphate, 214
Tryptophan, 11
Tybamate, 39
Tyramine, 11
Tyrosine, 11

U

n-Undecane, 142, 145
Undecanoic acid, 84-85
  silyl derivative of, 85
Undecyl acetate, 5

V

n-Valeraldehyde, 5, 94, 137
Valeric acid, 7, 75, 77, 97

n-Valeronitrile, 95, 137, 140
Valine, 11, 101
Vanillic acid, 8, 98
Vanillin, 8, 98, 431
Veratrole, 161
Vinyl acetate, 94
Vinylacetylene, 153
Vinyl chloride, 257
2-Vinyl-3-hydroxy-5-pentylbenzoic acid, 422, 424
  methyl derivative of, 422, 424
1-Vinylnaphthalene, 185
2-Vinylnaphthalene, 185
Vinylphenol, 144
p-Vinylphenol, 417
  silyl derivative of, 417
2-Vinylpyridine, 13, 142
3-Vinylpyridine, 9, 13, 19, 20, 100, 101, 104
4-Vinylpyridine, 13
Violaxanthin, 3
Vitamin K_1, 4

X

Xeanthin, 3
Xylan, 10
m-Xylene, 92, 136, 137, 145, 187, 226
o-Xylene, 92, 136, 137, 145, 187, 226
p-Xylene, 92, 137, 145, 226
2,3-Xylenol, 98
2,4-Xylenol, 98
2,5-Xylenol, 98
2,6-Xylenol, 98
3,4-Xylenol, 98
3,5-Xylenol, 98
Xylose, 10
2,3-Xylyl methyl ether, 248
2,4-Xylyl methyl ether, 248
2,5-Xylyl methyl ether, 247, 248
2,6-Xylyl methyl ether, 248
3,4-Xylyl methyl ether, 248
3,5-Xylyl methyl ether, 247, 248

n-Valeronitrile, 95, 197, 199
Valine, 11, 191
Vanillin, acid, 2, 45
/anillin, 4, 58, 59
Venetnater 181
vinyl acetate, 58
Vinylacetylene, 108
Vinyl chloride, 257
5-Vinyl-8-hetroxy-δ-carbbon-
  acid acid, 432, 434
methyl derivative of, 432, 434
4-Vinylbispirketone, 188
p-Vinylphenone hom, 142
Vinylphenol, 194
o-Vinylphenol, 271
ethyl derivative of, 473
2-Vinylpyridine, 30, 142
2-Vinylpyridine, 1, 6., 14, 26,
  109-10), 134
4-Vinylpyridine, 13
Viridicatin, 8
Vitamin K, 2

X

Xanthone, 2
Xolan, 70
m-Xylene, 92, 136, 137, 186, 187,
  188
o-Xylene, 92, 186, 137, 149, 181,
  186
p-Xylene, 92, 187, 188, 198
2,3-Xylenol, 98
2,4-Xylenol, 98
2,5-Xylenol, 77
2,6-Xylenol, 97
3,4-Xylenol, 98
3,5-Xylenol, 98
Xylose, 8
1,3-Xylyl methyl ether, 268
2,4-Xylyl methyl ether, 218
3,5-Xylyl methyl ether, 242, 243
2,6-Xylyl methyl ether, 218
3,4-Xylyl methyl ether, 242
3,5-Xylyl methyl ether, 247, 263

DATE DUE